Lecture Notes in Mathematics

Edited by A. Dold and B. Eckmann

Subseries: Nankai Institute of Mathematics, Tianjin, P.R. China
vol. 5

Adviser S.S. Chern, B.-j. Jiang

1369

Boju Jiang Chia-Kuei Peng
Zixin Hou (Eds.)

Differential Geometry and Topology

Proceedings of the Special Year at Nankai Institute of Mathematics, Tianjin, PR China, 1986–87

Springer-Verlag
Berlin Heidelberg New York London Paris Tokyo

Editors

Boju JIANG
Department of Mathematics, Peking University
Beijing 100871, P.R. China

Chia-Kuei PENG
Graduate School of University of Science and Technology of China
P.O. Box 3908
Beijing 10039, P.R. China

Zixin HOU
Department of Mathematics, Nankai University
Tianjin 300071, P.R. China

Mathematics Subject Classification (1980): 22E, 53A, 53C, 57M, 57R, 58G

ISBN 3-540-51037-0 Springer-Verlag Berlin Heidelberg New York
ISBN 0-387-51037-0 Springer-Verlag New York Berlin Heidelberg

Printed in Germany

Printing and binding: Druckhaus Beltz, Hemsbach/Bergstr.
2146/3140-543210

FOREWORD

The Nankai Institute of Mathematics held a Special Year in Geometry and Topology during the academic year 1986-1987. The program centered around invited series of lectures, listed on the next page. This volume contains several sets of notes from these lectures, along with articles submitted by the participants.

We would like to thank all the participants for their enthusiasm and cooperation. Our thanks are also due to those who offered courses in Fall 1986 which prepared the graduate students for the lectures. Finally, we wish to thank Mr. Zhang Shu-dong for smoothing the English of many articles.

For the editors,

Boju Jiang
Chia-Kuei Peng
Zixin Hou

June 1988

SERIES OF LECTURES

(in chronological order)

Fall 1986

S.S. Chern	Ten Lectures in Differential Geometry
R.J. Stern	Yang-Mills and 4-Manifolds
R.O. Wells, Jr.	Supermanifolds

Spring 1987

U. Simon	A Course on Affine Differential Geometry
R.L. Cohen	Immersions of Manifolds; Algebraic K-Theory and Groups of Diffeomorphisms of Manifolds
S. Murakami	Exceptional Simple Lie Groups and Related Topics in Recent Differential Geometry
W.S. Cheung	Exterial Differential Systems and Calculus of Variations
R.D. Edwards	Decomposition of Manifolds
N.H. Kuiper	Geometry in Curvature Theory and Tightness
J. Eells	Harmonic Maps between Spheres
R. Kirby	Topology of 4-Manifolds
P. May	Equivariant Homotopy Theory
R.S. Palais	Morse Theory
C.L. Terng	Geometry of Submanifolds
S. Helgason	Topics in Geometric Analysis; Lie Groups and Symmetric Spaces from a Geometric Viewpoint
R.F. Brown	Nielsen Fixed Point Theory and Parametrized Differential Equations
A. Granas	Fixed Point Theory and Applications to Analysis
S.Y. Cheng	Index of Minimal Hypersurfaces

TABLE OF CONTENTS

1. Thomas E. Cecil and S. S. Chern: 1
Dupin Submanifolds in Lie Sphere Geometry.

2. Chen Weihuan（陈维桓）： 45
The Mean Curvature of the Tubular Hypersurfaces in a Space of Constant Curvature.

3. Chen Xiuxiong and Peng Chia-Kuei（陈秀雄、彭家贵）： 63
Deformation of Surfaces Preserving Principal Curvatures.

4. Ralph L. Cohen and U. Tillmann: 71
Lectures on Immersion Theory.

5. Jiang Boju（姜伯驹）： 125
Surface Maps and Braid Equations I.

6. Li An-Min（李安民）： 142
Affine Maximal Surface and Harmonic Functions.

7. Li Bang-He and Tang Zizhou（李邦河、唐梓洲）： 152
Codimension 1 and 2 Immersions of Lens Spaces.

8. Li Bang-He and Xu Tao（李邦河、徐涛）： 164
On Third Order Nondegenerate Immersions and Maps of S^1 in R^2.

9. Ma Zhisheng（马志圣）： 176
Complete Surfaces in H^3 with a Constant Principal Curvature.

10. Shingo Murakami: 183
Exceptional Simple Lie Groups and Related Topics in Recent Differential Geometry.

11. Peng Chia-Kuei and Hou Zixin（彭家贵、侯自新）： 222
A Remark on the Isoparametric Polynomials of Degree 6.

12. Shen Chunli（沈纯理）： 225
On the Holomorphic Maps from Riemannian Surfaces to Grassmannians.

13. Shen Yibing (沈一兵): 235
Stability of Totally Real Minimal Submanifolds.

14. Udo Simon: 243
Dirichlet Problems and the Laplacian in Affine Hypersurface Theory.

15. Tai Hsin-Sheng (戴新生): 261
A Class of Symmetric Functions and Chern Classes of Projective Varieties.

16. Wang Shicheng (王诗宬): 275
Essential Invariant Circles of Surface Automorphisms of Finite Order.

17. Wu Yingqing (吴英青): 286
Jones Polynomal and the Crossing Number of Links.

18. Xiao Liang (肖良): 289
On Complete Minimal Surfaces with Parallel and Flat Ends.

19. Xin Yuan-Long (忻元龙): 295
Regularity of Harmonic Maps into Certain Homogeneous Spaces.

20. Yang Wenmao (杨文茂): 306
On Infinitesimal Deformations of Surfaces in E^3.

21. Yu Yanlin (虞言林): 322
Local Expressions of Classical Geometric Elliptic Operators.

22. Yu Yanlin (虞言林): 333
Volume of Geodesic Balls.

23. Zhang Shaoping (张少平): 339
On Complete Minimal Immersions χ: $RP^2-\{a,b\}\to R^3$ with Total Curvature -10π.

24. Zhang Weiping (张伟平): 351
Local Atiyah-Singer Theorem for Families of Dirac Operators.

DUPIN SUBMANIFOLDS IN LIE SPHERE GEOMETRY

Thomas E. Cecil and Shiing-Shen Chern

1. Introduction.

Consider a piece of surface immersed in three-dimensional Euclidean space E^3. Its normal lines are the common tangent lines of two surfaces, the focal surfaces. These focal surfaces may have singularities, and a classical theorem says that if the focal surfaces both degenerate to curves, then the curves are conics, and the surface is a cyclide of Dupin. (See, for example, [CR, pp. 151-166].) Equivalently, the cyclides can be characterized as those surfaces in E^3 whose two distinct principal curvatures are both constant along their corresponding lines of curvature.

The cyclides have been generalized to an interesting class of hypersurfaces in E^n, the Dupin hypersurfaces. Initially, a hypersurface M in E^n was said to be Dupin if the number of distinct principal curvatures (or focal points) is constant on M and if each principal curvature is constant along the leaves of its corresponding principal foliation. (See [CR], [Th], [GH].) More recently, this has been generalized to include cases where the number of distinct principal curvatures is not constant. (See [P3], [CC].)

The study of Dupin hypersurfaces in E^n is naturally situated in the context of Lie sphere geometry, developed by Lie [LS] as part of his work on contact transformations. The projectivized cotangent bundle PT^*E^n of E^n has a contact structure. In fact, if $x^1,\dots,x^n$ are the coordinates in E^n, the contact structure is defined by the linear differential form $dx^n-p_1dx^1-\dots-p_{n-1}dx^{n-1}$. Lie proved that the pseudo-group of all contact transformations carrying (oriented) hyperspheres in the generalized sense (i.e., including points and oriented hyperplanes) into hyperspheres is a Lie group, called the Lie sphere group, isomorphic to $O(n+1,2)/\pm I$, where $O(n+1,2)$ is the orthogonal group for an indefinite inner product on $\mathbb{R}^{n+3}$ with signature $(n+1,2)$. The Lie sphere group contains as a subgroup the Moebius group of

---

The first author was supported by NSF Grant No. DMS 87-06015, the second author by NSF Grant No. DMS 87-01609.

conformal transformations of E^n and, of course, the Euclidean group. Lie exhibited a bijective correspondence between the set of oriented hyperspheres in E^n and the points on the quadric hypersurface Q^{n+1} in real projective space P^{n+2} given by the equation $\langle x,x \rangle = 0$, where $\langle , \rangle$ is the inner product on $\mathbb{R}^{n+3}$ mentioned above. The manifold Q^{n+1} contains projective lines but no linear subspaces of P^{n+2} of higher dimension. The 1-parameter family of oriented spheres corresponding to the points of a projective line lying on Q^{n+1} consists of all oriented hyperspheres which are in oriented contact at a certain contact element on E^n. Thus, Lie constructed a local diffeomorphism between PT^*E^n and the manifold Λ^{2n-1} of projective lines which lie on Q^{n+1}.

An immersed submanifold $f:M^k \to E^n$ naturally induces a Legendre submanifold $\lambda:B^{n-1} \to \Lambda^{2n-1}$, where B^{n-1} is the bundle of unit normal vectors to f (take $B^{n-1} = M^{n-1}$ in the case $k = n-1$). This Legendre map λ has similarities with the familiar Gauss map, and like the Gauss map, it can be a powerful tool in the study of submanifolds of Euclidean space. In particular, the Dupin property for hypersurfaces in E^n is easily formulated in terms of the Legendre map, and it is immediately seen to be invariant under Lie sphere transformations.

The study of Dupin submanifolds has both local and global aspects. Thorbergsson [Th] showed that a Dupin hypersurface M with g distinct principal curvatures at each point must be taut, i.e., every nondegenerate Euclidean distance function $L_p(x) = |p-x|^2$, $p \in E^n$, must have the minimum number of critical points on M. Tautness was shown to be invariant under Lie transformations in our earlier paper [CC]. Using tautness and the work of Münzner [Mu], Thorbergsson was then able to conclude that the number g must be 1,2,3,4 or 6, as with an isoparametric hypersurface in the sphere S^n. The case g = 1 is, of course, handled by the well-known classification of umbilic hypersurfaces. Compact Dupin hypersurfaces with g=2 and g=3 were classified by Cecil and Ryan (see [CR, p. 168]) and Miyaoka [M1] respectively. In two recent preprints, Miyaoka [M2], [M3] has made further progress on the classification of compact Dupin hypersurfaces in the cases g=4 and g=6. Meanwhile, Grove and Halperin [GH] have determined several important topological invariants of compact Dupin hypersurfaces in the cases g=4 and g=6.

In this paper, we study Dupin hypersurfaces in the setting of Lie sphere geometry using local techniques. In Section 2, we give a brief introduction to Lie sphere geometry. In Section 3, we introduce the basic differential geometric notions: the Legendre map and the Dupin property. The case of E^3 is handled in Section 4, where we handle the case of g=2 distinct focal points for E^n. This was first done for n > 3 by Pinkall [P3]. Our main contribution lies in Section 5, where we treat the case E^4 by the method of moving frames. This case was also studied by Pinkall [P2], but our treatment seems to be more direct and differs from his in several essential points. It is our hope that this method will provide a framework and give some direction for the study of Dupin hypersurfaces in E^n for n > 4.

2. Lie Sphere Geometry.

We first present a brief outline of the main ideas in Lie's geometry of spheres in $\mathbb{R}^n$. This is given in more detail in Lie's original treatment [LS], in the book of Blaschke [B], and in our paper [CC].

The basic construction in Lie sphere geometry associates each oriented sphere, oriented plane and point sphere in $\mathbb{R}^n \cup \{\infty\} = S^n$ with a point on the quadric Q^{n+1} in projective space P^{n+2} given in homogeneous coordinates $(x_1,\dots,x_{n+3})$ by the equation

$$\langle x,x\rangle = -x_1^2 + x_2^2 + \dots + x_{n+2}^2 - x_{n+3}^2 = 0. \tag{2.1}$$

We will denote real (n+3)-space endowed with the metric (2.1) of signature (n+1,2) by $\mathbb{R}_2^{n+3}$.

We can designate the orientation of a sphere in $\mathbb{R}^n$ by assigning a plus or minus sign to its radius. Positive radius corresponds to the orientation determined by the field of inward normals to the sphere, while a negative radius corresponds to the orientation determined by the outward normal. (See Remark 2.1 below). A plane in $\mathbb{R}^n$ is a sphere which goes through the point ∞. The orientation of the plane can be associated with a choice of unit normal N. The specific correspondence between the points of Q^{n+1} and the set of oriented spheres, oriented planes and points in $\mathbb{R}^n \cup \{\infty\}$ is then given as follows:

Euclidean	Lie
Points: $u \in \mathbb{R}^n$	$\left[\left(\frac{1 + u\cdot u}{2}, \frac{1 - u\cdot u}{2}, u, 0\right)\right]$
∞	$[(1,-1,0,0)]$
(2.2)	
Spheres: Center p, signed radius r	$\left[\left(\frac{1 + p\cdot p - r^2}{2}, \frac{1 - p\cdot p + r^2}{2}, p, r\right)\right]$
Planes: $u\cdot N = h$, unit normal N	$[(h,-h,N,1)]$.

Here the square brackets denote the point in projective space P^{n+2} given by the homogeneous coordinates in the round brackets, and $u\cdot u$ is the standard Euclidean dot product in $\mathbb{R}^n$.

From (2.2), we see that the point spheres correspond to the points in the intersection of Q^{n+1} with the hyperplane in P^{n+2} given by the equation $x_{n+3} = 0$. The manifold of point spheres is called Moebius space.

A fundamental notion in Lie sphere geometry is that of oriented contact of spheres. Two oriented spheres S_1 and S_2 are in oriented contact if they are tangent and their orientations agree at the point of tangency. If p_1 and p_2 are the respective centers of S_1 and S_2, and r_1 and r_2 are the respective signed radii, then the condition of oriented contact can be expressed analytically by

$$|p_1 - p_2| = |r_1 - r_2|. \tag{2.3}$$

If S_1 and S_2 are represented by $[k_1]$ and $[k_2]$ as in (2.2), then (2.3) is equivalent to the condition

$$\langle k_1, k_2\rangle = 0. \tag{2.4}$$

In the case where S_1 and/or S_2 is a plane or a point in $\mathbb{R}^n$, oriented contact has the logical meaning. That is, a sphere S and plane π are in oriented contact if π is tangent to S and their orientations agree at the point of contact. Two oriented planes are in oriented contact if their unit normals are the same. They are in oriented contact at the point ∞. A point sphere is in oriented contact with a sphere or plane S if it lies on S, and ∞ is in oriented contact with each plane. In each case, the analytic condition for oriented contact is equivalent to (2.4) when the two "spheres" in question are represented in Lie coordinates as in (2.2).

Remark 2.1: In the case of a sphere $[k_1]$ and a plane $[k_2]$ as in (2.2), equation (2.4) is equivalent to $p \cdot N = h+r$. In order to make this correspond to the geometric definition of oriented contact, one must adopt the convention that the inward normal orientation of a sphere corresponds to positive signed radius. To get the outward normal orientation to correspond to positive radius, one should represent the plane by $[(-h,h,-N,1)]$ instead of $[(h,-h,N,1)]$. Then (2.4) becomes $p \cdot N = h-r$, which is the geometric formula for oriented contact with the outward normal orientation corresponding to positive signed radius.

Because of the signature of the metric (2.1), the quadric Q^{n+1} contains lines in P^{n+2} but no linear subspaces of higher dimension. A line on Q^{n+1} is determined by two points $[x]$, $[y]$ in Q^{n+1} satisfying $\langle x,y\rangle = 0$. The lines on Q^{n+1} form a manifold of dimension $2n-1$, to be denoted by Λ^{2n-1}. In $\mathbb{R}^n$, a line on Q^{n+1} corresponds to a 1-parameter family of oriented spheres such that any two of the spheres are in oriented contact, i.e., all the oriented spheres tangent to an oriented plane at a given point, i.e., an oriented contact element. Of course, a contact element can also be represented by an element of T_1S^n, the bundle of unit tangent vectors to the Euclidean sphere S^n in E^{n+1} with its usual metric. This is the starting point for Pinkall's [P3] considerations of Lie geometry.

A Lie sphere transformation is a projective transformation of P^{n+2} which takes Q^{n+1} to itself. Since a projective transformation takes lines to lines, a Lie sphere transformation preserves oriented contact of spheres. The group G of Lie sphere transformations is isomorphic to $O(n+1,2)/\{\pm I\}$, where $O(n+1,2)$ is the group of orthogonal transformations for the inner product (2.1). Moebius transformations are those Lie transformations which take point spheres to point spheres. The group of Moebius transformations is isomorphic to $O(n+1,1)/\{\pm I\}$.

3. Legendre Submanifolds.

Here we recall the concept of a Legendre submanifold of the contact manifold $\Lambda^{2n-1}(=\Lambda)$ using the notation of [CC]. In this section, the ranges of the indices are as follows:

$$1 \le A,B,C \le n+3, \qquad 3 \le i,j,k \le n+1. \tag{3.1}$$

Instead of using an orthonormal frame for the metric <,> defined by (2.1), it is useful to consider a <u>Lie frame</u>, that is, an ordered set of vectors Y_A in $\mathbb{R}^{n+3}_2$ satisfying

$$\langle Y_A, Y_B \rangle = g_{AB}, \tag{3.2}$$

with

$$(g_{AB}) = \begin{bmatrix} J & 0 & 0 \\ 0 & I_{n-1} & 0 \\ 0 & 0 & J \end{bmatrix}, \tag{3.3}$$

where I_{n-1} is the identity $(n-1) \times (n-1)$ matrix and

$$J = \begin{bmatrix} 0 & 1 \\ 1 & 0 \end{bmatrix}. \tag{3.4}$$

The space of all Lie frames can be identified with the orthogonal group $O(n+1,2)$, of which the Lie sphere group, being isomorphic to $O(n+1,2)/\{\pm I\}$, is a quotient group. In this space, we introduce the Maurer-Cartan forms

$$dY_A = \Sigma\, \omega_A^B\, Y_B. \tag{3.5}$$

Through differentiation of (3.2), we show that the following matrix of 1-forms is skew-symmetric

$$(\omega_{AB}) = \begin{bmatrix} \omega_1^2 & \omega_1^1 & \omega_1^i & \omega_1^{n+3} & \omega_1^{n+2} \\ \omega_2^2 & \omega_2^1 & \omega_2^i & \omega_2^{n+3} & \omega_2^{n+2} \\ \omega_j^2 & \omega_j^1 & \omega_j^i & \omega_j^{n+3} & \omega_j^{n+2} \\ \omega_{n+2}^2 & \omega_{n+2}^1 & \omega_{n+2}^i & \omega_{n+2}^{n+3} & \omega_{n+2}^{n+2} \\ \omega_{n+3}^2 & \omega_{n+3}^1 & \omega_{n+3}^i & \omega_{n+3}^{n+3} & \omega_{n+3}^{n+2} \end{bmatrix}. \tag{3.6}$$

Next, by taking the exterior derivative of (3.5), we get the Maurer-Cartan equations

$$d\omega_A^B = \sum_C \omega_A^C \wedge \omega_C^B. \tag{3.7}$$

In [CC], we then show that the form

$$\omega_1^{n+2} = \langle dY_1, Y_{n+3}\rangle$$

gives a contact structure on the manifold Λ.

Let $B^{n-1}(= B)$ be an (n-1)-dimensional smooth manifold. A <u>Legendre map</u> is a smooth map $\lambda : B \to \Lambda$ which annihilates the contact form on Λ, i.e., $\lambda^* \omega_1^{n+2} = 0$ on B. All of our calculations are local in nature. We use the method of moving frames and consider a smooth family of Lie frames Y_A on an open subset U of B, with the line $\lambda(b)$ given by $[Y_1(b), Y_{n+3}(b)]$ for each $b \in U$. The Legendre map λ is called a <u>Legendre submanifold</u> if for a generic choice of Y_1 the forms ω_1^i, $3 \leq i \leq n+1$, are linearly independent, i.e.,

(3.8) $$\wedge\, \omega_1^i \neq 0 \text{ on } U .$$

Here and later we pull back the structure forms to B^{n-1} and omit the symbols of such pull-backs for simplicity. Note that the Legendre condition is just

(3.9) $$\omega_1^{n+2} = 0 .$$

We now assume that our choice of Y_1 satisfies (3.8). By exterior differentiation of (3.9) and using (3.6), we get

(3.10) $$\Sigma\, \omega_1^i \wedge \omega_{n+3}^i = 0 .$$

Hence by Cartan's Lemma and (3.8), we have

(3.11) $$\omega_{n+3}^i = \Sigma\, h_{ij}\omega_1^j \text{ , with } h_{ij} = h_{ji} .$$

The quadratic differential form

$$II(Y_1) = \sum_{i,j} h_{ij}\omega_1^i\omega_1^j ,$$

defined up to a non-zero factor and depending on the choice of Y_1, is called the <u>second fundamental form</u>.

This form can be related to the well-known Euclidean second fundamental form in the following way. Let e_{n+3} be any unit timelike vector in $\mathbb{R}_2^{n+3}$. For each $b \in U$, let $Y_1(b)$ be the point of intersection of the line $\lambda(b)$ with the hyperplane $e^{\perp}_{n+3}$. Y_1 represents the locus of point spheres in the Moebius space $Q^{n+1} \cap e^{\perp}_{n+3}$, and we call Y_1 the <u>Moebius projection of λ</u> determined by e_{n+3}. Let e_1 and e_2 be unit timelike, respectively spacelike, vectors orthogonal to e_{n+3} and to each other, chosen so that Y_1 is not the point at infinity $[e_1-e_2]$ for any $b \in U$. We can represent Y_1 by the vector

$$(3.13) \qquad Y_1 = \frac{1 + f\cdot f}{2} e_1 + \frac{1 - f\cdot f}{2} e_2 + f,$$

as in (2.2), where $f(b)$ lies in the space $\mathbb{R}^n$ of vectors orthogonal to e_1, e_2 and e_{n+3}. We will call the map $f: B \to \mathbb{R}^n$ the <u>Euclidean projection of</u> λ determined by the ordered triple e_1, e_2, e_{n+3}. The regularity condition (3.8) is equivalent to the condition that f be an immersion on U into $\mathbb{R}^n$. For each $b \in U$, let $Y_{n+3}(b)$ be the intersection of $\lambda(b)$ with the orthogonal complement of the lightlike vector $e_1 - e_2$. Y_{n+3} is distinct from Y_1 and thus $\langle Y_{n+3}, e_{n+3}\rangle \neq 0$. So we can represent Y_{n+3} by a vector of the form

$$(3.14) \qquad Y_{n+3} = h(e_1 - e_2) + \xi + e_{n+3},$$

where $\xi: U \to \mathbb{R}^n$ has unit length and h is a smooth function on U. Thus, according to (2.2), $Y_{n+3}(b)$ represents the plane in the pencil of oriented spheres in $\mathbb{R}^n$ corresponding to the line $\lambda(b)$ on Q^{n+1}. Note that the condition $\langle Y_1, Y_{n+3}\rangle = 0$ is equivalent to $h = f\cdot\xi$, while the Legendre condition $\langle dY_1, Y_{n+3}\rangle = 0$ is the same as the Euclidean condition

$$(3.15) \qquad \xi \cdot df = 0 \; .$$

Thus, ξ is a field of unit normals to the immersion f on U. Since f is an immersion, we can choose the Lie frame vectors $Y_3, \ldots, Y_{n+1}$ to satisfy

$$(3.16) \qquad Y_i = dY_1(X_i) = (f\cdot df(X_i))(e_1 - e_2) + df(X_i), \; 3 \le i \le n+1 \; ,$$

for tangent vector fields $X_3, \dots, X_{n+1}$ on U. Then, we have

$$\omega_1^i(X_j) = \langle dY_1(X_j), Y_i \rangle = \langle Y_j, Y_i \rangle = \delta_{ij} . \tag{3.17}$$

Now using (3.14) and (3.16), we compute

$$\begin{aligned} \omega_{n+3}^i(X_j) &= \langle dY_{n+3}(X_j), Y_i \rangle = d\xi(X_j) \cdot df(X_i) \\ &= - df(AX_j) \cdot df(X_i) = -A_{ij} , \end{aligned} \tag{3.18}$$

where $A = [A_{ij}]$ is the Euclidean shape operator (second fundamental form) of the immersion f. But by (3.11) and (3.17), we have

$$\omega_{n+3}^i(X_j) = \Sigma h_{ik} \omega_1^k(X_j) = h_{ij} .$$

Hence $h_{ij} = -A_{ij}$, and $[h_{ij}]$ is just the negative of the Euclidean shape operator A of f.

Remark 3.1: The discussion above demonstrates how an immersion $f: B^{n-1} \to \mathbb{R}^n$ with field of unit normals ξ induces a Legendre submanifold $\lambda: B^{n-1} \to \Lambda$ defined by $\lambda(b) = [Y_1(b), Y_{n+3}(b)]$, for Y_1, Y_{n+3} as in (3.13), (3.14). Further, an immersed submanifold $f: M^k \to \mathbb{R}^n$ of codimension greater than one also gives rise to a Legendre submanifold $\lambda: B^{n-1} \to \Lambda$, where B^{n-1} is the bundle of unit normals to f in $\mathbb{R}^n$. As in the case of codimension one, $\lambda(b)$ is defined to be the line on Q^{n+1} corresponding to the oriented contact element determined by the unit vector b normal to f at the point $x = \pi(b)$, where π is the bundle projection from B^{n-1} to M^k.

As one would expect, the eigenvalues of the second fundamental form have geometric significance. Consider a curve $\gamma(t)$ on B. The set of points in Q^{n+1} lying on the lines $\lambda(\gamma(t))$ forms a ruled surface in Q^{n+1}. We look for the conditions that this ruled surface be developable, i.e., consist of tangent lines to a curve in Q^{n+1}. Let $rY_1 + Y_{n+3}$ be the point of contact. We have by (3.5) and (3.6)

$$d(rY_1 + Y_{n+3}) \equiv \sum_i (r\omega_1^i + \omega_{n+3}^i) Y_i, \ \mathrm{mod}\ Y_1, Y_{n+3} . \tag{3.19}$$

Thus, the lines $\lambda(\gamma(t))$ form a developable if and only if the tangent direction of $\gamma(t)$ is a common solution to the equations

$$(3.20) \qquad \sum_j (r\delta_{ij} + h_{ij})\omega_1^j = 0, \quad 3 \le i \le n+1 .$$

In particular, r must be a root of the equation

$$(3.21) \qquad \det(r\delta_{ij} + h_{ij}) = 0 .$$

By (3.11) the roots of (3.21) are all real. Denote them by $r_3, \ldots, r_{n+1}$. The points $r_i Y_1 + Y_{n+3}$, $3 \le i \le n+1$ are called the <u>focal points</u> or <u>curvature spheres</u> (Pinkall [P3]) on $\lambda(b)$. If Y_1 and Y_{n+3} correspond to an immersion $f: U \to \mathbb{R}^n$ as in (3.13) and (3.14), then these focal points on $\lambda(b)$ correspond by (2.2) to oriented spheres in $\mathbb{R}^n$ tangent to f at f(b) and centered at the Euclidean focal points of f. These spheres are called curvature spheres of f and the r_i are just the principal curvatures of f, i.e., eigenvalues of the shape operator A.

If r is a root of (3.21) of multiplicity m, then the equations (3.20) define an m-dimensional subspace T_r of T_bB, the tangent space to B at the point b. The space T_r is called a <u>principal space</u> of T_bB, the latter being decomposed into a direct sum of its principal spaces. Vectors in T_r are called <u>principal vectors</u> corresponding to the focal point $rY_1 + Y_{n+3}$. Of course, if Y_1 and Y_{n+3} correspond to an immersion $f: U \to \mathbb{R}^n$ as in (3.13) and (3.14), then these principal vectors are the same as the Euclidean principal vectors for f corresponding to the principal curvature r.

With a change of frame of the form

$$(3.22) \qquad Y_i^* = \sum_i c_i^j Y_j , \quad 3 \le i \le n+1 ,$$

where $[c_i^j]$ is an (n-1) x (n-1) orthogonal matrix, we can diagonalize $[h_{ij}]$ so that in the new frame, equation (3.11) has the form

$$(3.23) \qquad \omega_{n+3}^i = - r_i \omega_1^i , \quad 3 \le i \le n+1 .$$

Note that none of the functions r_i is ever infinity on U because of the assumption that (3.8) holds, i.e, Y_1 is not a focal point. By associating Y_1 to a Euclidean immersion f as in (3.13), we can apply results from Euclidean geometry to our situation. In particular, it follows from a result of Singley [S] on Euclidean shape operators that there is a dense open subset of B on which the number g(b) of distinct focal points on $\lambda(b)$ is locally constant. We will work exclusively on open subsets U of B on which g is constant. In that case, each distinct eigenvalue function $r:U \to \mathbb{R}$ is smooth (see Nomizu [N]), and its corresponding principal distribution is a smooth m-dimensional foliation, where m is the multiplicity of r (see [CR, p. 139]). Thus, on U we can find smooth vector fields $X_3, \ldots X_{n+1}$ dual to smooth 1-forms $\omega_1^3, \ldots, \omega_1^{n+1}$, respectively, such that each X_i is principal for the smooth focal point map $r_i Y_1 + Y_{n+3}$ on U. If $rY_1 + Y_{n+3}$ is a smooth focal point map of multiplicity m on U, then we can assume that

$$r_3 = \ldots = r_{m+2} = r \ . \tag{3.24}$$

By a different choice of the point at infinity, i.e., e_1 and e_2, if necessary, we can also assume that the function r is never zero on U, i.e., Y_{n+3} is not a focal point on U.

We now want to consider a Lie frame Y_A^* for which Y_1^* is a smooth focal point map of multiplicity m on U. Specifically, we make the change of frame

$$\begin{aligned} Y_1^* &= r\,Y_1 + Y_{n+3} \\ Y_2^* &= (1/r)Y_2 \\ Y_{n+2}^* &= Y_{n+2} - (1/r)Y_2 \\ Y_{n+3}^* &= Y_{n+3} \\ Y_i^* &= Y_i \ , \quad 3 \le i \le n+1 \ . \end{aligned} \tag{3.25}$$

We denote the Maurer-Cartan forms in this frame by θ_A^B. Note that

$$dY_1^* = d(rY_1 + Y_{n+3}) = (dr)Y_1 + rdY_1 + dY_{n+3} = \Sigma\, \theta_1^A Y_A^* \ . \tag{3.26}$$

By examining the coefficient of $Y_i^* = Y_i$ in (3.26), we see from (3.23) that

$$\theta_1^i = r\omega_1^i + \omega_{n+3}^i = (r-r_i)\omega_1^i, \quad 3 \leq i \leq n+1 \,. \tag{3.27}$$

From (3.24) and (3.27), we see that

$$\theta_1^a = 0 \,, \quad 3 \leq a \leq m+2 \,. \tag{3.28}$$

This equation characterizes the condition that a focal point map Y_1^* have constant multiplicity m on U.

We now introduce the concept of a Dupin submanifold and then see what further restrictions it allows us to place on the structure forms.

A connected submanifold $N \subset B$ is called a <u>curvature submanifold</u> if its tangent space is everywhere a principal space. The Legendre submanifold is called <u>Dupin</u> if for every curvature submanifold $N \subset B$, the lines $\lambda(b)$, $b \in N$, all pass through a fixed point, i.e., each focal point map is constant along its curvature submanifolds. This definition of "Dupin" is the same as that of Pinkall [P3]. It is weaker than the definition of Dupin for Euclidean hypersurfaces used by Thorbergsson [Th], Miyaoka [M1], Grove-Halperin [GH] and Cecil-Ryan [CR], all of whom assumed that the number g of distinct curvature spheres is constant on B. As we noted above, g is locally constant on a dense open subset of any Legendre submanifold, but g is not necessarily constant on the whole of a Dupin submanifold, as the example of the Legendre submanifold induced from a tube B^3 over a torus $T^2 \subset \mathbb{R}^3 \subset \mathbb{R}^4$ demonstrates (see Pinkall [P3], Cecil-Ryan [CR, p. 188]).

It is easy to see that the Dupin property is invariant under Lie sphere transformations as follows. Suppose that $\lambda : B \to \Lambda^{2n-1}$ is a Legendre submanifold and that α is a Lie transformation. The map $\alpha\lambda : B \to \Lambda^{2n-1}$ is also a Legendre submanifold with $\alpha\lambda(b) = [\alpha Y_1(b), \alpha Y_{n+3}(b)]$. Furthermore, if $k = rY_1 + Y_{n+3}$ is a curvature sphere of λ at a point $b \in B$, then since α is a linear transformation, αk is a curvature sphere of $\alpha\lambda$ at b with the same principal space as k. Thus λ and $\alpha\lambda$ have the same curvature submanifolds on B, and the Dupin property clearly holds for λ if and only if it holds for $\alpha\lambda$.

We now return to the calculation that led to equation (3.28). We have that $Y_1^* = rY_1 + Y_{n+3}$ is a smooth focal point map of multiplicity m on the open set U and its corresponding principal space is spanned by the vector fields $X_3, \ldots, X_{m+2}$. The Dupin condition that Y_1^* be constant along the leaves of T_r is simply

$$(3.29)\qquad dY_1^*(X_a) \equiv 0, \bmod Y_1^*, \ 3 \le a \le m+2 .$$

From (3.17), (3.27) and (3.28), we have that

$$(3.30)\qquad dY_1^*(X_a) = \theta_1^1(X_a)Y_1 + \theta_1^{n+3}(X_a)Y_{n+3}, \quad 3 \le a \le m+2.$$

Comparing (3.29) and (3.30), we see that

$$(3.31)\qquad \theta_1^{n+3}(X_a) = 0 \ , \ 3 \le a \le m+2 .$$

We now show that we can make one more change of frame and make $\theta_1^{n+3} = 0$. We can write the form θ_1^{n+3} in terms of the basis $\omega_1^3,\dots,\omega_1^{n+3}$ as

$$\theta_1^{n+3} = \Sigma\, a_i\omega_1^i .$$

From (3.31), we see that we actually have

$$(3.32)\qquad \theta_1^{n+3} = \sum_{b=m+3}^{n+1} a_b\omega_1^b .$$

Using (3.17), (3.27) and (3.32), we compute for $m+3 \le b \le n+1$,

$$(3.33)\qquad \begin{aligned} dY_1^*(X_b) &= \theta_1^1(X_b)Y_1^* + \theta_1^b(X_b)Y_b + \theta_1^{n+3}(X_b)Y_{n+3} \\ &= \theta_1^1(X_b)Y_1^* + (r-r_b)Y_b + a_b\, Y_{n+3} \\ &= \theta_1^1(X_b)Y_1^* + (r-r_b)(Y_b + (a_b/(r-r_b))Y_{n+3}). \end{aligned}$$

We now make the change of frame,

$$(3.34)\qquad \begin{aligned} &\bar{Y}_1 = Y_1^* \ , \ \bar{Y}_2 = Y_2^* \ , \\ &\bar{Y}_a = Y_a \ , \ 3 \le a \le m+2, \\ &\bar{Y}_b = Y_b + (a_b/r-r_b)Y_{n+3} \ , \ m+3 \le b \le n+1, \\ &\bar{Y}_{n+2} = -\sum_{b=m+3}^{n+1} (a_b/r-r_b)Y_b + Y_{n+2} - 1/2 \sum_{b=m+3}^{n+1} (a_b/r-r_b)^2\, Y_{n+3} \ , \\ &\bar{Y}_{n+3} = Y_{n+3}. \end{aligned}$$

Let α_A^B be the Maurer-Cartan forms for this new frame. We still have

$$\text{(3.35)} \qquad \alpha_1^a = \langle d\overline{Y}_1, \overline{Y}_a \rangle = \langle dY_1^*, Y_a \rangle = \theta_1^a = 0, \quad 3 \le a \le m+2 .$$

Furthermore, since $\overline{Y}_1 = Y_1^*$, the Dupin condition (3.29) still yields

$$\text{(3.36)} \qquad \alpha_1^{n+3}(X_a) = 0 \quad , \ 3 \le a \le m+2 .$$

Finally, for $m+3 \le b \le n+1$, we have from (3.33) and (3.34)

$$\text{(3.37)} \qquad \begin{aligned} \alpha_1^{n+3}(X_b) &= \langle d\overline{Y}_1(X_b), \overline{Y}_{n+2} \rangle \\ &= \langle \theta_1^1(X_b)\overline{Y}_1 + (r-r_b)\overline{Y}_b, \overline{Y}_{n+2} \rangle = 0 . \end{aligned}$$

From (3.36) and (3.37), we conclude that

$$\text{(3.38)} \qquad \alpha_1^{n+3} = 0 .$$

Thus, our main result of this section is that the assumption that the focal point map $\overline{Y}_1 = rY_1 + Y_{n+3}$ has constant multiplicity m and is constant along the leaves of its principal foliation T_r allows to produce a Lie frame $\overline{Y}_A$ whose structure forms satisfy

$$\text{(3.39)} \qquad \begin{aligned} \alpha_1^a &= 0 \quad , \ 3 \le a \le m+2 , \\ \alpha_1^{n+3} &= 0 . \end{aligned}$$

4. <u>Cyclides of Dupin</u>.

Dupin initiated the study of this subject in 1822 when he defined a cyclide to be a surface M^2 in E^3 which is the envelope of the family of spheres tangent to three fixed spheres. This was shown to be equivalent to requiring that both sheets of the focal set of M^2 in E^3 degenerate into curves. Then M^2 is the envelope of each of the two families of curvature spheres. The key step in the classical Euclidean proof (see, for example,

Eisenhart [E, pp. 312-314] or Cecil-Ryan [CR, pp. 151-166]) is to show that the two focal curves are a pair of so-called "focal conics" in E^3, i.e., an ellipse and hyperbola in mutually orthogonal planes such that the vertices of the ellipse are the foci of the hyperbola and vice-versa, or a pair of parabolas in orthogonal planes such the vertex of each is the focus of the other. This classical proof is local, i.e., one needs only a small piece of the surface to determine the focal conics and reconstruct the whole cyclide. Of course, envelopes of families of spheres can have singularities and some of the cyclides have one or two singular points in E^3. It turns out, however, that all of the different forms of cyclides in Euclidean space induce Legendre submanifolds which are locally Lie equivalent. In other words,they are all various Euclidean projections of one Legendre submanifold. Pinkall [P3] generalized this result to higher dimensional Dupin submanifolds. He defined a _cyclide of characteristic_ (p,q) to be a Dupin submanifold with the property that at each point it has exactly two distinct focal points with respective multiplicites p and q. He then proved the following.

Theorem 4.1: (Pinkall [P3]): (a) Every connected cyclide of Dupin is contained in a unique compact and connected cyclide of Dupin.
(b) Any two cyclides of the same characteristic are locally Lie equivalent, each being Lie equivalent to an open subset of a standard product of spheres in S^n.

In this section, we give a proof of Pinkall's result using the method of Lie frames. Let $\lambda: B \to \Lambda$ be the Dupin cyclide. The main step in the proof of the Theorem 4.1 is to show that the two focal point maps k_1 and k_2 from B to Q^{n+1} are such that the image $k_1(B)$ lies in the intersection of Q^{n+1} with a (p+1)-dimensional subspace E of P^{n+2} while $k_2(B)$ lies in the intersection of the (q+1)-dimensional subspace $E^{\perp}$ with Q^{n+1}. This generalizes the key step in the classical Euclidean proof that the two focal curves are focal conics. Once this fact has been established for k_1 and k_2, it is relatively easy to complete the proof of the Theorem.

We begin the proof by taking advantage of the results of the previous section. As we showed in (3.39), on any neighborhood U in B, we can find a local Lie frame, which we now denote by Y_A, whose Maurer-Cartan forms, now denoted ω_A^B, satisfy

$$(4.1)\qquad \omega_1^a = 0 \quad , \ 3 \le a \le p+2, \qquad \omega_1^{n+3} = 0 \ .$$

In this frame, Y_1 is a focal point map of multiplicity p from U to Q^{n+1}. By the hypotheses of Theorem 4.1, there is one other focal point of multiplicity $q = n-1-p$ at each point of B. By repeating the procedure used in constructing the frame Y_A, we can construct a new frame $\bar{Y}_A$ which has as $\bar{Y}_{n+3}$ the other focal point map $sY_1 + Y_{n+3}$, where s is a root of (3.21) of multiplicity q. The principal space T_s is the span of the vectors $X_{p+3},\ldots,X_{n+1}$ in the notation of the previous section. The fact that $\bar{Y}_{n+3}$ is a focal point yields

$$(4.2)\qquad \bar{\omega}_{n+3}^b = 0 \quad , \quad p+3 \le b \le n+1 \ ,$$

in analogy to (3.28). The Dupin condition analogous to (3.29) is

$$(4.3)\qquad d\bar{Y}_{n+3}(X_b) \equiv 0, \ \text{mod}\ \bar{Y}_{n+3} \ , \ p+3 \le b \le n+1 \ .$$

This eventually leads to

$$(4.4)\qquad \bar{\omega}_{n+3}^1 = 0 \ .$$

One can check that this change of frame does not affect condition (4.1). We now drop the bars and call this last frame Y_A with Maurer-Cartan forms ω_A^B satisfying,

$$(4.5)\qquad \begin{aligned} &\omega_1^a = 0 \quad , &&3 \le a \le p+2 \ , \\ &\omega_{n+3}^b = 0 \quad , &&p+3 \le b \le n+1 \ , \\ &\omega_1^{n+3} = 0 \quad , &&\omega_{n+3}^1 = 0 \ . \end{aligned}$$

Furthermore, the following forms are easily shown to be a basis for the cotangent space at each point of U,

$$(4.6)\qquad \{\omega_{n+3}^3,\ldots,\omega_{n+3}^{p+2},\omega_1^{p+3},\ldots,\omega_1^{n+1}\} \ .$$

We begin by taking the exterior derivative of the equations $\omega_1^a = 0$ and $\omega_{n+3}^b = 0$ in (4.5). Using (3.6), (3.7) and (4.5), we obtain

(4.7) $$0 = \omega_1^{p+3} \wedge \omega_{p+3}^a + \ldots + \omega_1^{n+1} \wedge \omega_{n+1}^a , \quad 3 \leq a \leq p+2 ,$$

(4.8) $$0 = \omega_{n+3}^3 \wedge \omega_3^b + \ldots + \omega_{n+3}^{p+2} \wedge \omega_{p+2}^b , \quad p+3 \leq b \leq n+1 .$$

We now show that (4.7) and (4.8) imply that

(4.9) $$\omega_b^a = 0 \quad , \quad 3 \leq a \leq p+2 \ , \ p+3 \leq b \leq n+1 .$$

To see this, note that since $\omega_b^a = -\omega_a^b$, each of the terms ω_b^a occurs in exactly one of the equations (4.7) and in exactly one of the equations (4.8). Equation (4.7) involves the basis forms $\omega_1^{p+3}, \ldots, \omega_1^{n+1}$, while equation (4.8) involves the basis forms $\omega_{n+3}^3, \ldots, \omega_{n+3}^{p+2}$. We now show how to handle the form ω_{p+3}^3; the others are treated in similar fashion. The equations from (4.7) and (4.8), respectively, involving $\omega_{p+3}^3 = -\omega_3^{p+3}$ are

(4.10) $$0 = \omega_1^{p+3} \wedge \omega_{p+3}^3 + \omega_1^{p+4} \wedge \omega_{p+4}^3 + \ldots + \omega_1^{n+1} \wedge \omega_{n+1}^3 ,$$

(4.11) $$0 = \omega_{n+3}^3 \wedge \omega_3^{p+3} + \omega_{n+3}^4 \wedge \omega_4^{p+3} + \ldots + \omega_{n+3}^{p+2} \wedge \omega_{p+2}^{p+3} .$$

We take the wedge product of (4.10) with $\omega_1^{p+4} \wedge \ldots \wedge \omega_1^{n+1}$ and get

$$0 = \omega_{p+3}^3 \wedge (\omega_1^{p+3} \wedge \omega_1^{p+4} \wedge \ldots \wedge \omega_1^{n+1}) ,$$

which implies that ω_{p+3}^3 is in the span of $\omega_1^{p+3}, \ldots, \omega_1^{n+1}$. On the other hand, taking the wedge product of (4.11) with $\omega_{n+3}^4 \wedge \ldots \wedge \omega_{n+3}^{p+2}$ yields

$$0 = \omega_{p+3}^3 \wedge (\omega_{n+3}^3 \wedge \ldots \wedge \omega_{n+3}^{p+2}) ,$$

and thus that ω_{p+3}^3 is in the span of $\omega_{n+3}^3, \ldots, \omega_{n+3}^{p+2}$. We conclude that $\omega_{p+3}^3 = 0$, as desired.

We next differentiate $\omega_1^{n+3} = 0$ and use (3.6), (3.9) and (4.5) to obtain

(4.12) $$0 = d\omega_1^{n+3} = \omega_1^{p+3} \wedge \omega_{p+3}^{n+3} + \ldots + \omega_1^{n+1} \wedge \omega_{n+1}^{n+3} .$$

This implies that

(4.13) $$\omega_b^{n+3} \in \mathrm{Span}\{\omega_1^{p+3}, \ldots, \omega_1^{n+1}\} , \quad p+3 \leq b \leq n+1.$$

Similarly, differentiation of $\omega_{n+3}^1 = 0$ yields

(4.14) $$0 = \omega_{n+3}^3 \wedge \omega_3^1 + \ldots + \omega_{n+3}^{p+2} \wedge \omega_{p+2}^1 ,$$

which implies that

(4.15) $$\omega_1^a \in \mathrm{Span}\{\omega_{n+3}^3, \ldots, \omega_{n+3}^{p+2}\} , \quad 3 \leq a \leq p+2 .$$

We next differentiate (4.9). Using the skew-symmetry relations (3.6) and equations (4.5) and (4.9), we see that all terms drop out except the following,

$$\begin{aligned} 0 = d\omega_b^a &= \omega_b^2 \wedge \omega_2^a + \omega_b^{n+3} \wedge \omega_{n+3}^a \\ &= -(\omega_a^1 \wedge \omega_1^b) + \omega_b^{n+3} \wedge \omega_{n+3}^a . \end{aligned}$$

Thus,

(4.16) $$\omega_a^1 \wedge \omega_1^b = \omega_b^{n+3} \wedge \omega_{n+3}^a , \quad 3 \leq a \leq p+2, \; p+3 \leq b \leq n+1 .$$

We now show that (4.16) implies that there is some function α on U such that

(4.17) $$\begin{aligned} \omega_a^1 &= \alpha\, \omega_{n+3}^a , \quad 3 \leq a \leq p+2 , \\ \omega_b^{n+3} &= -\alpha\, \omega_1^b , \quad p+3 \leq b \leq n+1 . \end{aligned}$$

To see this, note that for any a, $3 \leq a \leq p+2$, (4.15) gives

(4.18) $$\omega_a^1 = c_3\omega_{n+3}^3 + \ldots + c_{p+2}\omega_{n+3}^{p+2} \quad \text{for some } c_3, \ldots, c_{p+2} ,$$

while for any b, $p+3 \leq b \leq n+1$, (4.13) gives

$$(4.19) \qquad \omega_b^{n+3} = d_{p+3}\omega_1^{p+3} + \ldots + d_{n+1}\omega_1^{n+1} \text{ for some } d_{p+3}, \ldots, d_{n+1} .$$

Thus,

$$(4.20) \quad \omega_a^1 \wedge \omega_1^b = c_3\omega_{n+3}^3 \wedge \omega_1^b + \ldots + c_a\omega_{n+3}^a \wedge \omega_1^b + \ldots + c_{p+2}\omega_{n+3}^{p+2} \wedge \omega_1^b ,$$

$$(4.21) \quad \omega_b^{n+3} \wedge \omega_{n+3}^a = d_{p+3}\omega_1^{p+3} \wedge \omega_{n+3}^a + \ldots + d_b\omega_1^b \wedge \omega_{n+3}^a + \ldots + d_{n+1}\omega_1^{n+1} \wedge \omega_{n+3}^a .$$

From (4.16), we know that the right-hand sides of (4.20) and (4.21) are equal. But these expressions contain no common terms from the basis of 2-forms except those involving $\omega_{n+3}^a \wedge \omega_1^b$. Thence, all of the coefficients except c_a and d_b are zero, and we have

$$c_a\omega_{n+3}^a \wedge \omega_1^b = d_b\omega_1^b \wedge \omega_{n+3}^a = (-d_b)\, \omega_{n+3}^a \wedge \omega_1^b .$$

Thus $c_a = -d_b$. If we set $\alpha_a = c_a$ and $\mu_b = d_b$, we have shown that (4.18) and (4.19) reduce to

$$\omega_a^1 = \alpha_a\omega_{n+3}^a , \quad \omega_b^{n+3} = \mu_b\omega_1^b \text{ with } \mu_b = -\alpha_a .$$

This procedure works for any choice of a and b in the appropriate ranges. By holding a fixed and varying b, we see that all of the quantities μ_b are equal to each other and to $-\alpha_a$. Similarly, all of the quantities α_a are the same, and (4.17) holds. We now consider the expression (3.5) for dY_a, $3 \leq a \leq p+2$. We omit the terms which vanish because of (3.6), (4.5) or (4.9),

$$dY_a = \omega_a^1 Y_1 + \omega_a^3 Y_3 + \ldots + \omega_a^{p+2} Y_{p+2} + \omega_a^{n+2} Y_{n+2} + \omega_a^{n+3} Y_{n+3} .$$

Using (4.17) and the fact from (3.6) that $\omega_a^{n+2} = -\omega_{n+3}^a$, this becomes

$$(4.22) \qquad dY_a = \omega_{n+3}^a(\alpha Y_a - Y_{n+2}) + \omega_a^3 Y_3 + \ldots + \omega_a^{p+2} Y_{p+2} + \omega_a^{n+3} Y_{n+3} .$$

Similarly, for $p+3 \leq b \leq n+1$, we find

$$(4.23)\qquad dY_b = \omega_b^1 Y_1 + \omega_b^2(Y_2 + \alpha Y_{n+3}) + \omega_b^{p+3} Y_{p+3} + \ldots + \omega_b^{n+1} Y_{n+1} .$$

We make the change of frame

$$(4.24)\qquad \begin{aligned} &Y_2^* = Y_2 + \alpha Y_{n+3} , \quad Y_{n+2}^* = Y_{n+2} - \alpha Y_1 , \\ &Y_\beta^* = Y_\beta , \quad \beta \neq 2 \text{ or } n+2 . \end{aligned}$$

We now drop the asterisks but use the new frame. From (4.22) and (4.23), we see that in this new frame, we have

$$(4.25)\qquad dY_a = \omega_{n+3}^a(-Y_2) + \omega_a^3 Y_3 + \ldots + \omega_a^{p+2} Y_{p+2} + \omega_a^{n+3} Y_{n+3} ,$$

$$(4.26)\qquad dY_b = \omega_b^1 Y_1 + \omega_b^2 Y_2 + \omega_b^{p+3} Y_{p+3} + \ldots + \omega_b^{n+1} Y_{n+1} .$$

That is, in the new frame, we have

$$(4.27)\qquad \omega_a^1 = 0 , \quad 3 \leq a \leq p+2 ,$$

$$(4.28)\qquad \omega_b^{n+3} = 0 , \quad p+3 \leq b \leq n+1 .$$

Our goal now is to show that the two spaces

$$(4.29)\qquad E = \mathrm{Span}\{Y_1, Y_2, Y_{p+3}, \ldots, Y_{n+1}\} ,$$

and its orthogonal complement

$$(4.30)\qquad E^\perp = \mathrm{Span}\{Y_3, \ldots, Y_{p+2}, Y_{n+2}, Y_{n+3}\}$$

are invariant under d.

Concerning E, we have that $dY_b \in E$ for $p+3 \leq b \leq n+1$ by (4.26). Furthermore, (3.6), (3.9) and (4.5) imply that

$$dY_1 = \omega_1^1 Y_1 + \omega_1^{p+3} Y_{p+3} + \ldots + \omega_1^{n+1} Y_{n+1} ,$$

which is in E. Thus, it only remains to show that dY_2 is in E. To do this, we differentiate (4.27). As before, we omit terms which are zero because of (3.6), (4.5), (4.9) and (4.27). We see that formula (3.7) for $d\omega_a^1$ reduces to

$$(4.31) \qquad 0 = d\omega_a^1 = \omega_a^{n+2} \wedge \omega_{n+2}^1 = -\omega_{n+3}^a \wedge \omega_{n+2}^1 = \omega_{n+3}^a \wedge \omega_2^{n+3}, \ 3 \leq a \leq p+2 .$$

Similarly, by differentiating (4.28), we find that

$$0 = d\omega_b^{n+3} = \omega_b^2 \wedge \omega_2^{n+3} = -\omega_1^b \wedge \omega_2^{n+3}, \ p+3 \leq b \leq n+1 .$$

From this and (4.31), we see that the wedge product of ω_2^{n+3} with every form in the basis (4.16) is zero, and hence $\omega_2^{n+3} = 0$. Using this and the fact that $\omega_2^{n+2} = -\omega_{n+3}^1 = 0$, and that by (3.6) and (4.27),

$$\omega_2^a = -\omega_a^1 = 0 \quad , \quad 3 \leq a \leq p+2 ,$$

we have

$$dY_2 = \omega_2^2 Y_2 + \omega_2^{p+3} Y_{p+3} + \ldots + \omega_2^{n+1} Y_{n+1} ,$$

which is in E. So E is invariant under d and is thus a fixed subspace of P^{n+2}, independent of the choice of point of U. Obviously then, the space $E^{\perp}$ in (4.30) is also a fixed subspace of P^{n+2}.

Note that E has signature (1,q+1) as a vector subspace of $\mathbb{R}_2^{n+3}$, and $E^{\perp}$ has signature (1,p+1). Take an orthonormal basis $e_1, \ldots, e_{n+3}$ of $\mathbb{R}_2^{n+3}$ with e_1 and e_{n+3} timelike and

$$E = \mathrm{Span}\{e_1, \ldots, e_{q+2}\} , \ E^{\perp} = \mathrm{Span}\{e_{q+3}, \ldots, e_{n+3}\} .$$

Then $E \cap Q^{n+1}$ is given in homogeneous coordinates $(x_1, \ldots, x_{n+3})$ with respect to this basis by

$$(4.32) \qquad x_1^2 = x_2^2 + \ldots + x_{q+2}^2 , \ x_{q+3} = \ldots = x_{n+3} = 0 .$$

This quadric is diffeomorphic to the unit sphere S^q in the span E^{q+1} of the spacelike vectors $e_2,\dots,e_{q+2}$ with the diffeomorphism $\phi: S^q \to E \cap Q^{n+1}$ being given by

$$\phi(u) = [e_1 + u] \ , \ u \in S^q \ . \tag{4.33}$$

Similarly, $E^\perp \cap Q^{n+1}$ is the quadric given in homogeneous coordinates by

$$x_{n+3}^2 = x_{q+3}^2 + \dots + x_{n+2}^2 \ , \ x_1 = \dots = x_{q+2} = 0 \ . \tag{4.34}$$

$E^\perp \cap Q^{n+1}$ is diffeomorphic to the unit sphere S^p in the span E^{p+1} of $e_{q+3},\dots,e_{n+2}$ with the diffeomorphism $\psi: S^p \to E \cap Q^{n+1}$ being given by

$$\psi(v) = [v + e_{n+3}] \ , \ v \in S^p \ . \tag{4.35}$$

The focal point map Y_1 of our Dupin submanifold is constant on the leaves of the principal foliation T_r, and so Y_1 factors through an immersion of the q-dimensional space of leaves U/T_r into the q-sphere given by the quadric (4.32). Hence, the image of Y_1 is an open subset of this quadric. Similarly, Y_{n+3} factors through an immersion of the p-dimensional space of leaves U/T_s onto an open subset of the p-sphere given by the quadric (4.34). From this it is clear that the unique compact cyclide containing $\lambda: U \to \Lambda$ as an open submanifold is given by the Dupin submanifold $\bar{\lambda}: S^q \times S^p \to \Lambda$ with

$$\bar{\lambda}(u,v) = [k_1(u,v), k_2(u,v)] \ , \ (u,v) \in S^q \times S^p \ ,$$

where

$$k_1(u,v) = \phi(u) \quad \text{and} \quad k_2(u,v) = \psi(v) \ .$$

Geometrically, the image of $\bar{\lambda}$ consists of all lines joining a point on the quadric (4.32) to a point on the quadric (4.34).

Thus, any choice of (q+1)-plane E in P^{n+2} with signature (1,q+1) and corresponding orthogonal complement $E^{\perp}$ determines a unique compact cyclide of characteristic (p,q) and vice-versa. The local Lie equivalence of any two cyclides of the same characteristic is then clear.

From the standpoint of Euclidean geometry, if we consider the point spheres to be given by $Q^{n+1} \cap e^{\perp}_{n+3}$, as in Section 2, then the Legendre submanifold $\bar{\lambda}$ above is induced in the usual way from the unit normal bundle $B^{n-1} = S^q \times S^p$ of the standard embedding of S^q as a great q-sphere $E^{q+1} \cap S^n$, where E^{q+1} is the span of $e_2,\ldots,e_{q+2}$ and S^n is the unit sphere in E^{n+1}, the span of $e_2,\ldots,e_{n+2}$. The spheres S^q and $S^p = S^n \cap E^{p+1}$, where E^{p+1} is the span of $e_{q+3},\ldots,e_{n+2}$, are the two focal submanifolds in S^n of a standard product of spheres $S^p \times S^q$ in S^n (see [CR, p. 295]).

5. Dupin submanifolds for $n = 4$.

The classification of Dupin submanifolds induced from surfaces in $\mathbb{R}^3$ follows from the results of the last section. In his doctoral dissertation [P1] (later published as [P2]), U. Pinkall obtained a local classification up to Lie equivalence of all Dupin submanifolds induced from hypersurfaces in $\mathbb{R}^4$. As we shall see, this is a far more complicated calculation than that of the previous section, and as yet, no one has obtained a similar classification of Dupin hypersurfaces in $\mathbb{R}^n$ for $n \geq 5$. In this section, we will prove Pinkall's theorem using the method of moving frames.

We follow the notation used in Sections 3 and 4. We consider a Dupin submanifold

(5.1) $$\lambda : B \to \Lambda$$

where

(5.2) $$\dim B = 3\ , \quad \dim \Lambda = 7\ ,$$

and the image $\lambda(b)$, $b \in B$, is the line $[Y_1,Y_7]$ of the Lie frame $Y_1,\ldots Y_7$. We assume that there are three distinct focal points on each line $\lambda(b)$.

By (3.39), we can choose the frame so that

(5.3) $$\omega_1^3 = \omega_1^7 = 0\ , \qquad \omega_7^4 = \omega_7^1 = 0\ .$$

By making a change of frame of the form

$$Y_1^* = \alpha Y_1 \quad , \quad Y_2^* = (1/\alpha)Y_2$$

(5.4)

$$Y_7^* = \beta Y_7 \quad , \quad Y_6^* = (1/\beta)Y_6 \; ,$$

for suitable smooth functions α and β on B, we can arrange that Y_1+Y_7 represents the third focal point at each point of B. Then, using the fact that B is Dupin, we can use the method employed at the end of Section 3 to make a change of frame leading to the following equations similar to (3.39) (and to (5.3)),

(5.5)
$$\omega_1^5 + \omega_7^5 = 0 \; ,$$
$$\omega_1^1 - \omega_7^7 = 0 .$$

This completely fixes the Y_i, $i = 3,4,5$, and Y_1, Y_7 are determined up to a transformation of the form

(5.6)
$$Y_1^* = \tau\, Y_1 \; , \; Y_7^* = \tau\, Y_7 \; .$$

Each of the three focal point maps Y_1, Y_7, $Y_1 + Y_7$ is constant along the leaves of its corresponding principal foliation. Thus, each focal point map factors through an immersion of the corresponding 2-dimensional space of leaves of its principal foliation into Q^5. (See Section 4 of Chapter 2 of the book [CR] for more detail on this point.) In terms of moving frames, this implies that the forms ω_1^4, ω_1^5, ω_7^3 are linearly independent on B, i.e.,

(5.7)
$$\omega_1^4 \wedge \omega_1^5 \wedge \omega_7^3 \neq 0 \; .$$

This can also be seen by expressing the forms above in terms of a Lie frame $\overline{Y}_1, \ldots, \overline{Y}_7$, where $\overline{Y}_1$ satisfies the regularity condition (3.8), and using the fact that each focal point has multiplicity one. For simplicity, we will also use the notation

(5.8)
$$\theta_1 = \omega_1^4 \quad , \quad \theta_2 = \omega_1^5 \quad , \quad \theta_3 = \omega_7^3 \; .$$

Analytically, the Dupin conditions are three partial differential equations, and we are treating an over-determined system. The method of moving frames reduces the handling of its integrability conditions to a straightforward algebraic problem, viz. that of repeated exterior differentiations.

We begin by taking the exterior derivatives of the three equations $\omega_1^3 = 0$, $\omega_7^4 = 0$, $\omega_1^5 + \omega_7^5 = 0$. Using the skew-symmetry relations (3.6), as well as (5.3) and (5.5), the exterior derivatives of these three equations yield the system

$$\begin{aligned} 0 &= \omega_1^4 \wedge \omega_3^4 + \omega_1^5 \wedge \omega_3^5 , \\ 0 &= \omega_1^5 \wedge \omega_4^5 + \omega_7^3 \wedge \omega_3^4 , \\ 0 &= \omega_1^4 \wedge \omega_4^5 + \omega_7^3 \wedge \omega_3^5 . \end{aligned}$$

If we take the wedge product of the first of these with ω_1^4, we conclude that ω_3^5 is in the span of ω_1^4 and ω_1^5. On the other hand, taking the wedge product of the third equation with ω_1^4 yields that ω_3^5 is in the span of ω_1^4 and ω_3^7. Consequently, $\omega_3^5 = \rho\omega_1^4$ for some smooth function ρ on B. Similarly, one can show that there exist functions σ and τ such that $\omega_3^4 = \sigma\,\omega_1^5$ and $\omega_4^5 = \tau\,\omega_7^3$. Then, if we substitute these into the three equations above, we get that $\rho = \sigma = \tau$, and hence we have

$$\omega_3^5 = \rho\,\omega_1^4 , \quad \omega_3^4 = \rho\,\omega_1^5 , \quad \omega_4^5 = \rho\,\omega_7^3 . \tag{5.9}$$

Next we differentiate the equations $\omega_1^7 = 0$, $\omega_7^1 = 0$, $\omega_1^1 - \omega_7^7 = 0$. As above, use of the skew-symmetry relations (3.6) and the equations (5.3), (5.5) yields the existence of smooth functions a,b,c,p,q,r,s,t,u on B such that the following relations hold:

$$\begin{aligned} \omega_4^7 &= -\,\omega_6^4 = a\,\omega_1^4 + b\,\omega_1^5 , \\ \omega_5^7 &= -\,\omega_6^5 = b\,\omega_1^4 + c\,\omega_1^5 ; \end{aligned} \tag{5.10}$$

$$\begin{aligned} \omega_3^1 &= -\,\omega_2^3 = p\,\omega_7^3 - q\,\omega_1^5 , \\ \omega_5^1 &= -\,\omega_2^5 = q\,\omega_7^3 - r\,\omega_1^5 ; \end{aligned} \tag{5.11}$$

$$\omega_4^1 = -\omega_2^4 = b\,\omega_1^5 + s\,\omega_1^4 + t\,\omega_7^3 ,$$

(5.12)

$$\omega_6^3 = -\omega_3^7 = q\,\omega_1^5 + t\,\omega_1^4 + u\,\omega_7^3 .$$

We next see what can be deduced from taking the exterior derivatives of the equations (5.9)-(5.12). First, we take the exterior derivatives of the three basis forms ω_1^4, ω_1^5, ω_7^3. For example, using the relations that we have derived so far, we have from the Maurer-Cartan equation (3.7),

$$d\omega_1^4 = \omega_1^1 \wedge \omega_1^4 + \omega_1^5 \wedge \omega_5^4 = \omega_1^1 \wedge \omega_1^4 - \rho\omega_1^5 \wedge \omega_7^3 .$$

We obtain similar expressions for $d\omega_1^5$ and $d\omega_7^3$. When we use the forms $\theta_1,\theta_2,\theta_3$ defined in (5.8) for ω_1^4, ω_1^5,ω_7^3, we have

(5.13)

$$d\theta_1 = \omega_1^1 \wedge \theta_1 - \rho\,\theta_2 \wedge \theta_3 ,$$
$$d\theta_2 = \omega_1^1 \wedge \theta_2 - \rho\,\theta_3 \wedge \theta_1 ,$$
$$d\theta_3 = \omega_1^1 \wedge \theta_3 - \rho\,\theta_1 \wedge \theta_2 .$$

We next differentiate (5.9). We have $\omega_3^4 = \rho\omega_1^5$. On the one hand,

$$d\omega_3^4 = \rho\; d\omega_1^5 + d\rho \wedge \omega_1^5 .$$

Using the second equation in (5.13) with $\omega_1^5 = \theta_2$, this becomes

$$d\omega_3^4 = \rho\,\omega_1^1 \wedge \omega_1^5 - \rho^2\omega_7^3 \wedge \omega_1^4 + d\rho \wedge \omega_1^5 .$$

On the other hand, we can compute $d\omega_3^4$ from the Maurer-Cartan equation (3.7) and use the relationships that we have derived to find

$$d\omega_3^4 = (-p-\rho^2-a)(\omega_1^4 \wedge \omega_7^3) - q\,\omega_1^5 \wedge \omega_1^4 + b\,\omega_7^3 \wedge \omega_1^5 .$$

If we equate these two expressions for $d\omega_3^4$, we get

(5.14) $$(-p-a-2\rho^2)\,\omega_1^4 \wedge \omega_7^3 = (d\rho + \rho\,\omega_1^1 - q\,\omega_1^4 - b\,\omega_7^3) \wedge \omega_1^5 .$$

Because of the independence of ω_1^4, ω_1^5 and ω_7^3, both sides of the equation above must vanish. Thus, we conclude that

$$2\rho^2 = -a-p \ , \tag{5.15}$$

and that $d\rho + \rho\omega_1^1 - q\omega_1^4 - b\omega_7^3$ is a multiple of ω_1^5. Similarly, differentiation of $\omega_4^5 = \rho\omega_7^3$, yields the following analogue of (5.14),

$$(s-a-r + 2\rho^2)\ \omega_1^4 \wedge \omega_1^5 = (d\rho + \rho\ \omega_1^1 + t\ \omega_1^5 - q\ \omega_1^4) \wedge \omega_7^3 \ , \tag{5.16}$$

and differentiation of $\omega_3^5 = \rho\omega_1^4$ yields

$$(c+p+u - 2\rho^2)\ \omega_1^5 \wedge \omega_7^3 = (-d\rho - \rho\ \omega_1^1 - t\ \omega_1^5 + b\ \omega_7^3) \wedge \omega_1^4 \ . \tag{5.17}$$

In each of the equations (5.14), (5.16), (5.17) both sides of the equation must vanish. From the vanishing of the left-hand sides of the equations, we get the fundamental relationship,

$$2\rho^2 = -a-p = a+r-s = c+p+u \ . \tag{5.18}$$

Furthermore, from the vanishing of the right-hand sides of the three equations (5.14), (5.15) and (5.17), we can determine after some algebra that

$$d\rho + \rho\ \omega_1^1 = q\ \omega_1^4 - t\ \omega_1^5 + b\ \omega_7^3 \ . \tag{5.19}$$

The last equation shows the importance of the function ρ. Following the notation introduced in (5.8), we write (5.19) as

$$d\rho + \rho\omega_1^1 = \rho_1\theta_1 + \rho_2\theta_2 + \rho_3\theta_3, \tag{5.20}$$

where

$$\rho_1 = q,\ \rho_2 = -t,\ \rho_3 = b \ , \tag{5.21}$$

are the "covariant derivatives" of ρ.

Using the Maurer-Cartan equations, we can compute

$$\begin{aligned} d\omega_1^1 &= \omega_1^4 \wedge \omega_4^1 + \omega_1^5 \wedge \omega_5^1 \\ &= \omega_1^4 \wedge (b\,\omega_1^5 + t\,\omega_7^3) + \omega_1^5 \wedge (q\,\omega_7^3 - r\,\omega_1^5) \\ &= b\,\omega_1^4 \wedge \omega_1^5 + q\,\omega_1^5 \wedge \omega_7^3 - t\,\omega_7^3 \wedge \omega_1^4 \,. \end{aligned}$$

Using (5.8) and (5.21), this can be rewritten as

(5.22) $$d\omega_1^1 = \rho_3\,\theta_1 \wedge \theta_2 + \rho_1\,\theta_2 \wedge \theta_3 + \rho_2\,\theta_3 \wedge \theta_1 \,.$$

The trick now is to express everything in terms of $\bar{\rho}$ and its successive covariant derivatives.

We first derive a general form for these covariant derivatives. Suppose that σ is a smooth function which satisfies a relation of the form

(5.23) $$d\sigma + m\,\sigma\omega_1^1 = \sigma_1\theta_1 + \sigma_2\theta_2 + \sigma_3\theta_3$$

for some integer m. (Note that (5.19) is such a relationship for ρ with m=1.) By taking the exterior derivative of (5.23) and using (5.13) and (5.22) to express both sides in terms of the standard basis of two forms $\theta_1 \wedge \theta_2$, $\theta_2 \wedge \theta_3$ and $\theta_3 \wedge \theta_1$, one finds that the functions $\sigma_1, \sigma_2, \sigma_3$ satisfy equations of the form

(5.24) $$d\sigma_\alpha + (m+1)\sigma_\alpha\omega_1^1 = \sigma_{\alpha 1}\theta_1 + \sigma_{\alpha 2}\theta_2 + \sigma_{\alpha 3}\theta_3, \quad \alpha=1,2,3 \,,$$

where the coefficient functions $\sigma_{\alpha\beta}$ satisfy the commutation relations

(5.25) $$\begin{aligned} \sigma_{12} - \sigma_{21} &= -m\sigma\rho_3 - \rho\sigma_3 \,, \\ \sigma_{23} - \sigma_{32} &= -m\sigma\rho_1 - \rho\sigma_1 \,, \\ \sigma_{31} - \sigma_{13} &= -m\sigma\rho_2 - \rho\sigma_2 \,. \end{aligned}$$

In particular, from equation (5.20), we have the following commutation relations on ρ_1, ρ_2, ρ_3:

(5.26) $$\begin{aligned} \rho_{12} - \rho_{21} &= -2\rho\rho_3 \,, \\ \rho_{23} - \rho_{32} &= -2\rho\rho_1 \,, \\ \rho_{31} - \rho_{13} &= -2\rho\rho_2 \,. \end{aligned}$$

We next take the exterior derivatives of the equations (5.10)-(5.12). We first differentiate the equation

$$\omega^7_4 = a\,\omega^4_1 + b\,\omega^5_1 . \tag{5.27}$$

On the one hand, from the Maurer-Cartan equation (3.7) for $d\omega^7_4$, we have (by not writing those terms which have already been shown to vanish),

$$\begin{aligned} d\omega^7_4 &= \omega^2_4 \wedge \omega^7_2 + \omega^3_4 \wedge \omega^7_3 + \omega^5_4 \wedge \omega^7_5 + \omega^7_4 \wedge \omega^7_7 \\ &= -\,\omega^4_1 \wedge \omega^7_2 + (-\rho\,\omega^5_1) \wedge (-q\,\omega^5_1 - t\,\omega^4_1 - u\,\omega^3_7) \\ &\quad + \rho\,\omega^3_7 \wedge (b\,\omega^4_1 + c\,\omega^5_1) + (a\,\omega^4_1 + b\,\omega^5_1) \wedge \omega^1_1 . \end{aligned} \tag{5.28}$$

On the other hand, differentiation of the right-hand side of (5.27) yields

$$\begin{aligned} d\omega^7_4 &= da \wedge \omega^4_1 + a\,d\omega^4_1 + db \wedge \omega^5_1 + b\,d\omega^5_1 \\ &= da \wedge \omega^4_1 + a(\omega^1_1 \wedge \omega^4_1 - \rho\,\omega^5_1 \wedge \omega^3_7) \\ &\quad + db \wedge \omega^5_1 + b(\omega^1_1 \wedge \omega^5_1 - \rho\,\omega^4_1 \wedge \omega^3_7) . \end{aligned} \tag{5.29}$$

Equating (5.28) and (5.29) yields

$$\begin{aligned} &(da + 2a\,\omega^1_1 - 2b\rho\,\omega^3_7 - \omega^7_2) \wedge \omega^4_1 \\ &+(db + 2b\,\omega^1_1 + (a+u-c)\rho\,\omega^3_7) \wedge \omega^5_1 \\ &+ \rho t\,\omega^4_1 \wedge \omega^5_1 = 0 . \end{aligned} \tag{5.30}$$

Since $b = \rho_3$, it follows from (5.19) and (5.24) that

$$db + 2b\,\omega^1_1 = d\rho_3 + 2\rho_3\,\omega^1_1 = \rho_{31}\theta_1 + \rho_{32}\theta_2 + \rho_{33}\theta_3 . \tag{5.31}$$

By examining the coefficient of $\omega^5_1 \wedge \omega^3_7 = \theta_2 \wedge \theta_3$ in equation (5.30) and using (5.31), we get that

$$\rho_{33} = \rho(c-a-u) . \tag{5.32}$$

Furthermore, the remaining terms in (5.30) are

$$(5.33) \qquad (da + 2a\,\omega_1^1 - \omega_2^7 - 2\rho b\,\omega_7^3 - (\rho t + \rho_{31})\omega_1^5) \wedge \omega_1^4$$
$$+ \text{ terms involving } \omega_1^5 \text{ and } \omega_7^3 \text{ only.}$$

Thus, the coefficient in parentheses must be a multiple of ω_1^4, call it $\bar{a}\omega_1^4$. We can write this using (5.8) and (5.21) as

$$(5.34) \qquad da + 2a\,\omega_1^1 = \omega_2^7 + \bar{a}\theta_1 + (\rho_{31} - \rho\rho_2)\theta_2 + 2\rho\rho_3\theta_3 \; .$$

In a similar manner, if we differentiate

$$\omega_5^7 = b\,\omega_1^4 + c\,\omega_1^5 \; ,$$

we obtain,

$$(5.35) \qquad dc + 2c\,\omega_1^1 = \omega_2^7 + (\rho_{32} + \rho\rho_1)\theta_1 + \bar{c}\theta_2 - 2\rho\rho_3\theta_3 .$$

Thus, from the two equations in (5.10), we have obtained (5.32), (5.34) and (5.35). In completely analogous fashion, we can differentiate the two equations in (5.11) to obtain

$$(5.36) \qquad \rho_{11} = \rho(s+r-p) \; ,$$

$$(5.37) \qquad dp + 2p\,\omega_1^1 = -\,\omega_2^7 + 2\rho\rho_1\theta_1 + (-\rho_{13} - \rho\rho_2)\theta_2 + \bar{p}\theta_3 \; ,$$

$$(5.38) \qquad dr + 2r\,\omega_1^1 = -\,\omega_2^7 - 2\rho\rho_1\theta_1 + \bar{r}\theta_2 + (-\rho_{12} + \rho\rho_3)\theta_3 ,$$

while differentiation of (5.12) yields

$$(5.39) \qquad \rho_{22} + \rho_{33} = \rho(p-r-s) ,$$

$$(5.40) \qquad ds + 2s\,\omega_1^1 = \bar{s}\theta_1 + (\rho_{31} + \rho\rho_2)\theta_2 + (-\rho_{21} + \rho\rho_3)\theta_3 \; ,$$

$$(5.41) \qquad du + 2u\,\omega_1^1 = (-\rho_{23} - \rho\rho_1)\theta_1 + (\rho_{13} - \rho\rho_2)\theta_2 + \bar{u}\theta_3 \; .$$

In these equations, the coefficients $\bar{a}, \bar{c}, \bar{p}, \bar{r}, \bar{s}, \bar{u}$ remain undetermined. However, by differentiating (5.18) and using the appropriate equations among those involving these quantities above, one can show that

$$\bar{a} = -6\rho\rho_1 \quad , \quad \bar{c} = 6\rho\rho_2 \; ,$$

(5.42) $$\bar{p} = -6\rho\rho_3 \quad , \quad \bar{r} = 6\rho\rho_2 \; ,$$

$$\bar{s} = -12\rho\rho_1 \quad , \quad \bar{u} = 12\rho\rho_3 \; .$$

From equations (5.32), (5.36), (5.39) and (5.18), we easily compute that

(5.43) $$\rho_{11} + \rho_{22} + \rho_{33} = 0 \; .$$

Using (5.42), equations (5.40) and (5.41) can be rewritten as

(5.44) $$ds + 2s\,\omega_1^1 = -12\rho\rho_1\theta_1 + (\rho_{31} + \rho\rho_2)\theta_2 + (-\rho_{21} + \rho\rho_3)\theta_3 \; ,$$

(5.45) $$du + 2u\,\omega_1^1 = (-\rho_{23} - \rho\rho_1)\theta_1 + (\rho_{13} - \rho\rho_2)\theta_2 + 12\rho\rho_3\theta_3 \; .$$

By taking the exterior derivatives of these two equations and making use of (5.43) and of the commutation relations (5.25) for ρ and its various derivatives, one ultimately can show after a lengthy calculation that the following fundamental equations hold:

$$\rho\rho_{12} + \rho_1\rho_2 + \rho^2\rho_3 = 0 \; .$$

$$\rho\rho_{21} + \rho_1\rho_2 - \rho^2\rho_3 = 0 \; .$$

$$\rho\rho_{23} + \rho_2\rho_3 + \rho^2\rho_1 = 0 \; ,$$

(5.46) $$\rho\rho_{32} + \rho_2\rho_3 - \rho^2\rho_1 = 0 \; ,$$

$$\rho\rho_{31} + \rho_3\rho_1 + \rho^2\rho_2 = 0 \; ,$$

$$\rho\rho_{13} + \rho_3\rho_1 - \rho^2\rho_2 = 0 \; .$$

We now briefly outline the details of this calculation. By (5.44), we have

(5.47) $$s_1 = -12\rho\rho_1 \quad , \quad s_2 = \rho_{31} + \rho\rho_2 \quad , \quad s_3 = \rho\rho_3 - \rho_{21} \; .$$

The commutation relation (5.25) for s with m=2 gives

(5.48) $$s_{12} - s_{21} = -2s\rho_3 - \rho s_3 = -2s\rho_3 - \rho(\rho\rho_3 - \rho_{21}) \; .$$

On the other hand, we can directly compute by taking covariant derivatives of (5.47) that

(5.49) $$s_{12} - s_{21} = -12\rho\rho_{12} - 12\rho_2\rho_1 - (\rho_{311} + \rho_1\rho_2 + \rho\rho_{21}) .$$

The main problem now is to get ρ_{311} into a usable form. By taking the covariant derivative of the third equation in (5.26), we find

(5.50) $$\rho_{311} - \rho_{131} = -2\rho_1\rho_2 - 2\rho\rho_{21} .$$

Then using the commutation relation

$$\rho_{131} = \rho_{113} - 2\rho_1\rho_2 - \rho\rho_{12} ,$$

we get from (5.50)

(5.51) $$\rho_{311} = \rho_{113} - 4\rho_1\rho_2 - \rho\rho_{12} - 2\rho\rho_{21} .$$

Taking the covariant derivative of $\rho_{11} = \rho(s+r-p)$ and substituting the expression obtained for ρ_{113} into (5.51), we get

(5.52) $$\rho_{311} = \rho_3(s+r-p)-3\rho\rho_{21}-2\rho\rho_{12} + 8\rho^2\rho_3-4\rho_1\rho_2 .$$

If we substitute (5.52) for ρ_{311} in (5.49) and then equate the right-hand sides of (5.48) and (5.49), we obtain the first equation in (5.46). The cyclic permutations are obtained in a similar way from $s_{23} - s_{32}$, etc.

Our frame attached to the line $[Y_1, Y_7]$ is still not completely determined, viz., the following change is allowable:

(5.53) $$Y_2^* = \alpha^{-1}Y_2 + \mu Y_7 , \; Y_6^* = \alpha^{-1}Y_6 - \mu Y_1 .$$

The Y_i's, i = 3,4,5 being completely determined, we have under this change,

$$\omega_1^{4*} = \alpha\omega_1^4,\ \omega_1^{5*} = \alpha\omega_1^5,\ \omega_7^{3*} = \alpha\omega_7^3,$$
$$\omega_4^{7*} = \alpha^{-1}\omega_4^7 + \mu\omega_1^4,$$
$$\omega_3^{1*} = \alpha^{-1}\omega_3^1 - \mu\omega_7^3,$$

which implies that

$$a^* = \alpha^{-2}a + \alpha^{-1}\mu,$$
$$p^* = \alpha^{-2}p - \alpha^{-1}\mu.$$

We choose μ to make $a^* = p^*$. After dropping the asterisks, we have from (5.18) that

(5.54) $$a = p = -\rho^2,\ r = 3\rho^2 + s,\ c = 3\rho^2 - u.$$

Now using the fact that $a = p$, we can subtract (5.37) from (5.34) and get that

(5.55) $$\omega_2^7 = 4\rho\rho_1\theta_1 - ((\rho_{31}+\rho_{13})/2)\theta_2 - 4\rho\rho_3\theta_3.$$

We are finally in position to proceed toward the main results. Ultimately, we show that the frame can be chosen so that the function ρ is constant, and the classification naturally splits into the two cases $\rho = 0$ and $\rho \neq 0$.

<u>The case $\rho\neq 0$</u>.

We now assume that the function ρ is never zero on B. The following lemma is the key in this case. This is Pinkall's Lemma [P2, p. 108], where his function c is the negative of our function ρ. Since $\rho \neq 0$, the fundamental equations (5.46) allow one to express all of the second covariant derivatives $\rho_{\alpha\beta}$ in terms of ρ and its first derivatives ρ_α. This enables us to give a somewhat simpler proof than Pinkall gave for the lemma.

<u>Lemma 5.1</u>: **Suppose that ρ never vanishes on B. Then $\rho_1 = \rho_2 = \rho_3 = 0$ at every point of B.**

<u>**Proof:**</u> First note that if the function ρ_3 vanishes identically, then (5.46) and the assumption that $\rho \neq 0$ imply that ρ_1 and ρ_2 also vanish identically.

We now complete the proof of the lemma by showing that ρ_3 must vanish everywhere. This is accomplished by considering the expression $s_{12} - s_{21}$. By the commutation relations (5.25), we have

$$s_{12} - s_{21} = -2s\rho_3 - \rho s_3 .$$

By (5.46) and (5.47), we see that

$$\rho s_3 = \rho^2\rho_3 - \rho\rho_{21} = \rho_1\rho_2 ,$$

and so

$$s_{12} - s_{21} = -2s\rho_3 - \rho_1\rho_2 . \tag{5.56}$$

On the other hand, we can compute s_{12} directly from the equation $s_1 = -12\rho\rho_1$. Then using the expression for ρ_{12} obtained from (5.46), we get

$$\begin{aligned} s_{12} &= -12\rho_2\rho_1 - 12\rho\rho_{12} = -12(\rho_2\rho_1 + \rho\rho_{12}) \\ &= -12(\rho_2\rho_1 + (-\rho_2\rho_1 - \rho^2\rho_3)) = 12\rho^2\rho_3 . \end{aligned} \tag{5.57}$$

Next we have from (5.47),

$$s_2 = \rho_{31} + \rho\rho_2 .$$

Using (5.46), we can write

$$\rho_{31} = -\rho_3\rho_1\rho^{-1} - \rho\rho_2 ,$$

and thus

$$s_2 = -\rho_3\rho_1/\rho . \tag{5.58}$$

Then, we compute

$$s_{21} = -(\rho(\rho_3\rho_{11} + \rho_{31}\rho_1) - \rho_3\rho_1^2)/\rho^2 .$$

Using (5.36) for ρ_{11} and (5.46) to get ρ_{31}, this becomes

(5.59) $$s_{21} = -\rho_3(s+r-p) + 2\rho_3\rho_1^2\rho^{-2} + \rho_1\rho_2 .$$

Now equate the expression (5.56) for $s_{12} - s_{21}$ with that obtained by subtracting (5.59) from (5.57) to get

$$-2s\rho_3 - \rho_1\rho_2 = 12\rho^2\rho_3 + \rho_3(s+r-p) - 2\rho_3\rho_1^2\rho^{-2} - \rho_1\rho_2 .$$

This can be rewritten as

(5.60) $$0 = \rho_3(12\rho^2 + 3s+r-p - 2\rho_1^2\rho^{-2}) .$$

Using the expressions in (5.54) for r and p, we see that $3s+r-p = 4s + 4\rho^2$, and so (5.60) can be written as

(5.61) $$0 = \rho_3(16\rho^2 + 4s - 2\rho_1^2\rho^{-2}) .$$

Suppose that $\rho_3 \neq 0$ at some point $b \in B$. Then ρ_3 does not vanish on some neighborhood U of b. By (5.61), we have

(5.62) $$16\rho^2 + 4s - 2\rho_1^2\rho^{-2} = 0$$

on U. We now take the θ_2-covariant derivative of (5.62) and obtain

(5.63) $$32\rho\rho_2 + 4s_2 - 4\rho_1\rho_{12}\rho^{-2} + 4\rho_1^2\rho_2\rho^{-3} = 0 .$$

We now substitute the expression (5.58) for s_2 and the formula

$$\rho_{12} = -\rho_1\rho_2\rho^{-1} - \rho\rho_3$$

obtained from (5.46) into (5.63). After some algebra, (5.63) reduces to

$$\rho_2(32\rho^4 + 8\rho_1^2) = 0 .$$

Since $\rho \neq 0$, this implies that $\rho_2 = 0$ on U. But then the left side of the equation (5.46)

$$\rho\rho_{21} + \rho_1\rho_2 = \rho^2\rho_3$$

must vanish on U. Since $\rho \neq 0$, we conclude that $\rho_3 = 0$ on U, a contradiction to our assumption. Hence, ρ_3 must vanish identically on B and the lemma is proven.

We now continue with the case $\rho \neq 0$. According to Lemma 5.1, all the covariant derivatives of ρ are zero, and our formulas simplify greatly. Equations (5.32) and (5.36) give

$$c-a-u = 0 \ , \ s+r-p = 0 \ .$$

These combined with (5.54) give

$$c = r = \rho^2 \ , \ u = -s = 2\rho^2 \ . \tag{5.64}$$

By (5.55) we have $\omega_2^7 = 0$. So the differentials of the frame vectors can now be written

$$\begin{aligned}
dY_1 - \omega_1^1 Y_1 &= \omega_1^4 Y_4 + \omega_1^5 Y_5 \ , \\
dY_7 - \omega_1^1 Y_7 &= \omega_7^3 Y_3 - \omega_1^5 Y_5 \ , \\
dY_2 + \omega_1^1 Y_2 &= \rho^2(\omega_7^3 Y_3 + 2\omega_1^4 Y_4 + \omega_1^5 Y_5) \ , \\
dY_6 + \omega_1^1 Y_6 &= \rho^2(2\omega_7^3 Y_3 + \omega_1^4 Y_4 - \omega_1^5 Y_5) \ , \\
dY_3 &= \omega_7^3 Z_3 + \rho(\omega_1^5 Y_4 + \omega_1^4 Y_5) \ , \\
dY_4 &= -\omega_1^4 Z_4 + \rho(-\omega_1^5 Y_3 + \omega_7^3 Y_5) \ , \\
dY_5 &= \omega_1^5 Z_5 + \rho(-\omega_1^4 Y_3 - \omega_7^3 Y_4) \ ,
\end{aligned} \tag{5.65}$$

where

$$\begin{aligned}
Z_3 &= -Y_6 + \rho^2(-Y_1 - 2Y_7) \ , \\
Z_4 &= Y_2 + \rho^2(2Y_1 + Y_7) \ , \\
Z_5 &= -Y_2 + Y_6 + \rho^2(-Y_1 + Y_7) \ .
\end{aligned} \tag{5.66}$$

From this, we notice that

(5.67) $$Z_3 + Z_4 + Z_5 = 0 ,$$

so that the points Z_3, Z_4, Z_5 lie on a line.

From (5.20) and (5.22) and the lemma we see that

(5.68) $$d\rho + \rho\, \omega_1^1 = 0 \quad , \quad d\omega_1^1 = 0 .$$

We now make a change of frame of the form

(5.69) $$\begin{aligned} &Y_1^* = \rho Y_1,\ Y_7^* = \rho Y_7,\ Y_2^* = (1/\rho) Y_2,\ Y_6^* = (1/\rho) Y_6, \\ &Y_i^* = Y_i,\ i = 3,4,5. \end{aligned}$$

Then set

(5.70) $$\begin{aligned} &Z_i^* = (1/\rho) Z_i \quad , \quad i = 3,4,5 . \\ &\omega_1^{4*} = \rho\, \omega_1^4,\ \omega_1^{5*} = \rho\, \omega_1^5,\ \omega_7^{3*} = \rho\, \omega_7^3 . \end{aligned}$$

The effect of this change is to make $\rho^* = 1$ and $\omega_1^{1*} = 0$, for we can compute the following:

(5.71) $$\begin{aligned} dY_1^* &= \omega_1^{4*} Y_4 + \omega_1^{5*} Y_5 , \\ dY_7^* &= \omega_7^{3*} Y_3 - \omega_1^{5*} Y_5 , \\ dY_2^* &= \omega_7^{3*} Y_3 + 2\omega_1^{4*} Y_4 + \omega_1^{5*} Y_5 , \\ dY_6^* &= 2\omega_7^{3*} Y_3 + \omega_1^{4*} Y_4 - \omega_1^{5*} Y_5 , \\ dY_3 &= \omega_7^{3*} Z_3^* + \omega_1^{5*} Y_4 + \omega_1^{4*} Y_5 , \\ dY_4 &= - \omega_1^{4*} Z_4^* - \omega_1^{5*} Y_3 + \omega_7^{3*} Y_5 , \\ dY_5 &= \omega_1^{5*} Z_5 - \omega_1^{4*} Y_3 - \omega_7^{3*} Y_4 , \end{aligned}$$

with

(5.72) $$\begin{aligned} dZ_3^* &= 2(-2\omega_7^{3*} Y_3 - \omega_1^{4*} Y_4 + \omega_1^{5*} Y_5) , \\ dZ_4^* &= 2(\omega_7^{3*} Y_3 + 2\omega_1^{4*} Y_4 + \omega_1^{5*} Y_5) , \\ dZ_5^* &= 2(\omega_7^{3*} Y_3 - \omega_1^{4*} Y_4 - 2\omega_1^{5*} Y_5) , \end{aligned}$$

and

$$d\omega_1^{4*} = -\omega_1^{5*} \wedge \omega_7^{3*}, \text{ i.e.}, d\theta_1^* = -\theta_2^* \wedge \theta_3^*,$$
$$d\omega_1^{5*} = -\omega_7^{3*} \wedge \omega_1^{4*}, \text{ i.e.}, d\theta_2^* = -\theta_3^* \wedge \theta_1^*,$$
$$d\omega_7^{3*} = -\omega_1^{4*} \wedge \omega_1^{5*}, \text{ i.e.}, d\theta_3^* = -\theta_1^* \wedge \theta_2^*.$$

Comparing the last equation with (5.13), we see that $\omega_1^{1*} = 0$ and $\rho^* = 1$.

This is the final frame which we will need in this case $\rho \neq 0$. So, we again drop the asterisks.

We are now ready to prove Pinkall's classification result for the case $\rho \neq 0$ [P2, p. 117]. As with the cyclides, there is only one compact model, up to Lie equivalence. This is Cartan's isoparametric hypersurface M^3 in S^4. It is a tube of constant radius over each of its two focal submanifolds, which are standard Veronese surfaces in S^4. (See [CR, pp. 296-299] for more detail.) We will describe the Veronese surface after stating the theorem.

Theorem 5.2: (Pinkall [P2]): (a) Every connected Dupin submanifold with $\rho \neq 0$ is contained in a unique compact connected Dupin submanifold with $\rho \neq 0$.
(b) Any two Dupin submanifolds with $\rho \neq 0$ are locally Lie equivalent, each being Lie equivalent to an open subset of Cartan's isoparametric hypersurface in S^4.

Our method of proof differs from that of Pinkall in that we will prove directly that each of the focal submanifolds can naturally be considered to be an open subset of a Veronese surface in a hyperplane $P^5 \subset P^6$. The Dupin submanifold can then be constructed from these focal submanifolds.

We now recall the definition of a Veronese surface. First consider the map from the unit sphere $y_1^2 + y_2^2 + y_3^2 = 1$ in $\mathbb{R}^3$ into $\mathbb{R}^5$ given by

$$(5.73) \qquad (x_1, \ldots, x_5) = (2y_2y_3, 2y_3y_1, 2y_1y_2, y_1^2, y_2^2) .$$

This map takes the same value on antipodal points of the 2-sphere, so it induces a map $\phi: P^2 \to \mathbb{R}^5$. One can show by an elementary direct calculation that ϕ is an embedding of P^2 and that ϕ is substantial in $\mathbb{R}^5$, i.e., does not lie in any hyperplane. Any embedding of P^2 into P^5 which is projectively equivalent to ϕ is called a <u>Veronese surface</u>. (See Lane [L, pp. 424-430] for more detail.)

Let $k_1 = Y_1$, $k_2 = Y_7$, $k_3 = Y_1 + Y_7$ be the focal point maps of the Dupin submanifold $\lambda : B \to \Lambda$ with $\rho \neq 0$. Each k_i is constant along the leaves of its corresponding principal foliation T_i, so each k_i factors through an immersion ϕ_i of the 2-dimensional space of leaves B/T_i into P^6. We will show that each of these ϕ_i is an open subset of a Veronese surface in some $P^5 \subset P^6$.

We wish to integrate the differential system (5.71), which is completely integrable. For this purpose we drop the asterisks and write the system as follows:

(5.74)
$$\begin{aligned} dY_1 &= \theta_1 Y_4 + \theta_2 Y_5 , \\ dY_7 &= \theta_3 Y_3 - \theta_2 Y_5 , \\ dY_2 &= \theta_3 Y_3 + 2\theta_1 Y_4 + \theta_2 Y_5 , \\ dY_6 &= 2\theta_3 Y_3 + \theta_1 Y_4 - \theta_2 Y_5 , \\ dY_3 &= \theta_3 Z_3 + \theta_2 Y_4 + \theta_1 Y_5 , \\ dY_4 &= -\theta_1 Z_4 - \theta_2 Y_3 + \theta_3 Y_5 , \\ dY_5 &= \theta_2 Z_5 - \theta_1 Y_3 - \theta_3 Y_4 ; \end{aligned}$$

with

(5.75)
$$\begin{aligned} dZ_3 &= 2(-2\theta_3 Y_3 - \theta_1 Y_4 + \theta_2 Y_5) , \\ dZ_4 &= 2(\theta_3 Y_3 + 2\theta_1 Y_4 + \theta_2 Y_5) , \\ dZ_5 &= 2(\theta_3 Y_3 - \theta_1 Y_4 - 2\theta_2 Y_5) , \end{aligned}$$

where

(5.76)
$$\begin{aligned} d\theta_1 &= -\theta_2 \wedge \theta_3 , \\ d\theta_2 &= -\theta_3 \wedge \theta_1 , \\ d\theta_3 &= -\theta_1 \wedge \theta_2 , \end{aligned}$$

and

(5.77)
$$\begin{aligned} Z_3 &= -Y_1 - Y_6 - 2Y_7 , \\ Z_4 &= 2Y_1 + Y_2 + Y_7 , \\ Z_5 &= -Y_1 - Y_2 + Y_6 + Y_7 , \end{aligned}$$

so that

(5.78)
$$Z_3 + Z_4 + Z_5 = 0 .$$

Put

(5.79)
$$W_1 = -Y_1 + Y_6 - 2Y_7, \quad W_2 = -2Y_1 + Y_2 - Y_7 .$$

We find from (5.74) that

$$dW_1 = dW_2 = 0 , \tag{5.80}$$

so that the points W_1, W_2 are fixed. Their inner products are

$$\langle W_1, W_1 \rangle = \langle W_2, W_2 \rangle = -4, \ \langle W_1, W_2 \rangle = -2 , \tag{5.81}$$

and the line $[W_1, W_2]$ consists entirely of timelike points. Its orthogonal complement in $\mathbb{R}^7_2$ is spanned by Y_3, Y_4, Y_5, Z_4, Z_5. It consists entirely of spacelike points and has no point in common with Q^5. We will denote it as $\mathbb{R}^5$.

It suffices to solve the system (5.74) in $\mathbb{R}^5$ for Y_3, Y_4, Y_5, Z_4, Z_5. For we have

$$\begin{aligned} &d(Z_4 - Z_5 - 6Y_1) = 0 , \\ &d(Z_4 + 2Z_5 - 6Y_7) = 0 , \end{aligned} \tag{5.82}$$

so that there exist constant vectors C_1, C_2 such that

$$\begin{aligned} &Z_4 - Z_5 - 6Y_1 = C_1 , \\ &Z_4 + 2Z_5 - 6Y_7 = C_2 . \end{aligned} \tag{5.83}$$

Thus, Y_1 and Y_7 are determined by these equations, and then Y_2 and Y_6 are determined from (5.79). Note that C_1 and C_2 are timelike points and the line $[C_1, C_2]$ consists entirely of timelike points.

Equations (5.76) are the structure equations of SO(3). It is thus natural to take the latter as the parameter space, whose points are the 3x3 matrices

$$A = [a_{ik}] , \ 1 \leq i,j,k \leq 3 ,$$

satisfying

$$^tAA = A\,^tA = I , \ \det A = 1 . \tag{5.84}$$

The first equations above, when expanded, are

(5.85) $$\Sigma\, a_{ij}a_{ik} = \Sigma\, a_{ji}a_{ki} = \delta_{jk}\ .$$

The Maurer-Cartan forms of SO(3) are

(5.86) $$\alpha_{ik} = \Sigma\, a_{kj}da_{ij} = -\,\alpha_{ki}\ .$$

They satisfy the Maurer-Cartan equations

(5.87) $$d\alpha_{ik} = \Sigma\, \alpha_{ij} \wedge \alpha_{jk}\ .$$

If we set

(5.88) $$\theta_1 = \alpha_{23}\ ,\ \theta_2 = \alpha_{31}\ ,\ \theta_3 = \alpha_{12}\ ,$$

these equations reduce to (5.76). With the θ_i given by (5.88), we shall write down an explicit solution of (5.74).

Let E_A, $1 \le A \le 5$, be a fixed linear frame in $\mathbb{R}^5$. Let

(5.89) $$F_i = 2a_{i2}a_{i3}E_1+2a_{i3}a_{i1}E_2+2a_{i1}a_{i2}E_3+a_{i1}^2E_4+a_{i2}^2E_5,\ 1 \le i,j,k \le 3\ .$$

Since $$a_{i1}^2 + a_{i2}^2 + a_{i3}^2 = 1\ ,$$

we see from (5.73), with $y_j = a_{ij}$, that F_i is a Veronese surface for $1 \le i \le 3$. Using (5.85), we compute that

(5.90) $$F_1 + F_2 + F_3 = E_4 + E_5 = \text{constant}.$$

Since the coefficients in F_i are quadratic, the partial derivatives $\partial^2F_i/\partial a_{ij}\partial a_{ik}$ are independent of i. Moreover, the quantities G_{ik} defined below satisfy

(5.91) $$G_{ik} = \sum_j a_{ij}\frac{\partial F_k}{\partial a_{kj}} = G_{ki}\ .$$

We use these facts in the following computation:

$$dG_{ik} = \Sigma \frac{\partial F_k}{\partial a_{kj}} da_{ij} + \Sigma a_{ij} \frac{\partial^2 F_k}{\partial a_{kj}\partial a_{k\ell}} da_{k\ell}$$

(5.92)

$$= \Sigma \frac{\partial F_k}{\partial a_{kj}} da_{ij} + \Sigma a_{ij} \frac{\partial^2 F_i}{\partial a_{ij}\partial a_{i\ell}} da_{k\ell} = \Sigma \frac{\partial F_k}{\partial a_{kj}} da_{ij} + \Sigma \frac{\partial F_i}{\partial a_{ij}} da_{kj} ,$$

where the last step follows from the linear homogeneity of $\partial F_i/\partial a_{i\ell}$. In terms of α_{ij}, we have

$$dG_{ik} = \Sigma \frac{\partial F_k}{\partial a_{kj}} a_{\ell j}\alpha_{i\ell} + \Sigma \frac{\partial F_i}{\partial a_{ij}} a_{\ell j}\alpha_{k\ell} , \tag{5.93}$$

which gives, when expanded,

$$dG_{23} = 2(F_3-F_2)\theta_1 + G_{12}\theta_2 - G_{13}\theta_3 , \tag{5.94}$$

and its cyclic permutations.

On the other hand, by the same manipulation, we have

$$dF_i = \Sigma \frac{\partial F_i}{\partial a_{ij}} da_{ij} = \Sigma \frac{\partial F_i}{\partial a_{ij}} a_{kj}\alpha_{ik} , \tag{5.95}$$

giving

$$dF_1 = - G_{31}\theta_2 + G_{12}\theta_3 ,$$

and its cyclic permutations.

One can now immediately verify that a solution of (5.74) is given by

$$\begin{aligned} &Y_3 = G_{12} , \ Y_4 = - G_{23} , \ Y_5 = G_{31} , \\ &Z_3 = 2(F_2-F_1), \ Z_4 = 2(F_3-F_2), \ Z_5 = 2(F_1-F_3) . \end{aligned} \tag{5.96}$$

This is also the most general solution of (5.74), for the solution is determined up to a linear transformation, and our choice of frame E_A is arbitrary.

By (5.96), the functions Z_1, Z_2, Z_3 are expressible in terms of F_1, F_2, F_3, and then by (5.83), so also are Y_1, Y_7, Y_1+Y_7. Specifically, by (5.83), (5.90), and (5.96) we have

$$6Y_1 = Z_4 - Z_5 - C_1 = 2(-F_1 - F_2 + 2F_3) - C_1$$
$$= 6F_3 - 2(E_4 + E_5) - C_1 ,$$

so that the focal map Y_1, up to an additive constant vector, is the Veronese surface F_3. Similarly, the focal maps Y_7 and Y_1+Y_7 are the Veronese surfaces F_1 and $-F_2$, respectively, up to additive constants.

We see from (5.79) that

$$\langle Y_1, W_1 \rangle = 0, \ \langle Y_7, W_2 \rangle = 0, \ \langle Y_1+Y_7, W_1 - W_2 \rangle = 0 .$$

Thus Y_1 is contained in the Moebius space $\Sigma^4 = Q^5 \cap W_1^{\perp}$. Let $e_7 = W_1/2$ and $e_1 = (2W_2 - W_1)/\sqrt{12}$. Then e_1 is the unique unit vector on the timelike line $[W_1, W_2]$ which is orthogonal to W_1. In a manner similar to that of Section 3, we can write

$$Y_1 = e_1 + f ,$$

where f maps B into the unit sphere S^4 in the Euclidean space $\mathbb{R}^5 = [e_1, e_7]^{\perp}$ in $\mathbb{R}^7_2$. We call f the <u>spherical</u> <u>projection</u> of the Legendre map λ determined by the ordered pair $\{e_7, e_1\}$ (see [CC] for more detail). We know that f is constant along the leaves of the principal foliation T_1 corresponding to Y_1, and f induces a map $\tilde{f}: B/T_1 \to S^4$. By what we have shown above, $\tilde{f}$ must be an open subset a spherical Veronese surface.

Note that the unit timelike vector $W_2/2$ satisfies

$$W_2/2 = (\sqrt{3}/2)e_1 + (1/2)e_7 = \sin(\pi/3)e_1 + \cos(\pi/3)e_7 .$$

If we consider the points in the Moebius space Σ to represent point spheres in S^4, then as we show in [CC], points in $Q^5 \cap W_2^{\perp}$ represent oriented spheres in S^4 with oriented radius $-\pi/3$. In a way similar to that above, the second focal submanifold $Y_7 \subset Q^5 \cap W_2^{\perp}$ induces a spherical Veronese surface. When considered from the point of view of the Moebius space Σ, the points in Y_7 represent oriented spheres of radius $-\pi/3$ centered at points of this Veronese surface. These spheres must be in oriented contact with the point spheres of the first Veronese surface determined by Y_1 in $Q^5 \cap P^5$. Thus, the points in the second Veronese surface must lie at a distance $\pi/3$ along normal geodesics in S^4 to the first Veronese surface $\tilde{f}$. In fact (see, for example, [CR,pp. 296-299]), the set of all points in S^4 at distance $\pi/3$ from a spherical Veronese surface is another spherical Veronese surface.

Thus, with this choice of coordinates, the Dupin submanifold in question is simply an open subset of the Legendre submanifold induced in the standard way by considering B to be the unit normal bundle to the spherical Veronese embedding $\tilde{f}$ induced by Y_1. For values of $t = k\pi/3$, $k \in Z$, the parallel hypersurface at oriented distance t to $\tilde{f}$ in S^4 is a Veronese surface. For other values of t, the parallel hypersurface is an isoparametric hypersurface in S^4 with three distinct principal curvatures (Cartan's isoparametric hypersurface). All of these parallel hypersurfaces are Lie equivalent to each other and to the Legendre submanifolds induced by the Veronese surfaces.

The case $\rho = 0$.

We now consider the case where ρ is identically zero on B. It turns out that no new examples occur here, in that these Dupin submanifolds can all be constructed from Dupin cyclides by certain standard constructions. To make this precise, we recall Pinkall's [P3, p. 437] notion of reducibility. Our Dupin submanifold can be considered, as in Section 3, to have been induced from a Dupin hypersurface $M^3 \subset E^4$. The Dupin submanifold is reducible if M^3 is obtained from a Dupin surface $S \subset E^3 \subset E^4$ by one of the four following standard constructions.

i. M is a cylinder $S \times \mathbb{R}$ in E^4.

ii. M is the hypersurface of revolution obtained by revolving S about a plane π disjoint from S in E^3.

(5.97)

iii. Project S stereographically onto a surface $N \subset S^3 \subset E^4$. M is the cone $\mathbb{R}\cdot N$ over N.

iv. M is a tube of constant radius around S in E^4.

Pinkall proved [P3, p. 438] that the Dupin submanifold $\lambda : B \to \Lambda^7$ is reducible if and only if some focal point map is contained in a 4-dimensional subspace $P^4 \subset P^6$.

If ρ is identically zero on B, then by (5.20), all of the covariant derivatives of ρ are also equal to zero. From (5.21) and (5.54), we see that the functions in equations (5.10)-(5.12) satisfy

$$q = t = b = 0 \quad ,$$
$$a = p = 0 \ , \ r = s, \ c = -u.$$

Then from (5.55), we have that $\omega_2^7 = 0$. From these and the other relations among the Maurer-Cartan forms which we have derived, we see that the differentials of the frame vectors can be written

$$
\begin{aligned}
dY_1 - \omega_1^1 Y_1 &= \omega_1^4 Y_4 + \omega_1^5 Y_5 \ , \\
dY_7 - \omega_1^1 Y_7 &= \omega_7^3 Y_3 - \omega_1^5 Y_5 \ , \\
dY_2 + \omega_1^1 Y_2 &= s(-\ \omega_1^4 Y_4 + \omega_1^5 Y_5) \ , \\
dY_6 + \omega_1^1 Y_2 &= u(\omega_7^3 Y_3 + \omega_1^5 Y_5) \ , \\
dY_3 &= \omega_7^3(-Y_6 + uY_7) \ , \\
dY_4 &= \omega_1^4(sY_1 - Y_2) \ , \\
dY_5 &= \omega_1^5(-sY_1 - Y_2 + Y_6 - uY_7) \ .
\end{aligned}
\tag{5.98}
$$

Note that from (5.44), (5.45), we have

$$ds + 2s\,\omega_1^1 = 0 \quad , \quad du + 2u\,\omega_1^1 = 0 \ , \tag{5.99}$$

and from (5.13) that

$$d\theta_i = \omega_1^1 \wedge \theta_i \quad , \quad i = 1,2,3 \ . \tag{5.100}$$

From (5.22), we have that $d\omega_1^1 = 0$. Hence on any local disk neighborhood U in B, we have that

$$\omega_1^1 = d\sigma \quad , \tag{5.101}$$

for some smooth function σ on U. We next consider a change of frame of the form

$$\begin{aligned} &Y_1^* = e^{-\sigma} Y_1, \; Y_7^* = e^{-\sigma} Y_7, \; Y_2^* = e^{\sigma} Y_2, \; Y_6^* = e^{\sigma} Y_6 \; , \\ &Y_i^* = Y_i \quad , \quad i = 3,4,5 \; . \end{aligned} \tag{5.102}$$

The effect of this change is to make $\omega_1^{1*} = 0$ while keeping $\rho^* = 0$. If we set

$$\omega_1^{4*} = e^{-\sigma} \omega_1^4, \; \omega_1^{5*} = e^{-\sigma} \omega_1^5, \; \omega_7^{3*} = e^{-\sigma} \omega_7^3 \; ,$$

then we can then compute that from (5.98) that

$$\begin{aligned} dY_1^* &= \omega_1^{4*} Y_4 + \omega_1^{5*} Y_5 \; , \\ dY_7^* &= \omega_7^{3*} Y_3 - \omega_1^{5*} Y_5 \; , \\ dY_2^* &= s^* (- \omega_1^{4*} Y_4 + \omega_1^{5*} Y_5) \; , \\ dY_6^* &= u^* (\omega_7^{3*} Y_3 + \omega_1^{5*} Y_5) \; , \\ dY_3^* &= \omega_7^{3*} Z_3^*, \text{ where } Z_3^* = -Y_6^* - u^* Y_7^* \; , \\ dY_4^* &= \omega_1^{4*} Z_4^*, \text{ where } Z_4^* = s^* Y_1^* - Y_2^* \; , \\ dY_5^* &= \omega_1^{5*} Z_5^*, \text{ where } Z_5^* = -s^* Y_1^* - Y_2^* + Y_6^* - u^* Y_7^* \; , \end{aligned} \tag{5.103}$$

where

$$s^* = s e^{2\sigma} \quad , \; u^* = u e^{2\sigma} \; . \tag{5.104}$$

Using (5.99) and (5.104), we can then compute that

$$ds^* = 0 \quad , \; du^* = 0 \; , \tag{5.105}$$

i.e., s^* and u^* are constant functions on the local neighborhood U.

The frame (5.102) is our final frame, and we will now drop the asterisks in further references to (5.102)-(5.105). Since the functions s and u are now constant, we can compute from (5.103) that

$$
\begin{aligned}
dZ_3 &= -\,2u\,\omega_7^3\,Y_3\,, \\
dZ_4 &= 2s\,\omega_1^4\,Y_4\,, \\
dZ_5 &= 2(u-s)\omega_1^5\,Y_5\,.
\end{aligned}
\tag{5.106}
$$

From this we see that the following 4-dimensional subspaces,

$$
\begin{aligned}
&\mathrm{Span}\{Y_1, Y_4, Y_5, Z_4, Z_5\}\,, \\
&\mathrm{Span}\{Y_7, Y_3, Y_5, Z_3, Z_5\}\,, \\
&\mathrm{Span}\{Y_1+Y_7, Y_3, Y_4, Z_3, Z_4\}\,,
\end{aligned}
\tag{5.107}
$$

are invariant under exterior differentiation and are thus constant. Thus, each of the three focal point maps Y_1, Y_7 and Y_1+Y_7 is contained in a 4-dimensional subspace of P^6, and our Dupin submanifold is reducible in three different ways. Each of the three focal point maps is thus an immersion of the space of leaves of its principal foliation onto an open subset of a cyclide of Dupin in a space $\Sigma^3 = P^4 \cap Q^5$.

We state this result due to Pinkall [P2] as follows:

<u>Theorem 5.3:</u> Every Dupin submanifold with $\rho = 0$ is reducible. Thus, it is obtained from a cyclide in $\mathbb{R}^3$ by one of the four standard constructions (5.97).

Pinkall [P2, p. 111] then proceeds to classify Dupin submanifolds with $\rho = 0$ up to Lie equivalence. We will not prove his result here. The reader can follow his proof using the fact that his constants α and β are our constants s and -u, respectively.

REFERENCES

[B] W. Blaschke, <u>Vorlesungen über Differentialgeometrie</u>, Vol. 3, Springer, Berlin, 1929.

[CC] T. Cecil and S.S. Chern, <u>Tautness and Lie sphere geometry</u>, Math. Ann. 278 (1987), 381-399.

[CR] T. Cecil and P. Ryan, Tight and taut immersions of manifolds, Res. Notes Math. 107, Pitman, London, 1985.

[E] L. Eisenhart, A treatise on the differential geometry of curves and surfaces, Ginn, Boston, 1909.

[GH] K. Grove and S. Halperin, Dupin hypersurfaces, group actions and the double mappings cylinder, J. Differential Geometry 26 (1987), 429-459.

[L] E.P. Lane, A treatise on projective differential geometry, U. Chicago Press, Chicago, 1942.

[LS] S. Lie and G. Scheffers, Geometrie der Berührungstransformationen, Teubner, Leipzig, 1896.

[M1] R. Miyaoka, Compact Dupin hypersurfaces with three principal curvatures, Math. Z. 187 (1984), 433-452.

[M2] _____, Dupin hypersurfaces with four principal curvatures, Preprint, Tokyo Institute of Technology.

[M3] _____, Dupin hypersurfaces with six principal curvatures, Preprint, Tokyo Institute of Technology.

[Mu] H.F. Münzner, Isoparametrische Hyperflächen in Sphären, I and II, Math. Ann. 251 (1980), 57-71 and 256 (1981), 215-232.

[N] K. Nomizu, Characteristic roots and vectors of a differentiable family of symmetric matrices, Lin. and Multilin. Alg. 2 (1973), 159-162.

[P1] U. Pinkall, Dupin'sche Hyperflächen, Dissertation, Univ. Freiburg, 1981.

[P2] _____, Dupin'sche Hyperflächen in E^4, Manuscr. Math 51 (1985), 89-119.

[P3] _____, Dupin hypersurfaces, Math. Ann. 270 (1985), 427-440.

[S] D. Singley, Smoothness theorems for the principal curvatures and principal vectors of a hypersurface, Rocky Mountain J. Math., 5 (1975), 135-144.

[Th] G. Thorbergsson, Dupin hypersurfaces, Bull. Lond. Math. Soc. 15 (1983), 493-498.

Thomas E. Cecil
Department of Mathematics
College of the Holy Cross
Worcester, MA 01610

Shiing-Shen Chern
Department of Mathematics
University of California
Berkeley, CA 94720
and
Mathematical Sciences Research Institute
1000 Centennial Drive
Berkeley, CA 94720

THE MEAN CURVATURES ON THE TUBULAR HYPERSURFACES IN A SPACE OF CONSTANT CURVATURE

Chen Weihuan*

Recently the structure of focal sets of hypersurfaces in the space of constant curvature has been intensively studied (cf. [2], ch. 2), which is closely related to the geometry of the tubular hypersurfaces of submanifolds. In their discussion the spaces of constant curvature are usually considered as hypersurfaces in a Euclidean space or a pseudo-Euclidean space. In this paper, we shall first give the metric and the second fundamental form of the tubular hypersurface around a submanifold in the space of constant curvature using the technique of Jacobi fields, and then we shall give the formulas to the integral of mean curvatures over the tubular hypersurfaces, which are the generalizations of the well-known area formulas of the tubular hypersurfaces given by H. Weyl ([7]).

§1 The Normal Bundle of a Submanifold

Let N be an oriented n-dimensional manifold immersed in an oriented m-dimensional Riemannian manifold M, and its codimension $p=m-n$. Unless otherwise stated, we shall always agree on the ranges of indices as follows:

$$1\leqq i,j,k,l\leqq n, \quad n+1\leqq \alpha,\beta,\gamma,\delta\leqq m, \quad 1\leqq A,B,C,D\leqq m.$$

We denote $\langle\ ,\ \rangle$ the metric in M, and D the corresponding Levi-Civita connection. The normal bundle of N in M is denoted by $\nu(N)$, and the bundle projection is $\pi:\nu(N)\to N$. The normal connection on $\nu(N)$ induced from D is denoted by $D^{\perp}$. Thus the tangent space $T_\xi(\nu(N))$ to the normal bundle at $\xi\in\nu(N)$ can naturally be decomposed into the direct sum of its vertical subspace V_ξ and horizontal subspace H_ξ, where the vertical subspace V_ξ is exactly the tangent space to the fibre $\nu_{\pi(\xi)}(N)$ at ξ, so V_ξ can be identified with $\nu_{\pi(\xi)}$ itself (this identification is exactly the translation in the vector space $\nu_{\pi(\xi)}(N)$). Also the horizontal subspace H_ξ is isomorphic to $T_{\pi(\xi)}(N)$ under the tangent map $\pi_*: T_\xi(\nu(N)) \to T_{\pi(\xi)}(N)$.

In order to get a local coordinate system in the normal bundle, we take a locally trivial coordinate neighborhood $V\subset N$ for the bundle $\nu(N)$, in which the

---

* *This work is supported in part by the National Foundations of Science. The author would like to thank Nankai Institute of Mathematics for their hospitality during the preparation of this paper.*

coordinates are x^i, and the natural frame is $e_i = \frac{\partial}{\partial x^i}$. Meanwhile we take a smooth orthonormal frame field $\{e_\alpha\}$ along U such that $\{e_\alpha(x)\}$ form a basis of $\nu_x(N)$ for each $x \in U$ (In the following we shall always assume that $\{e_i\}$ and $\{e_i, e_\alpha\}$ are of the orientation coherent with N and M respectively). Putting

$$y^i(\xi) = x^i \circ \pi(\xi), \quad t^\alpha(\xi) = \langle \xi, \ e_\alpha(\pi(\xi))\rangle \tag{1.1}$$

for each $\xi \in \pi^{-1}(U)$, we get a local coordinate system (y^i, t^α) in $\pi^{-1}(U)$, and the natural frame field $\{\frac{\partial}{\partial y^i}, \frac{\partial}{\partial t^\alpha}\}$.

Obviously, the induced metric on N is given by

$$g_{ij} = \langle e_i, e_j\rangle . \tag{1.2}$$

along the submanifold N we let

$$\begin{aligned} dx &= \omega^A e_A , \\ De_A &= \omega_A^B e_B , \end{aligned} \tag{1.3}$$

then we have

$$\omega^i = dx^i, \ \omega^\alpha = 0. \tag{1.4}$$

By the Cartan's lemma we have

$$\omega_i^\alpha = h_{ij}^\alpha \omega^j , \quad h_{ij}^\alpha = h_{ji}^\alpha , \tag{1.5}$$

and

$$\omega_\alpha^i = -g^{ik} h_{kj}^\alpha \omega^j. \tag{1.6}$$

The second fundamental form of N is

$$II = h_{ij}^\alpha \omega^i \omega^j e_\alpha . \tag{1.7}$$

For $\xi = t^\alpha e_\alpha$ the shape operator S_ξ is an endomorphism of the tangent space $T_{\pi(\xi)}N$ defined by

$$S_\xi(e_i) = (D_{e_i}\xi)^T = -t^\alpha h_{ik}^\alpha g^{kj} e_j . \tag{1.8}$$

It should be noted that $(D_{e_i}\xi)^T$ is independent of the extension of the normal vector ξ along any curve tangent to e_i in N. Of course, ω_i^j are the forms of the Levi-Civita connection on N and ω_α^β, those of the induced normal connection in $\nu(N)$, i.e.

$$D^\perp e_\alpha = (De_\alpha)^\perp = \omega_\alpha^\beta e_\beta . \tag{1.9}$$

Evidently we have $\omega_\alpha^\beta + \omega_\beta^\alpha = 0$.

Lemma 1 Let $\xi = t^\alpha e_\alpha \in \pi^{-1}(U)$, X_i be the horizontal lift of e_i at ξ and X_α, be the vertical vector at ξ parallel to e_α in $\nu_{\pi(\xi)}(N)$, then we have

$$\begin{aligned}\left.\frac{\partial}{\partial y^i}\right|_\xi &= X_i + \Gamma^\alpha_{i\beta}\, t^\beta X_\alpha,\\ \left.\frac{\partial}{\partial t^\alpha}\right|_\xi &= X_\alpha,\end{aligned} \tag{1.10}$$

where $\omega^\alpha_\beta = \Gamma^\alpha_{i\beta} dx^i$.

Proof Consider an arbitrary smooth curve $\gamma(u)$ starting from $x_\circ = \pi(\xi)$ in U. The horizontal lift of $\gamma(u)$ passing through ξ is a curve in $\pi^{-1}(U)$ given by

$$\xi(u) = t^\alpha(u)\; e_\alpha(\gamma(u)),$$

where $t^\alpha(u)$ satisfy the following conditions:

$$\begin{cases}\dfrac{dt^\alpha(u)}{du} + t^\beta(u)\omega^\alpha_\beta(\gamma'(u)) = 0,\\[2mm] t^\alpha(0) = t^\alpha.\end{cases}$$

Hence the horizontal lift vector of $\gamma'(0)$ at ξ is

$$\xi'(0) = \left(\frac{dy^i(\xi(u))}{du}\frac{\partial}{\partial y^i} + \frac{dt^\alpha(\xi(u))}{du}\frac{\partial}{\partial t^\alpha}\right)_{u=0}$$

$$= \left(\frac{\partial}{\partial y^i} - t^\beta \Gamma^\alpha_{i\beta}\frac{\partial}{\partial t^\alpha}\right)\gamma^{i'}(0),$$

which implies the first equality in (1.10). The second holds true obviously. Q.E.D.

We denote the exponential map from the tangent bundle T(M) into M by Exp, and put

$$\exp_\nu = \mathrm{Exp}\big|_{\nu(N)}.$$

If the map $(\exp_\nu)_{*\xi}$ at $\xi \in \nu(N)$ is degenerate we call $\exp_\nu \xi$ a focal point of the submanifold N. For $\varepsilon>0$, let

$$\begin{aligned}T(\varepsilon) &= \{(x,\xi) \;:\; x \in N,\; \xi \in \nu_x(N),\; |\xi| \le \varepsilon\},\\ T_\varepsilon &= \exp_\nu T(\varepsilon).\end{aligned} \tag{1.11}$$

When N is compact, T_ε will not contain any focal point of N for ε sufficiently small. In this case, we call T_ε the ε-tube about N, and call $N_\varepsilon = \partial T_\varepsilon$ the ε-tubular hypersurface around N in M, which is obviously regular everywhere.

Now we want to attach a locally smooth frame field to the unit normal bundle $\nu'(N)$. In fact, we only need to assign a frame $\{x; a_A\}$ for each unit normal vector $v = \lambda^\alpha e_\alpha(x)$ where $x \in U$ and $\sum_\alpha(\lambda^\alpha)^2 = 1$ such that

$$a_i = e_i \,, \qquad a_\alpha = u_\alpha^\beta e_\beta \,, \tag{1.12}$$

where $(u_\alpha^\beta) \in SO(p)$, and $u_m^\beta = \lambda^\beta$, i.e. $a_m = v$. Evidently we can choose u_α^β smoothly depending on x^i and λ^α. Let

$$dx = \tau^A a_A, \qquad Da_A = \tau_A^B a_B \,. \tag{1.13}$$

Then we get

$$\begin{aligned} \tau^i &= \omega^i \,, & \tau^\alpha &= 0, \\ \tau_i^j &= \omega_i^j \,, & \tau_i^\beta &= \omega_i^\alpha u_\beta^\alpha \,, \\ \tau_\alpha^j &= \omega_\beta^j u_\alpha^\beta \,, & \tau_\alpha^\beta &= (du_\alpha^\gamma + u_\alpha^\delta \omega_\delta^\gamma) u_\beta^\gamma \,. \end{aligned} \tag{1.14}$$

Denote by $\{E_A(s)\}$ the frame parallel to $\{a_A\}$ along the geodisic $\gamma_v(s)=\exp_\nu(s\cdot v)$ with respect to the Levi-Civita connection in M, then we obtain a locally smooth frame field $\{E_A(\varepsilon)\}$ on the tubular hypersurface N_ε. The aim of §2 is to find its dual coframe. Before doing so we should recall some facts about the N-Jacobi field.

<u>Lemma 2</u> Given any $X \in T_\xi(\ (N))$ and $\xi=\varepsilon v$, there exists a N-Jacobi field A along γ_v such that $Y(\varepsilon)=(\exp_\nu)_{*\xi}X$. Moreover, this is exactly a Jacobi field satisfying the initial conditions:

$$Y(0) = \pi_*(X), \qquad (\frac{DY}{ds}(0))^\perp = \frac{1}{\varepsilon} X^v, \tag{1.15}$$

where X^v stands for the vertical component of X. In particular, such an N-Jacobi field is unique if ξ is not a focal point of N.

For the proof, the reader can be referred to [4].

Thus, the N-Jacobi field Y_i corresponding to the horizontal lift X_i of e_i at ξ is the Jacobi field satisfying

$$Y_i(0) = e_i, \qquad \frac{DY_i}{ds}(0) = S_v(e_i) = -\lambda^\alpha h_{ij}^\alpha g^{jk} e_k, \tag{1.16}$$

and the one corresponding to the vertical vector X_α is the Jacobi field Y_α satisfying

$$Y_\alpha(0) = 0, \qquad \frac{DY_\alpha}{ds}(0) = \frac{1}{\varepsilon} e_\alpha. \tag{1.17}$$

In this case we have

$$Y_i(\varepsilon) = (\exp_\nu)_{*\xi}(X_i), \qquad Y_\alpha(\varepsilon) = (\exp_\nu)_{*\xi}(X_\alpha). \tag{1.18}$$

$\tilde{X}_\alpha=u_\alpha^\beta X_\beta$, of course, is the vertical vector related to a_α, so the corresponding N-Jacobi field is $\tilde{Y}_\alpha = u_\alpha^\beta Y_\beta$. In particular

$$\tilde{Y}_m(\varepsilon) = (\exp_\nu)_{*\xi}(\tilde{X}_m) = E_m(\varepsilon). \tag{1.19}$$

§2 The Metric on the Tubular Hypersurface

By (1.11), (1.10), (1.14) and (1.18) we get

$$\begin{aligned} d(\exp_\nu(x,\xi)) &= (\exp_\nu)_{*\xi}\left(\frac{\partial}{\partial y^i}\right)dy^i+(\exp_\nu)_{*\xi}\left(\frac{\partial}{\partial t^\alpha}\right)dt^\alpha \\ &= (\exp_\nu)_{*\xi}(X_i)\omega^i+(\exp_\nu)_{*\xi}(X_\alpha)(dt^\alpha+t^\beta\omega_\beta^\alpha) \\ &= \tau^i Y_i(\varepsilon)+(d\varepsilon\cdot\lambda^\gamma+\varepsilon\cdot\tau_m^\gamma)\tilde{Y}_\gamma(\varepsilon), \end{aligned}$$

where $\xi = t^\alpha e_\alpha = \varepsilon\cdot\lambda^\alpha e_\alpha$, $\sum_\alpha(\lambda^\alpha)^2 = 1$. Restricting to N_ε we have $d\varepsilon = 0$, so

$$d\exp_\nu(x,\xi)\Big|_{N_\varepsilon} = \tau^i Y_i(\varepsilon)+\varepsilon\cdot\tau_m^\alpha\tilde{Y}_\alpha(\varepsilon). \qquad (2.1)$$

From the following proposition we know that $\tilde{Y}_m(\varepsilon)=E_m(\varepsilon)$ is the unit normal vector field to N .

<u>Proposition 1</u> $\xi=\varepsilon\cdot v(v \in \nu_x^1(N))$ is a focal point of N at x if and only if the N-Jacobi fields Y_i, $\tilde{Y}_\alpha$ along γ_v defined as in §1 satisfy

$$Y_1(\varepsilon)\wedge\cdots\wedge Y_n(\varepsilon)\wedge\tilde{Y}_{n+1}(\varepsilon)\wedge\cdots\wedge\tilde{Y}_{m-1}(\varepsilon) = 0. \qquad (2.2)$$

<u>Proof</u> Let J denote the Jacobian of map $\exp_\nu$ at ξ. Since

$$(\exp_\nu)_{*\xi}(X_i) = Y_i(\varepsilon), \qquad (\exp_\nu)_{*\xi}(X_\alpha) = Y_\alpha(\varepsilon)$$

and

$$\langle Y_i(\varepsilon),\tilde{Y}_m(\varepsilon)\rangle = \langle\tilde{Y}_\alpha(\varepsilon),\tilde{Y}_m(\varepsilon)\rangle = 0 \qquad (\alpha\neq m),$$

$$|\tilde{Y}_m(\varepsilon)| = |E_m(\varepsilon)| = 1,$$

we have

$$\begin{aligned} J &= \frac{|(\exp_\nu)_{*\xi}(X_1)\wedge\cdots\wedge(\exp_\nu)_{*\xi}(X_m)|}{|X_1\wedge\cdots\wedge X_m|} \\ &= C\cdot|Y_1(\varepsilon)\wedge\cdots\wedge Y_m(\varepsilon)| \\ &= C\cdot|Y_1(\varepsilon)\wedge\cdots\wedge Y_n(\varepsilon)\wedge\tilde{Y}_{n+1}(\varepsilon)\wedge\cdots\wedge\tilde{Y}_{m-1}(\varepsilon)|, \end{aligned}$$

where C is a non-zero constant. Therefore J = 0 if and only if (2.2) holds. QED.

Now we suppose that M is a space with constant curvature c, then the N-Jacobi field Y_i, $\tilde{Y}_\alpha$ along γ_v mentioned as in §1 can be explicitly expressed as

$$\begin{cases} Y_i(s) = (g_{ij}f_c(\varepsilon) - \lambda^\alpha h_{ij}^\alpha g_c(\varepsilon))g^{ik}E_k(s), \\ \tilde{Y}_\alpha(s) = \dfrac{g_c(s)}{\varepsilon}E_\alpha(s), \qquad n+1\leq\alpha\leq m-1 \\ \tilde{Y}_m(s) = \dfrac{s}{\varepsilon}E_m(s), \end{cases} \qquad (2.3)$$

where

$$g_c(s) = \begin{cases} \frac{1}{\sqrt{c}} \sin(\sqrt{c}\ s), & c>0, \\ s, & c=0, \\ \frac{1}{\sqrt{-c}} \mathrm{sh}(\sqrt{-c}\ s), & c<0, \end{cases} \tag{2.4}$$

$$f_c(s) = g_c'(s).$$

We put

$$h_c(s) = \frac{g_c(s)}{f_c(s)}. \tag{2.5}$$

Thus the formula (2.1) becomes

$$\mathrm{dexp}_\nu(x,\xi)\big|_{N_\varepsilon} = \tau^i(g_{ij}\cdot f_c(\varepsilon)-\lambda^\alpha h^\alpha_{ij} g_c(\varepsilon))g^{jk}E_k(\varepsilon)+g_c(\varepsilon)\tau^\alpha_m E_\alpha(\varepsilon). \tag{2.6}$$

Corollary 1 Let N be a submanifold immersed in the m-dimensional space M of constant curvature c, and $v \in v^1(N)$, then $\xi = \varepsilon\cdot v$ is a focal point of N if and only if ε satisfies the equation either $g_c(\varepsilon)=0$, or $\frac{f_c(\varepsilon)}{g_c(\varepsilon)}=k$ where k is a principal curvature of N with respect to v.

Proof Let k_i be the principal curvatures of N with respect to v, then under the assumption of M the condition (2.2) is equivalent to

$$[g_c(\varepsilon)]^{p-1}\cdot\prod_i(f_c(\varepsilon)-k_i g_c(\varepsilon)) = 0,$$

i.e. ε has to fill either $g_c(\varepsilon)=0$, or $\frac{f_c(\varepsilon)}{g_c(\varepsilon)} = k_i$ for some i. QED.

Assume that there is no focal point of N in the tuve T_ε about N, and N_ε is an immersed hypersurface in M. Let

$$\begin{aligned} \tilde{\omega}^k &= \tau^i(g_{ij} f_c(\varepsilon)-\lambda^\alpha h^\alpha_{ij} g_c(\varepsilon))g^{jk} = f_c(\varepsilon)\tau^k+g_c(\varepsilon)\tau^k_m, \\ \tilde{\omega}^\alpha &= g_c(\varepsilon)\tau^\alpha_m, \end{aligned} \tag{2.7}$$

then $\{\omega^A;\ 1\le A\le m-1\}$ is a coframe dual to $\{E_A(\varepsilon);\ 1\le A\le m-1\}$ on the tubular hypersurface N_ε. Obviously, if we take a locally smooth orthonormal frame $\{e_i\}$ on N in place of the natural frame $\{\frac{\partial}{\partial x^i}\}$, we still obtain the corresponding coframe on N_ε of the same form as in (2.7) with $g_{ij} = \delta_{ij}$. Now we sum up the above results as follows:

Proposition 2 Let N be an n-dimensional submanifold in the m-dimensional space of constant curvature c, and $\{x;\ e_i, e_\alpha\}$ an arbitrary smooth orthonormal frame field locally defined on N such that e_i are tangent and e_α are normal to N. For each $v \in v^1(N)$ we attach a frame $\{x;\ a_i, a_\alpha\}$ defined as in (1.12) and let it translate parallelly along the geodesic $\gamma_v(v=a_m)$ in the space M, then we obtain a local orthonormal frame field $\{\exp_v(\varepsilon\cdot v);E_A\}$ on the tubular hypersurface N_ε, in which E_m is exactly normal to N_ε. Moreover, the dual coframe is given by

$$\begin{aligned}\tilde{\omega}^k &= (f_c(\varepsilon)\delta_{kj} - g_c(\varepsilon)h^v_{kj})\omega^j,\\ \tilde{\omega}^\alpha &= g_c(\varepsilon)(d\lambda^\beta + \lambda^\gamma\omega^\beta_\gamma)u^\beta_\alpha,\end{aligned} \tag{2.8}$$

where $h^v_{ij}=\lambda^\alpha h^\alpha_{ij}$ is the second fundamental form of N with respect to $v=\lambda^\alpha e_\alpha$, and ω^β_γ are the forms of the induced connection in $v(N)$ under the local frame field $\{x;e_i,e_\alpha\}$.

Corollary 2 Under the assumption as in the proposition 2, the metric on the tubular hypersurface can be expressed as

$$I_\varepsilon = [f_c(\varepsilon)]^2 I - 2f_c(\varepsilon)g_c(\varepsilon)II^v + [g_c(\varepsilon)]^2|Dv|^2, \tag{2.9}$$

where I is the metric on N, $II^v=h^v_{ij}\omega^i\omega^j$ the second fundamental form of N with respect to v, and

$$Dv = \lambda^\alpha\omega^i_\alpha e_i + (d\lambda^\alpha + \lambda^\beta\omega^\alpha_\beta)e_\alpha. \tag{2.10}$$

The area element of N_ε at $\xi=\varepsilon\cdot v$ is

$$d\sigma_\varepsilon = [g_c(\varepsilon)]^{p-1}\det(f_c(\varepsilon)\delta_{ij} - g_c(\varepsilon)h^v_{ij})d\sigma \wedge dO_{p-1}, \tag{2.11}$$

where $d\sigma$ is the area element of N and dO_{p-1} that of unit sphere S^{p-1} in $\mathbb{R}^p$.

The formula (2.11) has been given in [4]. Here the formular (2.9) and (2.11) are the consequences of computing $\sum_A(\tilde{\omega}^A)^2$ and $\tilde{\omega}^1\wedge\cdots\wedge\tilde{\omega}^{m-1}$ respectively. It is worthwhile to note that the corollary 2 includes two special cases that the submanifold N could be a hypersurface or a single point, i.e. it gives the metric on a parallel hypersurface or a geodesic sphere about a point.

§3 The Second Fundamental Form on the Tubular Hypersurface

Lemma 3 Let N be an n-dimensional riemannian manifold M, and N_ε the ε-tubular hypersurface around N. Suppose that $v \in v^1(N)$, and $\xi=\varepsilon\cdot v$. Given any $Y \in T_{\gamma_v(\varepsilon)}(N_\varepsilon)$, then there exists a unique N-Jacobi field Y(s) along γ_v such that $Y(\varepsilon)=Y$, and

$$\tilde{S}(Y) = \left.\frac{DY(s)}{ds}\right|_{s=\varepsilon}, \tag{3.1}$$

where $\tilde{S}$ is the shape operator acting on the tangent space $T_{\gamma_v(\varepsilon)}(N_\varepsilon)$.

Proof Since there is no focal point in T_ε, the N-Jacobi field Y(s), as is stated above, exists uniquely. Suppose that we have a variation through geodesics perpendicular to N which yields the field Y(s) along γ_v:

$$F(s,u) = \exp_\gamma(s\cdot v(u)),$$

where $v(u)$ is a curve in $\nu^1(N)$ such that $v(0) = v$.

Putting

$$Y(s,u) = F_*(\frac{\partial}{\partial u}) = (\exp_\nu)_{*s\cdot v(u)}(\frac{\partial}{\partial u}(s\cdot v(u))),$$

$$V(s,u) = F_*(\frac{\partial}{\partial s}) = (\exp_\nu)_{*s\cdot v(u)}(v(u)),$$

we have

$$Y(s,0) = Y(s), \quad \text{and} \quad \gamma_v'(s) = V(s,0).$$

In particular

$$Y(\varepsilon,0) = Y, \quad V(\varepsilon,0) = \gamma_v'(\varepsilon).$$

Note that $F(\varepsilon,u)$ is a curve tangent to Y in N_ε, and $V(\varepsilon,u)$ a unit vector field normal to N_ε along $F(\varepsilon,u)$, hence

$$\tilde{S}(Y) = D_Y(V(\varepsilon,u)) = D_{Y(s,u)}(V(s,u))\Big|_{s=\varepsilon,\ u=0}$$

$$= D_{V(s,u)}(Y(s,u))\Big|_{\substack{s=\varepsilon\\ u=0}} = \frac{DY(s)}{ds}\Big|_{s=\varepsilon}. \qquad \text{QED.}$$

Proposition 3 Under the assumption as in the proposition 2 let

$$DE_m = \tilde{\omega}_m^A E_A, \tag{3.2}$$

then

$$\tilde{\omega}_m^i = -\tilde{h}_{ij}\tilde{\omega}^i = -c\cdot g_c(\varepsilon)\omega^i + f_c(\varepsilon)\lambda^\alpha\omega_\alpha^i,$$

$$\tilde{\omega}_m^\alpha = -\tilde{h}_{\alpha\beta}\tilde{\omega}^\beta = \frac{f_c(\varepsilon)}{g_c(\varepsilon)}\tilde{\omega}^\alpha, \tag{3.3}$$

$$\tilde{h}_{i\alpha} = \tilde{h}_{\alpha i} = 0.$$

Proof Let $Y_i(s), \tilde{Y}_\alpha(s)$ denote the N-Jacobi fields given by (2.3). Applying the lemma 3 to $Y_i(\varepsilon)$ and $\tilde{Y}_\alpha(\varepsilon)$ we get

$$(f_c(\varepsilon)\delta_{ij} - g_c(\varepsilon)\lambda^\alpha h_{ij}^\alpha)\tilde{S}(E_j) = (-c\cdot g_c(\varepsilon)\delta_{ij} - f_c(\varepsilon)\lambda^\alpha h_{ij}^\alpha)E_j, \tag{3.4}$$

$$\frac{g_c(\varepsilon)}{\varepsilon}\tilde{S}(E_\alpha) = \frac{f_c(\varepsilon)}{\varepsilon}E_\alpha, \tag{3.5}$$

in which we used the fact that

$$g_c'(s) = f_c(s), \quad f_c'(s) = -c\cdot g_c(s). \tag{3.6}$$

Multiplying the two sides of (3.4) by ω^i, and summing for i we obtain

$$\tilde{\omega}^i\tilde{S}(E_i) = -\tilde{\omega}^i(\tilde{h}_{ij}E_j+\tilde{h}_{i\alpha}E_\alpha) = -(c\cdot g_c(\varepsilon)\omega^j - f_c(\varepsilon)\lambda^\alpha\omega^j_\alpha)E_j ,$$

so

$$\tilde{\omega}^j_m = -c\cdot g_c(\varepsilon)\omega^j + f_c(\varepsilon)\lambda^\alpha\omega^j_\alpha ,$$

$$\tilde{h}_{i\alpha} = \tilde{h}_{\alpha i} = 0.$$

From (3.5) we get

$$\tilde{h}_{\alpha\beta} = -\frac{f_c(\varepsilon)}{g_c(\varepsilon)}\delta_{\alpha\beta} , \qquad \tilde{\omega}^\alpha_m = \frac{f_c(\varepsilon)}{g_c(\varepsilon)}\tilde{\omega}^\alpha . \qquad \text{QED.}$$

Corollary 3 In the space M of constant curvature c, the tubular hypersurface N_ε around submanifold N has the second fundamental form

$$\begin{aligned} II_\varepsilon = {} & c\cdot g_c(\varepsilon)f_c(\varepsilon)\cdot I+([f_c(\varepsilon)]^2-c\cdot[g_c(\varepsilon)]^2)II^v \\ & -g_c(\varepsilon)f_c(\varepsilon)|Dv|^2. \end{aligned} \tag{3.7}$$

Proof The conclusion follows immediatly from computing $II_\varepsilon = -(\tilde{\omega}^j_m\tilde{\omega}^j + \tilde{\omega}^\alpha_m\tilde{\omega}^\alpha)$. QED.

Remark From the second formulas of (3.3) we see that E_α are the principal directions corresponding to the principal curvature $-\frac{f_c(\varepsilon)}{g_c(\varepsilon)}$ on N_ε. If e_i is a principal direction of N with respect to the normal vector v, and the corresponging principal curvature is k_i, then E_i is a principal direction on N_ε at the point $\exp_v(\varepsilon\cdot v)$ corresponding to the principal curvature

$$\tilde{k}_i = \frac{c\cdot g_c(\varepsilon)+k_i\cdot f_c(\varepsilon)}{f_c(\varepsilon)-k_i\cdot g_c(\varepsilon)} .$$

§4 The Mean Curvature on the Tubular Hypersurface

Let N be an oriented submanifold immersed in the space M of constant curvature. We note that $N_{\varepsilon+\delta}$, the $(\varepsilon+\delta)$-tubular hypersurface around N, can be also regarded as a parallel hypersurface of N_ε. Hence we have two expressions for the area element on $N_{\varepsilon+\delta}$: on the one hand

$$\begin{aligned} \tilde{\tilde{\omega}}^i &= f_c(\varepsilon+\delta)\omega^i + g_c(\varepsilon+\delta)\lambda^\alpha\omega^i_\alpha , \\ \tilde{\tilde{\omega}}^\alpha &= g_c(\varepsilon+\delta)(d\lambda^\beta+\lambda^\gamma\omega^\beta_\gamma)u^\beta_\alpha ; \end{aligned} \tag{4.1}$$

on the other hand,

$$\tilde{\tilde{\omega}}^A = f_c(\delta)\tilde{\omega}^A + g_c(\delta)\tilde{\omega}^A_m . \tag{4.2}$$

According to (4.2), the area element of $N_{\varepsilon+\delta}$ is

$$\begin{aligned} d\sigma_{\varepsilon+\delta} &= \det(f_c(\delta)\cdot\delta_{AB} - g_c(\delta)\tilde{h}_{AB})d\sigma_\varepsilon \\ &= \sum_{r=o}^{m-1}(-1)^r\binom{m-1}{r}[f_c(\delta)]^{m-1-r}[g_c(\delta)]^r\tilde{H}_r d\sigma_\varepsilon , \end{aligned} \tag{4.3}$$

where $1\leq A, B\leq m-1$, and $\tilde{H}_r$ stands for the r-th mean curvature of N_ε, i.e., the r-th elementary symmetric function of the principal curvatures of N_ε divided by its term number $\binom{m-1}{r}$. We assume that $\tilde{H}_0=1$. By (4.1), however, we also have

$$d\sigma_{\varepsilon+\delta} = [g_c(\varepsilon+\delta)]^{p-1}\det(f_c(\varepsilon+\delta)\cdot\delta_{ij}-g_c(\varepsilon+\delta)h^{v}_{ij})d\sigma\wedge dO_{p-1}$$
$$= \sum_{q=0}^{n}(-1)^q\binom{n}{q}[f_c(\varepsilon+\delta)]^{n-q}[g_c(\varepsilon+\delta)]^{p+q-1}\cdot H_q(v)d\sigma\wedge dO_{p-1}\ , \tag{4.4}$$

where $H_q(v)$ is the q-th mean curvature of N with respect to the normal vector v.

The following lemma is due to W.Killing and H.Weyl ([6],[7]), but here we give a brief proof different from theirs.

Lemma 4 At each point $x \in N$, the average of $H_q(v)$ when v passes over the whole unit sphere S^{p-1} in the normal subspace $\nu_x(N)$ is

$$\frac{1}{O_{p-1}}\int_{S^{p-1}} H_q(v)dO_{p-1} = \begin{cases} \dfrac{1}{\binom{n}{2\mu}}\dfrac{K^c_{2\mu}}{p(p+2)\ldots(p+2\mu-2)}, & \text{if } q=2\mu,\\ 0, & \text{if } q \text{ is odd.}\end{cases} \tag{4.5}$$

where O_{p-1} is the total area of S^{p-1} and

$$K^c_{2\mu} = \frac{1}{2^{2\mu}\mu!}\cdot\delta^{i_1\cdots i_{2\mu}}_{j_1\cdots j_{2\mu}}K^c_{i_1i_2j_1j_2}\cdots K^c_{i_{2\mu-1}i_{2\mu}j_{2\mu-1}j_{2\mu}}$$
$$K^c_{ijkl} = R_{ijkl} - c(\delta_{ik}\delta_{jl}-\delta_{il}\delta_{jk}), \tag{4.6}$$

R_{ijkl} is the curvature tensor of N. When q=0, the right hand of (4.5) should be understood to be 1.

Proof Introducing a parameter τ, we have

$$(\omega^1+\tau\lambda^\alpha\omega^1_\alpha)\wedge\cdots\wedge(\omega^n+\tau\lambda^\alpha\omega^n_\alpha)$$
$$= \det(\delta_{ij}-\tau\cdot h^{v}_{ij})\omega^1\wedge\cdots\wedge\omega^n \tag{4.7}$$
$$= \sum_{q=0}^{n}(-1)^q\cdot\binom{n}{q}\tau^q\cdot H_q(v)d\sigma.$$

We can again expand it with λ^α and get

$$(\omega^1+\tau\lambda^\alpha\omega^1_\alpha)\wedge\cdots\wedge(\omega^n+\tau\lambda^\alpha\omega^n_\alpha)$$
$$= \sum_{q=0}^{n}\frac{\tau^q}{(n-q)!q!}\cdot\sum_{\substack{q_1+\cdots+q_p=q\\ q_\alpha\geq 0}}\frac{q!}{q_1!\cdots q_p!}(\lambda^{n+1})^{q_1}\cdots(\lambda^m)^{q_p}\Omega(q_1,\cdots q_p), \tag{4.8}$$

where

$$\Omega(q_1,\cdots,q_p) = \delta^{1\cdots\cdots n}_{i_1\cdots i_n}\underbrace{\omega^{i_1}_{n+1}\Lambda\cdots\Lambda\omega^{i_{q_1}}_{n+1}}_{q_1 \text{ factors}}\Lambda\cdots\Lambda\underbrace{\omega^{i_{q-q_p+1}}_{m}\Lambda\cdots\Lambda\omega^{i_q}_{m}}_{q_p \text{ factors}}\Lambda\omega^{i_{q+1}}\Lambda\cdots\Lambda\omega^{i_n}. \tag{4.9}$$

Comparing (4.7) and (4.8) we have

$$H_q(v)\, d\sigma = \frac{(-1)^q}{n!}\sum_{q_1+\cdots+q_p=q,\ q_\alpha\geqq 0}\frac{q!}{q_1!\cdots q_p!}(\lambda^{n+1})^{q_1}\cdots(\lambda^m)^{q_p}\Omega(q_1,\cdots,q_p). \tag{4.10}$$

It is easy to find that ([7])

$$\frac{1}{O_{p-1}}\int_{S^{p-1}}(\lambda^{n+1})^{q_1}\cdots(\lambda^m)^{q_p}dO_{p-1}$$

$$= \begin{cases} \dfrac{(q_1-1)!!\cdots(q_p-1)!!}{p(p+2)\ \cdots\ (p+q-2)}, & \text{if all of } q_\alpha \text{ are even,} \\ 0, & \text{otherwise.} \end{cases} \tag{4.11}$$

Here we agree on that $(q-1)!!=1$ if $q=0$, and that the right hand is equal to 1 if $q=0$. If $q=2\mu$, we integrate (4.10) over S^{p-1} and get

$$\left(\frac{1}{O_{p-1}}\int_{S^{p-1}}H_q(v)dO_{p-1}\right)d\sigma$$

$$= \frac{1}{\binom{n}{2\mu}}\sum_{\substack{\mu_1+\cdots+\mu_p=\mu\\ \mu_\alpha\geqq 0}}\frac{1}{(n-2\mu)!2^\mu\mu_1!\cdots\mu_p!}\cdot\frac{\Omega(2\mu_1,\cdots,2\mu_p)}{p(p+2)\cdots(p+2\mu-2)}.$$

Substituting into it the Gauss equations of N

$$\sum_\alpha \omega^i_\alpha \Lambda\, \omega^j_\alpha = \tfrac{1}{2}K^c_{ijkl}\omega^k \Lambda\, \omega^l,$$

we obtain

$$\frac{1}{(n-2\mu)!2^\mu}\sum_{\substack{\mu_1+\cdots+\mu_p=\mu\\ \mu_\alpha\geqq 0}}\frac{\Omega(2\mu_1,\cdots,2\mu_p)}{\mu_1!\cdots\mu_p!} = K^c_{2\mu}d\sigma,$$

so (4.5) holds. If q is odd, the integral is zero due to(4.11). QED.

Suppose that N is compact and

$$M^c_r(\varepsilon) = \int_{N_\varepsilon}\tilde{H}_r d\sigma_\varepsilon.$$

From (4.3) and (4.4) we have

$$\sum_{r=0}^{m-1}(-1)^r\binom{m-1}{r}[f_c(\delta)]^{m-1-r}[g_c(\delta)]^r M^c_r(\varepsilon) =$$

$$= \sum_{\mu=o}^{[\frac{n}{2}]} \frac{O_{p-1} \cdot I_{2\mu}^{c}}{p(p+2)\cdots(p+2\mu-2)} [f_c(\varepsilon+\delta)]^{n-2\mu} [g_c(\varepsilon+\delta)]^{p+2\mu-1} , \tag{4.13}$$

where

$$I_{2\mu}^{c} = \int_M K_{2\mu}^{c} d\sigma . \tag{4.14}$$

Theorem Let N be an oriented compact n-dimensional submanifold in the m-dimensional space M of constant curvature c, then the integral of the r-th mean curvature on the tubular hypersuface N_ε is

$$M_r^c(\varepsilon) = \frac{O_{p-1} \cdot [f_c(\varepsilon)]^{m-1}}{\binom{m-1}{r}} \sum_{\mu=o}^{[\frac{n}{2}]} \frac{I_{2\mu}^{c}}{p(p+2)\cdots(p+2\mu-2)} \cdot \sum_{\alpha=o}^{n-2\mu} (-1)^{r-\alpha} c^{\alpha} \binom{n-2\mu}{\alpha} \left\{ {p+2\mu-1 \atop r-\alpha} \right\} [h_c(\varepsilon)]^{p+2\mu-1-r+2\alpha} \tag{4.15}$$

where for any integer a>0, and b we define

$$\left\{ {a \atop b} \right\} = \begin{cases} 0, & \text{if } b<0 \text{ or } b>a, \\ \binom{a}{b}, & \text{if } 0 \leqq b \leqq a. \end{cases} \tag{4.16}$$

Proof By (2.4) we have

$$f_c(\varepsilon+\delta) = f_c(\varepsilon) f_c(\delta) - c \cdot g_c(\varepsilon) g_c(\delta),$$

$$g_c(\varepsilon+\delta) = f_c(\varepsilon) g_c(\delta) + g_c(\varepsilon) f_c(\delta) ,$$

hence (4.13) becomes

$$\sum_{r=o}^{m-1} (-1)^r \binom{m-1}{r} [h_c(\delta)]^r M_r^c(\varepsilon)$$

$$= \sum_{\mu=o}^{[\frac{n}{2}]} \frac{O_{p-1} \cdot I_{2\mu}^{c}}{p(p+2)\cdots(p+2\mu-2)} \cdot \sum_{\alpha=o}^{n-2\mu} \sum_{\beta=o}^{p+2\mu-1} (-c)^{\alpha} \binom{n-2\mu}{\alpha} \binom{p+2\mu-1}{\beta} \cdot$$

$$\cdot [f_c(\varepsilon)]^{n-2\mu-\alpha+\beta} [g_c(\varepsilon)]^{p+2\mu-1-\beta+\alpha} [h_c(\delta)]^{\alpha+\beta}$$

$$= \sum_{r=o}^{m-1} [h_c(\delta)]^r \cdot \sum_{\mu=o}^{[\frac{n}{2}]} \frac{O_{p-1} \cdot [f_c(\varepsilon)]^{m-1} \cdot I_{2\mu}^{c}}{p(p+2)\cdots\cdot(p+2\mu-2)} \cdot$$

$$\cdot \sum_{\alpha=o}^{n-2} (-1)^{\alpha} \cdot c^{\alpha} \cdot \binom{n-2\mu}{\alpha} \left\{ {p+2\mu-1 \atop r-\alpha} \right\} [h_c(\varepsilon)]^{p+2\mu-1-r+2\alpha} .$$

Since $h_c(\delta)$ is a strict increasing function at $\delta=0$, the above identity about δ implies that, for each r, the coefficients of $[h_c(\delta)]^r$ must be equal.

QED.

Corollary 4 In any space M of constant curvature, the integral of the r-th mean curvature ($0\leqq r\leqq m-1$) of the ε-tubular hypersurface around an oriented compact submanifold N is independent of the isometric deformation of N in M.

Remark (1) If c=0, and $r\leqq p-1$, then (4.15) is reduced to

$$M_r^0(\varepsilon) = (-1)^r\cdot O_{p-1}\cdot\sum_{\mu=o}^{[\frac{n}{2}]}\frac{I_{2\mu}^0}{p(p+2)\cdots(p+2\ \ -2)}\cdot\frac{\binom{p+2\mu-1}{r}}{\binom{m-1}{r}}\cdot\varepsilon^{p-1-r+2\mu} \tag{4.17}$$

When $r\geqq p$, the lower bound of μ in the summation should be replaced by $[\frac{r-p}{2}]+1$. In particular, if n is odd and r=m-1, it will be zero.

(2) If r=0 we have

$$M_0^c(\varepsilon) = O_{p-1}\cdot\sum_{\mu=o}^{[\frac{n}{2}]}\frac{I_{2\mu}^c}{p(p+2)\cdots(p+2\mu-2)}\,[f_c(\varepsilon)]^{n-2\mu}\cdot[g_c(\varepsilon)]^{p+2\mu-1}\ , \tag{4.18}$$

which has been obtained in [4] and is the famous area formula given by H.Weyl ([7]) if c=0 or c>0.

References

[1] R.L.Bishop and R.T.Crittenden, Geometry of Manifolds, Academic Press, New York, 1964.

[2] T.E.Cecil and R.J.Ryan, Tight and Taut Immersions of Manifolds, Pitman Publishing Inc., 1985.

[3] Chen Weihuan, Some integral formulas on submanifolds in Euclidean space, Proc. of the 1980 D.D.Beijing Symp., 1127-1140, Science Press, Beijing, China, 1982.

[4] Chen Weihuan, On the volumes of tubes in space forms, Acta Mathematica Sinica, 31(1988), 164-171 (in Chinese).

[5] E.Heintze and H.Karcher, A general comparison theorem with applications to volume estimates for submanifolds, Ann. Scient. Ec. Norm. Sup., 11(1978), 451-470.

[6] W.Killing, Die nicht-euclidischen Raumformen in analytische Behandlung, Teubner, Leipzig, 1885.

[7] H.Weyl, On the volume of tubes, Amer. J. Math., 61(1939), 461-472.

[8] Wu Guanglei, On the n-dimensional submanifolds in 2n-dimensional Euclidean space, Acta Scient. Natur. Univ. Pekin., 3(1957), 61-77(in Chinese).

Chen Weihuan
Department of Mathematics
Peking University
Beijing 100871
China

Deformation of Surfaces Preserving Principal Curvatures

Chen Xiuxiong   Peng Chia-Kuei
Graduate School of University of Science and Technology of China

§0. Introduction

The isometric deformation of surfaces preserving principal curvatures was first studied by O.Bonnet [1]. He proved that a surface of constant mean curvature can be isometrically deformed preserving the mean curvatures.

Recently S.S.Chern [2] has studied such deformations for surfaces of non-constant mean curvature and showed that they turn out to be W-surfaces.

In this paper we will show that the mean curvature of such W-surfaces satisfy an ordinary differential equation of third order. As an application, we also find that surfaces with constant Gaussian curvature admitting such deformation should have zero Gaussian curvature. This result has been proved by A.G.Colares and K.Kenmotsu [4].

§1. Notations

In this section we will develop the local theory of surfaces in R^3.

Our formulas and notations are same as S.S.Chern,[2] for which the reader is referred there if necessary.

Suppose M is an oriented surface in E^3, of sufficient smoothness and without umbolics. Then there is a well-defined field over M of orthonomal frames x, e_1, e_2, e_3 such that $x \in M$, e_3 is the unit normal vector of M at x, and e_1 and e_2 are along the principal directions. Hence we have

$$\begin{cases} dx = & \omega_1 e_1 + \omega_2 e_2 \\ de_1 = & \omega_{12} e_2 + \omega_{13} e_3 \\ de_2 = & -\omega_{12} e_1 + \omega_{23} e_3 \\ de_3 = & -\omega_{13} e_1 - \omega_{23} e_2 \end{cases} \tag{1}$$

where ω_1 and ω_2 are one-forms. Our choice of the frames allows us to set:

$$\begin{cases} \omega_{12} = & h\omega_1 + k\omega_2 \\ \omega_{13} = & a\omega_1, \quad \omega_{23} = c\omega_2, \qquad a > c. \end{cases} \tag{2}$$

Consequently a and c are both principal curvatures at x. As usual, we denote the mean curvature and Gaussian Curvature by

$$H = \frac{1}{2}(a + c), \qquad K = ac \tag{3}$$

Now we introduce functions u and v by

$$dH \overset{\text{def.}}{=} \sqrt{H^2 - K}(u\omega_1 + v\omega_2) \tag{4}$$

Then through direct computation we have

$$d\log\sqrt{H^2 - K} = (u - 2k)\omega_1 - (v - 2h)\omega_2 \tag{5}$$

For our treatment, we introduce the following forms:

$$\theta_1 = u\omega_1 + v\omega_2, \qquad \theta_2 = -v\omega_1 + u\omega_2, \tag{6}$$

$$\alpha_1 = u\omega_1 - v\omega_2, \qquad \alpha_2 = v\omega_1 + u\omega_2. \tag{7}$$

If $H \neq const$, the quadratic differential form

$$d\hat{s}^2 = \theta_1^2 + \theta_2^2 = \alpha_1^2 + \alpha_2^2 = |\nabla H|^2 ds^2/(H^2 - k) \tag{8}$$

(where ∇ and Δ denote gradiant and Laplacian operators respectively) defines a conformal metric on M.

We find it convenient to make use of the Hodge $*$ operator such that

$$*\omega_1 = \omega_2, \quad *\omega_2 = -\omega_1, \quad *^2 = -1 \quad \text{(on one forms)}. \tag{9}$$

Thus we have

$$*\theta_1 = \theta_2, \qquad *\theta_2 = -\theta_1. \tag{10}$$

$$*\alpha_1 = \alpha_2, \qquad *\alpha_2 = -\alpha_1. \tag{11}$$

Using these notations, we may rewrite Equations (4) and (5) as follows,

$$dH = \sqrt{H^2 - K}\theta_1 \tag{4a}$$

$$d \log \sqrt{H^2 - K} = \alpha_1 + 2 * \omega_{12}. \tag{5a}$$

According to S.S.Chern[2], we have the following lemma.

Lemma 1: A surface M admits a non-trival isometric deformation that keeps the principal curvatures fixed iff

$$\begin{cases} d\alpha_1 = 0 \\ d\alpha_2 = \alpha_1 \wedge \alpha_2 \qquad \text{or} \end{cases} \tag{12}$$

$$\alpha_{12} = \alpha_2.$$

From now on, we will only consider the surface where $H \neq const$ and $|\nabla H| \neq 0$ since it is only locally taken into account.

§2. The properties of surfaces

In this section, we wish to prove the following theorem

Theorem 1: If M admits a non-trival isometric deformation preserving the principal curvature, then M satisfies:

(1) M is a W-surface;

(2) the metric $ds_{-1}^2 = |\nabla H|^2 ds^2/(H^2 - K)$ has Gaussian curvature -1;

(3) the metric $ds_0^2 = (H^2 - K)^2 ds^2/|\nabla H|^2$ has Gaussian curvature 0;

(4) $|\nabla H|$ and ΔH are functions of H.

Remark: (1) and (2) are due to S.S.Chern[2], to which the interested reader is referred.

Proof: (1) we have

$$\left(d\log\frac{|\nabla H|^2}{H^2-K}\right)\wedge dH = 0.$$

(See [2], Eq.(41) and Eq.(6a).)

Since M is a W-surface,

$$d(H^2-K)\wedge dH = 0. \tag{14}$$

Thus $d|\nabla H|^2\wedge dH = 0$.

That is exactly the first half of (4), and the other half can be proved as follows.

$$dH = \sqrt{H^2-K}\theta_1. \tag{4a}$$

Hence

$$\begin{aligned} d\sqrt{H^2-K} =& (\Delta H)\theta_1 + \sqrt{H^2-K} * \theta_{12} \quad \text{or} \\ d\log\sqrt{H^2-K} =& \frac{\Delta H}{\sqrt{H^2-K}}\theta_1 + *\theta_{12}. \end{aligned}$$

Differentiating the equations above, and by $d * \theta_{12} = 0 = d\theta_1$ (see [2], Eq.(39)) we have

$$\begin{aligned} 0 =& d\left(\frac{\Delta H}{\sqrt{H^2-K}}\theta_1\right) = d\left(\frac{\Delta H}{\sqrt{H^2-K}}\right)\wedge\theta_1 \quad \text{or} \\ 0 =& d\left(\frac{\Delta H}{\sqrt{H^2-K}}\right)\wedge dH = d(\Delta H)\wedge dH, \end{aligned}$$

which imply the property (4).

(2) Applying $*$-operator to (5a) and differentiating it, we have

$$\Delta\log\sqrt{H^2-K} = 2K + \frac{|\nabla H|^2}{H^2-K}. \tag{15}$$

And from prop(2), we have

$$\Delta\log\frac{|\nabla H|}{\sqrt{H^2-K}} = K + \frac{|\nabla H|^2}{H^2-K}. \tag{16}$$

Subtracting Eq(15) from Eq(16), we get

$$\Delta\log\frac{|\nabla H|}{H^2-K} = -K \tag{17}$$

which is just (3). Thus we complete the proof of the theorem.

§3. Isothermal coordinates in M

In this section, we will use Theorem 1 to give an isothermal coordinate system in M.

It is known that ds_0^2 has a Gaussian Curvature equal to 0, which allows us to set

$$ds_0^2 = du^2 + dv^2.$$

For convinence, we denote the gradiant and Laplacian operators in ds_0^2, ds_{-1}^2 and ds^2 by ∇_0 and Δ_0, ∇_{-1} and Δ_{-1}, and ∇ and Δ respectively. Then we have the following equations:

$$ds_{-1}^2 = F^2 ds_0^2, \qquad F^2 = |\nabla H|^4/(H^2 - K)^3 \tag{18}$$

$$ds^2 = |\nabla H|^2 ds_0^2/(H^2 - K)^2 \tag{19}$$

$$|\nabla_0 H|^2 = |\nabla H|^2 \cdot \frac{|\nabla H|^2}{(H^2 - K)^2} = |\nabla H|^4/(H^2 - K)^2 \tag{20}$$

$$\Delta_0 H = (\Delta H) \cdot \frac{|\nabla H|^2}{(H^2 - K)^2}. \tag{21}$$

From Eq(20) and Eq(21) we learn that $|\nabla_0 H|$ and $\Delta_0 H$ remain the functions of H respectively. Now ds_0^2 is, however, plane metric, so we hope to get some fine properties of H. To this end we need the following lemma.

Lemma 2: If a function g in plane has $|\nabla_0 g|^2 \equiv 1$, and $\Delta_0 g \equiv V(g)$, then the curves of $g = const$ are parallel lines or concentric circles.

Proof: Suppose the plane coordinate is (x, y). On the curve $P: \; g \equiv 0$, we have

$$g_x^2 + g_y^2 = 1, \qquad g_{xx} + g_{yy} = const.$$

The normal vector of P is

$$N = (g_x, g_y),$$

and the tangent vector of P is

$$T = (-g_y, g_x) = (\dot{x}, \dot{y}),$$

where $x = x(s), y = y(s)$,

$$\dot{x} = \frac{dx}{ds}, \qquad \dot{y} = \frac{dy}{ds},$$

s being the arc length parameter of the curve P.

In the following we will show that the curvature of P is constant. In fact,

$$\begin{aligned} K =& (\frac{dN}{ds}, T) = (g_{xx}\dot{x} + g_{yy}\dot{y})\dot{x} + (g_{yx}\dot{x} + g_{xx}\dot{y})\dot{y} \\ =& g_{xx}g_y^2 - 2g_{xy}g_xg_y + g_x^2 g_{xy} \\ =& \Delta_0 g - (g_x^2 g_{xx} + 2g_{xy}g_xg_y + g_y^2 g_{yy}) \\ =& \Delta_0 g - (\nabla_0|\nabla_0 g|^2, \nabla_0 g) = const. \end{aligned}$$

So curve P is a round circle or a straight line. Moreover, according to [3], we know that $P_c : g \equiv c$ are equidistant and parallel. Thus follows our lemma.

Remark: If we replace $|\nabla_0 g| \equiv 1$ by $|\nabla_0 g| = U(g)$ a function of g, the lemma still hold true. Since one can use a function w instead of g such that $|\nabla_0 w(g)| \equiv 1$ $\Delta_0(w(g)) = \tilde{v}(w(g))$. It is clear that the conclusion in independent of the choics of w.

Applying the above remark to function H and metric ds_0^2 leads to

Lemma 3. Choose a frame in ds_0^2 (if necessary) then the mean curvature is either

$$\begin{aligned} &(i) H = H(u) \quad \text{or} \\ &(ii) H = H(\sqrt{u^2+v^2}). \end{aligned} \tag{22}$$

Remark: $|\nabla H|^2/(H^2-K)$ ds_0^2 is just the isothermal coordinate in M if we choose the frame as we did in lemma 3.

§4. The Main Theorem.

In this section, we will find Function F defined in (18), then derive out the ordinary equation satisfied by H and K, and determine the angle from the isothermal coordinate to the principal curvature frame.

(1) Suppose $(x, \tilde{e_1}, \tilde{e_2})$ is the frame corresponding to the isothermal coordinate and the coframe is $(x, e^f du, e^f dv)$ where

$$ds^2 = \frac{|\nabla H|^2}{(H^2-K)^2} ds_0^2 \overset{\text{def.}}{=} e^{2f}(du^2+dv^2), \quad f = f(H).$$

and set

$$\begin{aligned} -(de_3, \tilde{e_1}) &= h_{11}\tilde{\omega_1} + h_{12}\tilde{\omega_2} \\ -(de_3, \tilde{e_2}) &= h_{21}\tilde{\omega_1} + h_{22}\tilde{\omega_2} \\ \tilde{\omega_1} = e^f du, &\quad \tilde{\omega_2} = e^f dv \end{aligned} \tag{23}$$

$$\begin{pmatrix} \tilde{e_1} \\ \tilde{e_2} \end{pmatrix} = \begin{pmatrix} \cos\theta & \sin\theta \\ -\sin\theta & \cos\theta \end{pmatrix} \begin{pmatrix} e_1 \\ e_2 \end{pmatrix}.$$

Then we have

$$\begin{aligned} h_{11} &= G\cos 2\theta + H, \\ h_{12} = h_{21} &= G\sin 2\theta, \\ h_{22} &= -G\cos 2\theta + H. \end{aligned} \tag{24}$$

where $G = \frac{a-c}{2}$, $H = \frac{a+c}{2}$.

We are now going to deal with Gauss-Codazzi equations. Because of Eq.(22), we may consider them in two cases.

<u>Case 1</u>: $H = H(\sqrt{u^2+v^2})$.

Letting $\gamma = \sqrt{u^2+v^2}$ and $\tau = \operatorname{arctg}\frac{v}{u}$, we have

$$H = H(\sqrt{u^2+v^2}) = H(\gamma), \quad F = F(\gamma), \quad G = G(\gamma), \quad f = f(\gamma); \tag{25}$$

$$\tilde{\omega_{12}} = *df = f'\frac{udv - vdu}{\gamma}. \tag{26}$$

The Codazzi equation is

$$\begin{cases} h_{112} = & h_{121} \\ h_{212} = & h_{221} \end{cases} \tag{27}$$

Using Eqs.(25) and (26), and through direct computation we get $h_{112}, h_{121}, h_{212}$ and h_{221}. Substituting them into Eq.(27), we get

$$(2\theta)'_\tau = -(\ln F)'\gamma + F \cdot \gamma \cos(2\tau - 2\theta), \tag{32}$$

$$(2\theta)'_\gamma = F \sin(2\tau - 2\theta) \tag{33}$$

where $\gamma e^{i\tau} = u + iv$.

Since $ds_{-1}^2 = F^2 ds_0^2$ has a Gaussian curvature equal to -1, we have

$$\Delta_0 \ln F = F^2, \qquad F = F(\gamma) \tag{34}$$

namely

$$(\ln F)'' + (\ln F)' \frac{1}{\gamma} = F^2. \tag{35}$$

Using (35), we find $(2\theta)''_{\gamma\tau} - (2\theta)''_{\tau\gamma} = -F\cos(2\tau - 2\theta) \neq 0$. So Eqs.(32) and (33) are not integrable. Hence it implies that $H = H(\gamma)$ is impossible. So the only remaining case is

Case 2: $H = H(u)$.

In this case, $G = G(u)$, $K = K(u)$, $f = f(u)$, $F = F(u)$

$$\tilde{\omega}_{12} = f'(u)dv, \quad \tilde{\omega}_1 = e^f du, \quad \tilde{\omega}_2 = e^f dv.$$

Similarly, from the Codazzi equation we get the following:

$$\theta'_u = -F \sin 2\theta \tag{36}$$

$$\theta'_v = F \cos 2\theta - (\ln F)'. \tag{37}$$

We have the Gaussian Curvature equal to -1, $ds_{-1}^2 = F^2 ds_0^2$. Consequently

$$\Delta_0 \ln F = F^2 \tag{38}$$

namely

$$(\ln F)'' = F^2. \tag{39}$$

Finding F from the above equation, we get

$$F^2 = \begin{cases} \frac{1}{u^2} \\ \frac{\lambda^2}{\sin^2 \lambda u} \\ \frac{\lambda^2}{\operatorname{sh}^2 \lambda u} \end{cases} . \tag{40}$$

Finding θ from the equations (37), (38) and (39), we get

$$\operatorname{tg}\theta = \begin{cases} \frac{v+s}{u} & \text{when } F^2 = \frac{1}{u^2}, \\ \operatorname{tg}\frac{u}{2}\operatorname{th}(\frac{v}{2}+s) & \text{when } F^2 = \frac{1}{\sin^2 u}, \\ \operatorname{cth}\frac{u}{2}\operatorname{tg}(v+s) & \text{when } F^2 = \frac{1}{\operatorname{sh}^2 u}. \end{cases}$$

where θ depends on an arbitrary constant s.

Since $ds^2 = |\nabla H|^2 ds_0^2/(H^2-K)^2$ has Gaussian curvature K, we obtain

$$\Delta_0 \log\left(\frac{(H^2-K)^2}{|\nabla H|^2}\right)\cdot\frac{(H^2-K)^2}{|\nabla H|^2} = 2(H^2-(H^2-K)). \tag{42}$$

Eq.(20) can be rewritten as:

$$H'^2 = \frac{|\nabla H|^2}{(H^2-K)^2}|\nabla H|^2 = F^2(H^2-K). \tag{43}$$

While eq.(42) rewritten as

$$\left(\ln\frac{H'}{F^2}\right)''\frac{H'}{F^2} = 2\left(H^2-\frac{H'^2}{F^2}\right). \tag{44}$$

This is an ordinary differential equation of order 3 satisfied by H. Since F is given in eq.(40), the Gaussian Curvature can be obtained by

$$K = H^2 - \frac{H'^2}{F^2}. \tag{45}$$

From the above discussion comes our main theorem.

Theorem 2. If a surface M admits a nontrival isometric deformation preserving mean curvature, then there exists a conformal metric (on M) $ds_0^2 = du^2+dv^2$, and the angle from the frame corresponding to ds_0^2 to the principal curvature frame is:

$$\operatorname{tg}\theta = \begin{cases} (v+s)/u & \text{when } F^2 = 1/u^2, \\ \operatorname{ctg}\frac{u}{2}\operatorname{th}(\frac{v}{2}+s) & \text{when } F^2 = 1/\sin^2 u, \\ \operatorname{cth}\frac{u}{2}\operatorname{tg}(v+s) & \text{when } F^2 = 1/\operatorname{sh}^2 u. \end{cases}$$

(Here s is an arbitrary parameter, which really gives the isometric deformation). Moreover, the mean curvature H and Gaussian Curvature K on M satisfy

$$\begin{cases} (\ln\frac{H'}{F^2})''\frac{H'}{F^2} = 2(H^2-\frac{H'^2}{F^2}), \\ K = H^2 - \frac{H'^2}{F^2}. \end{cases}$$

where $F^2 = u^{-2}, \sin^{-2}u, \operatorname{sh}^{-2}u$.

Obviously, if M satisfies the above condition, then M admits an isometric deformation preserving the principal curvature.

Remark: To the surface M in Theorem 2, if we set a, or ac or a/c to be constant, it is easily that $K = 0$ or $H \equiv const$. This implies the result of [4].

Remark: To every function H satisfying Eq.(44), we can construct a surface M because of the codazzi equation. In this case, the Gauss equation hold true for the surface. Furthermore, we can construct an isometric deformation that keeps the mean curvature by Eq.(41) (where θ depends on a constant s). So to every solution of Eq.(44), there exists a corresponding surface M.

Question: Does there exist a solution of Equation (44) such that K is not identically zero?

Reference

[1] O.Bonnet, Memoire sur la theorie des surfaces applicables, J.Ec Polyt., 42(1867) 79-92.

[2] S.S.Chern, Deformation of surfaces preserving principal curvatures. Differential geometry and complex anlysis p.155-p.163 H.E.Rauch Memorial volume 1985.

[3] R.Courant and D.Hilbert, Methods of Mathematical Physics Vol.II p.89-, 1962.

[4] A.Gervasio Colares and Katsue Kenmotsu, Isometric Deformation of surfaces in R^3 preserving the Mean Curvature Function, preprint. 1987.

[5] I.M.Roussos, Principal curvature preserving isometries of surfaces in ordinary space. Priprint Univ. of Minnesota 1986.

[6] E.Cartan, Sur les couples de surfaces applicable avec conservation des courbures principales. Bull. Sci. Math. 66(1942) 1-30 or Oeuvres Completes, Partie III, Vol.2 1591-1620.

Graduate School of University of Science
and Technology of China
Beijing, China

Lectures on Immersion Theory

RALPH L. COHEN AND ULRIKE TILLMANN

Department of Mathematics
Stanford University

These notes are intended to be an exposition of several aspects of immersion theory and some related homotopy theory. They grew out of a short course given on this subject by the first author during April and May, 1987, at the Nankai Institute of Mathematics in Tianjin, China.

The modern topological theory of immersions of compact manifolds began with the seminal work of H. Whitney [48] who proved that every compact, C^∞ n - dimensional manifold immerses in $\mathbf{R}^{2n-1}$ and embeds in $\mathbf{R}^{2n}$. In the late 1950's, a powerful theory for the study of immersions was developed by Hirsch and Smale [30]. Roughly speaking this theory says that up to isotopy, studying immersions is equivalent to studying normal bundles. When combined with Steenrod's classification of vector bundles [43] this theory gives a translation of immersion theory from differential topology to homotopy theory.

For example, let M^n be a compact n - dimensional manifold, and let

$$e : M^n \hookrightarrow \mathbf{R}^{n+K}$$

be an embedding into some large dimensional Euclidean space. Let

$$\nu_M \longrightarrow M^n$$

be the K - dimensional normal bundle of this embedding. For K sufficiently large, all such embeddings are isotopic and so the isomorphism class of ν_M is independent of the embedding. By Hirsch - Smale theory, one has the following:

THEOREM. *M^n immerses in $\mathbf{R}^{n+q}$ if and only if the K - dimensional bundle ν_M has $K-q$ linearly independent cross sections.*

By Steenrod's classification theorem, the question of the existence of linearly independent cross sections of a vector bundle can be interpreted as a question about the existence of a certain homotopy lifting of the classifying map of the bundle. In particular, it is well known that the primary (cohomological) obstructions to the existence of $K-q$ cross sections of ν_M are the mod 2 Stiefel - Whitney characteristic classes $w_i(\nu_M)$, for $i > q$.

In a paper appearing in 1960 [36], Massey proved the following.

During the preparation of this work the first author was supported by an NSF grant and an NSF - P.Y.I award.

THEOREM. *Let M^n and $\nu_M \longrightarrow M^n$ be as above. Then*

$$w_i(\nu_M) = 0$$

for $i > n - \alpha(n)$, where $\alpha(n)$ is the number of ones in the binary expansion of n.

Moreover it is easy to find examples of manifolds M^n with $w_{n-\alpha(n)} \neq 0$ (we will give one in chapter *II*) and hence Massey's result is best possible. In view of this theorem it is natural to conjecture the following result (commonly referred to as the *immersion conjecture*) which was proved in [24].

THEOREM. *Every compact, C^∞, n - dimensional manifold immerses in $\mathbf{R}^{2n-\alpha(n)}$.*

The work in [24] was the final step in a long program developed in large part by E.H. Brown Jr. and F.P. Peterson. Essential contributions made to this program by many mathematicians. These contributions are scattered throughout the literature over a span of almost 25 years. It is the goal of these notes (as it was of the Nankai Institute lectures) to give a unified account of many of these results, and to outline the proof of the conjecture. These notes are meant to be self contained and complete in the sense that where full proofs are not given, precise references are. These notes are entirely expository. There are no new results here, although there are points of view of certain topics that we believe are new.

This paper is organized as follows. In chapter *I* we give background information on Hirsch - Smale theory, cobordism theory, and the Steenrod algebra. At the end of this chapter we give a proof of a theorem of R.L. Brown [15] that says that every n - manifold is cobordant to one that immerses in $\mathbf{R}^{2n-\alpha(n)}$. Our proof follows Brown's ideas, but is simplified by the use of an easy to immerse set of generators for the cobordism ring. In chapter *II* we give a description of E. Brown and F. Peterson's calculation of the ideal of relations among the normal Stiefel - Whitney classes of n - manifolds. We then study in detail a curious relationship between their results and the homology of Artin's braid groups. We then study certain homotopy theoretic relationships between the braid groups and spaces first constructed by Brown and Gitler, and then used by Brown and Peterson to give a kind of Thom space analogue of the immersion conjecture. We give a proof of that result at the end of chapter *II*. In chapter *III* we outline the proof of the immersion conjecture. This involved a type of obstruction theory which studies when a certain construction on Thom spaces is induced on the vector bundle level. Chapter *III* starts with a description of this *de-Thom-ification* obstruction theory, and then describes in some detail how it was used to prove the immersion conjecture.

Throughout the paper all manifolds will be compact, C^∞, and unless stated otherwise, without boundary. All (co)homology will be taken with $\mathbf{Z}_2$ coefficients.

The first author would like to take this opportunity to thank Professor S.S. Chern for inviting him to the Nankai Institute, and Professors Jiang Boju, Zhou Xueguang, and the staff of the institute for their kind hospitality during his visit.

Chapter I
Preliminaries

The primary goal of this chapter is to give a rapid review of some basic material concerning vector bundles, cobordism theory, and related homotopy theory. In section one we describe how, using Smale - Hirsch immersion theory and Steenrod's bundle classification theorem, questions concerning immersions of manifolds can be interpreted homotopy theoretically.

In section two, we recall the basic notions of cobordism theory and the Steenrod algebra. The goal of this section is to prove the immersion conjecture up to cobordism; that every compact n - manifold is cobordant to one that immerses in $\mathbf{R}^{2n-\alpha(n)}$. This result (theorem 1.30) was originally proved by R.L. Brown [15].

Before proving this theorem we need to recall the basics of cobordism theory, including the Thom - Pontryagin theorem which identifies the cobordism ring η_* as the homotopy groups of the Thom spectrum **MO**, and Thom's calculation of this ring. In order to describe these results we review some basic information about the Steenrod algebra and Eilenberg - MacLane spectra.

Using Thom's criteria to determine when a manifold M^n represents an indecomposable element in the cobordism ring η_*, we construct a particularly simple set of ring generators, each being a certain iterated twisted product of spheres. These generators have easily studied immersion properties from which we readily derive theorem 1.30.

§1 Immersion theory and homotopy theory.

Let M^n and N^{n+k} be closed manifolds, and let TM and TN denote their tangent bundles.

DEFINITION 1.1. *$f : M \to N$ is an immersion if its differential $D_x f : T_x M \to T_{f(x)} N$ is a vector space monomorphism for every $x \in M$.*

Recall that by the Implicit Function Function Theorem, every immersion is locally an embedding (i.e. is locally one - to - one).

Let $Imm(M, N)$ denote the space of immersions with the C^∞-topology (see for example [31]), and let $Mono(TM, TN)$ denote the space of vector bundle monomorphisms between their tangent bundles. By taking the differential we have a mapping

$$D : Imm(M^n, N^{n+k}) \longrightarrow Mono(TM^n, TN^{n+k})$$

The following result, due to Smale in the case when M and N are both spheres and due to Hirsch [30] in the general case, is one of the most fundamental theorems in immersion theory.

During the preparation of this work the first author was supported by an NSF grant and an NSF - P.Y.I award.

THEOREM 1.2. *$D : Imm(M^n, N^{n+k}) \longrightarrow Mono(TM, TN)$ is a weak homotopy equivalence. That is, D induces an isomorphism in homotopy groups:*

$$D_* : \pi_*(Imm(M,N)) \xrightarrow{\cong} \pi_*(Mono(TM,TN)).$$

In particular, if one considers what this theorem says about the path components of these function spaces, we see that every vector bundle monomorphism $F : TM \to TN$ is homotopic through vector bundle monomorphisms to the derivative Df of some immersion $f : M^n \hookrightarrow N^{n+k}$.

We now take $N = S^{n+k}$. For dimensional reasons it is clear that immersions of M^n into S^{n+k} are equivalent to immersions of M^n into $\mathbf{R}^{n+k}$. Since the tangent bundle of $\mathbf{R}^{n+k}$ is trivial we have that $Imm(M^n, S^{n+k})$ is weakly homotopy equivalent to $Mono(TM, \epsilon^{n+k})$, where ϵ^{n+k} denotes the trivial $(n+k)$ - dimensional bundle.

For $F \in Mono(TM, \epsilon^{n+k})$, let ν_F denote the k - dimensional orthogonal complement bundle of the subbundle of ϵ^{n+k} given by the image of F. ν_F is well defined up to isomorphism. We therefore have

$$TM^n \oplus \nu_F \cong \epsilon^{n+k}.$$

The following is an easy corollary of theorem 1.2.

COROLLARY 1.3. *There exists a k - dimensional bundle $\nu \to M^n$ such that*

$$\nu \oplus TM \cong \epsilon^{n+k}$$

if and only if there exists an immersion $f : M \hookrightarrow \mathbf{R}^{n+k}$ so that $\nu_{Df} \cong \nu$.

Thus we have translated the immersion problem (does there exist an immersion $f : M \hookrightarrow \mathbf{R}^{n+k}$?) into an equivalent vector bundle problem (does there exist a k - dimensional vector bundle $\nu \to M$ such that $\nu \oplus TM \simeq \epsilon^{n+k}$?). We now recall recall Steenrod's classification theorem so that we may interpret this vector bundle problem as a problem in homotopy theory.

Let G be a topological group, and let EG denote a contractible space equipped with a free G - action. Basic obstruction theory shows that there always exists such a space EG and its equivariant homotopy type is well defined. Let

$$BG = EG/G$$

be the orbit space. The homotopy type of BG is well defined. The principal bundle

$$\gamma^G : EG \longrightarrow BG$$

is referred to as the *universal principal G - bundle* and BG the *classifying space* of G because of the following classification theorem of Steenrod [43].

If X and Y are two topological spaces with basepoints, let $[X, Y]$ denote the set of homotopy classes of continuous, basepoint preserving maps from X to Y.

THEOREM 1.4. *Let X be of the (based) homotopy type of a C.W. complex. Then the set $[X, BG]$ is in bijective correspondence with the isomorphism classes of principal G - bundles over X. This correspondence is given by sending a map $f : X \longrightarrow BG$ to the pull - back $f^*(\gamma^G)$ of the universal G - bundle.*

We now list some well-known examples.

EXAMPLE 1.5: $G = \mathbf{Z}_2$. $\mathbf{Z}_2$ acts freely on S^n, the n-sphere, by the antipodal map $x \rightarrow -x$. S^n is embedded as the equator in S^{n+1}, and as such is contractible in S^{n+1}. Hence, $E\mathbf{Z}_2 = S^\infty = \lim_{n\to\infty} S^n$ is a contractible space with free $\mathbf{Z}_2$-action. Thus $B\mathbf{Z}_2 = S^\infty/\mathbf{Z}_2 = \mathbf{R}P^\infty$ (the infinite real projective space).

EXAMPLE 1.6: $G = S^1$. $S^1 \subset \mathbf{C}$ acts freely on the unit sphere $S^{2n-1} \subset \mathbf{C}^n$ by scalar multiplication. Then, as above, $ES^1 = S^\infty = \lim_{n\to\infty} S^{2n-1}$ and $BS^1 = S^\infty/S^1 = \mathbf{C}P^\infty$.

EXAMPLE 1.7: $G = Gl_n(\mathbf{R})$ or $G = O(n)$. $Gl_n(\mathbf{R})$ is the group of invertible $n \times n$ matrices, and $O(n)$ denotes the corresponding orthogonal group. Let $V_{n,k}$ be the space of vector space monomorphisms

$$V_{n,k} = Mono(\mathbf{R}^n, \mathbf{R}^{n+k}),$$

and let $V_{n,k}^{<,>} \subset V_{n,k}$ denote those monomorphisms that preserve the Euclidean metric. These are the well known Stiefel manifolds. $Gl_n(\mathbf{R})$ acts freely on $V_{n,k}$ by precomposition. Similarly, $O(n)$ acts freely on $V_{n,k}^{<,>}$. It is straightforward to see that the respective limit spaces $V_{n,\infty} = \lim_{k\to\infty} V_{n,k}$ and $V_{n,\infty}^{<,>}$ are both contractible (see, for example [43]). Hence, $EGl_n(\mathbf{R}) = V_{n,\infty}$ and $EO(n) = V_{n,\infty}^{<,>}$. It is easy to see now that in both cases the orbit space is the space of n-dimensional subspaces in $\mathbf{R}^\infty$. That is $BGl_n(\mathbf{R}) = BO(n) = G_{n,\infty}(\mathbf{R})$, the infinite real Grassmanian manifold.

EXAMPLE 1.8: $G = GL_n(\mathbf{C})$ or $G = U(n)$. Again, $Gl_n(\mathbf{C})$ is the group of invertible, and $U(n)$ is the group of unitary $n \times n$ matrices. By a similar construction as in the real case, $BGl_n(\mathbf{C}) = BU(n) = G_{n,\infty}(\mathbf{C})$, the infinite complex Grassmanian manifold.

This classification of principal G-bundles gives us the desired classification of vector bundles, as follows.

Given a principal $GL_n(\mathbf{R})$-bundle $E \rightarrow X$, there is an induced vector bundle

$$E \times_{Gl_n(\mathbf{R})} \mathbf{R}^n \longrightarrow X.$$

Moreover it is well known that every vector bundle is induced in this way by a unique (up to isomorphism) principal $Gl_n(\mathbf{R})$ - bundle [43]. This bijective correspondence between principal $Gl_n(\mathbf{R})$ - bundles and n - dimensional vector bundles implies the following corollary to theorem 1.4.

COROLLARY 1.9. *The set $[X, BO(n)]$ is in bijective correspondence with the isomorphism classes of n-dimensional vector bundles over X.*

We will now use this classification of vector bundles to study the problem of immersing manifolds in Euclidean space.

Let

$$e : M \hookrightarrow \mathbf{R}^{n+k}$$

be an embedding into large dimensional Euclidean space. By transversality theory, for k sufficiently large all such embeddings are isotopic. Thus, the induced k - dimensional normal bundles are isomorphic. We denote their isomorphism class by ν_M^k. Notice that composition with the canonical inclusion

$$\tilde{e} : M \hookrightarrow \mathbf{R}^{n+k} \hookrightarrow \mathbf{R}^{n+k+m}$$

yields the relation

$$\nu^{k+m} = \nu^k \oplus \epsilon^m.$$

By corollary 1.9, this stable isomorphism class of vector bundles (*the stable normal bundle*) is classified by a map

$$\nu_M : M^n \longrightarrow BO$$

where $BO = lim_{n\to\infty} BO(n)$. The following, which is an easy consequence of 1.3 and 1.9, completes the translation of the problem of immersing manifolds into Euclidean space into a homotopy lifting problem.

COROLLARY 1.10. *There exists an immersion $f : M^n \hookrightarrow \mathbf{R}^{n+q}$ if and only if there is a map $\nu : M^n \to BO(q)$ making the following diagram homotopy commutative:*

$$\begin{array}{ccc} M^n & \xrightarrow{\nu} & BO(q) \\ \| & & \downarrow \\ M^n & \xrightarrow{\nu_M} & BO. \end{array}$$

This was the setting in which the immersion conjecture was proved, and will be studied in this paper.

THEOREM 1.11 [24]. *Let M^n be any compact n - manifold. Then there is a map*

$$\tilde{\nu}_M : M^n \longrightarrow BO(n - \alpha(n))$$

that lifts (up to homotopy) the stable normal bundle map

$$\nu_M : M^n \longrightarrow BO.$$

The goal of the next section is to prove a theorem of R.L. Brown [15] saying that every n - manifold is cobordant to one that satisfies this lifting property.

§2 Cobordism theory and the immersion conjecture.

All homology and cohomology will be taken with $\mathbf{Z}_2$ - coefficients.

Thom Spaces and Spectra.

Let $\varsigma^k \to X$ be a k - dimensional vector bundle over a space X. If ς^k has a metric, let $D(\varsigma^k)$ be the unit disk bundle and $S(\varsigma^k)$ be its boundary sphere bundle. Then we define the *Thom space* of ς^k as the quotient space

$$T(\varsigma^k) = D(\varsigma^k)/S(\varsigma^k).$$

Notice that if X is compact then $T(\varsigma^k)$ is homeomorphic to the one point compactification of the total space of ς^k.

The following is the classical Thom isomorphism theorem.

THEOREM 1.12. *Let ς^k be a k - dimensional vector bundle over a space X. The Thom space $T(\varsigma^k)$ satisfies the following properties.*

(1)

$$H^k(T(\varsigma^k)) \cong \mathbf{Z}_2$$

generated by a class $u_k \in H^k(T(\varsigma^k))$ called the Thom class.

(2) *For every $n \geq 0$, there is an isomorphism of $H^n(X)$ and $H^{n+k}(T(\varsigma^k))$ given by the cup product with u_k:*

$$\cup u_k : H^n(X) \xrightarrow{\cong} H^{n+k}(D(\varsigma^k), S(\varsigma^k)) \cong H^{n+k}(T(\varsigma^k)).$$

As one might imagine, the most important examples of Thom spaces are the Thom spaces of the universal bundles

$$\gamma^k \longrightarrow BO(k)$$

described above. We denote the Thom space $T(\gamma^k)$ by $MO(k)$.

Now observe that the pull - back of γ^{k+1} under the canonical inclusion

$$i : BO(k) \longrightarrow BO(k+1)$$

is simply the $(k+1)$ - dimensional bundle $\gamma^k \oplus \epsilon^1$. The Thom space of this bundle is the suspension, $\Sigma MO(k)$. Thus on the Thom space level i induces a map

$$i_* : \Sigma MO(k) \longrightarrow MO(k+1).$$

These maps give the collection of spaces $\{MO(k)\}$ the structure of a *spectrum.*

DEFINITION 1.13. *A spectrum $\mathbf{E}$ is a sequence of spaces $\{E_n\}$, together with maps $e_n : \Sigma E_n \to E_{n+1}$. $\mathbf{E}$ is called an Ω - spectrum if the adjoint maps $\tilde{e}_n : E_n \to \Omega E_{n+1}$ are homotopy equivalences, where ΩE_n is the based loop space of E_n.*

We will always denote spectra with bold face letters. The reader is referred to [1] for a description of the category of spectra. In particular the homotopy and homology groups of a spectrum $\mathbf{E}$ are defined by

$$\pi_q \mathbf{E} = \lim_{k \to \infty} \pi_{q+k} E_k$$

$$H_q \mathbf{E} = \lim_{k \to \infty} H_{q+k} E_k$$

where the limits are defined using the structure maps e_n and the suspension homomorphisms.

EXAMPLE 1.14: For a space X its *suspension spectrum* is the spectrum

$$\Sigma^\infty X = \{\Sigma^n X, id\}.$$

Notice that by the suspension isomorphism, $H_*(\Sigma^\infty X) = H_*(X)$. $\pi_*(\Sigma^\infty X)$ are the stable homotopy groups of X.

As described in [1], given a spectrum $\mathbf{E}$, one can suspend or desuspend $\mathbf{E}$, and study homotopy classes of maps of any given degree from spaces to $\mathbf{E}$. In particular $\mathbf{E}$ defines a generalized homology and cohomology theory as follows. Namely, given a space X define[1 2]

$$\mathbf{E}^q(X) = [X, \mathbf{E}]^q = \lim_{k\to\infty} [\Sigma^k X, E_{k+q}]$$

$$\mathbf{E}_q(X) = \pi_q(\mathbf{E} \wedge X_+) = \lim_{k\to\infty} \pi_{q+k}(E_k \wedge X_+).$$

By theorems of Brown [8] and Whitehead [47], $\mathbf{E}^*(\_)$ and $\mathbf{E}_*(\_)$ are generalized cohomology and homology theories, and every such theory can be represented by a spectrum in the above way. The coefficient groups of $\mathbf{E}^*(\_)$ and $\mathbf{E}_*(\_)$ are given by

$$\mathbf{E}_q(point) = \mathbf{E}^q(point) = \pi_q(\mathbf{E})$$

EXAMPLE 1.15: Let $\mathbf{K(G)} = \{K(G,n), e_n\}$ be the Eilenberg-MacLane spectrum. Recall $K(G,n)$ is the unique space (up to homotopy) whose homotopy groups are equal to G in dimension n, and zero otherwise. Since $\pi_{q+1}K(G,n) = \pi_q \Omega K(G,n)$, by the uniqueness of the Eilenberg-MacLane spaces there is a homotopy equivalence $\tilde{e}_{n-1} : K(G, n-1) \to \Omega K(G,n)$. The structure maps $e_{n-1} : \Sigma K(G, n-1) \to K(G,n)$ defined to be the adjoint of $\tilde{e}_{n-1}$, make $\mathbf{K(G)}$ into an Ω - spectrum.

The following is a classical result of Whitehead [47].

$$\mathbf{K(G)}^q(X) = [X, \mathbf{K(G)}]^q = H^q(X, G)$$

$$\mathbf{K(G)}_q(X) = \pi_q(\mathbf{K(G)} \wedge X_+) = H_q(X, G)$$

EXAMPLE 1.16: Let $\mathbf{MO} = \{MO(n), e_n\}$ be defined as above. By the Thom isomorphism theorem we have that

$$H^*(\mathbf{MO}) \cong H^*(BO) \cong \mathbf{Z}_2[w_1, w_2, \cdots]$$

where $w_i \in H^*(BO)$ is the ith Stiefel - Whitney class (see [39]). As we will explain below, the generalized cohomology theory associated with the Thom spectrum $\mathbf{MO}$ is cobordism theory.

The spectra $\mathbf{K(Z_2)}$ and $\mathbf{MO}$ are examples of *ring spectra* [1]. Namely, they come equipped with pairings

$$c : \mathbf{K(Z_2)} \wedge \mathbf{K(Z_2)} \longrightarrow \mathbf{K(Z_2)}$$

and

---

[1] $X \wedge Y = (X \times Y)/(X \vee Y)$

[2] $X_+ = X \cup \{disjoint\ basepoint\}$

$$\mu : \mathbf{MO} \wedge \mathbf{MO} \longrightarrow \mathbf{MO}$$

that satisfy certain associativity and unital properties. The map c is induced by the cup product pairings $K(\mathbf{Z}_2, q) \wedge K(\mathbf{Z}_2, r) \to K(\mathbf{Z}_2, q + r)$ and the map μ is induced on the Thom space level by the natural inclusions $O(r) \times O(m) \hookrightarrow O(r + m)$.

Cobordism and the Thom-Pontryagin Theorem.

Two closed, n-dimensional manifolds M_1^n and M_2^n are said to be *cobordant* if their disjoint union is the boundary of a $(n + 1)$-dimensional manifold W^{n+1}. This clearly defines an equivalence relation on n-dimensional manifolds. Let η_n then be the set of cobordism classes of n-dimensional manifolds, and define $\eta_* = \bigoplus \eta_n$. The following basic results concerning cobordism theory are due to Thom [46]. A particularly nice account of this theory and its generalizations is given in [45].

THEOREM 1.17. *η_* is a commutative graded $\mathbf{Z}_2$-algebra, where addition is induced by disjoint union ($\sqcup$), and multiplication by Cartesian product ($\times$).*

The following Thom - Pontryagin theorem may be viewed as the fundamental theorem of cobordism theory.

THEOREM 1.18. *There is a natural isomorphism of rings*

$$\phi : \eta_* \xrightarrow{\cong} \pi_* \mathbf{MO}$$

where the ring structure of $\pi_*\mathbf{MO}$ is induced by the ring spectrum structure described above.

We will not give a proof of this theorem (see for example [46]) but we will describe the map ϕ.

Let M^n represent an element in η_n. M^n can be embedded in $\mathbf{R}^{n+k} \subset S^{n+k}$ for some large k. Any tubular neighborhood of $M^n \subset S^{n+k}$ is homeomorphic to the open disc bundle of the normal bundle $D(\nu^k) \to M^n$. Collapsing the compliment of $D(\nu^k)$ in S^{n+k} to a point, the resulting quotient space can be identified with the Thom space $T(\nu^k)$. This defines a map

$$\tau : S^{n+k} \longrightarrow T(\nu^k)$$

which we will refer to as the *Thom - Pontryagin collapse* map. The element

$$\phi([M]) \in \pi_n(\mathbf{MO})$$

is represented by the composition

$$S^{n+k} \longrightarrow T(\nu^k) \longrightarrow MO(k)$$

where the second map is induced by the classifying map of ν^k.

We now generalize the notion of cobordism to *cobordism of a space* X. Consider the set of pairs (M^n, f) of a closed n-manifold M^n and a continuous map $f : M^n \to X$. Two pairs (M_1^n, f_1) and (M_2^n, f_2) are cobordant if there is a $(n+1)$ - manifold W^{n+1} with boundary $\partial W^{n+1} = M_1^n \sqcup M_2^n$ and a continuous map $F : W^{n+1} \to X$ such that the restriction of F to ∂W^{n+1} is $f_1 \sqcup f_2$. Notice that if X is a point , we are reduced to our previous notion of cobordism. Again, the set of equivalence classes of n-manifolds, denoted by $\eta_n(X)$, forms a group under disjoint union. Furthermore, $\eta_*(X) = \bigoplus \eta_n(X)$ becomes a graded η_*- module via the action $([M^r], [N^s, f]) \to [M^r \times N^s, f \circ pr_2]$.

Given another space Y, a continuous map $g : X \to Y$ induces a map in cobordism $g_* : \eta_*(X) \to \eta_*(Y)$ mapping $[M^n, f] \to [M^n, g \circ f]$. Thus cobordism can be thought of as a graded functor from the category of topological spaces to the category of groups. Indeed, $\eta_*(\_)$ defines a generalized homology theory. In this context theorem 1.18 states that its coefficient group is given by $\pi_*\mathbf{MO}$. Furthermore, a generalization of theorem 1.18 yields that cobordism theory is the generalized homology theory represented by the spectrum $\mathbf{MO}$ (see [45]).

The Steenrod Algebra.

In order to understand cobordism theory better, one needs to understand the homotopy type of the spectrum $\mathbf{MO}$. To describe this, we begin by recalling some basic information about the mod 2 Steenrod algebra A. We refer the reader to [44] and [40] for the details of these results.

Recall that the Steenrod squares Sq^i, $i \geq 0$, satisfy the following axioms.

AXIOMS 1.19.

(1) *Sq^i defines a natural transformation of abelian group valued functors*

$$H^n(\_; \mathbf{Z}_2) \to H^{n+i}(\_; \mathbf{Z}_2).$$

(2) $Sq^0 = 1$
(3) $Sq^i x = 0$ *for dim* $x < i$.
(4) $Sq^i x = x^2$ *for dim* $x = i$.
(5) *Cartan formula:* $Sq^i(xy) = \sum_j (Sq^j x)(Sq^{i-j} y)$
(6) *Sq^1 is the Bockstein homomorphism of the coefficient sequence* $0 \to \mathbf{Z}_2 \to \mathbf{Z}_4 \to \mathbf{Z}_2 \to 0$.
(7) *Adem relations: For* $a < 2b$, $Sq^a Sq^b = \sum_j \binom{b-j-1}{a-2j} Sq^{a+b-j} Sq^j$, *where the binomial coefficients are taken mod 2.*

Axioms (6) and (7) can be shown to be consequences of axioms (1)-(5).

EXAMPLE 1.20: Let $\varsigma^k \to X$ be a k-dimensional vector bundle over a space X. Let $u_k \in H^k(T(\varsigma^k))$ be its Thom class. Then the following equation may be taken as the definition of the ith Stiefel-Whitney class of ς^k

$$w_i(\varsigma^k) \cup u_k = Sq^i(u_k).$$

The *mod* 2 *Steenrod Algebra* A is now defined to be the graded $\mathbf{Z}_2$-algebra generated by the Sq^i's subject to the Adem relations. A typical element in A is the sum of elements of the form $Sq^{i_1} \cdots Sq^{i_k} = Sq^I$, where I is the sequence of integers $(i_1, i_2, \cdots, i_k)$. I is called *admissible* if each $i_s \geq 2i_{s+1}$.

For any space X, $H^*(X)$ has the structure of an A-module by axiom 1.19 (1). Exploring this module structure on the cohomology of the n - fold infinite real projective space $(\mathbf{R}P^\infty)^n = BO(1)^n$, one proves

THEOREM 1.21.

(1) *$\{Sq^I : I$ admissible$\}$ is a $\mathbf{Z}_2$-vector space basis of A.*
(2) *$\{Sq^i : i = 2^r\}$ generates A as a $\mathbf{Z}_2$-algebra.*

One defines a diagonal map

$$\Delta : A \to A \otimes A$$

to be the map of algebras induced by the Cartan formula

$$Sq^i \to \sum_j Sq^j \otimes Sq^{i-j}.$$

Δ makes A into a Hopf algebra, which Milnor [38] showed was cocommutative.

More specifically, let A^* be the dual of A, that is $(A^*)_k = Hom((A)_k; \mathbf{Z}_2)$. Then Δ induces a multipication $\Delta^* : A^* \otimes A^* \to A^*$, making A^* into an algebra. In a very important work [38], Milnor proved that A^* is a polynomial algebra. More precisely,

$$A^* = \mathbf{Z}_2[\xi_1, .., \xi_k, ..] \tag{1.22}$$

where the generators ξ_k are dual to the elements Sq^{I_k} with $I = (2^{k-1}, 2^{k-2}, \cdots, 2, 1)$, with respect to the basis of admissible monomials. Notice that dim $\xi_k = 2^k - 1$.

The Steenrod algebra A has an alternative description in view of Cartan and Serre's calculation of the cohomology of the mod 2 - Eilenberg - MacLane spaces (see, for example [40]). In terms of spectra, this calculation yields

$$A \cong H^*(\mathbf{K}(\mathbf{Z}_2)) = [\mathbf{K}(\mathbf{Z}_2), \mathbf{K}(\mathbf{Z}_2)]^*.$$

Dually $A^* \cong H_*(\mathbf{K}(\mathbf{Z}_2))$ and the multiplication in A^* is the ring spectrum multiplication: $\mathbf{K}(\mathbf{Z}_2) \wedge \mathbf{K}(\mathbf{Z}_2) \to \mathbf{K}(\mathbf{Z}_2)$.

We now recall Thom's calculations of the unoriented cobordism ring η_* and of $\eta_*(\_)$. First by comparing $MO(n)$ with $MO(1) \wedge \cdots \wedge MO(1)$ (n factors) a straightforward calculation shows that

$$H_*(\mathbf{MO}) = \mathbf{Z}_2[\sigma_i, i \geq 0]$$

where dim $\sigma_i = i$. Since H^* is an A-algebra, its dual H_* is an A^*-coalgebra. By equation 1.22, as graded $\mathbf{Z}_2$-vector spaces

$$H_*(\mathbf{MO}) = \mathbf{Z}_2[\sigma_{2^j-1}] \otimes \mathbf{Z}_2[\sigma_i, i \neq 2^j - 1] \cong A^* \otimes \mathbf{Z}_2[\sigma_i, i \neq 2^j - 1].$$

Through further analysis of the A^* - comodule structure, one can show that this vector space isomorphism can be chosen to be one of A^* - comodules. Hence by duality, there is an isomorphism of A - modules

$$H^*(\mathbf{MO}) = A \otimes \mathbf{Z}_2[\sigma_i, i \neq 2^j - 1] = \bigoplus_\omega \Sigma^{|\omega|} A.$$

Here ω runs through all words in $\{\sigma_i, i \neq 2^i - 1\}$, and $\Sigma^{|\omega|}$ indicates a dimension shift : $(\Sigma^{|\omega|}A)^n = A^{n-|\omega|}$, where $|\omega|$ is the degree of ω. Now recall that $A = H^*(\mathbf{K}(\mathbf{Z}_2))$. Hence, there is an A-module isomorphism

$$H^*(\mathbf{MO}) = \bigoplus_\omega \Sigma^{|\omega|} H^*(\mathbf{K}(\mathbf{Z}_2)) = H^*(\bigvee_\omega \Sigma^{|\omega|}\mathbf{K}(\mathbf{Z}_2)).$$

Let $b_\omega \in H^{|\omega|}(\mathbf{MO})$ be a free generator over A corresponding to the index ω. We may think of b_ω as a map $b_\omega : \mathbf{MO} \to \Sigma^{|\omega|}\mathbf{K}(\mathbf{Z}_2)$. Then $b_\omega^* : \Sigma^{|\omega|}A = H^*(\Sigma^{|\omega|}\mathbf{K}(\mathbf{Z}_2)) \to H^*(\mathbf{MO})$ is given by $\Sigma^{|\omega|}a \to a(b_\omega)$. By taking the wedge of these maps we recover the following theorem of Thom.

THEOREM 1.23. *There is a homotopy equivalence of spectra*

$$\mathbf{MO} = \bigvee_\omega \Sigma^{|\omega|}\mathbf{K}(\mathbf{Z}_2).$$

COROLLARY 1.24. a. *η_* is a graded polynomial algebra over $\mathbf{Z}_2$ generated by one class σ_i in every dimension i, so long as i is not of the form $2^r - 1$. That is,*

$$\eta_* \cong \mathbf{Z}_2[\sigma_i : i \neq 2^r - 1].$$

b. Given any space X,

$$\eta_*(X) \cong \eta_* \otimes H_*(X).$$

PROOF: Part (a) follows from the Thom-Pontryagin theorem and theorem 1.23, since clearly $\pi_*(\bigvee_\omega \Sigma^{|\omega|}\mathbf{K}(\mathbf{Z}_2)) = \mathbf{Z}_2[\sigma_i, i \neq 2^j - 1]$, multiplication being introduced by the ring spectrum multiplication in $\mathbf{K}(\mathbf{Z}_2)$.

For part (b) recall that cobordism is the homology theory associated to the spectrum $\mathbf{MO}$ and that ordinary homology with $\mathbf{Z}_2$-coefficients is associated to the Eilenberg-MacLane spectrum $\mathbf{K}(\mathbf{Z}_2)$. The isomorphism follows then from theorem 1.23.

Generators for η_*.

In Thom's calculation, he also described how to determine when a given manifold M represents an indecomposable element in η_*. This criterion can be described as follows.

Let

$$\mu_n : BO(1)^n = BO(1) \times \cdots \times BO(1) \longrightarrow BO(n)$$

be the classifying map for the (external) Whitney sum of the n canonical line bundles over $BO(1)^n$. In cohomology this induces a ring homomorphism

$$\mu_n^* : \mathbf{Z}_2[w_1, \cdots, w_n] = H^*(BO(n); \mathbf{Z}_2) \longrightarrow H^*(BO(1); \mathbf{Z}_2)^{\otimes n} = \mathbf{Z}_2[t_1, \cdots, t_n]$$

where w_j is the jth Stiefel Whitney class, and where t_i is the one dimensional ring generator of the cohomology of the ith factor in the product $BO(1)^n$. Since t_i is the first Stiefel Whitney class of the corresponding line bundle, the Cartan formula allows the following calculation of μ_n^*. (A beautiful account of this calculation is in [39].)

THEOREM 1.25. *The homomorphism*

$$\mu_n^*: \mathbf{Z}_2[w_1, \cdots, w_n] \longrightarrow \mathbf{Z}_2[t_1, \cdots t_n]$$

is injective. The image of μ_n^* is the subalgebra of all symmetric polynomials in the variables $t_1, \cdots t_n$. (Recall that a symmetric polynomial is one whose values remain invariant under permutations of the variables.) In particular the image of w_i is the ith elementary symmetric function. Said another way,

$$\mu_n^*(1 + w_1 + \cdots + w_n) = \prod_{1 \leq i \leq n} (1 + t_i).$$

Indeed the representation of $H^*(BO(n); \mathbf{Z}_2)$ as symmetric polynomials was a crucial step in the calculation of $H^*(MO; \mathbf{Z}_2)$ as module over the Steenrod algebra A.

Let $I = (i_1, \cdots i_q)$ be a sequence of integers between one and n. Let $s_I \in \mathbf{Z}_2[t_1, \cdots, t_n]$ be the smallest homogeneous symmetric polynomial with $t_1^{i_1} t_2^{i_2} \cdots t_q^{i_q}$ as a summand. For example, if $n = 3$ and $I = (1, 1)$, then

$$s_I = t_1 t_2 + t_1 t_3 + t_2 t_3.$$

Since each s_I is a symmetric polynomial, by the above theorem, it determines a unique class in $H^*(BO(n); \mathbf{Z}_2)$. Let $s_i = s_{(i)}$. The following was an important calculational result in Thom's paper [46].

THEOREM 1.26. *For M^m a closed manifold and $s_I \in H^*(BO(n); \mathbf{Z}_2)$ as above, let $s_I(M) \in H^*(M; \mathbf{Z}_2)$ represent s_I applied to the stable normal bundle of M. Then M^m represents an indecomposable element of the cobordism ring η_* if and only if the characteristic number*

$$\langle s_m(M^m), [M^m] \rangle$$

is nonzero.

We now use this theorem to describe a particularly nice set of ring generators for η_*. To describe this set of generators, consider the following construction. Given a manifold M^m, consider the $2m+k$ dimensional product manifold $S^k \times M^m \times M^m$. This manifold has a free involution ($\mathbf{Z}_2$ - action) given by permuting the coordinates of $M \times M$ and by the antipodal action on S^k. We define the *twisted product of M by S^k* to be the orbit space of this involution, which we denote by

$$D_k(M) = S^k \times_{\mathbf{Z}_2} M^m \times M^m.$$

Observe that $D_k(point)$ is the projective space $\mathbf{R}P^k$.

The characteristic classes of the tangent and normal bundles of such twisted product manifolds $D_k(M)$ were studied in detail by Dold [26] and by R.L. Brown [15] in terms of the characteristic classes of M. A corollary of these calculations is the following.

Let

$$n = 2^{i_1} + 2^{i_2} + \cdots + 2^{i_r}, \qquad \text{where} \quad i_1 < \cdots < i_r$$

be the binary expansion of n. In particular we have $\alpha(n) = r$. We therefore may write

$$n = 2^{i_1} + 2k$$

where, if k is nonzero, then $k = 2^{i_2 - 1} + \cdots + 2^{i_r - 1}$, and so $\alpha(k) = \alpha(n) - 1$.

THEOREM 1.27. *(Dold, R.L. Brown) Let $n = 2^{i_1} + 2k$ as above. Let M^k be a k - dimensional manifold such that $\langle s_k(M^k), [M^k]\rangle \neq 0$. We then have*

$$\langle s_n(D_{2^{i_1}}(M^k)), [D_{2^{i_1}}(M^k)]\rangle \neq 0.$$

We now inductively define manifolds B^n as follows. For $n = 0$, $B^0 = \mathit{point}$. Now assume that for each $q < n, q \neq 2^r - 1$, we've defined a manifold B^q of dimension q. For $n \neq 2^r - 1$, with binary expansion $n = 2^{i_1} + \cdots + 2^{i_r}, i_1 < \cdots i_r$, set $n = 2^{i_1} + 2k$ as above. Define

$$B^n = D_{2^{i_1}}(B^k).$$

By the previous theorem and induction we see that each B^n represents an indecomposable element in the cobordism ring η_*. Let $b_n \in \eta_*$ denote the class represented by B^n. Then by corollary 1.24, we've therefore proved the following.

COROLLARY 1.28. $\eta_* \cong \mathbf{Z}_2[b_n : n \neq 2^r - 1]$.

By the definition of the B_n's we therefore have the following:

COROLLARY 1.29. *Every closed manifold is cobordant to a disjoint union of products of iterated twisted products of a point by spheres.*

Immersion Conjecture upto Cobordism.

With this explicit description of the unoriented cobordism ring, we can easily recover the following theorem of R.L. Brown [15].

THEOREM 1.30. *Every closed, n - dimensional manifold is cobordant to one that immerses in $\mathbf{R}^{2n-\alpha(n)}$.*

PROOF: We first show that every B^n immerses in $\mathbf{R}^{2n-\alpha(n)}$. We do this by induction on n.

If $\alpha(n) = 1$, i.e. $n = 2^i$ for some i, then $B^n = \mathbf{R}P^{2^i}$, which, by Whitney's theorem immerses in $\mathbf{R}^{2^{i+1}-1} = \mathbf{R}^{2n-\alpha(n)}$. So now assume $\alpha(n) \geq 2$.

If we write $n = 2^{i_1} + 2k$ as above, then by definition,

$$B^n = S^{2^{i_1}} \times_{\mathbf{Z}_2} B^k \times B^k.$$

By inductive assumption, we have an immersion

$$f_k : B^k \hookrightarrow \mathbf{R}^{2k-\alpha(k)} = \mathbf{R}^{2k-\alpha(n)+1}.$$

This then induces an immersion

$$1 \times f_k \times f_k : B^n = S^{2^{i_1}} \times_{\mathbf{Z}_2} B^k \times B^k \hookrightarrow S^{2^{i_1}} \times_{\mathbf{Z}_2} \mathbf{R}^{2k-\alpha(n)+1} \times \mathbf{R}^{2k-\alpha(n)-1}.$$

The target of this immersion is the total space of a vector bundle over $\mathbf{R}P^{2^{i_1}}$. Now it is a well known fact using standard bundle theory and the tubular neighborhood theorem, that the total space of any smooth vector bundle over a closed q - dimensional manifold immerses into Euclidean space of codimension q. Thus we have an immersion

$$e : S^{2^{i_1}} \times_{\mathbf{Z}_2} \mathbf{R}^{2k-\alpha(n)+1} \times \mathbf{R}^{2k-\alpha(n)+1} \hookrightarrow \mathbf{R}^{2^{i_1+1}+4k-2\alpha(n)+2}.$$

Now $2^{i_1+1}+4k-2\alpha(n)+2 = 2n-2\alpha(n)+2 \leq 2n-\alpha(n)$, because $\alpha(n) \geq 2$. Thus if we compose these two immersions together with the natural embedding of $\mathbf{R}^{2n-2\alpha(n)+2} \subset \mathbf{R}^{2n-\alpha(n)}$, we get an immersion

$$f_n : B^n \hookrightarrow \mathbf{R}^{2n-\alpha(n)}.$$

We now use the number theoretic fact that

$$\alpha(n+m) \leq \alpha(n) + \alpha(m) \tag{1.31}$$

to observe that if we take products of the above immersions, then every product

$$B^{i_1} \times \cdots \times B^{i_r}$$

immerses in $\mathbf{R}^{2m-\alpha(m)}$, where $m = i_1 + \cdots + i_r$. Now since, by the above results, every closed manifold is cobordant to a disjoint union of such products, the theorem is proved.

COROLLARY 1.32. *The homomorphism of stable homotopy groups*

$$\pi^s_{q+n-\alpha(n)} MO(n-\alpha(n)) \longrightarrow \pi_q \mathbf{MO}$$

is surjective for $q \leq n$.

PROOF: Let $\beta : S^q \to \mathbf{MO}$ be an element in $\pi_q\mathbf{MO}$. By the Thom-Pontryagin theorem, β represents a q - manifold M^q. Theorem 1.30 allows us to choose M^q such that it immerses in $\mathbf{R}^{2q-\alpha(q)}$. By equation 1.31, $\alpha(n) - \alpha(q) \leq \alpha(n-q) \leq n-q$, and so $q - \alpha(q) \leq n - \alpha(n)$. Thus, in particular M^q immerses in $\mathbf{R}^{q+(n-\alpha(n))}$. Then, by construction, β is given by the composition

$$\beta : S^{q+(n-\alpha(n))} \xrightarrow{\tau} T(\nu^{n-\alpha(n)}) \longrightarrow MO(n-\alpha(n)).$$

Chapter II
Relations among Characteristic Classes, Braids, and Thom Spaces

In this chapter we will describe the progress that was made toward the proof of the immersion conjecture during the roughly fifteen year time span from the early 1960's until the late 1970's. The progress made during this period was not only crucial to the proof of the conjecture, but also had dramatic applications to many other problems in algebraic topology. Some of these applications were discussed in [23].

The main explicit goal of this chapter will be to outline the proof of a theorem of Brown and Peterson [12] that can be viewed as a Thom space analogue of the immersion conjecture. To state the theorem precisely, we need to adopt some notation and terminology.

Let

$$\varsigma^k \longrightarrow X$$

be a k - dimensional vector bundle over a space X. Let $T(\varsigma^k)$ denote the Thom space of ς^k. So for example, if ς^k is given a Riemannian metric, then $T(\varsigma^k)$ is the associated disk bundle with the boundary sphere bundle identified to a point. We use bold face $\mathbf{T}(\varsigma^{\mathbf{k}})$ to denote the associated *Thom spectrum* of ς^k. That is, $\mathbf{T}(\varsigma^{\mathbf{k}})$ is the formal k - fold desuspension of $T(\varsigma^k)$. In the language of spectra we have

$$\mathbf{T}(\varsigma^{\mathbf{k}}) = \Sigma^{-k}\Sigma^{\infty}T(\varsigma^k)$$

where $\Sigma^{\infty}Y$ is the suspension spectrum of a space Y. Thus the Thom isomorphism

$$\Phi : H^*(X;\mathbf{Z}_2) \longrightarrow H^*(\mathbf{T}(\varsigma^{\mathbf{k}});\mathbf{Z}_2)$$

has no dimension shift. In particular the Thom class lies in dimension zero:

$$u \in H^0(\mathbf{T}(\varsigma^{\mathbf{k}});\mathbf{Z}_2).$$

Important examples of these types of Thom spectra include $\mathbf{MO(n)}$, the Thom spectrum corresponding to the Thom space $MO(n)$ of the universal n - plane bundle over $BO(n)$, and $\mathbf{T}(\nu_{\mathbf{M}})$, the Thom spectrum of the stable normal bundle over a manifold M.

The theorem of Brown and Peterson that we will outline in this chapter is the following.

THEOREM 2.1. *Let M^n be a compact n - manifold. Let*

$$\nu_M : M^n \longrightarrow BO$$

classify the stable normal bundle of M^n. Then there is a map of spectra

$$\tilde{t}(\nu_M) : \mathbf{T}(\nu_{\mathbf{M}}) \longrightarrow \mathbf{MO(n - \alpha(n))}$$

making the following diagram of spectra homotopy commute:

$$\begin{array}{ccc} \mathbf{T}\nu_\mathbf{M} & \xrightarrow{\tilde{t}(\nu_M)} & \mathbf{MO}(\mathbf{n}-\alpha(\mathbf{n})) \\ {\scriptstyle =}\downarrow & & {\scriptstyle i}\downarrow \\ \mathbf{T}\nu_\mathbf{M} & \xrightarrow[t(\nu_M)]{} & \mathbf{MO} \end{array}$$

where $t(\nu_M)$ *is the map of Thom spectra induced by* ν_M, *and* i *is induced by the inclusion* $BO(n-\alpha(n)) \hookrightarrow BO$.

The proof of this theorem involves several steps. The first is the primarily algebraic step of computing all the relations among the Stiefel Whitney classes of n - manifolds. This was done by Brown and Peterson in a paper appearing in 1964 [11]. It turns out that this calculation has a curious relationship with the homology of Artin's braid groups as carried out by F. Cohen [19]. In section one we will describe the Brown - Peterson calculation in terms of certain modules over the Steenrod algebra A. We will also outline the relevant Steenrod algebra calculations. In section 2 we will study Artin's braid groups and explore the relationship between their cohomologies and the Steenrod algebra. The relationship that we will describe is implicit in the literature, although the point of view we will take may be of some independent interest. We will also study the relationship between the braid groups and Brown - Gitler spectra. These are spectra constructed originally by Brown and Gitler in 1973. They have powerful homotopy theoretic properties and have proved to be perhaps the most important by-product of the immersion conjecture technology. In section 3 we will use these spectra to describe Brown and Peterson's argument for the proof of theorem 2.1. In this chapter all cohomology will be taken with $\mathbf{Z}_2$ coefficients.

§1 Relations among Stiefel Whitney Classes.

As was described in the introduction, the primary obstrucions to immersing a manifold in codimension k - Euclidean space are the normal Stiefel - Whitney classes, $\bar{w}_i(M) = w_i(\nu_M)$ for $i > k$. Using a formula of Wu [49] concerning how the tangential and normal Stiefel - Whitney classes of a manifold are related, Massey proved the following in a paper appearing in 1960 [36].

THEOREM 2.2. *Let* M^n *be a compact* n *- manifold. Then*

$$\bar{w}_i(M^n) = 0, \qquad \textit{for} \quad i > n - \alpha(n).$$

If $n = 2^i$ then $\alpha(n) = 1$ and a simple calculation yields that $\bar{w}_{2^i-1}(\mathbf{R}P^{2^i}) \neq 0$. This in particular says that $\mathbf{R}P^{2^i}$ does not immerse in $\mathbf{R}^{2^i-2}$ and hence Whitney's immersion result is best possible.

In general, write n in its dyadic expansion:

$$n = 2^{i_1} + 2^{i_2} + \cdots + 2^{i_k},$$

where the i_j's are distinct, and so $\alpha(n) = k$. Let

$$M^n = \mathbf{R}P^{2^{i_1}} \times \cdots \times \mathbf{R}P^{2^{i_k}}.$$

Then using the Cartan formula for Stiefel Whitney classes and the fact that

$$n - \alpha(n) = \sum_{j=1}^{k} (2^{i_j} - 1),$$

one sees that $\bar{w}_{n-\alpha(n)} \neq 0$ and hence M^n does not immerse in $\mathbf{R}^{2n-\alpha(n)-1}$. Thus the immersion conjecture (that every n - manifold immerses in $\mathbf{R}^{2n-\alpha(n)}$) is best possible.

Now recall that $H^*(BO(k)) \cong \mathbf{Z}_2[w_1, \cdots, w_k]$ and that $H^*(BO) \cong \mathbf{Z}_2[w_i : i \geq 1]$. Thus Massey's calculation describes what polynomial generators among the normal Stiefel - Whitney classes of n - manifolds vanish. In order to look for other possible obstructions, it is important to compute the set of all polynomials among the Stiefel - Whitney classes that vanish on the normal bundles of all n - manifolds. These polynomials form an ideal

$$I_n \in H^*(BO) \cong \mathbf{Z}_2[w_i : i \geq 1].$$

This ideal was computed explicitly by Brown and Peterson [11] in a paper appearing in 1964. Rather than describe I_n explicitly we will describe its Thom isomorphic image

$$\Phi(I_n) \subset H^*(\mathbf{MO})$$

where $\mathbf{MO}$ is the Thom spectrum of the universal stable vector bundle over BO. Actually we will describe the quotient group $H^*(\mathbf{MO})/\Phi(I_n)$. To describe this recall Thom's calculation (see chapter I).

$$H^*(\mathbf{MO}) \cong \bigoplus_{\omega} \Sigma^{|\omega|} A \qquad \text{as } A \text{ - modules}$$

where A is the mod 2 - Steenrod algebra, and where the indexing set of the direct sum is all sequences $\omega = (j_1, \cdots, j_r)$ of positive integers, none of which are of the form $2^k - 1$. Brown and Peterson's result is the following:

THEOREM 2.3 [11].

$$H^*(\mathbf{MO})/\Phi(I_n) \cong \bigoplus_{|\omega| \leq n} A/J_{[(n-|\omega|)/2]}$$

where $J_k \subset A$ is the left ideal

$$J_k = A\{\chi(Sq^i) : i > k\}$$

and χ is the canonical antiautomorphism of A.

Recall that χ is an antiautomorphism in the sense that it is an isomorphism of $\mathbf{Z}_2$ - vector spaces and has the property that

$$\chi(ab) = \chi(b)\chi(a) \tag{2.4}$$

for any $a, b \in A$. To recall the definition of χ, recall that the Steenrod algebra A is a Hopf algebra under the diagonal map

$$\Delta : A \longrightarrow A \otimes A$$

defined to be the map of algebras induced by the Cartan formula

$$\Delta(Sq^k) = \sum_{i \le k} Sq^i \otimes Sq^{k-i}.$$

Now suppose $a \in A$ and that $\Delta(a) = \sum_i a_i \otimes b_i$ where a_i has dimension i. Then χ is defined recursively by the rule

$$\sum_i a_i \chi(b_i) = 0.$$

It is easy to see from this definition that

$$\chi^2 = 1.$$

Let M^n be a closed n - manifold with stable normal bundle ν_{M^n} having Thom spectrum $\mathbf{T}\nu_{M^n}$. Let $u_M \in H^0(\mathbf{T}\nu_M)$ be the Thom class. Define the ideal $\tilde{J}(M^n) \subset A$ by

$$\tilde{J}(M^n) = \{a \in A : au_M = 0\}.$$

Finally we define the ideal $\tilde{J}_n$ to be the intersection

$$\tilde{J}_n = \bigcap_{M^n} \tilde{J}(M^n) \tag{2.5}$$

where the intersection is taken over all closed n - manifolds M^n. The following is the main calculational result needed to prove theorem 2.3.

THEOREM 2.6.

$$\tilde{J}_n = J_{[n/2]} = A\{\chi(Sq^i) : 2i > n\}.$$

PROOF: Let M^n be a closed n - manifold. Consider the following composite isomorphism

$$D : H_q(M^n) \xrightarrow{p.d} H^{n-q}(M^n) \xrightarrow{\Phi} H^{n-q}(\mathbf{T}(\nu_{M^n}))$$

where the map $p.d$ is Poincare duality and the map Φ is the Thom isomorphism. It is well known that the isomorphism D is induced by the Spanier - Whitehead duality between a manifold and the Thom space of its stable normal bundle. (See [4] for example.) In particular notice that if $[M^n] \in H_n(M^n)$ is the fundamental class, we then have that

$$D([M^n]) = u_M \in H^0(\mathbf{T}(M^n)).$$

Now let $a \in A$ be a cohomology operation. If a has dimension i it induces a natural transformation of cohomology

$$a : H^q \longrightarrow H^{q+i}$$

for all q. Using the universal coefficient theorem its dual homomorphism may be viewed as a natural transformation of homology

$$a_* : H_q \longrightarrow H_{q-i}$$

for all q. The following relationship between the duality homomorphism D and the canonical antiautomorphism χ was first worked out by Wu [49]:

LEMMA 2.7 [49]. *For $a \in A$ of dimension i we have*

$$D(\chi(a)_*([M^n])) = a(D([M^n])) = au_{M^n} \in H^i(\mathbf{T}(\nu_{M^n}).$$

COROLLARY 2.8. $a \in \tilde{J}_n$ *if and only if*

$$\chi(a) : H^{n-i}(M^n) \longrightarrow H^n(M^n)$$

is the zero homomorphism for every closed n - manifold M^n.

PROOF: Suppose there is a class $y \in H^{n-i}(M^n)$ such that $\chi(a)(y) \neq 0$ in $H^n(M^n) = \mathbf{Z}_2$. We then have

$$\begin{aligned} 0 &\neq \langle \chi(a)(y), [M^n] \rangle \\ &= \langle y, \chi(a)_*[M^n] \rangle. \end{aligned}$$

But since D is an isomorphism, this means that $D(\chi(a)_*([M^n])) \neq 0$. By lemma 2.7 this implies that $au_{M^n} \neq 0$ and hence $a \notin \tilde{J}_n$. The converse is proved in precisely the same way.

This corollary can be strengthened as follows.

LEMMA 2.9. *Let $a \in A$ have dimension i. Then $a \in \tilde{J}_n$ if and only if*

$$\chi(a) : H^{n-i}(X) \longrightarrow H^n(X)$$

is zero for every space X.

PROOF: In view of corollary 2.8, to prove this lemma it is sufficient to show that if there is a class $x \in H^q(X)$ and a cohomology operation $b \in A$ of dimension i with $b(x) \neq 0$, then there exists a closed manifold M^n of dimension $n = q + i$ and a class $z \in H^q(M^n)$ with $b(z) \neq 0$.

So suppose $x \in H^q(X)$ is such that $b(x) \neq 0$. By the universal coefficient theorem there is a class $y \in H_{q+i}(X)$ so that

$$\langle b(x), y \rangle \neq 0.$$

Now by Thom's cobordism theorem, the unoriented bordism

$$\eta_*(X) \cong \eta_* \otimes H_*(X).$$

This implies that every element of homology is representable by a manifold. So in particular there is a manifold M^n of dimension $n = q + i$ and a map

$$f : M^n \longrightarrow X$$

such that in homology, $f_*([M^n]) = y \in H_n(X)$.

Now define $z \in H^q(M^n)$ by the formula

$$z = f^*(x) \in H^q(M^n).$$

Then by the naturality of cohomology operations,

$$\begin{aligned} \langle b(z), [M^n] \rangle &= \langle f^*(b(x)), [M^n] \rangle \\ &= \langle b(x), f_*([M^n]) \rangle \\ &= \langle b(x), y \rangle \\ &\neq 0. \end{aligned}$$

Thus $b(z) \neq 0$ and hence the lemma is proved.

To compute the ideal $\tilde{J}_n$, we recall the notion of the *excess* of a cohomology operation (see [44]).

DEFINITION 2.10. *Let $\alpha \in A$ have dimension i. We say that α has excess $= k$, written $e(\alpha) = k$, if the following two conditions hold:*

(1) *For every space Y and integer $q < k$,*

$$\alpha : H^q(Y) \longrightarrow H^{q+i}(Y)$$

is the zero homomorphism

(2) *There exists a space X and a class $x \in H^k(X)$ such that*

$$\alpha(x) \neq 0.$$

Thus the excess of a cohomology operation is the smallest dimension in which that operation acts nontrivially. Now it is well known that $e(Sq^k) = k$ and more generally, if $I = (i_1, \cdots, i_r)$ is an admissible sequence (i.e. $i_j \geq 2i_{j+1}$ for $j = 1, \cdots, r$), then

$$e(Sq^I) = i_1 - i_2 - \cdots - i_r. \tag{2.11}$$

(See [44].)

Now fix an integer n and let $\alpha \in A$ have dimension i. Then by the definition of excess,

$$\alpha : H^{n-i}(X) \longrightarrow H^n(X)$$

is the zero homomorphism for every space X if and only if

$$e(\alpha) > n - i \qquad \text{i.e} \qquad e(\alpha) + i > n.$$

This observation together with lemma 2.8 implies the following:

COROLLARY 2.12. *$\alpha \in A$ lies in the ideal $\tilde{J}_n$ if and only if*

$$e(\chi(\alpha)) + dim(\alpha) > n$$

or equivalently, since $\chi^2 = 1$, we have that $\chi(\alpha) \in \tilde{J}_n$ if and only if

$$e(\alpha) + dim(\alpha) > n.$$

Formula 2.11 and corollary 2.12 allow us to calculate the ideal $\tilde{J}_n$. To do this, note that since χ is a vector space isomorphism, then

$$\{\chi(Sq^I) : \text{I is admissible}\}$$

forms a $\mathbf{Z}_2$ - vector space basis for A. Thus by 2.12

$$\{\chi(Sq^I) : \text{I is admissible and} \quad e(Sq^I) + dim(Sq^I) > n\}$$

forms as $\mathbf{Z}_2$ vector space basis for $\tilde{J}_n$. Now by formula 2.11 and the fact that if $I = (i_1, \cdots, i_r)$ then $dim(Sq^I) = i_1 + \cdots + i_r$, we see that

$$e(Sq^I) + dim(Sq^I) = 2i_i.$$

Thus we've proven that

$$\{\chi(Sq^I) : I = (i_1, \cdots, i_r) \quad \text{is admissible and} \quad 2i_1 > n\}$$

is a $\mathbf{Z}_2$ - vector space basis for the ideal $\tilde{J}_n$. Now using the fact that $\chi(ab) = \chi(b)\chi(a)$ it is easy to see that $\tilde{J}_n$ is the left ideal

$$\tilde{J}_n = A\{\chi(Sq^i) : 2i > n\} = J_{[n/2]}$$

which proves theorem 2.6.

We end this section with a calculation that uses theorem 2.6 to recover Massey's result (theorem 2.2).

By the above calculation, we see that the quotient module $A/J_{[n/2]}$ has a vector space basis

$$\{\chi(Sq^I) : I = (i_1, \cdots, i_r) \quad \text{is admissible and} \quad 2i_1 \leq n\}.$$

This basis is graded by dimension, and by the admissibility requirements one sees immediately that this is a finite basis. Furthermore an easy exercise with the admissibility requirements in this basis shows that the admissible sequence in this basis having highest dimension is

$$I_n = (n', n^{(2)}, n^{(3)}, \cdots, 1)$$

where $n' = [n/2]$, and $n^{(q)}$ is recursively defined by $n^{(q)} = [n^{(q-1)}/2]$. A similarly easy exercise shows that the dimension of Sq^{I_n} is $n - \alpha(n)$. These observations imply the following:

COROLLARY 2.13. *If $a \in A$ has $dim(a) > n - \alpha(n)$ then $a \in J_{[n/2]} = \tilde{J}_n$.*

Thus in particular $Sq^i \in \tilde{J}_n$ for every $i > n - \alpha(n)$. By the definition of $\tilde{J}_n$ this implies that $Sq^i(u_{M^n}) = 0$ for every n - manifold M^n and every $i > n - \alpha(n)$. But since $Sq^i(u_{M^n})$ is the Thom isomorphic image of the normal Stiefel Whitney class $\tilde{w}_i(M^n)$, Massey's theorem (theorem 2.2) follows.

§2 Braid groups and the Steenrod algebra.

In this section we study the homology of Artin's braid groups and their relation with the Steenrod algebra. The braid groups were originally defined by Artin in [3] and were used by Alexander and others in the 1920's to study knot theory [2]. Since then they have had many applications in the study of mapping class groups of surfaces (see [5]). Very recently the dramatic work of V. Jones [28] has again made use of the braid groups in studying knot theory. This time it was the representation theory of braid groups that yielded polynomial invariants of knots and links.

During the past ten to fifteen years Artin's braid groups have also had some striking applications to algebraic topology. They have been used in studying both unstable and stable homotopy theory [18, 22, 35], as well as immersion theory. An elegant exposition of the use of braids in classical homotopy theory was given by F. Cohen in [18]. The homology of the braid groups was worked out by Fuks, Milgram, and F. Cohen [27, 37, 19]. Their cohomology as modules over the Steenrod algebra was worked out implicitly by F. Cohen in [19] and explicitly by Mahowald [35] at the prime 2 and by R. Cohen [22] at odd primes. In this section we will describe their homology in terms of explicit geometric constructions (products and twisted products) , and then observe that the Eilenberg - MacLane spectra have analogous constructions. This was essentially the approach taken in [22] and [9]. The upshot will be the description of the cohomology of a Thom space of a certain representation of the braid groups as the cyclic modules A/J_k over the Steenrod algebra. These Thom spaces have certain important homotopy theoretic properties that are important in immersion theory. These properties were first established by Brown and

Peterson [13] by comparing these Thom spaces to certain spectra constructed by Brown and Gitler [10] that were known to have these properties. These properties were later established directly by Brown and Cohen [9]. Besides their importance in immersion theory, these Brown - Gitler spectra and braid group Thom spaces have played very important roles in many of the recent advances in homotopy theory. A description of some of these applications was given in [23].

We will begin our study with a description of the braid groups themselves. As above, all homology and cohomology will be taken with $\mathbf{Z}_2$ - coefficients.

We denote Artin's braid group on k strings by β_k. An element $b \in \beta_k$ can be thought of as a configuration of k - stings, connecting two sets of k fixed points, each set lying in parallel planes in $\mathbf{R}^3$. Thus one can picture $b \in \beta_k$ as follows:

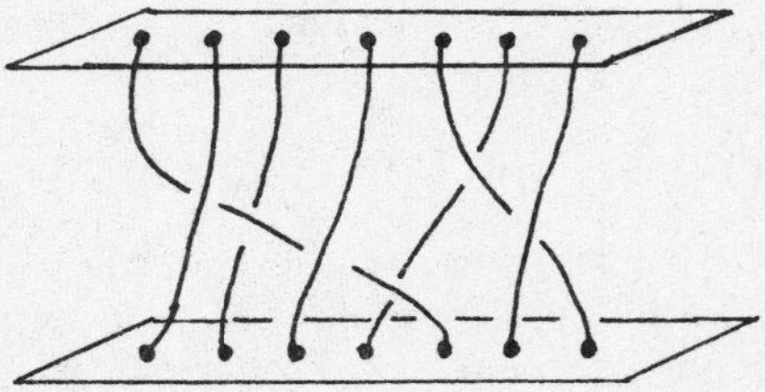

Actually elements of β_k are isotopy classes of such configurations. The group multiplication in β_k is given by juxtaposition of braids. The most direct way of making this definition precise is by viewing β_k as the fundamental group of a certain configuration space. This is done as follows.

Let

$$F_k = \{(t_1, \cdots, t_k) \in (\mathbf{R}^2)^k : t_i \neq t_j \quad \text{if} \quad i \neq j\}.$$

Notice that the symmetric group on k - letters Σ_k acts freely on F_k by permuting the coordinates. We denote the orbit space by B_k:

$$B_k = F_k / \Sigma_k.$$

DEFINITION 2.14. *Artin's braid group on k strings is defined to be the fundamental group,*

$$\beta_k = \pi_1(B_k).$$

Notice that this definition agrees with the conceptual description given above. That is, by use of the Σ_k - covering space

$$F_k \longrightarrow B_k,$$

a loop in B_k can be represented by a path in F_k whose endpoints differ by a permutation. A path in F_k is a one parameter family of k - distinct points in $\mathbf{R}^2$ and thus represents a braid. The homotopy relation in the definition of π_1 is the braid isotopy relation.

Now it is not difficult to see that B_k is in fact an Eilenberg - MacLane space (see [41] for example) and hence

$$B_k = K(\beta_k, 1).$$

Thus the (co)homology of the group β_k is given by the (co)homology of the space B_k. In this paper we will only be concerned with the mod 2 (co)homology of these groups. In order to describe this homology we need no establish some structure on the braid groups β_k and their classifying spaces B_k. For example, notice that there is a natural inclusion

$$i_k : \beta_k \hookrightarrow \beta_{k+1}$$

given by adding on a trivial $(k+1)^{st}$ string:

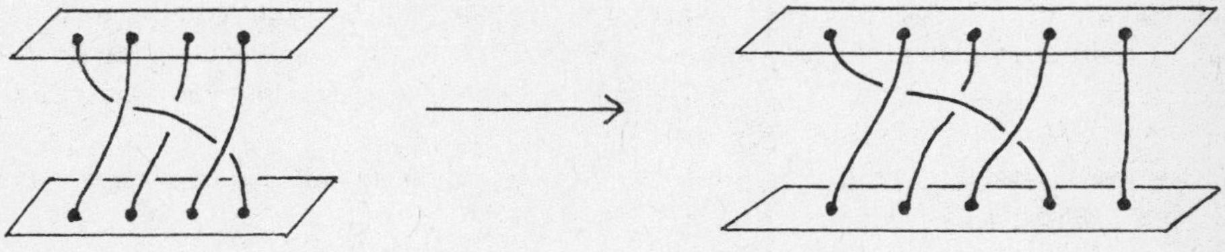

This induces a map on the level of Eilenberg - MacLane spaces $B_k \to B_{k+1}$. Let β_∞, the *infinite braid group* be the direct limit

$$\beta_\infty = \lim_{k \to \infty} \beta_k$$

and similarly $B_\infty = \varinjlim B_k$. We will describe, in a geometric fashion, the following calculation of F. Cohen:

THEOREM 2.15.

(1) *Each inclusion*

$$i_k : H_*(B_k) \longrightarrow H_*(B_{k+1})$$

is a monomorphism.

(2) $H_*(B_\infty)$ *is a polynomial algebra*

$$H_*(B_\infty) \cong \mathbf{Z}_2[x_1, \cdots, x_i, \cdots]$$

where the dimension of the generator x_i *is* $2^i - 1$.

As mentioned above, we wish to geometrically describe this calculation. We begin by describing the product structure in $H_*(B_\infty)$. This product structure is induced by pairings

$$\beta_k \times \beta_r \longrightarrow \beta_{k+r}$$

defined by placing the braid on k strings adjacent (and disjoint from) the braid on r strings, thus realizing a braid on $k+r$ strings. This pairing is seen in the following picture:

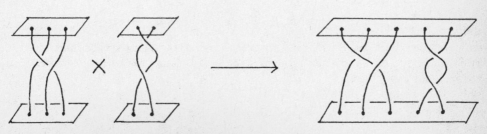

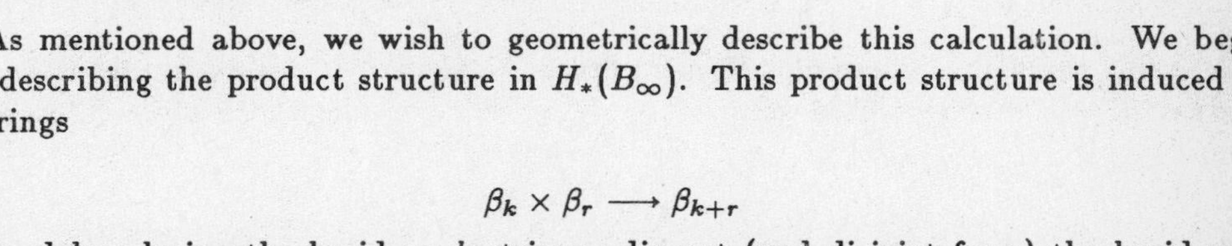

This pairing induces maps of classifying spaces

$$B_k \times B_r \longrightarrow B_{k+r}$$

which in the limit induces maps

$$B_\infty \times B_\infty \longrightarrow B_\infty .$$

A description of this map of classifying spaces is given in [41]. This pairing induces the product in homology reflected in theorem 2.15 part (2). Thus to complete our geometric description of the algebra $H_*(B_\infty) = \mathbf{Z}_2[x_i]$, we need to describe, geometrically, the generators x_i. We do this inductively on i.

First observe that the braid group on two strings, β_2 is an infinite cyclic group, with generator σ defined to be a single half twist:

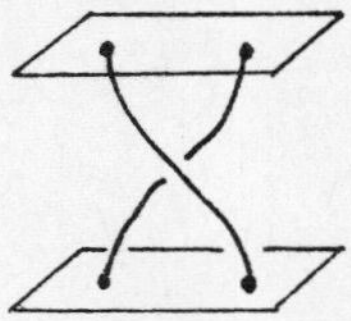

Thus the classifying space B_2 is naturally homotopy equivalent to the circle S^1. This homotopy equivalence is given by the map

$$h : S^1 \longrightarrow B_2$$

defined by $h(t) = (0, t) \in B_2$. The generator $x_1 \in H_*(B_\infty)$ is defined to be the image under the inclusion

$$i_* : H_1(B_2) \longrightarrow H_1(B_\infty)$$

of the generator of $H_1(S^1) \cong H_1(B_2)$.

To define the higher dimensional generators, we make use of the extended or *twisted* product structure of the braid groups. To define this consider the semidirect product group

$$\mathbf{Z} \tilde{\times} \beta_k \times \beta_k$$

defined to be the extension

$$1 \longrightarrow \beta_k \times \beta_k \longrightarrow \mathbf{Z} \tilde{\times} \beta_k \times \beta_k \longrightarrow \mathbf{Z} \longrightarrow 1$$

given by letting $\mathbf{Z}$ act on $\beta_k \times \beta_k$ by permuting coordinates. This action therefore factors through an action of $\mathbf{Z}_2$ via the projection $\mathbf{Z} \to \mathbf{Z}_2$. The *twisted* products referred to above are homomorphisms

$$\xi_k : \mathbf{Z} \tilde{\times} \beta_k \times \beta_k \longrightarrow \beta_{2k}$$

extending the pairing $\beta_k \times \beta_k \longrightarrow \beta_{2k}$ defined above. ξ_k is defined by associating to a triple (n, b_1, b_2), where $n \in \mathbf{Z}$ and $b_1, b_2 \in \beta_k$, the braid on $2k$ strings defined by twisting b_1 around the braid b_2 by n half twists:

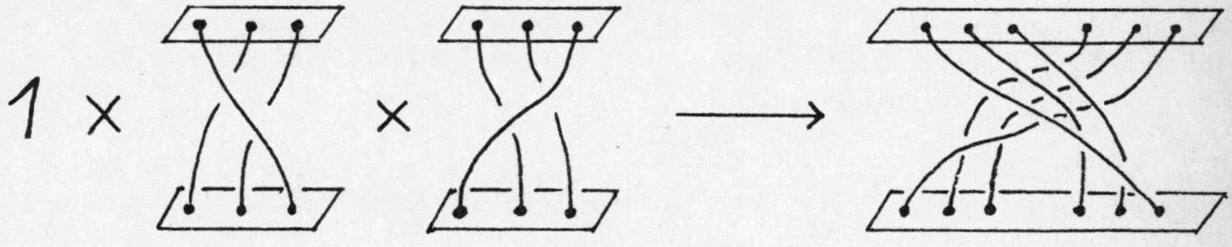

The classifying space of $\mathbf{Z}\tilde{\times}\beta_k \times \beta_k$ is clearly the space $S^1 \times_{\mathbf{Z}_2} B_k \times B_k$, and these twisted products induce classifying maps

(2.16) $$\xi_k : S^1 \times_{\mathbf{Z}_2} B_k \times B_k \longrightarrow B_{2k}.$$

and in the limit

$$\xi : S^1 \times_{\mathbf{Z}_2} B_\infty \times B_\infty \longrightarrow B_\infty$$

that extends the pairings $B_\infty \times B_\infty \longrightarrow B_\infty$ described above.

These twisted pairings allow us to define a homology operation

(2.17) $$Q : H_r(B_k) \longrightarrow H_{2r+1}(B_{2k})$$

by the rule

$$Q(x) = (\xi_k)_*(e_1 \otimes x \otimes x) \in H_{2r+1}(B_{2k})$$

where $e_1 \in H_1(S^1)$ is the generator. This operation is a type of Araki - Kudo or Dyer - Lashof homology operation. They were defined on the homology of braid groups by F. Cohen in [19]. We now define

$$x_2 = Q(x_1) \in H_3(B_4)$$

and inductively we define

$$x_i = Q(x_{i-1}) \in H_{2^i-1}(B_{2^i}).$$

By abuse of notation we also let x_i denote the image of these classes in $H_*(B_\infty)$. In [19] F. Cohen proved that these classes form polynomial generators of $H_*(B_\infty)$. Thus in some sense $H_*(B_\infty)$ is the universal commutative algebra generated by the twisted product operation Q.

Now Cohen also described the homology of the finite braid groups $H_*(B_k)$. To do this we define a weight to a monomial in the x_i's defined to be the number of strings in the braid group necessary to define the monomial by the above constructions. So for example we needed the braid group on 2 stings β_2 to define x_1 and so x_1 has weight = 2. More generally, the above constructions show us that the weight function is defined by the rules

(1) $wt(1) = 0$
(2) $wt(x_i) = 2^i$, and
(3) $wt(xy) = wt(x) + wt(y)$.

The following calculation of F. Cohen [19] gives a beautiful correspondence between the geometric constructions (i.e the product and twisted product) described above, and the algebra of the homology:

THEOREM 2.18.
$$H_*(B_k) \hookrightarrow H_*(B_\infty) = \mathbf{Z}_2[x_1, \cdots, x_i, \cdots]$$
is spanned by those monomials of weight $\leq k$.

We end our discussion of the braid groups by observing that the monomial of weight $= k$ of maximal dimension is easily seen to be

$$x_{i_1} \cdots x_{i_r}$$

where $k = 2^{i_1} + \cdots + 2^{i_r}$ is the dyadic expansion of k. Thus $\alpha(k) = r$. The dimension of this monomial is $k - \alpha(k)$. This proves the following.

COROLLARY 2.19. *B_k has the same mod 2 homotopy type as a C.W complex of dimension $k - \alpha(k)$.*

Observe that the homology of the infinite braid group,

$$H_*(B_\infty) \cong \mathbf{Z}_2[x_1, \cdots, x_i, \cdots]$$

is formally isomorphic to the dual of the Steenrod algebra,

$$A^* \cong \mathbf{Z}_2[\xi_1, \cdots, \xi_i, \cdots].$$

We will now describe how this isomorphism actually points to a much deeper, geometric relationship between the braid groups and the Steenrod algebra. Our first step in doing this is showing that, like $H_*(B_\infty)$, the dual of the Steenrod algebra A^* also can be viewed as the universal commutative algebra generated by a twisted product operation.

To describe this operation, let

$$c : \mathbf{K}(\mathbf{Z}_2) \wedge \mathbf{K}(\mathbf{Z}_2) \longrightarrow \mathbf{K}(\mathbf{Z}_2)$$

be the cup product pairing as described in chapter I. Alternatively, c can be viewed as the generator of the cohomology group

$$c \in H^0(\mathbf{K}(\mathbf{Z}_2) \wedge \mathbf{K}(\mathbf{Z}_2)) = \mathbf{Z}_2.$$

As was pointed out in chapter I, this pairing induces the multiplication in

$$H_*(\mathbf{K}\mathbf{Z}_2)) = A^* = \mathbf{Z}_2[\xi_1, \cdots, \xi_i, \cdots].$$

Now consider the twisted product spectrum

$$S^1 \wedge_{\mathbf{Z}_2} \mathbf{K}(\mathbf{Z}_2) \wedge \mathbf{K}(\mathbf{Z}_2).$$

It is easy to see that the cohomology class c extends to a cohomology class (namely the generator)

$$\varsigma \in H^0(S^1 \wedge_{\mathbf{Z}_2} \mathbf{K}(\mathbf{Z}_2) \wedge \mathbf{K}(\mathbf{Z}_2)) = \mathbf{Z}_2.$$

Viewing ς as a map

$$\varsigma : S^1 \wedge_{\mathbf{Z}_2} \mathbf{K}(\mathbf{Z}_2) \wedge \mathbf{K}(\mathbf{Z}_2) \longrightarrow \mathbf{K}(\mathbf{Z}_2),$$

we can use it to define a homology operation

$$Q : A_r^* = H_r(\mathbf{K}(\mathbf{Z}_2)) \longrightarrow H_{2r+1}(\mathbf{K}(\mathbf{Z}_2))$$

by the rule

$$Q(x) = \varsigma_*(e_1 \otimes x \otimes x).$$

Compare this formula with formula (2.17). It was shown in [9] how this operation can be used to define the entire dual of the Steenrod algebra. Namely, analogous to how we defined the generators $x_i \in H_*(B_\infty)$, we define classes $t_i \in H_*(\mathbf{K}(\mathbf{Z}_2)) = A^*$ as follows. We define

$$t_1 \in A_1^* \cong \mathbf{Z}_2$$

to be the generator. Notice that t_1 is dual to Sq^1 and is therefore equal to the Milnor generator ξ_1. We then inductively define

$$t_i = Q(t_{i-1}) \in A_{2^i-1}^*.$$

It was observed in [9] that

$$t_i = \chi^*(\xi_i)$$

where $\chi^* : A^* \longrightarrow A^*$ is the dual of the antiautomorphism χ studied above. Since χ is a homomorphism of coalgbras, χ^* is an isomorphism of algebras. Therefore, since the ξ_i's are ring generators, so are the t_i's. We therefore have that

$$A^* \cong \mathbf{Z}_2[t_1, \cdots, t_i, \cdots]$$

is, like $H_*(B_\infty)$, the universal commutative algebra generated by the twisted product operation Q.

We now identify the submodule of A^* that takes the role of $H_*(B_k) \subset H_*(B_\infty)$. To do this we define, like above, a weight function on A^*. We define it by the rules

(1) $wt(1) = 0$
(2) $wt(t_i) = 2^i$, and
(3) $wt(xy) = wt(x) + wt(y)$.

This defines a weight function on monomials in $\mathbf{Z}_2[t_1, \cdots, t_i, \cdots] = A^*$. In analogy to theorem 2.6 we define submodules $M_k^* \subset A^*$ by the following rule.

DEFINITION 2.20.

$$M_k^* \subset \mathbf{Z}_2[t_1, \cdots, t_i, \cdots]$$

is the submodule spanned by those monomials of weight $\leq k$.

The following was proved in [9].

THEOREM 2.21. *There is a natural isomorphism*

$$M_k^* \cong (A/J_{[k/2]})^* \subset A^*.$$

PROOF: Recall from section 1 that

$$A/J_{[k/2]} = \{a \in A : dim(a) + e(\chi(a)) \leq k\}.$$

Consider the Milnor basis $A^* = \mathbf{Z}_2[\xi_1, \cdots, \xi_i, \cdots]$. For a monomial $\xi^R \subset A^*$, we define

$$e(\xi^R) = e(Sq^R)$$

where Sq^R is the class in the corresponding dual basis for A. It is now easy to verify that for every monomial t^R in the t_i's,

$$wt(t^R) = dim(\xi^R) + e(\xi^R).$$

The theorem now follows.

We now relate the braid groups and the modules $M_k = A/J_{[k/2]}$ even more closely as follows. Consider the representation γ_k of the braid group β_k defined to be the composition

$$\gamma_k : \beta_k \xrightarrow{\sigma_k} \Sigma_k \xrightarrow{j_k} O(k)$$

where σ_k sends a braid to the corresponding permutation of the endpoints of the strings, and where j_k is the representation of Σ_k as the permutation matrices. Notice that γ_k induces the following bundle on the classifying space B_k:

$$\gamma_k : F_k \times_{\Sigma_k} \mathbf{R}^{\mathbf{k}} \longrightarrow F_k/\Sigma_k = B_k$$

which has Thom space

$$T\gamma_k = (F_k)_+ \wedge_{\Sigma_k} S^k$$

where $(F_k)_+$ denotes F_k together with a disjoint basepoint. This bundle and its Thom space have received a great deal of study over the years [13, 17, 21, 22, 35]. In particular one has the following homology calculation which was implicit in [35], but carried out explicitly in [9].

Let

$$\gamma : \beta_\infty \longrightarrow O$$

be the limit of the γ_k's. Let $\mathbf{T}\gamma$ denote the corresponding Thom spectrum.

PROPOSITION 2.22. *Let $\Phi : H_*(B_\infty) \longrightarrow H_*(\mathbf{T}\gamma)$ be the Thom isomorphism. Let $u_* : H_*(\mathbf{T}\gamma) \longrightarrow A^*$ be the dual of the action of A on the Thom class u. We then have*

$$(u_* \circ \Phi)(x_i) = t_i.$$

As corollaries of this result and the above calculations, we get the following, which were originally done by Mahowald in [35].

COROLLARY 2.23. *$H^*(\mathbf{T}\gamma) \cong A$ as modules over A, and so there is an equivalence of spectra*

$$\mathbf{T}\gamma \simeq \mathbf{K}(\mathbf{Z}_2).$$

COROLLARY 2.24.

$$H^*(\mathbf{T}\gamma_\mathbf{k}) \cong A/J_{[k/2]}$$

as modules over A.

The homotopy types of the Thom spectra $\mathbf{T}\gamma_\mathbf{k}$ have proved very interesting. In particular the following was proved in [9].

THEOREM 2.25. *Let X be any space. Let $\tilde{k}$ denote the integer $\tilde{k} = 2[k/2] + 1$. Then the 2 - primary Adams spectral sequence [1] for computing $\pi_*(\mathbf{T}\gamma_\mathbf{k} \wedge X)$ has the following properties:*

(1) $E_2^{0,t} = H_t(X)$, *for* $t \leq \tilde{k}$.
(2) *The differentials* $d_r : E_r^{s,t} \longrightarrow E_r^{s+r,t+r-1}$ *are zero for* $t - s \leq \tilde{k}$ *and* $r \geq 2$.

It was also proved in [9] that the properties listed in (2.24) and (2.25) characterize the 2 - local homotopy type of $\mathbf{T}\gamma_k$. Notice that as a corollary of of 2.25 we get the following.

COROLLARY 2.26. *Let $u_* : \mathbf{T}\gamma_\mathbf{k} \longrightarrow \mathbf{K}(\mathbf{Z}_2)$ represent the generator of $H^*(\mathbf{T}\gamma_\mathbf{k})$ as a module over the Steenrod algebra. Then if X is any space,*

$$u_* : \pi_q(\mathbf{T}\gamma_\mathbf{k} \wedge X_+) \longrightarrow \pi_q(\mathbf{K}(\mathbf{Z}_2) \wedge X_+) \cong H_q(X)$$

is onto for $q \leq \tilde{k}$.

Spectra satisfying the properties described in (2.24) and (2.26) were originally constructed by Brown and Gitler in [10]. The fact that the spectra $\mathbf{T}\gamma_\mathbf{k}$ are equivalent to the original Brown - Gitler spectra was proved by Brown and Peterson in [13]. The Adams spectral sequence properties given in (2.25) were implicit in the work of Brown and Gitler, but were proved explicitly, (and without reference to the original Brown - Gitler spectra) for the braid Thom spectra in [9]. Since the original work of Brown and Gitler, there have been many different constructions and descriptions of their spectra [13, 20, 21, 29, 33]. They have had many applications in homotopy theory. One of their first applications, however, was in Brown and Peterson's proof of the Thom space analogue of the immersion conjecture (theorem 2.1) whose proof we give in the next section.

§3 Proof of theorem 2.1.

We now have the machinery necessary to prove theorem 2.1, the Thom spectrum analogue of the immersion conjecture. This was proved by Brown and Peterson in [12]. We will give essentially their argument here, but our notation will be somewhat different.

We begin by recalling the Thom splitting of the spectrum **MO**:

$$\mathbf{MO} \simeq \bigvee_{\omega} \Sigma^{|\omega|}\mathbf{K}(\mathbf{Z}_2).$$

Using the braid Thom spectra $\mathbf{T}\gamma_k$ studied in the last section, we define a new spectrum

DEFINITION 2.27. *Let* $\mathbf{MO}/\mathbf{I_n}$ *be the spectrum*

$$\mathbf{MO}/\mathbf{I_n} = \bigvee_{|\omega|\leq n} \Sigma^{|\omega|}\mathbf{T}\gamma_{n-|\omega|}$$

and let

$$\rho : \mathbf{MO}/\mathbf{I_n} \longrightarrow \mathbf{MO}$$

be defined using the above splittings and the generating maps

$$u : \mathbf{T}\gamma_k \longrightarrow \mathbf{K}(\mathbf{Z}_2)$$

of $H^*(\mathbf{T}\gamma_k)$ *as cyclic* A *- modules.*

The name for this spectrum was chosen for the following reason.

LEMMA 2.28.

(1)
$$H^*(\mathbf{MO}/\mathbf{I_n}) \cong H^*(\mathbf{MO})/\Phi(I_n)$$
as A *- modules, where* $\Phi(I_n)$ *is the submodule described in section one.*

(2) *The map*
$$\rho^* : H^*(\mathbf{MO}) \longrightarrow H^*(\mathbf{MO}/\mathbf{I_n})$$
is the natural projection.

PROOF: This follows immediately from theorem 2.3 and corollary 2.24.

The spectrum $\mathbf{MO}/\mathbf{I_n}$ is a universal spectrum for Thom spectra of normal bundles on n - manifolds in the following sense.

THEOREM 2.29. *Let* M^n *be an* n *- manifold with stable normal bundle* ν_M. *Let*

$$t(\nu_M) : \mathbf{T}\nu_{\mathbf{M}} \longrightarrow \mathbf{MO}$$

be the map of Thom spectra induced by ν_M. *Then there is a map of spectra*

$$\tau_M : \mathbf{T}\nu_{\mathbf{M}} \longrightarrow \mathbf{MO}/\mathbf{I_n}$$

making the following diagram of spectra homotopy commute:

$$\begin{array}{ccc} \mathbf{T}\nu_{\mathbf{M}} & \xrightarrow{\tau_M} & \mathbf{MO/I_n} \\ {\scriptstyle =}\downarrow & & \downarrow{\scriptstyle \rho} \\ \mathbf{T}\nu_{\mathbf{M}} & \xrightarrow[t(\nu_M)]{} & \mathbf{MO} \end{array}$$

The following is the other important property of $\mathbf{MO/I_n}$ necessary to prove theorem 2.1.

THEOREM 2.30. *There is a map of spectra*

$$\rho_n : \mathbf{MO/I_n} \longrightarrow \mathbf{MO}(n - \alpha(n))$$

making the following diagram homotopy commute.

$$\begin{array}{ccc} \mathbf{MO/I_n} & \xrightarrow{\rho_n} & \mathbf{MO}(n - \alpha(n)) \\ {\scriptstyle =}\downarrow & & \downarrow{\scriptstyle i} \\ \mathbf{MO/I_n} & \xrightarrow[\rho]{} & \mathbf{MO}. \end{array}$$

Notice that these two results together imply theorem 2.1 because we can define $\tilde{t}(\nu_M)$ to be the composition

$$\tilde{t}(\nu_M) : \mathbf{T}\nu_{\mathbf{M}} \xrightarrow{\tau_M} \mathbf{MO/I_n} \xrightarrow{\rho_n} \mathbf{MO}(n - \alpha(n)).$$

We are therefore reduced to proving theorems 2.29 and 2.30.

PROOF OF 2.29: Let ω be an appropriate indexing sequence for the cobordism ring, and hence for the splitting of $\mathbf{MO}$ given above. Consider the cohomology class $j_\omega \in H^{|\omega|}(\mathbf{T}\nu_{\mathbf{M}}))$ given by the composition

$$j_\omega : \mathbf{T}\nu_{\mathbf{M}} \xrightarrow{t(\nu_M)} \mathbf{MO} \xrightarrow{s_\omega} \Sigma^{|\omega|}\mathbf{K}(\mathbf{Z}_2)$$

where s_ω is the projection onto the split wedge summand corresponding to ω. By the definition of the splitting of spectrum $\mathbf{MO/I_n}$, to prove this result it is sufficient to prove that j_ω lifts to a map

$$\tilde{j}_\omega : \mathbf{T}\nu_{\mathbf{M}} \longrightarrow \Sigma^{|\omega|}\mathbf{T}\gamma_{n-|\omega|}.$$

Moreover, since $\mathbf{T}\nu_{\mathbf{M}}$ is an n - dimensional spectrum, we may assume, without loss of generality, that $|\omega| \leq n$.

Now by the Spanier - Whitehead duality between M^n and $\mathbf{T}\nu_{\mathbf{M}}$, $j_\omega \in H^{|\omega|}(\mathbf{T}\nu_{\mathbf{M}})$ corresponds to a class

$$g_\omega \in H_{n-|\omega|}(M^n).$$

By corollary 2.26, g_ω lifts to a class

$$\tilde{g}_\omega \in \pi_{n-|\omega|}(\mathbf{T}\gamma_{n-|\omega|} \wedge M^n_+).$$

Using Spanier - Whitehead duality again, $\tilde{g}_\omega$ corresponds to a map of spectra

$$\tilde{j}_\omega : \mathbf{T}\nu_{\mathbf{M}} \longrightarrow \Sigma^{|\omega|}\mathbf{T}\gamma_{n-|\omega|}.$$

By the naturality of Spanier - Whitehead duality, $\tilde{j}_\omega$ lifts j_ω. This completes the proof of theorem 2.29.

We now complete the proof of theorem 2.1 by proving theorem 2.30.

PROOF OF 2.30: We begin by recalling some information about the Thom splitting of **MO**. Let ω be an indexing sequence representing a monomial $b_\omega : S^{|\omega|} \to \mathbf{MO}$ in the cobordism ring η_*. Recall that the inclusion

$$i_\omega : \Sigma^{|\omega|}\mathbf{K}(\mathbf{Z}_2) \longrightarrow \mathbf{MO}$$

in the Thom splitting is given by a composition of the form

$$i_\omega : S^{|\omega|} \wedge \mathbf{K}(\mathbf{Z}_2) \xrightarrow{b_\omega \wedge i_1} \mathbf{MO} \wedge \mathbf{MO} \xrightarrow{\mu} \mathbf{MO} \tag{2.31}$$

where $i_1 : \mathbf{K}(\mathbf{Z}_2) \to \mathbf{MO}$ is the inclusion of the bottom dimensional summand, and where μ is the ring spectrum multiplication induced by the Whitney sum pairings

$$BO(r) \times BO(k) \longrightarrow BO(r+k).$$

By the construction of the spectrum $\mathbf{MO}/\mathbf{I_n}$ and the map $\rho : \mathbf{MO}/\mathbf{I_n} \to \mathbf{MO}$, to prove theorem 2.30 it is sufficient to prove that each of the compositions

$$\rho_\omega : S^{|\omega|} \wedge \mathbf{T}\gamma_{n-|\omega|} \xrightarrow{1 \wedge u} S^{|\omega|} \wedge \mathbf{K}(\mathbf{Z}_2) \xrightarrow{i_\omega} \mathbf{MO}$$

lifts to $\mathbf{MO}(n - \alpha(n))$.

Now by the cobordism immersion result, (1.32), the homotopy class

$$b_\omega : S^{|\omega|} \longrightarrow \mathbf{MO}$$

lifts to a class

$$\tilde{b}_\omega : S^{|\omega|} \longrightarrow \mathbf{MO}(|\omega| - \alpha(|\omega|)).$$

Also by the cohomology calculation (2.24) and corollary (2.13), the spectrum $\mathbf{T}\gamma_k$ has the 2 - local homotopy type of a complex of dimension $k - \alpha(k)$, and hence by obstruction theory, the composition

$$\mathbf{T}\gamma_k \xrightarrow{u} \mathbf{K}(\mathbf{Z}_2) \xrightarrow{i_1} \mathbf{MO}$$

has a lifting to a map

$$\tilde{u}_k : \mathbf{T}\gamma_k \longrightarrow \mathbf{MO}(k - \alpha(k)).$$

We now define $\tilde{\rho}_\omega : S^{|\omega|} \wedge \mathbf{T}\gamma_{n-|\omega|} \longrightarrow \mathbf{MO}(n - \alpha(n))$ to be the composition

$$\tilde{\rho}_\omega : S^{|\omega|} \wedge \mathbf{T}\gamma_{n-|\omega|} \xrightarrow{\tilde{b}_\omega \wedge \tilde{u}_{n-|\omega|}} \mathbf{MO}(|\omega| - \alpha(|\omega|)) \wedge \mathbf{MO}(n - |\omega| - \alpha(n - |\omega|))$$
$$\xrightarrow[\mu]{} \mathbf{MO}(n - \alpha(|\omega|) - \alpha(n - |\omega|)) \longrightarrow \mathbf{MO}(n - \alpha(n))$$

where the last map is the inclusion that exists because

$$\alpha(k) + \alpha(r) \geq \alpha(k + r).$$

By the construction of the maps involved, and by formula (2.31) for i_ω, it is straightforward to check that $\tilde{\rho}_\omega$ is an appropriate lifting of ρ_ω. As was argued above, this is what was needed to complete the proof of theorem 2.30 and thereby complete the proof of theorem 2.1.

Chapter III
De-Thom-ification Obstruction Theory and the Immersion Conjecture

In this section we outline the proof of the immersion conjecture. The program for its proof was developed by Brown and Peterson. The idea is to *de-Thom-ify* both the statement and the proof of theorem 2.1 (the Thom space analogue of the immersion conjecture).

As was seen in chapter *II* §3 , the proof of theorem 2.1 consisted of essentially two parts. The first part, (theorem 2.29) established the existence of a spectrum, $\mathbf{MO/I_n}$, which had certain universal properties with respect to liftings of the classifying maps of Thom spectra of normal bundles of n - manifolds. The second part (theorem 2.30) was a proof of the fact that the canonical map from this universal spectrum, $\rho : \mathbf{MO/I_n} \longrightarrow \mathbf{MO}$, lifts to $\mathbf{MO}(n-\alpha(n))$. The proof of the immersion conjecture amounted to showing that each of these results about Thom spectra actually is induced on the vector bundle level. Namely, the following two results were proved in [14] and [24] respectively.

THEOREM 3.1. *There is a space BO/I_n and a map*

$$\rho : BO/I_n \longrightarrow BO$$

that satisfies the following properties:

(1)
$$H^*(BO/I_n; \mathbf{Z}_2) \cong H^*(BO; \mathbf{Z}_2)/I_n$$

and ρ induces the natural projection in cohomology. Here I_n is the ideal of relations among the Stiefel - Whitney classes discussed in chapter II.

(2) *There is a natural equivalence of spectra*

$$\mathbf{T}\rho \simeq \mathbf{MO/I_n}$$

where $\mathbf{T}\rho$ is the Thom spectrum associated with the map $\rho : BO/I_n \to BO$.

(3) *The stable normal bundle map*

$$\nu_M : M^n \longrightarrow BO$$

of any n - manifold M^n can be factored up to homotopy as a composition

$$\nu_M : M^n \xrightarrow{\tilde{\nu}_M} BO/I_n \xrightarrow{\rho} BO.$$

THEOREM 3.2. *The map $\rho : BO/I_n \longrightarrow BO$ lifts (up to homotopy) to a map*

$$\rho_n : BO/I_n \longrightarrow BO(n - \alpha(n)).$$

Notice that just as theorems 2.29 and 2.30 together implied the Thom spectrum analogue of the immersion conjecture, theorems 3.1 and 3.2 together imply the immersion conjecture itself. Namely, given an n - manifold M^n, these results allow the construction of a composition

$$M^n \xrightarrow{\tilde{\nu}_M} BO/I_n \xrightarrow{\rho_n} BO(n-\alpha(n))$$

which lifts the stable normal bundle map $\nu_M : M^n \to BO$.

In view of theorems 2.29 and 2.30 it is natural that one of the main techniques in the proofs of theorems 3.1 and 3.2 is a type of obstruction theory that identifies the obstructions to a Thom spectrum level construction being induced by constructions on the vector bundle level. In particular one needs to understand when liftings of maps between Thom spectra are induced by liftings of bundle maps. In section one we will examine this *de-Thom-ification obstruction theory* and show how it, together with the Adams spectral sequence results concerning the braid Thom spectra $\mathbf{T}\gamma_k$ (theorem 2.25) was used by Brown and Peterson to prove theorem 3.1. In section two we will describe how this obstruction theory as well as a general study of the homotopy type of the spaces BO/I_n done in [24] was used to prove theorem 3.2.

As above, all (co)homology will be taken with $\mathbf{Z}_2$ coefficients.

§1 De-Thom-ifications and the construction of BO/I_n.

The goal of this section is to describe Brown and Peterson's construction of the space BO/I_n and their proof of theorem 3.1 [14]. In order to do that we begin by describing certain obstructions to de-Thom-ifying lifts of maps between Thom spectra. By use of Postnikov towers, the basic such obstruction will occur when one is trying to de-Thom-ify a lift of a map between Thom spectra

$$\phi : \mathbf{T}\zeta_1 \longrightarrow \mathbf{T}\zeta_2$$

to a spectrum $\mathbf{T}'$ obtained from $\mathbf{T}\zeta_2$ by killing certain cohomology classes. In order to understand the obstructions to de-Thom-ifying such lifts, one needs to understand the effect on the cohomology of a Thom spectrum when a cohomology class in the base space is killed. This problem was studied by Mahowald [34], Browder [7], and Brown and Peterson [14]. We will present the exposition of this theory given in [24].

Suppose $f : B \longrightarrow BO$ is a map which induces an isomorphism in homotopy groups through dimension k. Let V be a graded $\mathbf{Z}_2$-vector space with $V_q = 0$ for $q \leq k$, and let $\mathbf{K(V)}$ be the corresponding Eilenberg - MacLane spectrum of type $\mathbf{K}(\mathbf{Z}_2)$ with the property that

$$\pi_*(\mathbf{K(V)}) \cong V.$$

Represent $\mathbf{K(V)}$ as an Ω - spectrum $\{K(V)_q\}$. Let

$$\gamma : B \longrightarrow K(V)_1$$

represent a sum of cohomology classes and let B' be the homotopy fiber of γ. Thus we have a two - stage Postnikov system

$$\begin{array}{ccccc} B' & \xrightarrow{i} & B & \xrightarrow{f} & BO \\ & & \big\downarrow{\scriptstyle \gamma} & & \\ & & K(V)_1. & & \end{array}$$

Let B/B' denote the mapping cone of i. Notice that there is a canonical factorization of γ through a map

$$\tilde{\gamma} : B/B' \longrightarrow K(V)_1.$$

Let $\mathbf{T}$ and $\mathbf{T}'$ denote the Thom spectra of the stable bundles classified by f and $f \circ i$ respectively. The cohomology $H^*(\mathbf{T}/\mathbf{T}')$ can, in a range of dimensions, be described as follows.

Let $A(BO)$ be the semi - tensor product of the Steenrod algebra A with $H^*(BO)$. That is

$$A(BO) = A \otimes H^*(BO)$$

with the algebra structure defined by

$$(a \otimes u)(b \otimes v) = \sum_i ab'_i \otimes (\chi(b''_i)u)v$$

where if Δ is the Cartan diagonal map $\Delta : A \longrightarrow A \otimes A$, then $\Delta(b) = \sum_i b'_i \otimes b''_i$. As in [14], we denote $a \otimes u$ by $a \circ u$.

Notice that if $\varsigma \longrightarrow X$ is any vector bundle, $H^*(X)$ has an obvious $H^*(BO)$ - module structure induced by the classifying map of ς. This structure induces an $A(BO)$ - module structure on the cohomology of the Thom spectrum

$$A(BO) \otimes H^*(\mathbf{T}\varsigma) \longrightarrow H^*(\mathbf{T}\varsigma)$$

given by

$$(a \circ u)(\phi(x)) = a(\phi(u \cup x))$$

where $x \in H^*(X)$ and $\phi : H^*(X) \xrightarrow{\cong} H^*(\mathbf{T}\varsigma)$ is the Thom isomorphism.

Now consider the homomorphism

$$\psi : (A(BO) \otimes V)^q \longrightarrow H^{q+1}(\mathbf{T}/\mathbf{T}')$$

given by

$$\psi(a \circ u \otimes v) = a(u \cup \phi(\tilde{\gamma}^*(v_1)))$$

where $v_1 \in H^*(K(V)_1)$ corresponds to $v \in V$ and where here ϕ denotes the relative Thom isomorphism. In [14] Brown and Peterson proved the following.

THEOREM 3.3. *The map*

$$\psi : (A(BO) \otimes V)^q \longrightarrow H^{q+1}(\mathbf{T}/\mathbf{T}')$$

is an isomorphism for $q \leq 2k$.

We observe that this theorem can be used to describe obstructions to de-Thom -ifying spectra and maps between them. For example if

$$\theta : \mathbf{T}\varsigma_1 \longrightarrow \mathbf{T}\varsigma_2$$

is a map between Thom spectra, then an obvious necessary condition to de-Thom-ify θ, i.e to construct a map of vector bundles $f : \varsigma_1 \longrightarrow \varsigma_2$ inducing θ, is that

$$\theta^* : H^*(\mathbf{T}\varsigma_2) \longrightarrow H^*(\mathbf{T}\varsigma_1)$$

be a homomorphism of $A(BO)$ - modules. Thus $A(BO)$ - linearity is the basis of our de-Thom-ification obstruction theory. In fact theorem 3.3 will imply that in a certain setting $A(BO)$ module - structure describes the complete obstructions to de-Thom-ifying. More precisely we have the following. First we make a definition.

DEFINITION 3.4. *Let* $h : X \longrightarrow BO$ *have Thom spectrum* **TX**. *Then a map of spectra* $\alpha : \mathbf{Z} \longrightarrow \mathbf{TX}$ *is said to de-Thom-ify through dimension* m *if there is a space* Y *and a map*

$$g : Y \longrightarrow X$$

so that if **TY** *denotes the Thom spectrum of* $h \circ g$, *then there is an* m *- connected map of spectra*

$$\kappa : \mathbf{TY} \longrightarrow \mathbf{Z}$$

that makes the following diagram of spectra homotopy commute:

$$\begin{array}{ccc} \mathbf{TY} & \xrightarrow{\kappa} & \mathbf{Z} \\ {\scriptstyle Tg}\downarrow & & \downarrow{\scriptstyle \alpha} \\ \mathbf{TX} & \xrightarrow[=]{} & \mathbf{TX}. \end{array}$$

THEOREM 3.5. *Let* $f : B \longrightarrow BO$ *be a* k *- connected map as above, with Thom spectrum* **T**. *Let* $\alpha : \mathbf{Z} \longrightarrow \mathbf{T}$ *be a* k *- connected map from another spectrum* **Z**. *Then* α *de-Thom-ifies through dimension* $2k$ *if and only if the cohomology of the mapping cone*

$$H^*(\mathbf{T}/\mathbf{Z})$$

is, through dimension $2k$, *a free module over* $A(BO)$.

PROOF: The necessity of of this condition is given by theorem 3.3. We will now show that it is sufficient.

Let V be the graded $\mathbf{Z}_2$ vector space generated by a minimal generating set of $H^*(\mathbf{T}/\mathbf{Z})$ as a free $A(BO)$ - module, through dimension $2k$. Let $\mathbf{K(V)}$ be the corresponding Eilenberg - MacLane spectrum. Let

$$\gamma : \mathbf{T} \longrightarrow \mathbf{K}(\mathbf{V})$$

be the composition $\mathbf{T} \to \mathbf{T}/\mathbf{Z} \xrightarrow{j} \mathbf{K}(\mathbf{V})$ where j represents the cohomology classes representing this generating set. Let

$$\gamma : B \longrightarrow K(V)$$

also represent the Thom isomorphic image of $\gamma \in H^*(\mathbf{T})$. We define the space Y to be the homotopy fiber of γ, and

$$g : Y \longrightarrow B$$

to be the inclusion of this fiber. The fact that there is an $2k$ - connected map

$$\kappa : \mathbf{TY} \longrightarrow \mathbf{Z}$$

that lifts $Tg : \mathbf{TY} \longrightarrow \mathbf{T}$ follows from theorem 3.3 and simple obstruction theory.

We now show how Brown and Peterson used this obstruction theory to construct the spaces BO/I_n and to prove theorem 3.1.

Consider the spectrum $\mathbf{MO}/\mathbf{I_n}$ studied in the last section. As we saw, this spectrum is a wedge of suspensions of the braid Thom spectra

$$\rho : \mathbf{MO}/\mathbf{I_n} \longrightarrow \mathbf{MO}$$

which, when viewed as a sum of cohomology classes (since $\mathbf{MO}$ is a wedge of $\mathbf{K}(\mathbf{Z}_2)$'s) is the wedge of the Thom classes for each $\mathbf{T}\gamma_k$. Moreover this map induces a surjection in cohomology and therefore can be used as the first step in building an Adams resolution for $\mathbf{MO}/\mathbf{I_n}$ (i.e a tower, which when one applies homotopy groups yields the Adams spectral sequence). In [14] Brown and Peterson constructed a particular Adams resolution

$$\begin{array}{ccccccccccc} \mathbf{MO}/\mathbf{I_n} & \to & \cdots & \to & \mathbf{T_{i+1}} & \to & \mathbf{T_i} & \to & \cdots & \to & \mathbf{T_0} = \mathbf{MO} \\ & & & & & & \downarrow{\scriptstyle q_i} & & & & \downarrow{\scriptstyle q_0} \\ & & & & & & \mathbf{L_i} & & & & \mathbf{L_0} \end{array} \tag{3.6}$$

where each $\mathbf{T_{i+1}} \longrightarrow \mathbf{T_i} \xrightarrow{q_i} \mathbf{L_i}$ is a cofibration sequence of spectra with $\mathbf{L_i}$ a wedge of Eilenberg - MacLane spectra. This Adams resolution was built as a wedge of Adams resolutions for each of the relevant braid Thom spectra (Brown - Gitler spectra) $\mathbf{T}\gamma_k$. These resolutions were built using the *Lambda algebra* of Bousfield, Curtis, et al. [6] The main property of this tower is essentially a translation of the statement about the differentials in the Adams spectral sequence given in theorem 2.25. Namely, this resolution satisfies the following:

LEMMA 3.7. *Let X be any space. Then*

$$(q_i)_* : \pi_k(\mathbf{T_i} \wedge X) \longrightarrow \pi_k(\mathbf{L_i} \wedge X)$$

is the zero homomorphism for $k \leq n$.

Probably the most remarkable property of this resolution is the following calculation for which we refer the reader to [14].

LEMMA 3.8. *Each map $\mathbf{T_{i+1}} \longrightarrow \mathbf{T_i}$ is $[n/2]$ - connected and each $H^*(\mathbf{L_i})$ is, through dimension n, a free $A(BO)$ - module.*

Thus by theorem 3.5 and its proof, tower (3.6) for $\mathbf{MO/I_n}$ de-Thom-ifies! That is, there is a tower of fibrations

$$\begin{array}{ccccccccccc} \rightarrow & \cdots & \rightarrow & Y_{i+1} & \rightarrow & Y_i & \rightarrow & \cdots & \rightarrow & Y_0 & = & BO \\ & & & & & \downarrow{\scriptstyle k_i} & & & & \downarrow{\scriptstyle k_0} & & \\ & & & & & K_i & & & & K_0 & & \end{array} \tag{3.9}$$

where the Thom spectra $\mathbf{TY_i}$ have the same n - dimensional homotopy type as the spectra $\mathbf{T_i}$, and where each K_i is an Eilenberg - MacLane space whose homotopy groups form the graded $\mathbf{Z}_2$ - vector space generated by a minimal set of $A(BO)$ - module generators of $H^*(\mathbf{L_i})$.

BO/I_n was defined to be the n - dimensional skeleton of the inverse limit of the spaces Y_i in this tower. (See [14] to make this precise.)

$$\rho : BO/I_n \longrightarrow BO$$

is defined to be the composition

$$BO/I_n \longrightarrow \cdots \longrightarrow Y_i \longrightarrow \cdots \longrightarrow Y_0 = BO.$$

There is an obvious abuse of notation here. This map $\rho : BO/I_n \longrightarrow BO$ is the de-Thom-ification of the map $\rho : \mathbf{MO/I_n} \longrightarrow \mathbf{MO}$ studied above.

The fact that BO/I_n satisfies properties (1) and (2) of theorem 3.1 follows by the properties of de-Thom-ification described above. We now verify that it satisfies property (3).

Let M^n be an n manifold, and let

$$\nu_M : M^n \longrightarrow BO$$

classify its stable normal bundle. Suppose inductively that there is a homotopy lifting of ν_M to a map

$$\nu_i : M^n \longrightarrow Y_i$$

in this tower. The obstruction to lifting ν_i to a map

$$\nu_{i+1} : M^n \longrightarrow Y_{i+1}$$

is the composite map (cohomology class)

$$M^n \xrightarrow{\nu_i} Y_i \xrightarrow{k_i} K_i.$$

But by the definition of the map k_i and the above mentioned de-Thom-ification properties, to prove this cohomology class is zero, it is sufficient to prove that the composition

$$\mathbf{T}\nu_{\mathbf{M}} \xrightarrow{T\nu_i} \mathbf{T_i} \xrightarrow{q_i} \mathbf{L_i}$$

is zero. (Notice that we can use $\mathbf{T_i}$ instead of $\mathbf{TB_i}$ because $\mathbf{T}\nu_{\mathbf{M}}$ is an n - dimensional spectrum.) Now by Spanier - Whitehead duality, this follows because the homomorphism

$$\pi_n(\mathbf{T_i} \wedge M^n_+) \longrightarrow \pi_n(\mathbf{L_i} \wedge M^n_+)$$

is zero, by lemma 3.7.

This completes the inductive step, and hence we can lift the map ν_M all the way up the tower to produce a lifting

$$\tilde{\nu}_M : M^n \longrightarrow BO/I_n$$

satisfying theorem 3.1.

Notice that in the above proof we actually showed that *any* lifting of the stable normal bundle map of an n - manifold to the ith stage of the tower lifts to the $(i+1)^{st}$ stage. This is a very strong homotopy theoretic property of this tower which was important in the proof of theorem 3.2 [24]. In fact this tower has this lifting property for the following somewhat larger class of bundles.

DEFINITION 3.10. *Let X be any finite C.W. complex and* $h : X \longrightarrow BO$ *any map. The pair (X,h) is said to be quasi - normal of dimension n if there exists an n - dimensional manifold* M^n *and a map* $g : M^n \longrightarrow X$ *satisfying*

(1) *The composition*
$$h \circ g : M^n \longrightarrow BO$$
classifies the stable normal bundle of M^n, *and*

(2) $g^* : H^*(X) \longrightarrow H^*(M)$ *is injective.*

The following was proved in [24, Cor. 2.7] using an easy generalization of the above argument.

THEOREM 3.11. *If* (X,h) *is quasi - normal of dimension* n, *then* $h : X \longrightarrow BO$ *lifts to* BO/I_n. *In fact if* $h_{i-1} : X \longrightarrow Y_{i-1}$ *is any lifting of* h *to the* $(i-1)^{st}$ *stage of tower (3.9), then* h_{i-1} *lifts to a map* $h_i : X \longrightarrow Y_i$.

Besides actual normal bundles of n - manifolds, the following are examples of quasi - normal bundles of dimension n.

(1) $(BO/I_n, \rho)$. This is quasi-normal because of the lifting property in theorem 3.1 and the fact that by the definition of I_n, every class in $H^*(BO/I_n)$ is represented by a characteristic class which is nonzero on the normal bundle of some n - manifold. (See [24] for a complete proof.)

(2) $(BO/I_r \times BO/I_{n-r}, \rho \times \rho)$, for every $r \geq 0$. This yields pairings

$$BO/I_r \times BO/I_{n-r} \longrightarrow BO/I_n$$

that lift the Whitney sum pairing $BO \times BO \longrightarrow BO$.

(3) The braid space $(B_n = K(\beta_n, 1), \gamma_n)$ studied in chapter II. The fact that the braid spaces are quasi-normal was proved by Brown and Peterson in [14] (although they did not use this language). This yields a lifting

$$\tilde{\gamma}_n : B_n \longrightarrow BO/I_n$$

of γ_n. These liftings, as well as the pairings $BO/I_r \times BO/I_{n-r} \longrightarrow BO/I_n$ were very important in the proof of theorem 3.2 given in [24].

§2 The lifting of $\mathbf{BO/I_n}$.

The object of this chapter is to outline the arguments in [24] used to prove that

$$\rho : BO/I_n \longrightarrow BO$$

lifts to $BO(n - \alpha(n))$ (theorem 3.2), and thus complete the proof of the immersion conjecture.

Now theorem 2.30 says that the Thom spectrum analogue of this theorem is true (i.e that $\mathbf{MO/I_n} \longrightarrow \mathbf{MO}$ lifts to $\mathbf{MO(n - \alpha(n))}$). Thus the idea was to use the de-Thom-ification obstruction theory described in the last section to prove that this result de-Thom-ifies. The main difficulty in doing so was that the space BO/I_n was constructed as the inverse limit of a tower (3.9) and although this construction implies certain homotopy theoretic properties (e.g theorem 3.11), we did not have a concrete, point - set description of BO/I_n to use to examine its lifting properties. Nonetheless, the following construction of a space X_n related to BO/I_n suggested that these lifting properties could be studied from an obstruction theoretic point of view.

Recall that the spectrum $\mathbf{MO/I_n}$ is the wedge of the braid Thom spectra,

$$\mathbf{MO/I_n} = \bigvee_{\omega} S^{|\omega|} \wedge \mathbf{T}\gamma_{n-|\omega|}$$

and the map $\mathbf{MO/I_n} \longrightarrow \mathbf{MO}$ was given by the wedge of the compositions

$$S^{|\omega|} \wedge \mathbf{T}\gamma_{n-|\omega|} \xrightarrow{b_\omega \wedge \gamma_*} \mathbf{MO} \wedge \mathbf{MO} \xrightarrow{\mu} \mathbf{MO}$$

where γ_* is the map of Thom spectra induced by the classifying map of the bundle $\gamma_{n-|\omega|} \longrightarrow B_{n-|\omega|}$.

Now by the cobordism immersion result, (theorem 1.30) the cobordism class

$$b_\omega : S^{|\omega|} \longrightarrow \mathbf{MO}$$

is represented by a manifold M_ω that immerses in $\mathbf{R}^{2|\omega|-\alpha(|\omega|)}$. Let

$$\nu_\omega : M_\omega \longrightarrow BO(|\omega| - \alpha(|\omega|))$$

classify the normal bundle of such an immersion. Notice that the class b_ω is given by the composition

$$b_\omega : S^{|\omega|} \xrightarrow{\tau} \mathbf{T}\nu_\omega \longrightarrow \mathbf{MO}(|\omega| - \alpha(|\omega|)) \hookrightarrow \mathbf{MO}$$

where τ is the Thom - Pontrjagin collapse map described in chapter *I*.

Now consider the braid bundle

$$\gamma_k \longrightarrow B_k$$

studied in chapter *II*. As observed there, B_k has the 2 - local homotopy type of a $(k - \alpha(k))$ - dimensional C.W. complex, and hence the classifying map of γ_k factors through a map, which by abuse of notation we call

$$\gamma_k : B_k \longrightarrow BO(k - \alpha(k)).$$

Now define the space X_n as follows.

DEFINITION 3.12. *Let X_n be the space*

$$X_n = \coprod_{|\omega| \le n} M_\omega \times B_{n-|\omega|}$$

The space X_n has the following properties, which begins to suggest the possibility of lifting $\rho : BO/I_n \longrightarrow BO$.

PROPOSITION 3.13. *There are maps $f_n : X_n \longrightarrow BO(n - \alpha(n))$ and $g_n : X_n \longrightarrow BO/I_n$ that satisfy the following properties:*

(1) *The following diagram homotopy commutes*

$$\begin{array}{ccc} X_n & \xrightarrow{f_n} & BO(n - \alpha(n)) \\ {\scriptstyle g_n}\downarrow & & \downarrow{\scriptstyle i} \\ BO/I_n & \xrightarrow[\rho]{} & BO. \end{array}$$

(2) *If $\mathbf{TX_n}$ is the Thom spectrum of the composition $i \circ f_n \simeq \rho \circ g_n$, then there is a splitting map of spectra*

$$\sigma_n : \mathbf{MO/I_n} \longrightarrow \mathbf{TX_n}.$$

That is, $1 \simeq Tg_n \circ \sigma_n : \mathbf{MO/I_n} \longrightarrow \mathbf{TX_n} \longrightarrow \mathbf{MO/I_n}$.

PROOF:

$$f_n : X_n \longrightarrow BO(n - \alpha(n))$$

is defined to be the disjoint union of the compositions

$$M_\omega \times B_{n-|\omega|} \xrightarrow{\nu_\omega \times \gamma_{n-|\omega|}} BO(|\omega| - \alpha(|\omega|)) \times BO(n - |\omega| - \alpha(n - |\omega|))$$
$$\underset{\mu}{\longrightarrow} BO(n - \alpha(|\omega|) - \alpha(n - |\omega|)) \hookrightarrow BO(n - \alpha(n)).$$

To define $g_n : X_n \longrightarrow BO/I_n$, observe that by theorem 3.11 and the examples of quasi - normal bundles given after it, there are liftings

$$\tilde{\nu}_\omega : M_\omega \longrightarrow BO/I_{|\omega|}$$

of the stable normal bundle map for M_ω,

$$\tilde{\gamma}_k : B_k \longrightarrow BO/I_k$$

of the stable bundle represented by γ_k, and

$$\tilde{\mu} : BO/I_r \times BO/I_{n-r} \longrightarrow BO/I_n$$

lifting the Whitney sum pairing $\mu : BO \times BO \longrightarrow BO$. We therefore define

$$g_n : X_n \longrightarrow BO/I_n$$

to be the disjoint union of the compositions

$$M_\omega \times B_{n-|\omega|} \xrightarrow{\tilde{\nu}_\omega \times \tilde{\gamma}_k} BO/I_{|\omega|} \times BO/I_{n-|\omega|} \xrightarrow{\tilde{\mu}} BO/I_n.$$

We now leave it for the reader to check that the diagram in the statement of the proposition homotopy commutes.

To prove the second part of the proposition, notice that, by construction, there is a natural splitting of Thom spectra

$$\mathbf{TX_n} \simeq \bigvee_\omega \mathbf{T}\nu_\omega \wedge \mathbf{T}\gamma_{n-|\omega|}.$$

We define the splitting map $\sigma_n : \mathbf{MO/I_n} \longrightarrow \mathbf{TX_n}$ to be the wedge of the maps

$$\tau \wedge 1 : S^{|\omega|} \wedge \mathbf{T}\gamma_{n-|\omega|} \longrightarrow \mathbf{T}\nu_\omega \wedge \mathbf{T}\gamma_{n-|\omega|}.$$

We again leave it to the reader to check that σ_n is in fact a splitting of Tg_n.

Notice that there were many choices in the definition of the space X_n and the maps $f_n : X_n \longrightarrow BO(n - \alpha(n))$ and $g_n : X_n \longrightarrow BO/I_n$. In particular one could have chosen different manifolds M_ω and different immersions. Also there was choice involved in the definitions of the liftings $\tilde{\nu}_\omega$, $\tilde{\gamma}_k$, and $\tilde{\mu}$. A major step in the proof of the immersion conjecture was showing that it would be implied by the following refinement of proposition 3.13. Philosophically it will say that if one can make these choices concerning X_n satisfy certain homotopy theoretic properties, then the obstructions to lifting $\rho : BO/I_n \longrightarrow BO$ to $BO(n-\alpha(n))$ will necessarily vanish. These properties were described in [24] in terms of the homotopy pull - back of $i : BO(n-\alpha(n)) \longrightarrow BO$ along the map $\rho : BO/I_n \longrightarrow BO$.

More precisely, define the pull - back space P_n as the space

$$P_n = \{(x, y, \alpha) \in BO/I_n \times BO(n-\alpha(n)) \times BO^I : \alpha(0) = i(y) \quad \text{and} \quad \alpha(1) = \rho(x)\}.$$

Here $X^I = Map(I, X)$ is the path space of X.

Define maps $P_n \longrightarrow BO/I_n$ and $P_n \longrightarrow BO(n-\alpha(n))$ by projecting onto the first and second coordinates respectively. Observe that

$$\begin{array}{ccc} P_n & \longrightarrow & BO(n-\alpha(n)) \\ \downarrow & & \downarrow i \\ BO/I_n & \xrightarrow[\rho]{} & BO \end{array}$$

is a homotopy pull - back (or homotopy cartesian) diagram. The following was the main lemma in [24] and should be viewed as a refinement of proposition 3.13.

LEMMA 3.14. *There exists a space X_n together with a map $h_n : X_n \longrightarrow P_n$ satisfying the following properties:*

(1) *If f_n and g_n are the compositions*

$$f_n : X_n \xrightarrow{h_n} P_n \longrightarrow BO(n-\alpha(n))$$

and

$$g_n : X_n \xrightarrow{h_n} P_n \longrightarrow BO/I_n$$

then there is a splitting map of Thom spectra

$$\sigma_n : \mathbf{MO/I_n} \longrightarrow \mathbf{TX_n}.$$

That is, $1 \simeq Tg_n \circ \sigma_n : \mathbf{MO/I_n} \longrightarrow \mathbf{TX_n} \longrightarrow \mathbf{MO/I_n}$.

(2) *The follwoing diagram of Thom spectra homotopy commutes:*

$$\begin{array}{ccc} \mathbf{TX_n} & \xrightarrow{Th_n} & \mathbf{TP_n} \\ {\scriptstyle Tg_n}\downarrow & & \uparrow{\scriptstyle Th_n} \\ \mathbf{MO/I_n} & \xrightarrow[\sigma_n]{} & \mathbf{TX_n}. \end{array}$$

Notice that by the pull - back property, part (1) of this lemma is implied by proposition 3.13. Part (2) of this lemma is the refinement of that proposition necessary to prove the following somewhat stronger version of theorem 3.2.

THEOREM 3.15. *Let $h_n : X_n \longrightarrow P_n$ be as in lemma 3.14. Then there is a homotopy lifting $\rho_n : BO/I_n \longrightarrow BO(n-\alpha(n))$ of $\rho : BO/I_n \longrightarrow BO$ that makes the following diagram of Thom spectra homotopy commute:*

$$\begin{array}{ccc} \mathbf{TX_n} & \xrightarrow{Tf_n} & \mathbf{MO(n-\alpha(n))} \\ {\scriptstyle Tg_n}\downarrow & & \uparrow{\scriptstyle T\rho_n} \\ \mathbf{MO/I_n} & \xrightarrow[=]{} & \mathbf{MO/I_n}. \end{array}$$

The proof of the immersion conjecture in [24] thereby split into two major parts. The first was to prove that lemma 3.14 implies theorem 3.15. The second was to prove lemma 3.14. We now proceed to outline the arguments in each of these steps.

We first discuss the ideas in the proof that lemma 3.14 implies theorem 3.15. (In the notation of [24] lemma 3.14 was lemma B and theorem 3.15 was theorem A.) As one might imagine, de-Thom-ification obstruction theory was used heavily in this argument. However there was one other ingredient that heretofore we have not discussed. That is the following stable analogue of theorem 3.2.

PROPOSITION 3.16. *There is a map of suspension spectra*

$$r_n : \Sigma^\infty BO/I_n \longrightarrow \Sigma^\infty BO(n-\alpha(n))$$

which homotopy lifts the stable map

$$\Sigma^\infty \rho : \Sigma^\infty BO/I_n \longrightarrow \Sigma^\infty BO.$$

PROOF: The obstruction to the existence of the map r_n is the composition

$$BO/I_n \xrightarrow{\rho} BO \longrightarrow BO/BO(n-\alpha(n)).$$

Now it is rather easy to show that $BO/BO(n-\alpha(n))$ has the same n - dimensional homotopy type as a product of mod 2 - Eilenberg - MacLane spaces. This was shown in [24] using a splitting theorem of Snaith [42] (See theorem 3.17 below). Since BO/I_n has the homotopy type of an n - dimensional complex, the above obstruction is entirely cohomological. But the fact that all cohomological obstructions vanish follows from Massey's calculation (theorem 2.2) and the definition of the ideal $I_n \subset H^*(BO)$.

The following refinement of the splitting theorem of Snaith [42] mentioned above was also an important ingredient in this argument.

THEOREM 3.17. *There is a map of suspension spectra*

$$\kappa : \Sigma^\infty BO \longrightarrow \Sigma^\infty BO(n-\alpha(n))$$

so that the composition

$$\Sigma^\infty BO(n-\alpha(n)) \xrightarrow{i} \Sigma^\infty BO \xrightarrow{\kappa} \Sigma^\infty BO(n-\alpha(n))$$

is (stably) homotopic to the identity.

An immediate observation following from this splitting theorem is the following. If Y is a space let

$$QY = \lim_{m\to\infty} \Omega^m \Sigma^m Y.$$

Thus QY is the zero space in the Ω - spectrum corresponding to $\Sigma^\infty Y$. The adjoint of the map κ in 3.17 is a map of spaces

$$\kappa : BO \longrightarrow QBO(n-\alpha(n)).$$

One then easily sees that the adjoint of any stable lifting r_n satisfying 3.16 is given by the composition

$$r_n : BO/I_n \xrightarrow{\rho} BO \xrightarrow{\kappa} QBO(n-\alpha(n)).$$

The way these results were used the the proof of theorem 3.15 (assuming lemma 3.14) was as follows.

The map $h_n : X_n \longrightarrow P_n$ to the pull - back is equipped with a canonical homotopy between the maps

$$i \circ f_n : X_n \longrightarrow BO(n-\alpha(n)) \longrightarrow BO$$

and

$$\rho \circ g_n : X_n \longrightarrow BO/I_n \longrightarrow BO.$$

This homotopy is given by the composition

$$E \circ (h_n \times 1) : X_n \times I \longrightarrow P_n \times I \longrightarrow BO$$

where $E : P_n \times I \longrightarrow BO$ is the canonical homotopy given by the formula

$$E((x,y,\alpha),t) = \alpha(t).$$

Let $BO/I_n \cup_{g_n} X \times I$ be the mapping cylinder of $g_n : X_n \longrightarrow BO/I_n$ The homotopy $E \circ (h_n \times 1)$ defines a map

$$\tilde{E} : BO/I_n \cup_{g_n} X \times I \longrightarrow BO$$

which, when restricted to BO/I_n is ρ, and when restricted to $X_n = X_n \times \{1\} \subset X_n \times I$ is the composition

$$X_n \xrightarrow{f_n} BO(n-\alpha(n)) \xrightarrow{i} BO.$$

Now let

$$H : BO/I_n \cup_{g_n} X \times I \longrightarrow QBO(n-\alpha(n))$$

be the composition $H = \kappa \circ \tilde{E}$. Thus the restriction of H to BO/I_n is the map $r_n : BO/I_n \longrightarrow QBO(n-\alpha(n))$, and the restriction of H to X_n is the composition

$$X_n \xrightarrow{f_n} BO(n-\alpha(n)) \hookrightarrow QBO(n-\alpha(n)).$$

The following lemma (lemma 1.9 of [24]) relates these stable constructions to Thom spectrum constructions.

LEMMA 3.18. *The map of pairs*

$$H : (BO/I_n \cup_{g_n} X \times I, \quad X_n) \longrightarrow (QBO(n-\alpha(n)), \quad BO(n-\alpha(n)))$$

has trivial Thom-ification. That is, the induced map of quotients of Thom spectra

$$\mathbf{T(BO/I_n \cup_{g_n} X \times I)/TX_n \longrightarrow T(QBO(n-\alpha(n)))/MO(n-\alpha(n))}$$

is null homotopic.

This result was used to set up an inductive argument using a (modified) Postnikov tower for the map $i : BO(n-\alpha(n)) \longrightarrow BO$:

$$BO(n-\alpha(n)) \to \cdots \to Y_i \to Y_{i-1} \to \cdots \to Y_0 = BO.$$

The inductive assumptions were essentially the following (see [24, 1.11] for a precise statement):

There exists a homotopy lifting $\rho_i : BO/I_n \longrightarrow Y_i$ of ρ satisfying the following properties.

(1) The diagram

$$\begin{array}{ccc} BO/I_n \sqcup X_n & \xrightarrow{\rho_i \sqcup f_{n,i}} & Y_i \\ \downarrow & & \downarrow \\ BO/I_n \cup_{g_n} X \times I & \xrightarrow[H]{} & QBO(n-\alpha(n)) \end{array}$$

commutes, where $f_{n,i}$ is the composition

$$X_n \xrightarrow{f_n} BO(n-\alpha(n)) \longrightarrow Y_i$$

and where the vertical map in this diagram is the composition

$$Y_i \longrightarrow BO \xrightarrow{\kappa} QBO(n-\alpha(n)).$$

(2) The induced map of pairs

$$(BO/I_n \cup_{g_n} X \times I, \quad BO/I_n \sqcup X_n) \longrightarrow (QBO(n-\alpha(n)), \quad Y_i)$$

has trivial Thom-ification.

Roughly speaking, in order to complete the inductive step, it was shown that the obstruction to doing so is a map from the Thom spectrum $\mathbf{T(BO/I_n \cup_{g_n} X \times I)}$ to a product of Eilenberg - MacLane spectra whose cohomology is, in a range of dimensions, a free $A(BO)$ - module. This freeness allowed us to change the map ρ_i if necessary, so that the obstruction became zero. See [24, §1] for details.

We end this section with an outline of the proof of lemma 3.14 (the last step in the proof of the immersion conjecture).

As one might guess, this lemma was proved by induction on n. So assume that for $k \leq n-1$ there exists spaces X_k and maps $h_k : X_k \longrightarrow P_k$ and $\sigma_k : \mathbf{MO/I_k} \longrightarrow \mathbf{TX_k}$ satisfying the requirements of lemma 3.14. We furthermore assume that each of these spaces X_k is of the form described above. Namely,

$$X_k = \coprod_{|\omega| \leq k} M_\omega \times B_{k-|\omega|}.$$

The space X_n will be of the same form. We first define the subspace $X_n^{(n-1)}$ to be the disjoint union

$$X_n^{(n-1)} = \coprod_{|\omega| \leq n-1} M_\omega \times B_{n-|\omega|}$$

where M_ω is the manifold that was used in the definition of $X_{|\omega|}$, which by induction makes sense since we are assuming that $|\omega| \leq n-1$.

Notice that to construct X_n out of $X_n^{(n-1)}$ we need to add on n - dimensional manifolds. That is, X_n will be of the form

$$X_n = X_n^{(n-1)} \sqcup \coprod_{|\omega|=n} M_\omega$$

where the union is taken over all monomials ω in the cobordism ring of dimension n.

Let ω be such a monomial, and let $l(\omega)$ be the length of ω. So ω represents a decomposable monomial if and only if $l(\omega) \geq 2$. Notice furthermore that by the calculation of the cobordism ring, there is at most one indecomposable monomial in dimension n and this occurs if and only if n is not of the form $2^r - 1$.

Let ω represent the decomposable monomial $b_1^{i_1} \cdots b_r^{i_r}$ of dimension n. Since it is decomposable, each generator b_j in this monomial has dimension $j < n$, so by induction we have already chosen a manifold M_j representing b_j in the definition of X_j. We therefore define M_ω to be the product manifold

$$M_\omega = M_1^{i_1} \times \cdots \times M_r^{i_r}.$$

where the superscripts refer to cartesian products. We may therefore define the space

$$\tilde{X}_n = X_{n-1} \sqcup \coprod_{l(\omega) \geq 2} M_\omega$$

where the union is taken over all decomposable monomials in the cobordism ring of dimension n. If n is of the form $2^r - 1$, then we will have

$$X_n = \tilde{X}_n.$$

If n is not of the form $2^r - 1$, X_n will be formed out of $\tilde{X}_n$ by adding on a manifold (yet to be defined) M_n representing an indecomposable element of dimension n in the cobordism ring. Before we describe how M_n was defined, we first show how the restriction of h_n to $\tilde{X}_n$ was defined.

By the pull - back property, the existence of the pairings

$$\tilde{\mu} : BO/I_r \times BO/I_{n-r} \longrightarrow BO/I_n$$

defines pairings of the pull - backs

$$\nu : P_r \times P_{n-r} \longrightarrow P_n.$$

The map

$$\tilde{h}_n : \tilde{X}_n \longrightarrow P_n$$

is defined in terms of these pairings, as follows.

Let

$$M_\omega \times B_{n-|\omega|} \subset X_n^{(n-1)} \subset \tilde{X}_n.$$

The restriction of $\tilde{h}_n$ to $M_\omega \times B_{n-|\omega|}$ is given by the composition

$$M_\omega \times B_{n-|\omega|} \subset X_{|\omega|} \times X_{n-|\omega|} \xrightarrow{h_{|\omega|} \times h_{n-|\omega|}} P_{|\omega|} \times P_{n-|\omega|} \xrightarrow{\nu} P_n.$$

Now let $\omega = (i_1, \cdots, i_r)$ be a decomposable monomial of dimension n. We define the restriction of $\tilde{h}_n$ to M_ω to be the composition

$$M_\omega = M_1^{i_1} \times \cdots \times M_r^{i_r} \xrightarrow{h_{i_1} \times \cdots \times h_{ri_r}} P_{i_1} \times \cdots \times P_{ri_r} \xrightarrow{\nu} P_n.$$

This then defines

$$\tilde{h}_n : \tilde{X}_n \longrightarrow P_n.$$

In [24] it was observed that we had to choose the pairings $\tilde{\mu}$ and ν to satisfy some mild conditions which are not difficult. The reader is referred to [24] for details. In any case in dimensions n of the form $2^r - 1$ (so that there are no indecomposables in the cobordism ring of dimension n), we let $X_n = \tilde{X}_n$, and $h_n = \tilde{h}_n$. The splitting map σ_n is defined in terms of the Thom - Pontryagin collapse maps as described above. In this case it was then easy to verify that the triple (X_n, h_n, σ_n) satisfied the hypotheses of lemma 3.14.

We now explain what was done to complete the inductive step in the proof of lemma 3.14 in the case when n is not of the form $2^r - 1$. In this case we need to construct a suitable n - dimensional manifold representing an indecomposable class in the cobordism ring. This was actually done in a rather round about way, as follows.

First, it was shown, using a standard, easy obstruction theoretic argument that there exists a subcomplex

$$\tilde{B}O/I_n \subset BO/I_n$$

that satisfies the following properties.

(1) BO/I_n is formed out of $\tilde{B}O/I_n$ by attaching one n - dimensional disk

$$BO/I_n = \tilde{B}O/I_n \cup_\alpha D^n.$$

(2) The attaching map α has trivial Thom-ification. That is, on the Thom spectrum level we have

$$\mathbf{MO/I_n} = \tilde{\mathbf{MO}}/\mathbf{I_n} \vee S^n$$

where $\tilde{\mathbf{MO}}/\mathbf{I_n}$ is the Thom spectrum of the restriction of $\rho : BO/I_n \longrightarrow BO$ to $\tilde{BO}/I_n$. Furthermore, the sphere S^n in this splitting represents, via the map $\rho : \mathbf{MO/I_n} \longrightarrow \mathbf{MO}$, an indecomposable element of the cobordism ring.

It was then shown, using elementary obstruction theory, that

$$\tilde{h}_n : \tilde{X}_n \longrightarrow P_n$$

naturally factors through $\tilde{P}_n$, defined to be the restriction of the fibration $P_n \longrightarrow BO/I_n$ to $\tilde{BO}/I_n \subset BO/I_n$. One can then prove that the analogue of lemma 3.14 holds, with $\tilde{X}_n$, $\tilde{h}_n$, and $\tilde{P}_n$ replacing X_n, h_n, and P_n, respectively. Next one observes that the same argument used to show that lemma 3.14 implies theorem 3.15, implies the analogue of 3.15 with $\tilde{BO}/I_n$ replacing BO/I_n. Thus we have a lifting

$$\tilde{\rho}_n : \tilde{BO}/I_n \longrightarrow BO(n - \alpha(n))$$

satisfying certain Thom spectrum level properties. In particular it is easily seen that on the Thom spectrum level

$$\tilde{\rho}_n : \tilde{\mathbf{MO}}/\mathbf{I_n} \longrightarrow \mathbf{MO(n - \alpha(n))}$$

extends over $\mathbf{MO/I_n}$, and that stably,

$$\Sigma^\infty \tilde{BO}/I_n \longrightarrow \Sigma^\infty BO(n - \alpha(n))$$

extends over $\Sigma^\infty BO/I_n$. Moreover, it was shown that these extensions could be made compatible, in an appropriate sense. Thus the setting is ripe for the use of de-Thom-ification obstruction theory. This was done using the Postnikov tower

$$BO(n - \alpha(n)) \longrightarrow \cdots \longrightarrow Y_i \longrightarrow Y_{i-1} \longrightarrow \cdots \longrightarrow Y_0 = BO$$

studied above. This was probably the most technically complicated argument of the paper. The outcome was a lifting

$$\rho_n : BO/I_n \longrightarrow BO(n - \alpha(n))$$

of $\rho : BO/I_n \longrightarrow BO$ that extends $\tilde{\rho}_n : \tilde{BO}/I_n \longrightarrow BO(n - \alpha(n))$.

To complete the inductive step in the proof of lemma 3.14 we need to construct an appropriate indecomposable manifold of dimension n. This was done as follows.

Consider the stable map

$$j : S^n \longrightarrow \mathbf{MO/I_n}$$

given by the splitting $\mathbf{MO/I_n} = \tilde{\mathbf{MO}}/\mathbf{I_n} \vee S^n$ mentioned above. By standard Thom - Pontryagin cobordism arguments, j represents an n - dimensional manifold M_n, and a map

$$\phi : M_n \longrightarrow BO/I_n.$$

Furthermore, by property (2) of the space $\tilde{B}O/I_n$ given above, M_n represents an indecomposable element in the cobordism ring.

Now by the pull-back property, the lifting $\rho_n : BO/I_n \longrightarrow BO(n-\alpha(n))$ defines a section $s : BO/I_n \longrightarrow P_n$ of the fibration $P_n \longrightarrow BO/I_n$. We may then define

$$X_n = \tilde{X}_n \sqcup M_n$$

and

$$h_n : X_n \longrightarrow P_n$$

is defined to be $\tilde{h}_n$ when restricted to $\tilde{X}_n$, and when restricted to M_n it is defined to be the composition

$$M_n \xrightarrow{\phi} BO/I_n \xrightarrow{s} P_n.$$

As above, the Thom spectrum splitting

$$\sigma_n : \mathbf{MO/I_n} \longrightarrow \mathbf{TX_n}$$

is defined via the Thom - Pontryagin collapse maps. It is then easily verified that the triple (X_n, h_n, σ_n) satisfies the properties stated in lemma 3.14.

This then completed the inductive step in the proof of lemma 3.14, which was the final step in the proof of the immersion conjecture.

REFERENCES

1. J.F. Adams, "Stable Homotopy and Generalized Homology," Mathematical Lecture Notes, University of Chicago, 1971.
2. J.W. Alexander, *Topological invariants of knots and links*, Trans. A.M.S. **30** (1928), 275–306.
3. E.Artin, *Theorie der Zöpfe*, Hamburg Abh. **4** (1925), 47–72.
4. M.F. Atiyah, *Thom complexes*, Proc. Lond. Math. Soc. (3) **11** (1961), 291–310.
5. J. Birman, "Braids, Links, and Mapping Class Groups," Annals of Math. Studies 82, Princeton Univ. Press, 1974.
6. A. Bousfield, E. Curtis, D. Kan, D. Quillen, D. Rector, and J. Schlesinger, *The mod p lower central series and the Adams spectral sequence*, Topology **5** (1966), 331–342.
7. W. Browder, *The Kervaire invariant of framed manifolds and its generalizations*, Annals of Math. **90** (1969), 157–186.
8. E.H. Brown, *Cohomology theories*, Annals of Math. **75** (1962), 467–484.
9. E.H. Brown and R.L. Cohen, *The Adams spectral sequence of $\Omega^2 S^3$ and Brown-Gitler spectra*, Annals of Math. Studies **113** (1987), 101–125.
10. E.H. Brown and S. Gitler, *A spectrum whose cohomology is a certain cyclic module over the Steenrod algebra*, Topology **12** (1973), 283–295.
11. E.H. Brown and F.P. Peterson, *Relations among characteristic classes I*, Topology **3** (1964), 39–52.
12. ——, *On immersions of n-manifolds*, Advances in Math. **24** (1977), 74–77.
13. ——, *On the stable decomposition of $\Omega^2 S^{r+2}$*, Trans. A.M.S. **243** (1978), 287–298.
14. ——, *A universal space for normal bundles of n-manifolds*, Comment. Math. Helv. **54** (1979), 405–430.
15. R.L. Brown, *Immersions and embeddings up to cobordism*, Canad. J. Math. (6) **23** (1971), 1102–1115.
16. S. Bullett, *Braid orientations and Stiefel-Whitney classes*, Quart. J. Math. Oxford **2** (1981), 267–285.
17. F.R. Cohen, *Braid orientations and bundles with flat connections*, Invent. Math. **46** (1978), 99–110.

18. ——, *Artin's braid groups and classical homotopy theory*, Contemp. Math. **44** (1985), 207–219.
19. F.R. Cohen, T. Lada, and J.P. May, "The Homology of Iterated Loop Spaces," Lecture Notes 533, Springer Verlag, New York, 1976.
20. R.L. Cohen, *The geometry of* $\Omega^2 S^3$ *and braid orientations*, Invent. Math. **54** (1979), 53-67.
21. ——, *Representations of Brown-Gitler spectra*, Proc. Top. Symp. at Siegen, 1979, Lecture Notes 788, Springer Verlag, New York (1980), 399–417.
22. ——, *Odd primary infinite families in stable homotopy theory*, Memoirs of A.M.S. **242** (1981).
23. ——, *The homotopy theory of immersions*, Proc. Int. Cong. of Math., Warszawa 1982 **1** (1984), 627–640.
24. ——, *The immersion conjecture for differentiable manifolds*, Annals of Math. **122** (1985), 237–328.
25. R.L. Cohen, J.D.S. Jones, and M.Mahowald, *The Kervaire invariant of immersions*, Inven. Math. **79** (1985), 95–123.
26. A. Dold, *Erzeugende der Thomschen Algebra* η_*, Math. Zeit. **65** (1956), 25–35.
27. D.B. Fuks, *Cohomologies of the braid groups mod 2*, Functional Anal. and its Applic.. **4** (1970), 143–151.
28. V.F.R. Jones, *A polynomial invariant for knots via von Neumann algebras*, Bull. A.M.S. **12** (1985), 103–111.
29. P.G. Goerss, *A direct construction for the duals of Brown- Gitler spectra*, Indiana J. of Math. **34** (1985), 733–751.
30. M.W. Hirsch, *Immersions of manifolds*, Trans. A.M.S. **93** (1959), 242–276.
31. M.W. Hirsch, "Differential Topology," Springer Verlag, New York, 1976.
32. D. Husemoller, "Fibre Bundles," Springer Verlag, New York, 1966.
33. J. Lannes and S. Zarati, *Sur les foncteurs derives de la destbilisation*, C.R. Acad. Sci. Paris **296** (1983), 573–576.
34. M. Mahowald, *On obstruction theory in orientable fibre bundles*, Trans. A.M.S. **110** (1964), 315–349.
35. ——, *A new infinite family in* ${}_2\pi_*^s$, Topology **16** (1977), 249–256.
36. W.S. Massey, *On the Stiefel-Whitney classes of a manifold*, Amer. J. Math. **82** (1960), 92–102.
37. R.J. Milgram, *Iterated loop spaces*, Annals Math. **84** (1966), 386–403.
38. J.W. Milnor, *The Steenrod algebra and its dual*, Annals of Math. (2) **67** (1958), 150–171.
39. J.W. Milnor and J.D. Stasheff, "Characteristic Classes," Princeton University Press, New Jersey, 1974.
40. R. Mosher and M. Tangora, "Cohomology Operations and Applications in Homotopy Theory," Harper and Row, New York, 1968.
41. J.P. May, "The geometry of iterated loop spaces," Lecture Notes 271, Springer Verlag, New York, 1972.
42. V. Snaith, *Algebraic cobordism and K-theory*, Memoirs of A.M.S. **221** (1979).
43. N. Steenrod, "The Topology of Fibre Bundles," Princeton University Press, New Jersey, 1951.
44. N.E. Steenrod and D.B.A. Epstein, "Cohomology Operations," Princeton University Press, New Jersey, 1962.
45. R.E. Stong, "Notes on Cobordism Theory," Princeton University Press, New Jersey, 1968.
46. R. Thom, *Quelques propertés globales des varietës differentiables*, Comment. Math. Helv. **28** (1954), 17–86.
47. G.W. Whitehead, *Generalized homology theories*, Trans. A.M.S. **102** (1962), 227–283.
48. H. Whitney, *The singularities of a smooth n-manifold in (2n-1)-space*, Annals of Math. **45** (1944), 247–293.
49. W.T. Wu, *Classes caractéristiques et i-carrés d'une varieté*, C.R. Acad. Sci. Paris **230** (1950), 508–511.

Surface maps and braid equations, I

BOJU JIANG

Peking University

and

Nankai Institute of Mathematics

§0. Introduction.

Dimension two is the special dimension in fixed point theory. For selfmaps of compact manifolds of other dimensions, the Nielsen fixed point theory gives us the best lower bound for the number of fixed points in the homotopy class. Counter-examples on surfaces have been discovered recently ([J2],[J3]), arousing new interest in this classical subject.

In the present paper, an algebraic approach is proposed to fixed point theory of surface maps. The fixed point problem is shown to be equivalent to a certain equation in the pure 2-braid group of the surface. For planar surface this equation was studied by Zhang [Z]. We apply commutator calculus to analyse the equation. The first stage of commutator analysis is abelianization. The invariants of Nielsen theory, e.g. the Reidemeister trace, arise as just the abelian obstructions. In a sequal, we will show that by a deeper commutator analysis, new algebraic information will be obtained. The known examples, verified previously with *ad hoc* techniques, can now be done by routine (though sometimes tedious) computation.

§1. An algebraic formulation of the fixed point problem.

Notations and conventions.

Let M be a connected compact surface, with $\pi_2(M) = 0$. Let Δ be the diagonal in $M \times M$. A pure 2-braid in M is a homotopy class of loops in $M \times M - \Delta$. Let x_1, x_2 be two distinct preassigned points in $\text{int}M = M - \partial M$. The fundamental group $\pi_1(M \times M - \Delta, (x_1, x_2))$ is the pure 2-braid group of M. Let $i : M \times M - \Delta \to M \times M$ be the inclusion. Let $p_k : (y_1, y_2) \mapsto y_k$, $k = 1, 2$, be the projections of $M \times M$ onto the factors, and let $i_1 : y \mapsto (y, x_2)$ and $i_2 : y \mapsto (x_1, y)$ be the inclusions of the factors.

Let $U \subset M$ be an oriented chart homeomorphic to $\mathbf{R}^2$ and containing x_1, x_2. In $U - x_1$ there is a loop at x_2 going around x_1 once in the possitive sense of U. Its i_2-image in $M \times M - \Delta$ represents a braid $B \in \pi_1(M \times M - \Delta, (x_1, x_2))$.

For brevity we make the following convention about base points: Unless otherwise specified, for subspaces of $M \times M$ we take (x_1, x_2) as base point. For subspaces of M we take x_1, or take x_2 if x_1 is not in the subspace.

Suppose ∂M has n components, $n \geq 0$. With arbitrarily chosen base points and orientations they are denoted $S_1, \ldots, S_n$. Let D be a disk in U with x_1 in ∂D but x_2 not in D. Orienting ∂D in the possitive sense of U we get a loop T.

Let $f : M \to M$ be a map with $f(x_1) = x_2$ such that the fixed point set $\text{Fix}(f)$ is finite and contained in $\text{int}M$. Let $\bar{f} : M \to M \times M$ denote the graph of f, i.e. $\bar{f}(x) = (x, f(x))$. For $i = 1, \ldots, n$, let a_i be a path in $M - \text{Fix}(f)$ from x_1 to the base point on S_i. Let w_i be the loop $a_i S_i a_i^{-1}$ in M. Then the loop $\bar{f} \circ w_i$ is in $M \times M - \Delta$ and represents a braid $\sigma_i \in \pi_1(M \times M - \Delta, (x_1, x_2))$.

Partially supported by a TWAS grant.

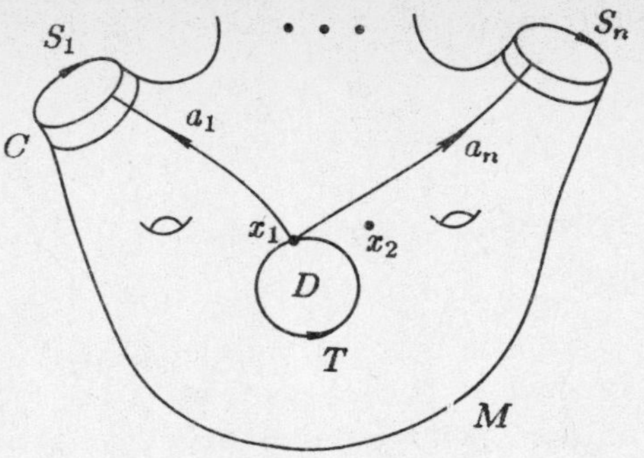

Figure 1

THEOREM 1.1. *Let $f : M \to M$ be as above. Let $k \geq 0$, $i_1, \ldots, i_k \in \mathbf{Z}$. Then the following two conditions are equivalent:*

(I) There exists a map $g \simeq f : M \to M$, with k fixed points of indices $i_1, \ldots i_k$ respectively.

(II) There exists a homomorphism

$$\phi : \pi_1(M - \mathrm{int}D, x_1) \to \pi_1(M \times M - \Delta, (x_1, x_2)),$$

and elements

$$u_i \in \mathrm{Ker}\,(i_\pi : \pi_1(M \times M - \Delta) \to \pi_1(M \times M)), \quad i = 1, \ldots, n,$$

$$v_j \in \mathrm{Ker}\,(p_{1\pi} : \pi_1(M \times M - \Delta) \to \pi_1(M)), \quad j = 1, \ldots, k,$$

such that the diagram

$$\begin{array}{ccc} \pi_1(M - \mathrm{int}D) & \xrightarrow{\phi} & \pi_1(M \times M - \Delta) \\ {\scriptstyle i_\pi}\downarrow & & \downarrow{\scriptstyle i_\pi} \\ \pi_1(M) & \xrightarrow{\bar{f}_\pi} & \pi_1(M \times M) \end{array}$$

commutes, and that

$$\phi([w_i]) = u_i \sigma_i u_i^{-1}, \qquad \text{for } i = 1, \ldots, n,$$

and

$$\phi([T]) = v_1 B^{i_1} v_1^{-1} \ldots v_k B^{i_k} v_k^{-1}.$$

Remark. In (II) the restriction $v_j \in \mathrm{Ker}\, p_{1\pi}$ can be replaced by $v_j \in \pi_1(M \times M - \Delta)$ when M is orientable. This is also true for nonorientable M if we allow the fixed point indices to differ from i_j by a minus sign. See the Corollary in the next section.

PROOF: *(I) implies (II).* According to the Lemma in [J2], we may assume that $g \simeq f$ rel $\partial M \cup x_1$. Without loss we assume all fixed points of g are in $\mathrm{int}D$.

(i) Let $\phi = \bar{g}_\pi : \pi_1(M - \mathrm{int}D) \to \pi_1(M \times M - \Delta)$. Then obviously the diagram in (II) commutes.

(ii) $\phi[w_i]$ is represented by the loop

$$\bar{g} \circ w_i \simeq (\bar{g} \circ a_i)(\bar{f} \circ a_i)^{-1}(\bar{f} \circ w_i)(\bar{f} \circ a_i)(\bar{g} \circ a_i)^{-1}.$$

Hence $\phi[w_i] = u_i \sigma_i u_i^{-1}$, where u_i is the braid represented by $(\bar{g} \circ a_i)(\bar{f} \circ a_i)^{-1}$. It is readily seen that $u_i \in \mathrm{Ker}\, i_\pi$.

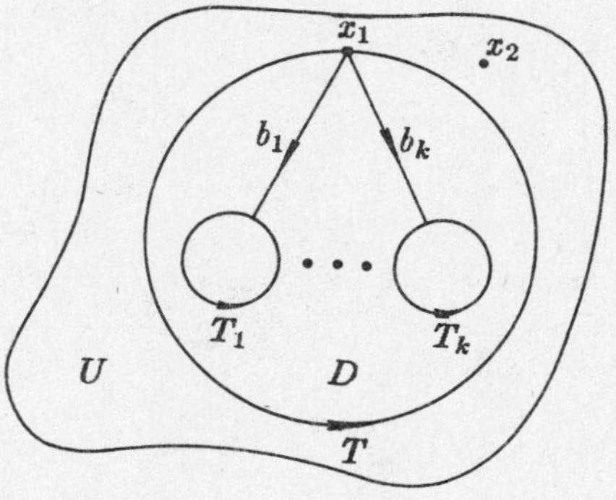

Figure 2

(iii) Let the fixed points of g be $y_1, \ldots, y_k$. For $j = 1, \ldots, k$, let D_j be mutually disjoint small disks centred at y_j such that $g(D_j) \subset U$. Draw arcs b_j from x_1 to ∂D_j, and let T_j be the loop of oriented ∂D_j, as in Fig.2. Then $\phi[T]$ is represented by $\prod \bar{g} \circ (b_j T_j b_j^{-1})$. For each j take a path c_j in $U \times U - \Delta$ joining the end points of $\bar{g} \circ b_j$ so that $p_1 \circ c_j = b_j$. Let $v_j = [(\bar{g} \circ b_j)c_j^{-1}] \in \pi_1(M \times M - \Delta)$. Then $[\bar{g} \circ (b_j T_j b_j^{-1})] = v_j[c_j(\bar{g} \circ T_j)c_j^{-1}]v_j^{-1}$ and $p_{1\pi}(v_j) = 1$. Now $[c_j(\bar{g} \circ T_j)c_j^{-1}] \in \pi_1(U \times U - \Delta) = \pi_1(\mathbf{R}^2 \times \mathbf{R}^2 - \Delta)$, the infinite cyclic group generated by B. Since the index of the fixed point y_j is i_j, we have $[c_j(\bar{g} \circ T_j)c_j^{-1}] = B^{i_j}$. Hence the formula for $\phi[T]$.

(II) implies (I). We shall construct the map g in several steps. Take a collar C on ∂M so small that it avoids any fixed point of f. Let M_0 be the closure of $M - C$. It is a deformation retract of M. By deforming f we may assume f maps M_0 into M_0. We may also assume the a_i's cross themselves transversely, intersect C nicely, touch D only at x_1, and avoid x_2. Let A be the union of the paths $a_i \cap M_0$, $i = 1, \ldots, k$.

(i) Define $g : C \cup A \cup D \to M$.

Let $g = f$ on $C \cup x_1$. Since $u_i \in \mathrm{Ker}\, i_\pi$, we can extend g over each a_i without fixed points (with value x_2 except in a small neighborhood of C, to avoid problems at the crossings of A) so that the loop $(\bar{g} \circ a_i)(\bar{f} \circ a_i)^{-1}$ represents u_i and $g \simeq f : C \cup A \to M$ rel $C \cup x_1$.

For $j = 1, \ldots, k$, take D_j and b_j as in Fig.2. Since $v_j \in \mathrm{Ker}\, p_{1\pi}$, we can define g on each b_j so that $g \circ b_j$ is in $M_0 - D$ and $(\bar{g} \circ b_j)(i_1 \circ b_j)^{-1}$ represents v_j.

On D_j, the map g has now been defined only at one point on ∂D_j, with value x_2. Extend it to $g : D_j \to U$ such that it has only one fixed point y_j with index i_j, and it sends ∂D_j outside of D. Since g maps $\bigcup(\partial D_j \cup b_j)$ out of D, we can further extend it over D without introducing new fixed points. We can even extend the previous homotopy to a homotopy $g \simeq f : C \cup A \cup D \to M$ rel $C \cup x_1$, because D is contractible to x_1.

Note that this construction guarantees $\bar{g}_\pi[w_i] = \phi[w_i]$ and $\bar{g}_\pi[T] = \phi[T]$.

(ii) Extend the homotopy $\bar{g} \simeq \bar{f} : C \cup A \cup D \to M \times M$ to a homotopy $h \simeq \bar{f} : M \to M \times M$.

From the above construction, we have a commutative diagram

$$\begin{array}{ccc}
\pi_1(C \cup A \cup \partial D) & \xrightarrow{\bar{g}_\pi} & \pi_1(M \times M - \Delta) \\
\downarrow & & \| \\
\pi_1(M - \mathrm{int} D) & \xrightarrow{\phi} & \pi_1(M \times M - \Delta)
\end{array}$$

where the vertical arrow is induced by inclusion. By obstruction theory (cf. [H], Prop.11.1), the map $\bar{g} \mid C \cup A \cup \partial D$ can be extended to a map $h : M - \mathrm{int} D \to M \times M - \Delta$ with $h_\pi = \phi$. Combining this map with the graph $\bar{g} \mid D : D \to M \times M$ we get a map $h : M \to M \times M$.

Comparing the commutative diagram

$$\begin{array}{ccc}
\pi_1(M - \mathrm{int} D) & \xrightarrow{\phi} & \pi_1(M \times M - \Delta) \\
{\scriptstyle i_\pi}\downarrow & & \downarrow{\scriptstyle i_\pi} \\
\pi_1(M) & \xrightarrow{h_\pi} & \pi_1(M \times M)
\end{array}$$

with the one in hypothesis (II), we see $h_\pi = \bar{f}_\pi : \pi_1(M) \to \pi_1(M \times M)$ because the vertical arrows are surjective. Again by obstruction theory (cf. [H] Prop.17.1, and $\pi_2(M \times M) = \pi_2(M) \oplus \pi_2(M) = 0$ by hypothesis), our previous homotopy $\bar{g} \simeq \bar{f} : C \cup A \cup D \to M \times M$ rel $C \cup x_1$ can be extended to a homotopy $h \simeq \bar{f} : M \to M \times M$ rel $C \cup x_1$. Without loss we may assume this homotopy sends M_0 into $M_0 \times M$.

(iii) Deform h to the graph of the desired $g : M \to M$.

Since M_0 is in the interior of M, the projection $p_1 : M_0 \times M - \Delta \to M_0$ is a fiber bundle and hence has the homotopy lifting property. The above homotopy projects to a homotopy $p_1 \circ h \simeq$ inclusion $: M_0 - \mathrm{int} D \to M_0$ rel $\partial M_0 \cup A \cup \partial D$. The latter lifts to a homotopy $h \simeq h' : M_0 - \mathrm{int} D \to M_0 \times M - \Delta$ rel $\partial M_0 \cup A \cup \partial D$. This h' is the graph of a map $g' : M_0 - \mathrm{int} D \to M$ with no fixed point, and $\bar{g}' = h = \bar{g}$ on $\partial M_0 \cup A \cup \partial D$. So g' extends our previous g to a map $g : M \to M$ with only k fixed points $y_1, \ldots, y_k$. The homotopy $h \simeq h'$ extends to a homotopy $h \simeq \bar{g} : M \to M \times M$ rel $C \cup A \cup D$. This shows $\bar{g} \simeq \bar{f}$, hence $g \simeq f : M \to M$. ∎

§2. Presentation of the braid group.

We use the group theoretic notation $(u, v) = u^{-1} v^{-1} u v$ and $u^{kv} = v^{-1} u^k v$ for group elements u, v and integer k. For brevity we denote

$$G = \pi_1(M \times M - \Delta, (x_1, x_2)),$$
$$H = \mathrm{Ker}\,(p_{1\pi} : \pi_1(M \times M - \Delta, (x_1, x_2)) \to \pi_1(M, x_1)),$$
$$K = \mathrm{Ker}\,(i_\pi : \pi_1(M \times M - \Delta, (x_1, x_2)) \to \pi_1(M \times M, (x_1, x_2))).$$

The group G is the pure 2-braid group of M. Its structure is well known, see e.g. [S]. We will give it a more convenient presentation.

Remove a disk from the surface M to get the surface shown in Fig.3 with $n+1>0$ boundary components. In the standard form of bounded surfaces, it is a disk with g pairs of linked flat bands, h twisted bands, and n separated flat bands. The surface M is obtained from this surface by identifying the outer boundary to a point. We can always assume at least one of g, h is 0.

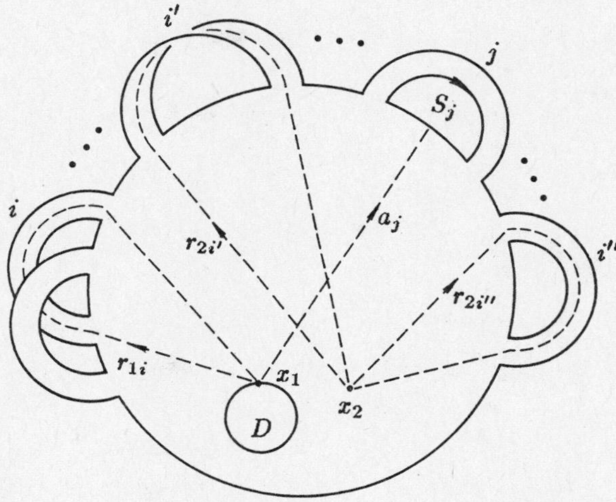

Figure 3

Let us define the generators. The braid B is already defined. For $k=1,2$ and $j=1,\ldots,2g+h+n$, let r_{kj} be the loop in M based at x_k and going through the j-th band in the fashion shown in Fig.3. Its i_k-image represents the braid $\rho_{kj}\in G$.

THEOREM 2.1. *The group G has generators*

$$B, \rho_{1i}, \rho_{2i}, \qquad \text{for } i=1,\ldots,2g+h+n,$$

and the following relations:

$$\text{(2-1)} \qquad \rho_{1i}^{-1}B\rho_{1i}=\begin{cases} B^{-1}\rho_{2i}B\rho_{2i}^{-1}B, & \text{if band } i \text{ flat,}\\ B^{-1}\rho_{2i}B^{-1}\rho_{2i}^{-1}B, & \text{otherwise;}\end{cases}$$

$$\text{(2-2)} \qquad \rho_{1i}^{-1}\rho_{2i}\rho_{1i}=\begin{cases} B^{-1}\rho_{2i}B, & \text{if band } i \text{ flat,}\\ B^{-1}\rho_{2i}, & \text{otherwise;}\end{cases}$$

when $i<j$ then

$$\text{(2-3)} \qquad \rho_{1i}^{-1}\rho_{2j}\rho_{1i}=\begin{cases} B^{-1}\rho_{2j}, & \text{if bands } i,j \text{ linked,}\\ \rho_{2j}, & \text{otherwise;}\end{cases}$$

when $i > j$ then

(2-4) $$\rho_{1i}^{-1}\rho_{2j}\rho_{1i} = \begin{cases} (B,\rho_{2i}^{-1})\rho_{2j}B, & \text{if bands } i,j \text{ linked,} \\ B^{-1}\rho_{2i}B^{-1}\rho_{2i}^{-1}\rho_{2j}\rho_{2i}B\rho_{2i}^{-1}B, & \text{if band } i \text{ twisted,} \\ (B,\rho_{2i}^{-1})\rho_{2j}(\rho_{2i}^{-1},B), & \text{otherwise;} \end{cases}$$

when M orientable,

$$\begin{aligned} B &= (\rho_{11}^{-1},\rho_{12})\dots(\rho_{1,2g-1}^{-1},\rho_{1,2g})\rho_{1,2g+1}\cdots\rho_{1,2g+n} \\ &= (\rho_{21}^{-1},\rho_{22})\dots(\rho_{2,2g-1}^{-1},\rho_{2,2g})\rho_{2,2g+1}\cdots\rho_{2,2g+n}, \end{aligned}$$

or, when M nonorientable,

$$\begin{aligned} B &= \rho_{11}^2\cdots\rho_{1h}^2\rho_{1,h+1}\cdots\rho_{1,h+n} \\ &= \rho_{21}^2\cdots\rho_{2h}^2\rho_{2,h+1}\cdots\rho_{2,h+n}. \end{aligned}$$

Introducing the notation

$$\begin{aligned} R(z_1,\dots,z_{2g+h+n}) =& (z_1^{-1},z_2)\dots(z_{2g-1}^{-1},z_{2g})z_{2g+1}^2\cdots z_{2g+h}^2 \\ & \cdot z_{2g+h+1}\cdots z_{2g+h+n}, \end{aligned}$$

we have

(2-5) $$B = R(\rho_{11},\dots,\rho_{1,2g+h+n}) = R(\rho_{21},\dots,\rho_{2,2g+h+n}).$$

PROOF: The validity of these relations can be verified directly by pictures.

Consider the Fadell-Neuwirth exact sequence [FN]

$$1 \to \pi_1(M - x_1, x_2) \xrightarrow{i_{2\pi}} \pi_1(M \times M - \Delta, (x_1,x_2)) \xrightarrow{p_{1\pi}} \pi_1(M, x_1) \to 1.$$

The subgroup $H = i_{2\pi}\pi_1(M - x_1, x_2)$ has generators B and ρ_{2j} for $j = 1,\dots,2g+h+n$, with one relation

$$B = R(\rho_{21},\dots,\rho_{2,2g+h+n}).$$

The group $\pi_1(M,x_1)$ has as generators the $p_{1\pi}$-image of ρ_{1i}, $i = 1,\dots,2g+h+n$, with the relation

$$p_{1\pi}R(\rho_{11},\dots,\rho_{1,2g+h+n}) = 1.$$

Now apply the presentation theorem for group extensions ([Jo] p.187). ∎

For reference we list some other useful relations. The symbol $\rightleftharpoons$ means 'commutes with'.

(2-1') $$\rho_{1i}B\rho_{1i}^{-1} = \begin{cases} \rho_{2i}^{-1}B\rho_{2i}, & \text{if band } i \text{ flat,} \\ \rho_{2i}^{-1}B^{-1}\rho_{2i}, & \text{otherwise;} \end{cases}$$

(2-2') $$\rho_{1i}\rho_{2i}\rho_{1i}^{-1} = \begin{cases} \rho_{2i}^{-1}B\rho_{2i}B^{-1}\rho_{2i}, & \text{if band } i \text{ flat,} \\ \rho_{2i}^{-1}B^{-1}\rho_{2i}^2, & \text{otherwise;} \end{cases}$$

when $i < j$ then

$$\text{(2-3')} \qquad \rho_{1i}\rho_{2j}\rho_{1i}^{-1} = \begin{cases} \rho_{2i}^{-1} B \rho_{2i}\rho_{2j}, & \text{if bands } i, j \text{ linked,} \\ \rho_{2j}, & \text{otherwise;} \end{cases}$$

when $i > j$ then

$$\text{(2-4')} \qquad \rho_{1i}\rho_{2j}\rho_{1i}^{-1} = \begin{cases} (\rho_{2i}, B^{-1})\rho_{2j}\rho_{2i}^{-1}B^{-1}\rho_{2i}, & \text{if bands } i, j \text{ linked,} \\ \rho_{2i}^{-1}B^{-1}\rho_{2i}B^{-1}\rho_{2j}B\rho_{2i}^{-1}B\rho_{2i}, & \text{if band } i \text{ twisted,} \\ (\rho_{2i}, B^{-1})\rho_{2j}(B^{-1}, \rho_{2i}), & \text{otherwise;} \end{cases}$$

$$\text{(2-6)} \qquad \rho_{1i} \rightleftharpoons \begin{cases} B^{-1}\rho_{2i}, & \text{if band } i \text{ flat,} \\ B^{-1}\rho_{2i}^2, & \text{otherwise;} \end{cases}$$

when $i < j$ then

$$\text{(2-7)} \qquad \rho_{1i} \rightleftharpoons \begin{cases} \rho_{2j}^{-1}\rho_{2i}\rho_{2j}, & \text{if bands } i, j \text{ linked,} \\ \rho_{2j}, & \text{otherwise;} \end{cases}$$

when $i > j$ then

$$\text{(2-8)} \qquad \rho_{1i} \rightleftharpoons \begin{cases} B^{-1}\rho_{2j}\rho_{2i}\rho_{2j}^{-1}B, & \text{if bands } i, j \text{ linked,} \\ B^{-1}\rho_{2j}B, & \text{otherwise.} \end{cases}$$

COROLLARY 2.2. *For each $w \in G$, there is some $v \in H$ such that $w^{-1}Bw = v^{-1}Bv$ or $w^{-1}Bw = v^{-1}B^{-1}v$, according as $p_{1\pi}(w) \in \pi_1(M, x_1)$ preserves or reverses orientation.*

PROOF: Write w as a word in the generators and argue by induction on its length. Start with the relation (2-1). ∎

PROPOSITION 2.3. *The group H has generators*

$$B, \rho_{2i}, \qquad \text{for } i = 1, \ldots, 2g + h + n,$$

and a single relation

$$B = R(\rho_{21}, \ldots, \rho_{2,2g+h+n}). \qquad ∎$$

The group H is in fact a free group with basis $\{\rho_{2i} \mid i = 1, \ldots, 2g + h + n\}$. When M has boundary, i.e. $n > 0$, a more convenient basis is $\{B, \rho_{2i} \mid i = 1, \ldots, 2g + h + n - 1\}$.

The group $K = \operatorname{Ker} i_\pi = H \cap \operatorname{Ker} p_{2\pi}$ is the normal subgroup generated by the element B. Let π be the factor group H/K which can naturally be identified with $\pi_1(M, x_2)$ via $p_{2\pi}$. We will use the following presentation for π: Let ρ_{2i} stand for $[r_{2i}]$ in $\pi = \pi_1(M, x_2)$. Then π has generators $\{\rho_{2i} \mid i = 1, \ldots, 2g + h + n\}$ and a single relation $R(\rho_{21}, \ldots, \rho_{2,2g+h+n}) = 1$. (When $n > 0$, π is free with basis $\{\rho_{2i} \mid i = 1, \ldots, 2g + h + n - 1\}$.) To get a basis for K, we need a minimal Schreier transversal (cf. [Jo] p.11).

PROPOSITION 2.4. *The group K has a free basis $\{sBs^{-1} \mid s \in S\}$ where S is a minimal Schreier transversal with respect to the projection $H \to \pi$.*

PROOF: A standard argument involving the Reidemeister-Schreier rewriting process for $K \subset H$ and the Dehn algorithm for π. ∎

For calculations in K, we extend the exponential notation $u^{kv} = v^{-1}u^k v$ by writing $B^u B^v = B^{u+v}$. This suggests the following notion and the 'logarithmic' notation.

Definition. Let Γ be a (multiplicative) group. The *free additive group on* Γ, written $\mathbf{F}[\Gamma]$, is the additive free group with basis Γ. Thus the elements of $\mathbf{F}[\Gamma]$ are integral linear combinations of elements of Γ, but the addition is *not* commutative. The negative of an element is obtained by not only changing the sign of the coefficients, but also reversing the order of the terms. The conjugate ξ^* of an element $\xi \in \mathbf{F}[\Gamma]$ is obtained by inverting the elements of Γ which appear in the terms, while retaining the order of the terms. Left and right multiplication by elements of Γ are defined naturally and are distributive. The abelianization of $\mathbf{F}[\Gamma]$ is the integral group-ring $\mathbf{Z}[\Gamma]$.

The 'exponential' homomorphisms $\beta : \mathbf{F}[G] \to K$ and $\beta : \mathbf{F}[H] \to K$ are defined by $g \mapsto B^g$. Suppose S is a minimal Schreier transversal with respect to the projection $H \to \pi$. Define $\beta_S : \mathbf{F}[\pi] \to K$ by $\alpha \mapsto B^{\bar{\alpha}}$, where $\bar{\alpha}$ is the representative of α in S. It is an isomorphism by the previous Proposition. Its inverse is the 'logarithm' $\lambda_S : K \to \mathbf{F}[\pi]$. We omit the subscript S if it is clear from the context or if it is immaterial. We define $\lambda^* : K \to \mathbf{F}[\pi]$ by $\lambda^*(u) = \lambda(u)^* =$ the conjugate of $\lambda(u)$, which is preferable to λ because of its connection with the Fox calculus (see Section 4).

Notation. Let $\widehat{G}$ denote the free group on the basis $\{B, \rho_{1i}, \rho_{2i} \mid i = 1, \ldots, 2g+h+n\}$; $\widehat{H}$ denote the free group on the basis $\{B, \rho_{2i} \mid i = 1, \ldots, 2g+h+n\}$; $\widehat{\pi}$ denote the free group on the basis $\{\rho_{2i} \mid i = 1, \ldots, 2g+h+n\}$. We will not distinguish between the notation for elements in the free groups $\widehat{G}$, $\widehat{H}$, $\widehat{\pi}$ and the corresponding elements in the groups G, H, π. The group in question should be clear from the context.

§3. Fixed point equation in the braid group.

We shall use the presentation of §2 to change the Condition (II) of §1 into an equation. The maps $f : M, x_1 \to M, x_2$ are classified by the homomorphisms

$$f_\pi : \pi_1(M, x_1) \to \pi_1(M, x_2).$$

Suppose

$$f_\pi : [r_{1i}] \mapsto f_i, \qquad i = 1, \ldots, 2g+h+n,$$

where each $f_i = f_i(\rho_{21}, \ldots, \rho_{2,2g+h+n}) \in \widehat{\pi}$ is a word in the letters $\{\rho_{2i} \mid i = 1, \ldots, 2g+h+n\}$. Since $R([r_{1i}]) = 1$ in $\pi_1(M, x_1)$, we must have $R(f_i) = 1$ in π. Hence $R(f_i)$, when regarded as an elemant of $\widehat{\pi}$, is a product of conjugates of $R(\rho_{2i})$. In other words, there exists $A^* \in \mathbf{F}[\widehat{\pi}]$ such that

$$R(f_1, \ldots, f_{2g+h+n}) = R(\rho_{21}, \ldots, \rho_{2,2g+h+n})^{A^*}$$

in $\widehat{\pi}$.

Let the paths a_i, for $i = 2g+h+1, \ldots, 2g+h+n$ (this is our new range of i corresponding to the boundary components), be as in Fig.3. Similar to Lemma 2 of [J3] we have

PROPOSITION 3.1. *For* $i = 2g+h+1, \ldots, 2g+h+n$,

$$\sigma_i = \rho_{1i} f_i' = f_i' \rho_{1i},$$

where

$$f_i' = f_i(B^{-1}\rho_{21}B, \ldots, B^{-1}\rho_{2,i-1}B, B^{-1}\rho_{2i}, \rho_{2,i+1}, \ldots, \rho_{2,2g+h+n}).$$

For brevity we will write

$$f_i' = f_i(B^{-1}\rho_{2s}B_{|s<i}, B^{-1}\rho_{2i}, \rho_{2s|s>i}).$$

PROOF: We can deform f so that it sends each a_i to x_2, while on each S_i it traces out a loop *disjoint* from w_i. Thus the two projections of $\bar{f} \circ w_i$ are disjoint, hence $\sigma_i = [\bar{f} \circ w_i] = [i_1 \circ p_1 \circ \bar{f} \circ w_i][i_2 \circ p_2 \circ \bar{f} \circ w_i] = \rho_{1i}[i_2 \circ f \circ w_i]$ and the two factors commute. Figure out the second factor to complete the formula. ▮

THEOREM 3.2. *The condition (II) is equivalent to*

(II′) There exist $u_i \in K$, *for* $i = 1, \ldots, 2g+h+n$, *and* $v_j \in H$, *for* $j = 1, \ldots, k$, *such that in* K *we have*

$$R(u_i\sigma_{i|i\leq 2g+h}, u_i\sigma_i u_i^{-1}{}_{|i>2g+h}) = v_1 B^{i_1} v_1^{-1} \ldots v_k B^{i_k} v_k^{-1},$$

where for all $1 \leq i \leq 2g+h+n$,

$$\sigma_i = f_i' \rho_{1i},$$

$$f_i' = f_i\left(\rho_{21}^{(i)}, \ldots, \rho_{2,2g+h+n}^{(i)}\right),$$

and

$$\rho_s^{(i)} = \begin{cases} B^{-1}\rho_{2s}B & \text{if } s<i \text{ unless bands } s=i-1 \text{ and } i \text{ linked,} \\ B^{-1}\rho_{2s} & \text{if } s=i \text{ or bands } s=i-1 \text{ and } i \text{ linked,} \\ \rho_{2s} & \text{if } s>i. \end{cases}$$

Remark. There is some freedom in choosing σ_i for $i \leq 2g+h$. One could use, say, $\rho_{1i}f_i$. Our choise here simplifies later computations.

PROOF: The fundamental group $\pi_1(M - \text{int}D)$ has generators

$$[r_{11}], \ldots, [r_{1,2g+h}], [w_{2g+h+1}], \ldots, [w_{2g+h+n}], [T],$$

and a single relation

$$R([r_{1i}]_{|i\leq 2g+h}, [w_i]_{|i>2g+h}) = [T].$$

Now describe the homomorphism ϕ in (II) in terms of its value on these generators, and interprete the commutative diagram into equations. ▮

Let us abbreviate the left hand side of the equation (II′) to L. For arbitrary $\{u_i \in K\}$, L is always in K. Each element of K can be written in the form of the right hand side of (II′), i.e. as a product of conjugates of B. We are thus led to the following algebraic notion.

Definition. Let F be a free group with a preferred basis $\{a_i\}$. Each element $x \in F$ can be written as a product of conjugates of powers of basis elements

$$x = y_1 a_{i(1)}^{k_1} y_1^{-1} \cdots y_m a_{i(m)}^{k_m} y_m^{-1},$$

where $k_1, \dots k_m \in \mathbf{Z}$ and $y_1, \dots, y_m \in F$. Such expressions are not unique, e.g. $a_1^{-1}a_2^{-1}a_1^2a_2 = a_2^{-a_1}a_1a_2 = a_1^{-1}a_1^{2a_2}$. The minimal value of m for such expressions will be called the *width* of x with respect to the basis $\{a_i\}$. It is a kind of measure of complexity for elements of F.

It is an interesting question how to calculate the width for $x \in F$. By abelianizing F, x being uniquely expressed as an integral linear combination of the basis elements, the number of terms in this expression is a lower bound for the width. A deeper commutator analysis may be helpful for a better estimate. The method of 'turning index' of Zhang [Z] seems more powerful.

Now come back to our free group K with the basis $\{sBs^{-1} \mid s \in S\}$ of §2. The width of an element $k \in K$ with respect to such a basis does not depend on the Schreier transversal S used. (Cf. Lemma 4.1.) So we call it the width of k.

In fixed point theory one is interested in finding $MF[f]$, the least number of fixed points in the homotopy class of f. With the above notions, this can be reformulated as an algebraic optimization problem.

PROPOSITION 3.3. *$MF[f]$ equals to the minimal width of*

$$L = R(u_i\sigma_i|_{i \le 2g+h}, u_i\sigma_i u_i^{-1}|_{i>2g+h}),$$

the minimum being taken over all $u_1, \dots, u_{2g+h+n} \in K$. Here the σ_i's are the constants determined by the map f as before. ∎

§4. Abelianization of K.

We will use the commutator calculus (see [MKS] Chapter 5) and the Fox calculus (see [B] Chapter 3 or [LS] I.10 and II.3). Let

$$K = K_1 \supset K_2 \supset K_3 \supset \dots,$$

where $K_m = (K_{m-1}, K)$, $m = 2, 3, \dots$, be the lower central series of K. The symbol $\equiv_m$ will denote the congruence modulo K_m.

In this section we prepare for the *mod* K_2 analysis. K/K_2 is the abelianization of K. So the isomorphism β_S between $\mathbf{F}[\pi]$ and K in §2 induces an isomorphism between $\mathbf{Z}[\pi]$ and K/K_2. The following easy lemma shows the latter is really independent of the Schreier transversal S. We therefore call it $\beta_2 : \mathbf{Z}[\pi] \to K/K_2$, and its inverse $\lambda_2 : K/K_2 \to \mathbf{Z}[\pi]$. We define $\lambda_2^* : K/K_2 \to \mathbf{Z}[\pi]$ by $\lambda_2^*(u) = \lambda_2(u)^* =$ the conjugate of $\lambda_2(u)$. We shall not hesitate to use the same symbol for an element of K and for its image in K/K_2. Thus the meaning of $\lambda_2^* k$ is clear for $k \in K$.

LEMMA 4.1. *Suppose $g, g' \in G$ and $g \equiv g'$ mod K. Then for any $k \in K$, the elements k^g and $k^{g'}$ are conjugate in K. In particular, $B^g \equiv_2 B^{g'}$.*

PROOF: Suppose $g = g'k'$ with $k' \in K$. Then $k^g = (k^{g'})^{k'}$. ∎

LEMMA 4.2. *Suppose $h \in H$. Then*

$$B^{h\rho_{1i}^{-1}} \equiv_2 B^{\pm\rho_{2i}h},$$

where the sign is minus iff the band i is twisted.

PROOF: By relation (2-1') we have $B^{\rho_{1i}^{-1}} = B^{\pm\rho_{2i}}$. It is clear that $\rho_{1i}h\rho_{1i}^{-1} \equiv h \bmod K$, because their $p_{2\pi}$-images are the same. So, by Lemma 4.1,

$$B^{h\rho_{1i}^{-1}} = \left(B^{\rho_{1i}^{-1}}\right)^{(\rho_{1i}h\rho_{1i}^{-1})} \equiv_2 \left(B^{\pm\rho_{2i}}\right)^h = B^{\pm\rho_{2i}h}. \qquad \blacksquare$$

Definition. Let $w = w(\rho_{21}, \cdots, \rho_{2,2g+h+n})$ be a word in $\hat{\pi}$. We define $\varepsilon(w)$ to be 1 or -1 according as $w \in \pi$ is orientation preserving or reversing. This ε is a homomorphism of $\hat{\pi}$ into the multiplicative group $\{1, -1\}$. It extends naturally to a ring homomorphism $\varepsilon : \mathbf{Z}[\hat{\pi}] \to \mathbf{Z}$.

Definition. In view of Lemmas 4.1 and 4.2, we define the left action of $\widehat{G}$ on $\mathbf{Z}[\hat{\pi}]$ as follows: each ρ_{2i} acts by left multiplication, B acts trivially, and each ρ_{1i} acts by $\rho_{1i}h = \varepsilon(\rho_{2i})h\rho_{2i}^{-1}$. So we can write $\lambda_2^*(gBg^{-1}) = g$ for all $g \in \widehat{G}$.

Thus, elements of $\mathbf{Z}[\widehat{G}]$ can represent elements of $\mathbf{Z}[\pi]$. Whenever an element of $\mathbf{Z}[\widehat{G}]$ appears as the result of computing some λ_2^*k, it should be interpreted in this way. In this context, B is the same as 1. Sometimes we use the evaluation notation $|_{B=1}$ to emphasize that B should be replaced by 1. Also note that in this context each ρ_{1i} becomes commutative with each ρ_{2j}. This fact is frequently used in computation.

By induction on the length of words, we have

LEMMA 4.3. *Suppose $g \in \widehat{G}$ is a word. Let $g_{(2)} \in \hat{\pi}$ be the word obtained from g by deleting the letters $\{B, \rho_{1i} | i = 1, \cdots, 2g+h+n\}$. Let $g_{(1)} \in \hat{\pi}$ be obtained from g by deleting the letters $\{B, \rho_{2i} | i = 1, \cdots, 2g+h+n\}$ and then replacing each ρ_{1i} with ρ_{2i}. Then*

$$g\xi = \varepsilon(g_{(1)})g_{(2)}\xi g_{(1)}^{-1}$$

for all $\xi \in \mathbf{Z}[\hat{\pi}]$. ∎

The following simple observation enables us to use the Fox calculus for computations.

LEMMA 4.4. *Suppose $g \in \widehat{G}$ belongs to the normal subgroup of $\widehat{G}$ generated by the element B. Then in $\mathbf{Z}[\pi]$ we have*

$$\lambda_2^* g = \left.\frac{\partial g}{\partial B}\right|_{B=1}.$$

PROOF: By linearity of both sides, it suffices to verify the formula for a basis of the normal subgroup generated by B. So suppose $g = B^{u^{-1}}$ with $u \in \widehat{G}$. Then

$$\frac{\partial g}{\partial B} = \frac{\partial}{\partial B}(uBu^{-1}) = u = \lambda_2^* g. \qquad \blacksquare$$

In some situations, it is more convenient to use the other commutator notation $[g, h] = ghg^{-1}h^{-1} = (g^{-1}, h^{-1})$.

LEMMA 4.5. *Suppose $g, g_1, g_2 \in \widehat{G}$ are words in the letters $\{\rho_{1i}\}$ and $h, h_1, h_2 \in \widehat{G}$ are words in the letters $\{B, \rho_{2i}\}$. Then $[g,h] \in G$ is in K, and $\lambda_2^*[g,h]$ is a derivation with respect to both variables g and h. That is,*

$$\lambda_2^*[g_1g_2, h] = \lambda_2^*[g_1, h] + g_1\lambda_2^*[g_2, h],$$

$$\lambda_2^*[g, h_1h_2] = \lambda_2^*[g_1, h_1] + h_1\lambda_2^*[g, h_2].$$

PROOF:

$$\begin{aligned}\lambda_2^*[g_1g_2, h] =& \lambda_2^*(g_1g_2hg_2^{-1}g_1^{-1}h^{-1})\\ =& \lambda_2^*(g_1(g_2hg_2^{-1}h^{-1})g_1^{-1}\cdot g_1hg_1^{-1}h^{-1})\\ =& g_1\lambda_2^*[g_2, h] + \lambda_2^*[g_1, h].\end{aligned}$$

The second formula is verified similarly. ∎

The Reidemeister conjugacy relation in π is the equivalence relation $\sim_R$ generated by $\alpha \sim_R f_i\alpha\rho_{2i}^{-1}$, for all $\alpha \in \pi$ and $i = 1, \cdots, 2g+h+n$. (See [W],[FH], also see [J1] for a dual definition.) Let π_R be the set of Reidemeister conjugacy classes, and let $\mathbf{Z}[\pi_R]$ be the free abelian group generated by π_R. There is a natural projection $\mathbf{Z}[\pi] \to \mathbf{Z}[\pi_R]$ which is nothing but identifying the elements of π of the same Reidemeister conjugacy class. We will write $\xi \equiv_R \eta$ if $\xi, \eta \in \mathbf{Z}[\pi]$ have the same projection in $\mathbf{Z}[\pi_R]$.

An easy corollary of Lemma 4.3 is the following

LEMMA 4.6. *Let $g \in \widehat{G}$. Suppose $w = w(\rho_{21}, \cdots, \rho_{2,2g+h+n})$ is a word in $\widehat{\pi}$ such that $g_{(1)} = w$ and $g_{(2)} = w(f_1, \cdots, f_{2g+h+n})$. Then for $\xi \in \mathbf{Z}[\pi]$*

$$g\xi \equiv_R \varepsilon(w)\xi. \qquad ∎$$

Alternatively, consider the *Reidemeister action*—the right $\widehat{\pi}$-action on $\mathbf{Z}[\pi]$ defined by

$$\eta \bullet w(\rho_{2s}) = w(f_s)^{-1}\eta w(\rho_{2s}) \qquad \text{for } \eta \in \mathbf{Z}[\pi] \text{ and } w \in \widehat{\pi}.$$

Thus π_R and $\mathbf{Z}[\pi_R]$ are respectively π and $\mathbf{Z}[\pi]$ modulo the Reidemeister action.

§5. The Reidemeister trace invariant.

We now do the *mod K_2* analysis for the fixed point equation (II′). Let L be the left hand side of the equation. Our aim is to evaluate λ_2^*L in $\mathbf{Z}[\pi_R]$. L can be written in the form

$$L = L'L''R(f_s)R(\rho_{1s}),$$

where

$$\begin{aligned}L' &= R(u_s\sigma_s|_{s\le 2g+h}, u_s\sigma_s u_s^{-1}|_{s>2g+h})R(\sigma_s)^{-1},\\ L'' &= R(\sigma_s)R(\rho_{1s})^{-1}R(f_s)^{-1}.\end{aligned}$$

Note that the u's only appear in L'.

Notation. We define inductively the segments $R_i = R_i(z_s)$ of the word $R = R(z_s)$ by

$$R_1 = 1,$$

$$R_{i+1} = \begin{cases} R_i z_i z_{i+1}^{-1} z_i^{-1} & \text{if } i \text{ odd} \leq 2g \\ R_i z_i & \text{if } i \text{ even} \leq 2g \\ R_i z_i^2 & \text{if } 2g < i \leq 2g+h \\ R_i z_i & \text{if } 2g+h < i \leq 2g+h+n. \end{cases}$$

For brevity we also denote

$$\begin{aligned} R_{1i} &= R_i(\rho_{1s}), \\ R_{2i} &= R_i(\rho_{2s}), \\ R_{fi} &= R_i(f_s), \\ R_{\sigma i} &= R_i(\sigma_s). \end{aligned}$$

Let

$$\xi_i = \lambda_2^* u_i \in \mathbf{Z}[\pi].$$

L' is a word in $\widehat{G}$ belonging to the normal subgroup generated by B.

$$L' = \prod_{\substack{\text{odd } i=1 \\ k=i+1}}^{i=2g-1} R_{\sigma i} u_i \sigma_i \sigma_k^{-1} u_k^{-1} \sigma_i^{-1} u_i^{-1} u_k \sigma_k R_{\sigma,k+1}^{-1}$$

$$\prod_{i=2g+1}^{2g+h} R_{\sigma i} u_i \sigma_i u_i \sigma_i R_{\sigma,i+1}^{-1} \cdot \prod_{i=2g+h+1}^{2g+h+n} R_{\sigma i} u_i \sigma_i u_i^{-1} R_{\sigma,i+1}^{-1} \cdot$$

Hence

$$\begin{aligned} \lambda_2^* L' &= \sum_{\substack{\text{odd } i \leq 2g \\ k=i+1}} R_{\sigma i} \left(1 - \sigma_i \sigma_k^{-1} \sigma_i^{-1}\right) \xi_i + R_{\sigma k} \left(1 - \sigma_i\right) \xi_k \\ &+ \sum_{2g < i \leq 2g+h} R_{\sigma i} \left(1 + \sigma_i\right) \xi_i + \sum_{i > 2g+h} R_{\sigma i} \left(1 - \sigma_i\right) \xi_i \\ &= \sum_i R_{\sigma i} \xi_i \bullet \left(1 - \tau_{2i}\right), \end{aligned}$$

where

$$\tau_{2i} = \begin{cases} R_{2i} \rho_{2i} \rho_{2,i+1} \rho_{2i}^{-1} R_{2i}^{-1} & \text{for odd } i < 2g, \\ R_{2i} \rho_{2,i-1}^{-1} R_{2i}^{-1} & \text{for even } i \leq 2g, \\ R_{2i} \rho_{2i}^{-1} R_{2i}^{-1} & \text{for } i > 2g \end{cases}$$

are elements in $\widehat{\pi}$.

Remark. It is not difficult to show by induction on k that, for even $k \leq 2g$ and for every $k > 2g$, the set of words $\{\tau_{21}, \cdots, \tau_{2k}\}$ can be obtained from $\{\rho_{21}, \cdots, \rho_{2k}\}$ by a regular Nielsen transformation (cf. [Jo] p.20). Therefore $\{\tau_{2i} | 1 \leq i \leq 2g+h+n\}$ is a basis for $\widehat{\pi}$.

PROPOSITION 5.1. *$\lambda_2^* L' \equiv_R 0$. Conversely, for every $\eta \in \mathbf{Z}[\pi]$ with $\eta \equiv_R 0$ there exist $\xi_1, \cdots, \xi_{2g+h+n} \in \mathbf{Z}[\pi]$ such that $\lambda_2^* L' = \eta$.*

PROOF: It is evident that $\lambda_2^* L'$ is 0 modulo the Reidemeister action. Conversely, suppose $\eta \equiv_R 0$. Since by the above Remark $\{\tau_{2i} | 1 \leq i \leq 2g+h+n\}$ is a basis for $\widehat{\pi}$, there are $\{\eta_i\}$ such that

$$\eta = \sum_i \eta_i \bullet (1 - \tau_{2i}).$$

Then $\lambda_2^* L' = \eta$ when $\xi_i = \eta_i \bullet R_{2i}$. ∎

For the calculation of $\lambda_2^* L''$ we need a few more lemmas.

LEMMA 5.2. *Suppose $g_1, \cdots, g_m \in \widehat{G}$ are words in the letters $\{\rho_{1i} \mid i = 1, ..., 2g+h+n\}$ and $h_1, \cdots, h_m \in \widehat{G}$ are words in the letters $\{B, \rho_{2i} \mid i = 1, ..., 2g+h+n\}$, such that $g_1 \cdots g_m = h_1 \cdots h_m|_{B=1} = 1$. Then $g_1 h_1 g_2 h_2 \cdots g_m h_m \in G$ is in K, and*

$$\begin{aligned}\lambda_2^*(g_1 h_1 \cdots g_m h_m) =& \lambda_2^*[g_1, h_1] + h_1 \lambda_2^*[g_1 g_2, h_2] + \cdots \\ &+ h_1 \cdots h_{m-2} \lambda_2^*[g_1 \cdots g_{m-1}, h_{m-1}] + \lambda_2^*(h_1 \cdots h_m).\end{aligned}$$

PROOF: Induction on m. Suppose it is true for $m = t$. Then when $m = t+1$ we have

$$\begin{aligned}g_1 h_1 \cdots g_t h_t g_{t+1} h_{t+1} =& g_1 h_1 \cdots g_{t-1} h_{t-1} (g_t g_{t+1})(h_1 \cdots h_{t-1})^{-1} \\ &\cdot h_1 \cdots h_{t-1} g_{t+1}^{-1} h_t g_{t+1} h_{t+1}.\end{aligned}$$

By inductive hypothesis,

$$\begin{aligned}&\lambda_2^*(g_1 h_1 \cdots g_{t+1} h_{t+1}) \\ &= \lambda_2^*[g_1, h_1] + \cdots + h_1 \cdots h_{t-2} \lambda_2^*[g_1 \cdots g_{t-1}, h_{t-1}] + \lambda_2^*(1) \\ &\quad + \lambda_2^*(h_1 \cdots h_{t-1} [g_{t+1}^{-1}, h_t](h_1 \cdots h_{t-1})^{-1}(h_1 \cdots h_{t-1}) h_t h_{t+1}) \\ &= \lambda_2^*[g_1, h_1] + \cdots + h_1 \cdots h_{t-2} \lambda_2^*[g_1 \cdots g_{t-1}, h_{t-1}] \\ &\quad + h_1 \cdots h_{t-1} \lambda_2^*[g_1 \cdots g_t, h_t] + \lambda_2^*(h_1 \cdots h_{t+1}).\end{aligned}$$

This completes the induction. ∎

LEMMA 5.3.

$$\lambda_2^*[g, B] = g - 1 \qquad \textit{for all } g \in \widehat{G}.$$

$$\lambda_2^*[\rho_{1i}, \rho_{2i}] = \begin{cases} \rho_{1i} - 1 & \textit{if band } i \textit{ flat} \\ \rho_{1i} & \textit{if band } i \textit{ twisted.} \end{cases}$$

When $i < j$,

$$\lambda_2^*[\rho_{1i}, \rho_{2j}] = \begin{cases} \rho_{1i} & \textit{if bands } i, j \textit{ linked} \\ 0 & \textit{otherwise.} \end{cases}$$

When $i > j$,

$$\lambda_2^*[\rho_{1i}, \rho_{2j}] = \begin{cases} (\rho_{1i} - 1)(1 - \rho_{2j}) - \rho_{2j} & \textit{if bands } i, j \textit{ linked} \\ (\rho_{1i} - 1)(1 - \rho_{2j}) & \textit{otherwise.} \end{cases}$$

PROOF: The first formula follows from the definition

$$\lambda_2^*[g, B] = \lambda_2^*(gBg^{-1}B^{-1}) = \lambda_2^*(gBg^{-1}) + \lambda_2^*(B^{-1}) = g - 1.$$

The rest are direct consequences of the relations (2-2')–(2-4') and Lemma 4.3. ∎

LEMMA 5.4.

$$\lambda_2^*[R_{1i}, \rho_{2j}] = \begin{cases} 0 & \text{if } i \le j \\ R_{1i} & \text{if } j < i \text{ and bands } j, i \text{ linked} \\ R_{1i}(1-\rho_{2j}) & \text{otherwise.} \end{cases}$$

PROOF: It is trivial when $i = 1$. Argue inductively on i on the basis of Lemma 5.3 and the inductive definition of R_i. ∎

We can write

$$L'' = R(f_s'\rho_{1s})R(\rho_{1s})^{-1}R(f_s)^{-1}.$$

By Lemma 5.2, we have

$$\begin{aligned}\lambda_2^* L'' = &\sum_{\text{odd } i \le 2g} (R_{fi}\lambda_2^*[R_{1i}, f_i'] + R_{fi}f_i\lambda_2^*[R_{1,i+1}\rho_{1i}, f_{i+1}'^{-1}] \\ &\qquad + R_{f,i+1}f_i\lambda_2^*[R_{1,i+1}, f_i'^{-1}f_{i+1}']) \\ &+ \sum_{2g<i\le 2g+h} (R_{fi}\lambda_2^*[R_{1i}, f_i'] + R_{fi}f_i\lambda_2^*[R_{1i}\rho_{1i}, f_i']) \\ &+ \sum_{i>2g+h} R_{fi}\lambda_2^*[R_{1i}, f_i'] \\ &+ \lambda_2^*(R(f_s')R(f_s)^{-1}).\end{aligned}$$

Applying Lemma 4.4 to the last term and Lemma 4.5 to the first sum then collecting the terms involving f_i', we see

$$\lambda_2^* L'' = S_1 + \cdots + S_{2g+h+n},$$

where for odd $i \le 2g$

$$\begin{aligned}S_i = &R_{fi}\left(\lambda_2^*[R_{1i}, f_i'] + \lambda_2^*(f_i'f_i^{-1})\right) \\ &- R_{f,i+1}\left(\lambda_2^*[R_{1,i+1}, f_i'] + \lambda_2^*(f_i'f_i^{-1})\right),\end{aligned}$$

for even $i \le 2g$

$$\begin{aligned}S_i = &R_{fi}\left(\lambda_2^*[R_{1i}, f_i'] + \lambda_2^*(f_i'f_i^{-1})\right) \\ &- R_{fi}f_{i-1}\left(\lambda_2^*[R_{1i}\rho_{1,i-1}, f_i'] + \lambda_2^*(f_i'f_i^{-1})\right),\end{aligned}$$

for $2g < i \le 2g + h$

$$\begin{aligned}S_i = &R_{fi}\left(\lambda_2^*[R_{1i}, f_i'] + \lambda_2^*(f_i'f_i^{-1})\right) \\ &+ R_{fi}f_i\left(\lambda_2^*[R_{1i}\rho_{1i}, f_i'] + \lambda_2^*(f_i'f_i^{-1})\right),\end{aligned}$$

and for $i > 2g + h$

$$S_i = R_{fi}\left(\lambda_2^*[R_{1i}, f_i'] + \lambda_2^*(f_i'f_i^{-1})\right).$$

LEMMA 5.5. *Let g be a word in the letters $\{\rho_{1s}\}$. Then*

$$\lambda_2^*[g, f_i'] + \lambda_2^*(f_i' f_i^{-1}) = \sum_j \frac{\partial f_i}{\partial \rho_{2j}} \left(\lambda_2^*[g, \rho_{2j}] + g \frac{\partial \rho_j^{(i)}}{\partial B} \right).$$

PROOF: Apply Lemmas 4.5 and 4.4. ∎

A straight forward computation using Lemmas 5.3–5.5 now shows that for all $1 \leq i \leq 2g + h + n$ we have the same formula

$$S_i = -R_{fi} \frac{\partial f_i}{\partial \rho_{2i}} R_{1i},$$

hence by Lemma 4.6

$$S_i \equiv_R -\frac{\partial f_i}{\partial \rho_{2i}}.$$

To sum up, we have got

PROPOSITION 5.6.

$$\lambda_2^* L'' \equiv_R -\sum_i \frac{\partial f_i}{\partial \rho_{2i}}. \qquad ∎$$

The factor $R(\rho_{1s})$ of L is easy to determine. By relation (2-5), $R(\rho_{1s}) = B$ in π. Hence $\lambda_2^* R(\rho_{1s}) = 1$.

It remains to investigate the factor $R(f_s)$. In §3 we already found some $A^* \in \mathbf{F}[\widehat{\pi}]$ such that $R(f_s) = R(\rho_{2s})^{A^*}$ in $\widehat{\pi}$. Hence in π we have $R(f_s) = R(\rho_{2s})^{A^*}$, thus $\lambda_2^* R(f_s) = A^{**} = A$.

This completes our computation:

THEOREM 5.7.

$$\lambda_2^* L \equiv_R 1 - \sum_i \frac{\partial f_i}{\partial \rho_{2i}} + A. \qquad ∎$$

Now compare our result with the Reidemeister trace invariant [W]. Fadell and Husseini ([FH] §2) called it the generalized Lefschetz-Hopf-Reidemeister number, and designed a special method for computing it on surfaces. Their formula looks much like our formula in the above Theorem. A closer look will convince us that they are indeed the same. For the convenience of the reader we include the following dictionary, in which the left column contains their notation, and the right ours.

$R(x_1, \cdots, x_k) \in F$	$R(\rho_{21}, \cdots, \rho_{2,2g+h+n}) \in \widehat{\pi}$
$\psi, w_i, R[\psi]$	f_π, f_i, π_R
w lifts to a cycle $\widetilde{w}$	$w = 1$ in π
$\partial\widetilde{u}$	$R(\rho_{2s})$
$\widetilde{w} = \gamma\partial\widetilde{u}, \quad \gamma \in F$	$w = \gamma R(\rho_{2s})\gamma^{-1}, \quad \lambda_2^* w = \gamma$
$\partial \widetilde{f}_{2\#}\widetilde{u} = A\partial\widetilde{u}$	$\lambda_2^* R(f_s) = A\lambda_2^* R(\rho_{2s}) = A$

Our final conclusion is

THEOREM 5.8. *$\lambda_2^* L \in \mathbf{Z}[\pi_R]$ is equal to the Reidemeister trace invariant.* ▮

We will write $\lambda_{2R}^* L$ when $\lambda_2^* L$ is considered in $\mathbf{Z}[\pi_R]$, i.e. as an integral linear combination of the Reidemeister conjugacy classes. The theorem says $\lambda_{2R}^* L$ carries all the information of Nielsen fixed point theory. Namely, the number of terms is equal to the Nielsen number of f, the coefficients are the indices of the fixed point classes, etc.

On the other hand, in view of Proposition 5.1, there is no loss of useful information under the projection $\mathbf{Z}[\pi] \to \mathbf{Z}[\pi_R]$, $\lambda_2^* L \mapsto \lambda_{2R}^* L$. The Reidemeister trace invariant is all that we can get from a *mod* K_2 analysis of the fixed point equation (II′).

§6. Remarks.

Fadell and Husseini constructed in [FH] a weaker (abelianized) obstruction theory for deforming a surface map into a fixed point free one, and identified their result with the Reidemeister invariant. Our approach is to construct the real obstruction, in the form of a braid equation, and then abelianize it. It reveals how far away the Reidemeister invariant is from the solution of the fixed point problem. In Part II we shall use *mod* K_3 analysis to get better knowledge about $MF[f]$.

REFERENCES

[B] Birman, J.S., "Braids, Links, and Mapping Class Groups," Princeton Univ. Press, Princeton, 1974.

[FH] Fadell, E., Husseini, S., *The Nielsen number on surfaces*, in "Topological Methods in Nonlinear Functional Analysis," Contemp. Math. vol.21, AMS, Providence, 1983, pp. 59–98.

[FN] Fadell, E., Neuwirth, L., *Configuration spaces*, Math. Scand. **10** (1962), 111–118.

[H] Hu, S.-T., *Extensions and classification of maps*, Osaka Math. J. **2** (1950), 165–209.

[J1] Jiang, B., "Lectures on Nielsen Fixed Point Theory," Contemp. Math. vol.14, AMS, Providence, 1983.

[J2] Jiang, B., *Fixed points and braids*, Invent. Math. **75** (1984), 69–74.

[J3] Jiang, B., *Fixed points and braids, II*, Math. Ann. **272** (1985), 249–256.

[Jo] Johnson, D.L., "Topics in the Theory of Group Presentations," Cambridge Univ. Press, London, 1980.

[LS] Lyndon, R.C., Schupp, P.E., "Combinatorial Group Theory," Springer, Berlin, Heidelberg, New York, 1977.

[MKS] Magnus, W., Karrass, A., Solitar, D., "Combinatorial Group Theory," Dover, New York, 1976.

[S] Scott, G.P., *Braid groups and the group of homeomorphisms of surfaces*, Proc. Camb. Phil. Soc. **68** (1970), 605–617.

[W] Wecken, F., *Fixpunktklassen, II*, Math. Ann. **118** (1942), 216–234.

[Z] Zhang, X.-G., *The least number of fixed points can be arbitrarily larger than the Nielsen number*, Acta Sci. Natur. Univ. Pekin. 1986:3, 15–25.

1980 *Mathematics subject classifications*: 55M20, 57M99

Department of Mathematics, Peking University, Beijing 100871, China

Affine Maximal Surfaces and Harmonic Functions

Li An-Min
Department of Mathematics, Sichuan University
Chengdu, P.R.China

The purpose of this paper is to study affine maximal surfaces. In section 2 we give an affine analogue of the Weierstrass representation of minimal surfaces in R^3. In section 3 we study the distribution of the affine normals of affine maximal surface and prove that a locally strongly convex, affine complete, affine maximal surface is an elliptic paraboloid if its affine normals omit 5 or more directions. In §4 we prove a theorem on harmonic functions by means of the affine Weierstrass representation and fundamental formulas in affine differential geometry.

This work was done while the author was a guest at TU Berlin, financed by a research grant of the Alexander von Humboldt-Stiftung. The author would like to express his sincere gratitude to Prof. Dr. U.Simon and Dr. M.Kozlowski for the discussions on this work and for their hospitality.

§1 Preliminaries

Let A^3 denote the unimodular affine space of dimension 3, i.e., the space with real coordinates x^1, x^2, x^3 and volume element $dv = dx^1 \wedge dx^2 \wedge dx^3$. Let M be a 2-dimensional, oriented connected manifold and $x : M \to A^3$ an immersion. Choose an affine frame field x, e_1, e_2, e_3 on M such that e_1 and e_2 are tangent to M at x, and

$$(e_1, e_2, e_3) = 1.$$

Throughout this paper we shall agree on the index ranges:

$$1 \leq i, j, k, \cdots \leq 2, \quad 1 \leq \alpha, \beta, \gamma, \cdots \leq 3.$$

We can wirte

$$\begin{aligned} dx &= \sum \omega^\alpha e_\alpha \\ de_\alpha &= \sum \omega_\alpha^\beta e_\beta \end{aligned} \tag{1}$$

The structure equation of A^3 gives

$$\begin{aligned} d\omega^\alpha &= \sum \omega^\beta \wedge \omega_\beta^\alpha \\ d\omega_\alpha^\beta &= \sum \omega_\alpha^\gamma \wedge \omega_\gamma^\beta. \end{aligned} \tag{2}$$

If we restrict the forms to the surface M, we have

$$\begin{aligned} &\omega^3 = 0 \\ &\sum \omega^i \wedge \omega_i^3 = 0. \end{aligned}$$

By Cartan's lemma we have

$$\omega_i^3 = \sum h_{ik}\omega^k, \qquad h_{ik} = h_{ki}. \tag{3}$$

We assume that $x(M)$ is non-degenerate, i.e., $\det(h_{ij}) \neq 0$. The quadratic differential form

$$II = |\det(h_{ij})|^{-\frac{1}{4}} \sum h_{ik}\omega^i\omega^k := \sum G_{ik}\omega^i\omega^k \tag{4}$$

is affinely invariant and called the Berwald-Blaschke metric. Since $x(M)$ is non-degenerate, we can suitably choose e_3 such that

$$\omega_3^3 + \frac{1}{4} d\log|\det(h_{ij})| = 0. \tag{5}$$

Under such a choice the vector

$$Y = |\det(h_{ij})|^{\frac{1}{4}} e_3 \tag{6}$$

is affinely invariant and called the affine normal vector.

Differentiating (5) we get

$$\sum \omega_3^i \wedge \omega_i^3 = 0$$

which gives

$$\omega_3^i = -\sum l^{ik}\omega_k^3, \qquad l^{ik} = l^{ki}.$$

The quadratic differential form

$$III = \sum l^{ik}\omega_i^3\omega_k^3 \tag{7}$$

is invariant under any change of frame keeping the affine normal fixed. The trace of III relative to II, i.e.,

$$H = \frac{1}{2}|\det(h_{ij})|^{\frac{1}{4}} \sum l^{ik}h_{ki} \tag{8}$$

is called the affine mean curvature. A surface is called affine maximal if $H \equiv 0$ on M.

Taking the exterior differential of (3) and defining h_{ijk} by

$$\sum h_{ijk}\omega^k = dh_{ij} + h_{ij}\omega_3^3 - \sum h_{ik}\omega_j^k - \sum h_{kj}\omega_i^k$$

we get

$$h_{ijk} = h_{ikj}. \tag{9}$$

The cublic form

$$\begin{aligned} A =& \frac{1}{2}|\det(h_{ij})|^{-\frac{1}{4}} \sum h_{ijk}\omega^i\omega^j\omega^k : \\ =& \sum A_{ijk}\omega^i\omega^j\omega^k \end{aligned} \tag{10}$$

is called the Fubini-Pick form, and the affine invariant

$$J := \frac{1}{2}\|A_{ijk}\|_G^2 = \frac{1}{2}\sum G^{im}G^{jn}G^{kl}A_{ijk}A_{mnl}, \tag{11}$$

the Pick invariant. Denote by R the scalar curvature with respect to the Berwald-Blaschke metric and we have [3]:

$$R = J + H. \tag{12}$$

Define the conormal vector field U by

$$\begin{aligned} \langle U, Y\rangle &=1 \\ \langle U, e_i\rangle &=0, \end{aligned} \tag{13}$$

where $\langle , \rangle : A^* \times A \to R$ denotes the standard scalar product, then we have [3]:

$$\Delta U = -2HU, \tag{14}$$

where Δ denotes the Laplacian with respect to the Berwald-Blaschke metric.

From (5) and (6) we have

$$dY = |\det(h_{ij})|^{\frac{1}{4}} \cdot \sum \omega_3^i e_i.$$

It follows from (13) that

$$\langle U, dY\rangle = 0,$$

i.e.,

$$\langle U, Y_i\rangle = 0. \tag{15}$$

From (1-6) we have

$$de_i = \sum \omega_i^j e_j + \sum G_{ik}\omega^k Y. \tag{16}$$

It follows that

$$\langle dU, e_i\rangle = -\langle U, de_i\rangle = -G_{ik}\omega^k,$$

i.e.,

$$\langle U_k, e_i\rangle = -G_{ik}. \tag{17}$$

In this paper we shall limit our consideration to those surfaces that are locally strongly convex, such that G_{ij} is positive definite.

§2. Affine analogue of the Weierstrass representation

Now we introduce a Euclidean scalar product '$\cdot$' in A^3, and denote $(A^3, \cdot)$ by R^3. Given a triple of functions

$$U = (U^1(u,v), U^2(u,v), U^3(u,v))$$

satisfying

$$\begin{aligned} U \times U_{uv} &=0 \\ (U, U_u, U_v) &>0 \qquad \forall u, v \in \quad \textit{textdomain} \end{aligned}$$

one can construct a hyperbolic (i.e., $\det(G_{ij}) < 0$) affine maximal surface [5]

$$x = \int U \times U_u du - u \times U_v dv.$$

Terng Chuu-Lian [4] pointed out (without proof) that, given a triple of harmonic functions U satisfying $(U, U_u, U_v) > 0$, one can construct a locally strongly convex, affine maximal surface. But her formula (44) in [4] should be corrected because of a misprint. In the following we shall derive the correct formula.†

Let $x : M \to R^3$ be a locally strongly convex surface. Choose the isothermal parameters u and v on M with respect to the Berwald-Blaschke metric, and let $e_1 = x_u, e_2 = x_v$. Then $G_{12} = G_{21} = 0, G_{11} = G_{22} > 0$. The formulas (13) and (17) in this case respectively become

$$U \cdot x_u = 0, \quad U_u \cdot x_u = -G_{11}, \quad U_v \cdot x_u = 0, \tag{18}$$

$$U \cdot x_v = 0, \quad U_u \cdot x_v = 0, \quad U_v \cdot x_v = -G_{22}, \tag{19}$$

$$U \cdot Y = 1, \quad U_u \cdot Y = 0, \quad U_v \cdot Y = 0. \tag{20}$$

It follows that

$$x_u = \lambda U \times U_v, \tag{21}$$

$$x_v = \mu U \times U_u, \tag{22}$$

$$(x_u, x_v, Y)(U, U_u, U_v) = G_{11}^2. \tag{23}$$

Since

$$(x_u, x_v, Y) = |\det(h_{ij})|^{\frac{1}{4}} = G_{11}, \tag{24}$$

we have

$$(U, U_u, U_v) = G_{11} > 0. \tag{25}$$

From (18) and (21) it follows that

$$-G_{11} = U_u \cdot x_u = \lambda(U, U_v, U_u) = -\lambda G_{11}.$$

Hence $\lambda = 1$. Similarly, we have $\mu = -1$. Consequently

$$x_u = U \times U_v,$$
$$x_v = -U \times U_u.$$

Thus we obtain the following formula:

$$x = \int U \times U_v du - U \times U_u dv. \tag{26}$$

---

†Remark. At the conference which was held in Oberwolfach, in November 1986, Calabi announced a complex representation of a locally strongly convex affine maximal surface, which is equivalent to the formula (27) in this paper.

If $x(M)$ is affine maximal, then

$$\Delta U = 0$$

where

$$\Delta = \frac{1}{(U, U_u, U_v)}\left(\frac{\partial^2}{\partial u^2} + \frac{\partial^2}{\partial v^2}\right).$$

It follows that $U^1(u,v), U^2(u,v)$ and $U^3(u,v)$ are harmonic functions.

Conversely, given a triple of harmonic functions $U = (U^1(u,v), U^2(u,v), U^3(u,v))$ defined on $\Omega \subset R^2$ satisfying $(U, U_u, U_v) > 0$ in Ω, we can construct a surface

$$x = \int_{(u_0,v_0)}^{(u,v)} U \times U_v du - U \times U_u dv, \tag{27}$$

where Ω is a simply connected domain and $(u_0, v_0), (u, v) \in \Omega$. It is well defined since

$$(U \times U_v)_v + (U \times U_u)_u = U \times (U_{uu} + U_{vv}) = 0.$$

Now let us prove that the surface M defined by (27) is a locally strongly convex affine maximal surface. From (27) we have

$$x_u = U \times U_v, \tag{28}$$

$$x_v = -U \times U_u, \tag{29}$$

$$x_u \times x_v = (U, U_u, U_v)U. \tag{30}$$

Let

$$e_3 = \frac{U}{(U, U_u, U_v)U \cdot U},$$

then $(x_u, x_v, e_3) = 1$, i.e., $\{x_u, x_v, e_3\}$ is a unimodular affine frame field. Let

$$x_{ij} = \sum \Gamma_{ij}^k x_k + h_{ij} e_3 \qquad 1 \le i, j, k \le 2.$$

From (28), (29) and (30) we have

$$h_{11} = h_{22} = (U, U_u, U_v)^2, \qquad h_{12} = h_{21} = 0.$$

Hence $x(M)$ is locally strongly convex and the Berwalk-Blaschke metric is

$$G_{ij} = |\det(h_{ij})|^{-\frac{1}{4}} h_{ij},$$

i.e.,

$$G_{11} = G_{22} = (U, U_u, U_v), \qquad G_{12} = G_{21} = 0.$$

The conormal vector is

$$[\det(h_{ij})]^{-\frac{1}{4}} x_u \times x_v = U.$$

Since $U^1(u,v), U^2(u,v)$ and $U^3(u,v)$ are harmonic functions, $x(M)$ is affine maximal.

Example 1. Take $U = (1, u, v)$ and we get

$$x = (\frac{1}{2}(u^2 + v^2), -u, -v),$$

which is an elliptic paraboloid.

Example 2. Take $U = (1, u^2 - v^2, v)$, then

$$(U, U_u, U_v) = 2u, \qquad (u > 0).$$

We get

$$x = (\frac{1}{3}u^3 + uv^2, -u, -2uv), \qquad u > 0.$$

Example 3. Take $U = (u, v, 2uv)$, then

$$(U, U_u, U_v) = -2uv, \qquad (u > 0, \;\; v < 0).$$

We get

$$x = (-\frac{2}{3}v^3, -\frac{2}{3}u^3, \frac{1}{2}(u^2 + v^2)) \qquad u > 0, \;\; v < 0.$$

From the point of view of local differential geometry the formula (27) gives all affine maximal surfaces. The following global problem is interesting:

Problem. Find an affine complete, affine maximal surface which is not an elliptic paraboloid.

This problem is equivalent to the following one

Problem. Find a triple of harmonic functions $U = (U_1(u,v), U_2(u,v), U_3(u,v))$, which doesn't lie on a plane, such that $(U, U_u, U_v) > 0$ on C and $(U, U_u, U_v)(du^2 + dv^2)$ is a complete metric on C.

§3. Gauss map

To study the distribution of the affine normals we need to define the Gauss map. We will first introduce an equivalence relation $\sim$ in $A^3 - \{0\} : (a^1, a^2, a^3) \sim (b^1, b^2, b^3)$ if and only if there is a real number $\lambda > 0$ such that $(b^1, b^2, b^3) = \lambda(a^1, a^2, a^3)$. Denote by Q the quotient space, by $[a] = \{y \in A^3 - \{0\} | y \sim a\}$ the equivalence class of a, and by $\Pi : A^3 - \{0\} \to Q$ the natural projection taking each $a \in A^3 - \{0\}$ to its equivalence class, i .e.,

$$\Pi(a) = [a].$$

With these notations we can define the Gauss map

$$\begin{aligned} g : M &\to Q \\ x &\mapsto [Y_x]. \end{aligned} \tag{31}$$

When an Euclidean metric is introduced in A^3, Q may be identified with S^2, and the Gauss map (31) may be identified with the following map

$$\begin{aligned} g' : M &\longrightarrow S^2 \\ x &\longmapsto Y_x/\|Y_x\| \qquad \text{where } \|Y_x\| = \sqrt{Y_x \cdot Y_x}. \end{aligned} \tag{32}$$

From now on we shall identify g with g'.

In the following we will give another explanation of the Gauss map (32). From $(U, U_u, U_v) > 0$ we know that $U : M \to R^3$ is an immersed surface with central normalization. From (20) it follows that

$$\begin{aligned} Y =& \lambda U_u \times U_v, \\ 1 =& Y \cdot U = \lambda(U, U_u, U_v). \end{aligned}$$

Consequently $\lambda > 0$ and

$$\frac{Y}{\|Y\|} = \frac{U_u \times U_v}{\|U_u \times U_v\|},$$

i.e., the Gauss map defined by (32) is just the classical Gauss map of the conormal immersion $U : M \to R^3$.

The problem how the surface $x(M)$ is determined by $U(M)$ looks interesting. It is easy to see that $x(M)$ is an improper affine sphere if and only if $U(M)$ lies on a plane. Furthermore, we have

Theorem 1. Let $x : M \to A^3$ be a locally strongly convex, affine complete, affine maximal surface. If $U(M)$ lies on a half space, then $x(M)$ is an elliptic paraboloid.

Proof. After a homogeneous coordinate transformation we can assume that $U^3 > c$, where c is a constant. By Liouville's theorem $U^3 = constant$. It follows that $x(M)$ is an improper affine sphere. Being affine complete, $x(M)$ must be an elliptic paraboloid.

Remark. It is easy to see that $U^3 > 0$ if $x(M)$ is a graph defined by $x^3 = f(x^1, x^2)$. Hence, Theorem 1 is a little more general than Calabi's theorem ([7]).

In the following we will give an affine analogue of Xavier's theorem of minimal surface in R^3.

Definition 1. A set of vectors in R^{n+1} (or C^{n+1}) is in general position if each subset of $n+1$ vectors is linearly independent. A set of points in S^n (or CP^n) is in general position if it is in general position as a set of vectors in R^{n+1} (or C^{n+1}).

Definition 2. A set of hyperplanes in CP^n is in general position if each subset of $n+1$ hyperplanes has no common point.

Lemma 1 (Borel's theorem). Let $f : C \to CP^n$ be a holomorphic curve not lying in a hyperplane of CP^n. For any $n+2$ hyperplanes in CP^n in general position, the image $f(C)$ meets one of them.

Lemma 2. Let $V(k), 1 \le k \le 5$, be 5 vectors in C^2 satisfying

$$rank\{V(i), V(j), V(k)\} = 2 \quad \forall \{i, j, k\} \subset \{1, 2, 3, 4, 5\}.$$

then there are at least 3 vectors in general position.

Proof. Since $rank\{V(1), V(2), V(3)\} = 2$, there are two linearly independent vectors in $\{V(1), V(2), V(3)\}$. Suppose that $V(1)$ and $V(2)$ are linearly independent. If $\{V(1), V(2), V(3)\}$ is not in general position, then $V(3)//V(2)$ or $V(3)//V(1)$. Suppose that $V(3)//V(1)$. If $\{V(1), V(2), V(4)\}$ is not in general position either, then $V(4)//V(1)$ or $V(4)//V(2)$. Since $rank\{V(1), V(3), V(4)\} = 2$, we have $V(4)//V(2)$. Thus we can prove that $\{V(1), V(2), V(5)\}$ is in general position. In fact, if $V(5)//V(1)$, we have $rank\{V(1), V(3), V(5)\} = 1$; if $V(5)//V(2)$ we have $rank\{V(1), V(4), V(5)\} = 1$. Both of them are impossible.

<u>Theorem 2</u>. Let $x : M \to A^3$ be a locally strongly convex, affine complete, affine maximal surface. If $g(M)$ omits 5 or more points in general position and its antipodal points, then $x(M)$ must be an elliptic paraboloid.

<u>Proof</u>. Choose the isothermal parameters u and v on M and deifne the map

$$f : M \longrightarrow CP^2$$
$$(u,v) \longmapsto (Z^1(u,v), Z^2(u,v), Z^3(u,v))$$

where (Z^1, Z^2, Z^3) are the homogeneous coordinates of CP^2, and

$$Z^k(u,v) = U^k_u - iU^k_v.$$

It is easily seen that f is well defined. Since $x(M)$ is affine maximal, $U^k (1 \le k \le 3)$ are harmonic functions, hence $Z^k(\xi)$, $\xi = u + iv$, are holomorphic functions. We assume that M is simply connected (otherwise we may pass to the universal covering surface of M). Since $x(M)$ is affine complete and $R \ge 0$, it is conformally equivalent to C. Suppose that $g(M)$ omits the points

$$\pm W(k) = \pm(W^1(k), W^2(k), W^3(k)) \in S^2, \quad 1 \le k \le 5,$$

then

$$W(k) \cdot Z^1(\xi) + W^2(k) \cdot Z^2(\xi) + W^3(k) \cdot Z^3(\xi) \ne 0, \quad 1 \le k \le 5.$$

This implies that $f(M)$ doesn't meet 5 planes:

$$P_k : \ W^1(k) \cdot Z^1 + W^2(k) \cdot Z^2 + W^3(k) \cdot Z^3 = 0, \quad 1 \le k \le 5.$$

It is obvious that $\{P_k, 1 \le k \le 5\}$ is in general position. By Lemma 1 $f(M)$ must lie in a plane $P: \ aZ^1 + bZ^2 + cZ^3 = 0$ of CP^2. After a Hermitian transformation

$$\begin{pmatrix} Z^{*1} \\ Z^{*2} \\ Z^{*3} \end{pmatrix} = L \begin{pmatrix} Z^1 \\ Z^2 \\ Z^3 \end{pmatrix}$$

we can assume that the equation of the plane P is $\overset{*}{Z}{}^1 = 0$. The map $f : M \to CP^2$ can be regarded as that $f : M \to CP^1$. Let

$$\overset{*}{W}(k) = \begin{pmatrix} W^{*1}(k) \\ W^{*2}(k) \\ W^{*3}(k) \end{pmatrix} = L \begin{pmatrix} W^1(k) \\ W^2(k) \\ W^3(k) \end{pmatrix}.$$

Then $\{\overset{*}{W}(k), 1 \le k \le 5\}$ is in general position, i.e.,

$$\det(\overset{*}{W}(i), \overset{*}{W}(j), \overset{*}{W}(k)) \ne 0, \quad \forall \{i,j,k\} \subset \{1,2,3,4,5\}. \tag{33}$$

Hence we have

$$\begin{aligned} &\bar{\overset{*}{W}}{}^1(k) \cdot \overset{*}{Z}{}^1(\xi) + \bar{\overset{*}{W}}{}^2(k) \cdot \overset{*}{Z}{}^2(\xi) + \bar{\overset{*}{W}}{}^3(k) \cdot \overset{*}{Z}{}^3(\xi) \\ =&W^1(k) \cdot Z^1(\xi) + W^2(k) \cdot Z^2(\xi) + W^3(k) \cdot Z^3(\xi) \ne 0, \quad \forall \xi \in C, \ 1 \le k \le 5. \end{aligned}$$

Since $\overset{*}{Z}{}^1(\xi) = 0$ we get

$$\bar{\overset{*}{W}}{}^2(k) \cdot \overset{*}{Z}{}^2(\xi) + \bar{\overset{*}{W}}{}^3(k) \cdot \overset{*}{Z}{}^3(\xi) \neq 0 \quad \forall \xi \in C, \quad 1 \leq k \leq 5.$$

i.e., $f(M)$ omits the points $(-\bar{\overset{*}{W}}{}^3(k), \bar{\overset{*}{W}}{}^2(k)), 1 \leq k \leq 5$. From (33) it follows that

$$rank \begin{pmatrix} -\bar{W^*}^3(i) & \bar{W^*}^2(i) \\ -\bar{W^*}^3(j) & \bar{W^*}^2(j) \\ -\bar{W^*}^3(k) & \bar{W^*}^2(k) \end{pmatrix} = 2 \quad \forall \{i,j,k\} \subset \{1,2,3,4,5\}.$$

By Lemma 2 there are at least three distinct points in $\{(-\bar{\overset{*}{W}}{}^3(k), \bar{\overset{*}{W}}{}^2(k)\}$. By Picard's theorem $f(M)$ is a point, i.e., $U(M)$ lies on a plane. Thus $x(M)$ is an improper affine sphere. Being affine complete, $x(M)$ must be an elliptic paraboloid.

§4. A theorem on harmonic functions

Theorem 3. Let $f(u,v), g(u,v), p(u,v)$ be harmonic functions defined over the whole R^2. If there is a constant $\varepsilon > 0$ such that

$$F = \begin{vmatrix} f & g & p \\ f_u & g_u & p_u \\ f_v & g_v & p_v \end{vmatrix} \geq \varepsilon > 0 \quad \forall (u,v) \in R^2,$$

then $f(u,v), g(u,v)$ and $p(u,v)$ are linear functions.

Proof. Let $U = f(u,v), g(u,v), p(u,v))$ and construct an affine maximal surface as (27). Then we have

$$R = H + J = J \geq 0.$$

The Blaschke metric is $F(du^2 + dv^2)$ and $-\frac{1}{2}\Delta \log F = R \geq 0$, i.e., $\log F$ is a superharmonic function. Since $F \geq \varepsilon > 0$ we have $F = const$. It follows that $R = J = 0$. Hence $x(M)$ is an elliptic paraboloid. After an affine transformation

$$\begin{pmatrix} x^{*1} \\ x^{*2} \\ x^{*3} \end{pmatrix} = \begin{pmatrix} b_{11} & b_{12} & b_{13} \\ b_{21} & b_{22} & b_{23} \\ b_{31} & b_{32} & b_{33} \end{pmatrix} \begin{pmatrix} x^1 \\ x^2 \\ x^3 \end{pmatrix} + \begin{pmatrix} C^1 \\ C^2 \\ C^3 \end{pmatrix}$$

the surface can be exprerssed by the following equation

$$\overset{*}{x}{}^3 = (\overset{*}{x}{}^1)^2 + (\overset{*}{x}{}^2)^2.$$

A direct calculation shows that the conormal vector field of the surface is

$$(-\sqrt{2}\overset{*}{x}{}^1, -\sqrt{2}\overset{*}{x}{}^2, 1)$$

and the Berwald-Blaschke metric is

$$\sqrt{2}((d\overset{*}{x}{}^1)^2 + (d\overset{*}{x}{}^2)^2).$$

That is to say, $\overset{*}{x}{}^1$ and $\overset{*}{x}{}^2$ are isothermal parameters. Hence the function $\varphi(\xi) = \overset{*}{x}{}^1(u,v) + \overset{*}{x}{}^2(u,v)i$ is a holomorphic or an anti-holomorphic function, i.e.,

$$\overset{*}{x}{}^1_u = \overset{*}{x}{}^2_v, \quad \overset{*}{x}{}^1_v = -\overset{*}{x}{}^2_u;$$

or

$$\overset{*}{x}{}^1_u = -\overset{*}{x}{}^2_v, \quad \overset{*}{x}{}^1_v = \overset{*}{x}{}^2_u.$$

Since

$$\begin{pmatrix} -\sqrt{-2}x^{*1} \\ -\sqrt{2}x^{*2} \\ 1 \end{pmatrix} = \begin{pmatrix} b_{11} & b_{12} & b_{13} \\ b_{21} & b_{22} & b_{23} \\ b_{31} & b_{32} & b_{33} \end{pmatrix} \begin{pmatrix} f \\ g \\ p \end{pmatrix}$$

and

$$\begin{vmatrix} f & g & p \\ f_u & g_u & p_u \\ f_v & g_v & p_v \end{vmatrix} = const.$$

we have

$$\begin{vmatrix} x^{*1}_u & x^{*2}_u \\ x^{*1}_v & x^{*2}_v \end{vmatrix} = const,$$

i.e.,

$$(\overset{*}{x}{}^1_u)^2 + (\overset{*}{x}{}^1_v)^2 = const.$$

By Liouville's theorem $\varphi'(\xi) = const$, i.e., $\varphi(\xi)$ is a linear function. Thus $f(u,v), g(u,v)$ and $p(u,v)$ are linear functions.

Remark. The condition $\begin{vmatrix} f & g & p \\ f_u & g_u & p_u \\ f_v & g_v & p_v \end{vmatrix} \geq \varepsilon > 0$ is necessary. For example, let $f = 1, g = e^u \cos v$ and $p = e^u \sin v$, then

$$\begin{vmatrix} f & g & p \\ f_u & g_u & p_u \\ f_v & g_v & p_v \end{vmatrix} = e^{2u} > 0.$$

References

1. Calabi, E., Improper Affine Hyperspheres of Convex Type and Generalization of a Theorem by K. Jorgens. Michigan Math. J. 5 105-126 (1958).

2. Chern, S.S., Affine Minimal Hypersurfaces. Proc. of the Japan- United States Seminar on Minimal Submanifold and Geodesics, 17-30 (1977).

3. Simon, U., Hypersurfaces in Equiaffine Differential Geometry. Geom. Dedicata. 17, 157-168 (1984).

4. Terng, Chuu-Lian, Affine Minimal Surfaces. Seminar on Minimal Submanifolds. Annals of Mathematics Studies 103, Princeton University Press, 1983.

5. Blaschke, W., Vorlesungen Uber Differential geometric II, Berlin 1923.

6. Osserman, R., A Survey of Minimal Surfaces, Van Nostrand, 1969.

7. Calabi, E., Hypersurfaces with maximal affinely invariant area. Amer. J. Math. 104, 91-126 (1982).

CODIMENSION 1 AND 2 IMMERSIONS OF LENS SPACES

Li Bang-He & Tang Zizhou

Abstract

The existence and classification problems of codimension 1 and 2 immersions of lens spaces in Euclidean spaces have been solved completely. Also, the ring structures of $\widetilde{KO}(L^n(p))$ for $n\leqq 3$ are determined.

Introduction

Let S^{2n+1} be the unit sphere in $(n+1)$-dimensional complex space $\mathbb{C}^{n+1}$ and $p>1$, an integer. Introduce an equivalence relation $\sim$ in S^{2n+1} as follows:

$$(Z_1,\cdots,Z_{n+1}) \sim (Z_1',\cdots,Z_{n+1}')$$

iff there exists an integer q such that $Z_k'=\exp(\frac{2\pi q\sqrt{-1}}{p})Z_k$, $k=1,2,\cdots,n+1$. Then we have the quotient space $L^n(p)=S^{2n+1}/\sim$ which is called an ordinary lens space. $L^n(p)$ is a $(2n+1)$-dimensional orientable smooth closed manifold and $L^n(2)=\mathbb{R}P^{2n+1}$, the $(2n+1)$-dimensional real projective space.

$L^n(p)$ immerses in $\mathbb{R}^{2n+2}$ iff $L^n(p)$ is a π-manifold. For $n=0$ or 1, $L^n(p)$ is always parallelizable; and if and only if $n=0,1$ or 3, $L^n(2)$ is a π-manifold (therefore, parallelizable) (cf[1]). Thus, the problem of existence of lens spaces is solved by

<u>Theorem 1</u>. Let $n\geqq 2$, $p\geqq 3$, then $L^n(p)$ is a π-manifold iff $p=n+1$ is a prime.

We conclude also that only $L^0(p)=S^1$, $L^3(2)=\mathbb{R}P^7$ and $L^1(p)$ (for any $p>1$) are parallelizable.

For the case of codimension 2, we have

<u>Theorem 2</u>. If $n\geqq 4$, then $L^n(p)$ immerses in $\mathbb{R}^{2n+3}$ iff $n+1=p$ or $n+2=p$ and p is a prime.

<u>Theorem 3</u>. $L^2(p)$ immerses in $\mathbb{R}^7$ iff $p=2^\alpha 3^\beta(3m_1+1)^{r_1}\cdots(3m_t+1)^{r_t}$ and $\alpha,\beta=0$ or 1; $L^3(p)$ immerses in $\mathbb{R}^9$ iff $p=2^\alpha(4m_1+1)^{r_1}\cdots(4m_t+1)^{r_t}$, $\alpha=0,1$, or 2, where $3m_i+1$ and $4m_j+1$ are the prime factor of p.

For determining codimension 2 immersions, we need the knowledge of $\widetilde{KO}(L^n(p))$ for $n=2,3$. So, in §1, we calculate the ring structures of $\widetilde{KO}(L^n(p))$ for $n=1,2,3$, which are presumably new (cf. [2],[3],[4]).

The classification of codimension 1 and 2 immersions is given in §5.

Usually, one uses K-theory to prove the non-existence of immersions. And here, we use K-theory to prove the existence of immersions. Another thing we find interesting in this paper is the use of number theory in both existence and classification problems.

<u>Remark</u>. The stable parallelizability for $L^n(p; a_0,a_1,\cdots,a_n)$ with p an odd prime and $L^3(p^m; a_0,a_1,a_2,a_3)$ with p a prime was determined respectively in [19] and [2] (p.214).

§1 The ring structure of $\widetilde{KO}(L^n(p))$ for $n \leqq 3$

Since $L^0(p)=S^1$, $\widetilde{KO}(L^0(p))=\mathbb{Z}_2$ is given. As is known to all,

$$H^i(L^n(p);\mathbb{Z}) = \begin{cases} \mathbb{Z}, & i=0,2n+1 \\ \mathbb{Z}_p, & i=2,4,\cdots,2n \\ 0, & \text{otherwise.} \end{cases}$$

$$H^i(L^n(p);\mathbb{Z}_2) = \begin{cases} \mathbb{Z}_2, & i=0,2n+1 \\ 0, & \text{otherwise} \end{cases}$$

if p is odd; and

$$H^i(L^n(p);\mathbb{Z}_2) = \mathbb{Z}_2, \quad 0 \leqq i \leqq 2n+1,$$

if p is even.

By the definitions of $L^n(p)$ and the complex projective space $\mathbb{C}P^n$, we see that there is a natural bundle projection $\pi: L^n(p) \to \mathbb{C}P^n$, with fiber S^1. Let η' be the canonical complex line bundle on $\mathbb{C}P^n$ and $\eta=\pi^*\eta'$, then we have $\sigma=\eta-1 \in \widetilde{K}(L^n(p))$, $r\sigma=r\eta-2 \in \widetilde{KO}(L^n(p))$, where r is the realification. Let $x=C_1(\eta)$ be the first Chern class of η, then x is a generator of $H^2(L^n(p),\mathbb{Z}) \cong \mathbb{Z}_p$ and x^i a generator of $H^{2i}(L^n(p),\mathbb{Z}) \cong \mathbb{Z}_p$. Denote by $\tau(L^n(p))$ the tangent bundle of $L^n(p)$, we have

<u>Lemma 1</u>. $\tau(L^n(p))\oplus 1=(n+1)r\eta$.

Since $\pi_1(BO) \cong \pi_2(BO)=\mathbb{Z}_2, \pi_3(BO)=0$, from obstruction theory, we see that

$$\# \widetilde{KO}(L^1(p)) \leqslant 1, \quad \text{if p odd,}$$
$$\# \widetilde{KO}(L^1(p)) \leqslant 4, \quad \text{if p even.}$$

In the case of $p=2\ell$ even, there isaa unique non-trivial real line bundle ρ since $H^1(L^n(p),\mathbb{Z}_2)=\mathbb{Z}_2$. By Lemma 3.1 in [3], the Euler class $\chi(2\rho)$ is ℓx. Since $\pi_1(BSO)=\pi_3(BSO)=0$, $\pi_2(BSO)=\mathbb{Z}_2$, there are just two elements in $\widetilde{KO}(L^1(2\ell))$ with the first Stiefel-Witney class w_1 vanishing, and distinguished by their second Stiefel-Whitney class $w_2 \in H^2(L^1(2\ell),\mathbb{Z}_2) \cong \mathbb{Z}_2$. The mod 2 homomorphism

$$H^2(L^1(2\ell);\mathbb{Z}) = \mathbb{Z}_{2\ell} \to H^2(L^1(2\ell),\mathbb{Z}_2)=\mathbb{Z}_2$$

is onto, so $w_2(r\sigma)\neq 0$. Let $k=\rho-1$, we then have

$$2k = \begin{cases} 0, & \text{if } \ell \text{ is even;} \\ r\sigma, & \text{if } \ell \text{ is odd.} \end{cases}$$

Thus, if ℓ is odd, k has order 4, and $\widetilde{KO}\ (L^1(2\ell)) \cong \mathbb{Z}_4$ generated by k. If ℓ is even, $\widetilde{KO}\ (L^1(2\ell))$ has at least 3 elements : $0, r\sigma$, and k, and $2r\sigma=2k=0$, hence $\widetilde{KO}\ (L^1(2\ell)) \cong \mathbb{Z}_2\oplus\mathbb{Z}_2$ generated by $r\sigma$ and k. Since $\rho^2=1$, we have

$$K^2 = \rho^2-2\rho+1 = -2k = \begin{cases} 0, & \text{if } \ell \text{ is even;} \\ r\sigma, & \text{if } \ell \text{ is odd.} \end{cases}$$

It is well-known that $C_1(\eta^2)=2C_1(\eta)=C_1(2\eta)$, hence

$$C_1((\eta-1)^2) = C_1(\eta^2-2\eta+1) = 0,$$

and $$w_2((r\sigma)^2) = w_2(r((\eta-1)^2)) = 0.$$

Therefore $(r\sigma)^2 = 0$. Now, in the case of ℓ even, we have

$$(r\sigma)k = (r\eta-2)(\rho-1) = (r\eta)\rho-r\eta-2k = (r\eta)\rho-r\eta.$$

By a formula in [6] (p.87), we see that

$$w_1(r\eta\otimes\rho) = w_1(r\eta)+2w_1(\rho) = 0,$$

$$w_2(r\eta\otimes\rho) = w_2(r\eta)+w_1(r\eta)w_1(\rho)+w_1(\rho)^2.$$

But $w_1(\rho)^2=w_2(2\rho)=0$, so $w_2(r\eta\otimes\rho)=w_2(r\eta)$ and $w_1((r\sigma)k)=w_2((r\sigma)k)=0$. This shows that $(r\sigma)k=0$. In this way, we have proved the following theorem.

Theorem 4. $\widetilde{KO}(L^1(2m+1))=0$; $\widetilde{KO}(L^1(4m+2))\cong\mathbb{Z}_4$ is generated by k and $k^2=2k=r\sigma$; $\widetilde{KO}(L^1(4m)) \cong \mathbb{Z}_2\oplus\mathbb{Z}_2$ is generated by $r\sigma$ and k with $(r\sigma)^2=k^2=(r\sigma)k=0$.

To determine $\widetilde{KO}(L^2(p))$ and $\widetilde{KO}(L^3(p))$, we notice first that $\Pi_{8m+5}(BO)=$ $=\Pi_{8m+6}(BO)=\Pi_{8m+7}(BO)=0$, and $L^n(p)$ has a standard CW-decomposition $L^n(p)=e^0\cup e^1\cup\cdots$ $\cdots\cup e^{2n+1}$ with each nonnegative $i\leq 2n+1$ an i-dimensional cell e^i such that $L^m(p)$ is the $(2m+1)$-skeleton of $L^n(p)$ for $m<n$. Thus, by obstruction theory, we have immediately the following

Lemma 2. J^*: $\widetilde{KO}(L^{4m+3}(p))\to\widetilde{KO}(L^{4m+2}(p))$ is an isomorphism, where j: $L^{4m+2}(p)\subsetneq L^{4m+3}(p)$ is the natural inclusion.

Therefore, we need only to calculate $\widetilde{KO}(L^3(p))$. By Mahammed [7], we have

Lemma 3. There is a ring isomorphism:

$$K(L^n(p)) \cong \mathbb{Z}[\eta]/\langle(\eta-1)^{n+1},\eta^p-1\rangle.$$

And by [3], we have

Lemma 4.

$$\#\,\tilde{K}(L^n(p)) = p^n.$$

Since $(\eta-1)^{n+1}=\sigma^{n+1}$, $\eta^p-1=\sum_{i=1}^{p}\binom{p}{i}\sigma^i$, it follows from Lemma 3 and 4 that

Lemma 5.
The group structure of $\tilde{K}(L^n(p))$ is given by $\mathbb{Z}^n=\{\sum_{i=1}^{n}a_i\sigma^i/a_i\in\mathbb{Z}^n\}$ modulo the subgroup generated by

$$\sum_{i=1}^{n-j}\binom{p}{i}\sigma^{i+j},\quad j=0,1,\cdots,n-1. \text{ (Notice: } \tbinom{s}{t}=0 \text{ if } t>s.)$$

Let c: $\widetilde{KO}(L^3(p))\to\tilde{K}(L^3(p))$ be the complexification which is a ring hommorphism. Since the conjugate complex bundle $\bar{\eta}=\eta^{p-1}$, we have $c(r\sigma)=\sigma+\bar{\sigma}=\sigma+\eta^{p-1}-1=$ $=\sigma+\sum_{i=1}^{p-1}\binom{p-1}{i}\sigma^i=p\sigma+\binom{p-1}{2}\sigma^2+\binom{p-1}{3}\sigma^3$. Thus, by $p\sigma+\binom{p}{2}\sigma^2+\binom{p}{3}\sigma^3=0$, we obtain

$$c(r\sigma) = [\tbinom{p-1}{2}-\tbinom{p}{2}]\sigma^2 + [\tbinom{p-1}{3}-\tbinom{p}{3}]\sigma^3.$$

Assume first $p>3$, then $c(r\sigma)=\sigma^2-\sigma^3-p\sigma^2-\binom{p}{2}\sigma^3+p\sigma^3$; but from Lemma 5, $p\sigma^2+\binom{p}{2}\sigma^3=0$, and $p\sigma^3=0$, so $c(r\sigma)=\sigma^2-\sigma^3$. An easy calculation shows that this is also true for $p=2,3$. Hence, we have

Lemma 6. $c(r\sigma)=\sigma^2-\sigma^3$ in $\tilde{K}(L^3(p))$

From Lemma 5, it is seen that σ^3 has order p. Now $pc(r\sigma)=p\sigma^2=\frac{1-p}{2}p\sigma^3$, so

$$pc(r\sigma) = 0, \quad \text{if } p \text{ is odd;}$$
$$2pc(r\sigma) = 0, \text{ if } p \text{ is even.}$$

By Lemma 5, it is easily seen that if

$mc(r\sigma)=m\sigma^2-m\sigma^3=0$, then m must be a multiple of p.

Hence, the order of $c(r\sigma)$ is p in the case of p odd. And in the case of p even,

$$pc(r\sigma)=(1-p)\frac{p}{2}\sigma^3 \neq 0$$

so the order of $c(r\sigma)$ in $\tilde{K}(L^3(p))$ is 2p. Thus we have proved

Lemma 6. The order of $c(r\sigma)$ in $\tilde{K}(L^3(p))$ is p if p is odd, and it is 2p if p is even.

Now, we are going to calculate $c(r(\eta^k-1))$ for $k=1,2,\cdots,p-1$. Since $r\eta^k=r\eta^{p-k}$, we need only to work with $k=1,2,\cdots,[\frac{p}{2}]$.

$$c(r(\eta^k-1)) = \eta^k-1+\eta^{p-k}-1 = \sum_{i=1}^{3}[\tbinom{k}{i}+\tbinom{p-k}{i}]\sigma^i$$
$$= p\sigma+\sum_{i=2}^{3}[\tbinom{k}{i}+\tbinom{p-k}{i}]\sigma^i.$$

Thus, by $p\sigma+\binom{p}{2}\sigma^2+\binom{p}{3}\sigma^3=0$, we have

$$c(r(\eta^k-1)) = \sum_{i=2}^{3}[\tbinom{k}{i}+\tbinom{p-k}{i}-\tbinom{p}{i}]\sigma^i .$$

If $k\geqq 2$ and $p-k\geqq 2$, then we have

$$\frac{1}{2}[k(k-1)+(p-k)(p-k-1)-p(p-1)] = k^2-pk.$$

Hence, in the case of k and $p-k\geqq 3$, we have by $p\sigma^2+\binom{p}{2}\sigma^3=0$,

$$c(r(\eta^k-1)) = k^2\sigma^2 + \frac{1}{6}[k(k-1)(k-2)+(p-k)(p-k-1)(p-k-2)$$
$$-p(p-1)(p-2)+3kp(p-1)]\sigma^3$$
$$= k^2(\sigma^2-\sigma^3)+[(k+1)p+k(k+1)/2]p\sigma^3 = k^2(\sigma^2-\sigma^3).$$

If $k\geqq 3$, then $[p/2]\geqq k$ implies $p\geqq 2k$, hence $p-k\geqq k\geqq 3$. Thus the only case left is $k=2$ and $p\geqq 4$. If $k=2$ and $p=4$, then we have

$$c(r(\eta^2-1)) = (k^2-pk)\sigma^2-\tbinom{p}{3}\sigma^3 = k^2\sigma^2-pk\sigma^2-p\sigma^3$$
$$= k^2\sigma^2+[kp(p-1)/2]\sigma^3 = k^2\sigma^2 = k^2(\sigma^2-\sigma^3), \text{ since } 4\sigma^3=0.$$

If $k=2$ and $p\geqq 5$, we have

$$c(r(\eta^2-1)) = (k^2-pk)\sigma^2+[\tbinom{p-k}{3}-\tbinom{p}{3}]\sigma^3$$
$$= 4\sigma^2-2p\sigma^2+ \frac{1}{6}[(p-2)(p-3)(p-4)-p(p-1)(p-2)]\sigma^3$$
$$= 4\sigma^2-2p\sigma^2-(p-2)^2\sigma^3 = 4\sigma^2-4\sigma^3.$$

So far, we have proved

Lemma 7. For any $k \geq 0$, in $\tilde{K}(L^3(p))$, we have

$$c(r(\eta^k-1) = k^2 c(r\sigma).$$

Since $\Pi_4(BSO) \cong \mathbb{Z}$ and $\Pi_5(BSO)=\Pi_6(BSO)=\Pi_7(BSO)=0$, from obstruction theory, it is seen that the subgroup $\widetilde{KSO}(L^3(p))$ of $\widetilde{KO}(L^3(p))$ consisting of elements with $w_1=0$ has order at most p if p is odd or 2p if p is even. Thus by Lemma 6,

$$\widetilde{KSO}(L^3(p)) \cong \begin{cases} \mathbb{Z}_p & \text{if p is odd,} \\ \mathbb{Z}_{2p} & \text{if p is even} \end{cases}$$

generated by $r\sigma$.

In the case of p odd, $\widetilde{KSO}(L^3(p))=\widetilde{KO}(L^3(p))$, since $H^1(L^3(p);\mathbb{Z}_2)=0$. And in the case of p even, $\widetilde{KSO}(L^3(p))$ has index 2 in $\widetilde{KO}(L^3(p))$. $k \in \widetilde{KSO}(L^3(p))$, and by Lemma 3.1 in [3], we see that $2k=r(\eta^\ell-1)$, where $p=2\ell$. Thus, by Lemma 7, and the fact that

$c\colon \widetilde{KSO}(L^3(p)) \to \tilde{K}(L^3(p))$ is a monomorphism, we see that $2k=\ell^2(r\sigma)$.

If $\ell=4m$, then $\ell^2=2mp$, so $2k=0$; if $\ell=4m+2$, then $\ell^2(r\sigma)=2\ell(r\sigma)$. Thus $\widetilde{KO}(L^3(2\ell))\cong$ $\cong\mathbb{Z}_{41} \oplus \mathbb{Z}_2$ generated by $r\sigma$ and k if $\ell=4m$; by $r\sigma$ and $k+\ell(r\sigma)$ if $\ell=4m+2$. In the case of ℓ odd, we have

$$2\left(k-\frac{\ell^2-1}{2}\,r\sigma\right) = r\sigma,$$

so $\widetilde{KO}(L^3(2\ell) \cong \mathbb{Z}_{81}$ generated by $k-\frac{\ell^2-1}{2}\,r\sigma$.

Now, let us take a look at the multiplication in $\widetilde{KO}(L^3(p))$. Since $c(r\sigma)^2 =$ $=(c(r\sigma))^2=(\sigma^2-\sigma^3)^2=0$, we have $(r\sigma)^2=0$. In the case of $p=2\ell$,

$$\begin{aligned} c((r\sigma)k) &= c(k)c(r\sigma) = (\eta^\ell-1)(\sigma^2-\sigma^3) = (\ell\sigma+\text{terms including } \sigma^2 \text{ or } \sigma^3)(\sigma^2-\sigma^3) \\ &= \ell\sigma^3+2\ell\sigma^2+\ell(2\ell-1)\sigma^3/2 = 2\ell\sigma^2 = 2\ell(\sigma^2-\sigma^3), \end{aligned}$$

so $(r\sigma)k=p(r\sigma)$. Obviously, $k^2=-2k$. In this way we have proved

Theorem 5. $\widetilde{KO}(L^3(p)) \cong \widetilde{KO}(L^2(p))$, $\widetilde{KO}(L^3(2m+1)) \cong \mathbb{Z}_{2m+1}$ is generated by $r\sigma$; $\widetilde{KO}(L^3(8m)) \cong \mathbb{Z}_{16m} \oplus \mathbb{Z}_2$ is generated by $r\sigma$ and k; $\widetilde{KO}(L^3(8m+4)) \cong \mathbb{Z}_{16m+8} \oplus \mathbb{Z}_2$ is generated by $r\sigma$ and $k+(4m+2)r\sigma$; $\widetilde{KO}(L^3(4m+2)) \cong \mathbb{Z}_{16m+8}$ is generated by $k-2m(m+1)r\sigma$. The multiplication is given by $(r\sigma)^2=0$, $k^2=-2k$, $(r\sigma)k=p(r\sigma)$.

Examples. $\widetilde{KO}(\mathbb{R}p^5)=\widetilde{KO}(\mathbb{R}p^7) \cong \mathbb{Z}_8$, $\widetilde{KO}(\mathbb{R}p^3) \cong \mathbb{Z}_4$, $\widetilde{KO}(L^3(6)) \cong \mathbb{Z}_{24}$.

2. The Proof of Theorem 1

Since $L^n(2)$ immerses in $\mathbb{R}^{2n+1}$ iff $n=0,1,3$ and when p is even, $L^n(2)$ is a covering space of $L^n(p)$, we have immediately the following lemma.

Lemma 8. If p is even, and $n\neq 0,1$ or 3, then $L^n(p)$ is not a π-manifold.

From Sjerve [8], we have

Lemma 9. Let m be odd, and any prime factor of it is greater than $n-2s+2$, then $L^n(m)$ immerses in $\mathbb{R}^{2n+2s}$ iff $\binom{n+i}{i}\equiv 0 \bmod m$, for any $s\leq i\leq[\frac{n}{2}]$.

It is easy to see that if q is an odd prime, then

(1) $\binom{q-1+i}{i}\equiv 0 \bmod q$, for $1\leq i\leq\frac{q-1}{2}$;

(2) $\binom{q-1+i}{i}\equiv q\not\equiv 0 \bmod q^u$, for any integer $u>1$.

Setting s=1 in Lemma 9, we have

Lemma 10. Let q be an odd prime and u>1 an integer, then

(1) $L^{q-1}(q)$ immerses in $\mathbb{R}^{2q}$;

(2) $L^{q-1}(q^u)$ does not immerse in $\mathbb{R}^{2q}$.

From [9], we have

Lemma 11. Let n>2 and q be odd prime, then if $L^n(q)$ immerses in $\mathbb{R}^{2n+2}$, we must have $n+1\equiv 0 \bmod q^{1+[n-2/q-1]}$.

Now, we prove the following

Lemma 12. Let q be an odd prime and $n\neq q-1$, $n\geq 2$, then $L^n(q)$ does not immerse in $\mathbb{R}^{2n+2}$.

Proof. Suppose $L^n(q)$ immerses in $\mathbb{R}^{2n+2}$, then by Lemma 11- we would have

$$n+1\equiv 0 \bmod q^{1+[n-2/q-1]}, \quad (*)$$

(*) together with the assumptions would imply $n+1\geq 2q>q+1$, hence $n-2>q-1$. Therefore, there would exist $k\geq 1$ and $0\leq r<p-1$ such that $n-2=k(q-1)+r$. Thus, we would get

$$q^{1+[n-2/p-1]} = q^{1+k} = (1+(q-1))^{1+k} \geq 1+(1+k)(q-1)+(q-1)^{1+k}$$

$$= q+k(q-1)+(q-1)^{1+k} > 3+k(q-1)+r = n+1,$$

a contradiction to (*). The Lemma is proved.

Lemma 13. Let $n\geq 2$ and p be odd, then if $L^n(p)$ is a π-manifold we must have $(n-p)=(q-1,q)$ with q a prime.

Proof. Let $p=q_1^{r_1}\cdots q_s^{r_s}$ be the standard decomposition of p, then if $L^n(p)$ is a π-manifold, so is $L^n(q_i)$ for $i=1,\cdots,s$. Thus, by Lemma 12, $n+1=q_i$, $i=1,\cdots,s$, and $p=q_1^k$, where $k=r_1+r_2+\cdots+r_s$. By Lemma 10, we conclude that k=1, and $(n,p)=(q_1-1,q_1)$, thus proving the lemma.

Lemma 14. Let p be even, and $p\neq 2$, then $L^3(p)$ is not a π-manifold.

Proof. If $L^3(p)$ is a π-manifold, then $\tau(L^3(p))\oplus 1$ is trivial. It follows from Lemma 1 that $4r\sigma=0$. From Lemma 6 it is seen that the order of $c(r\sigma)$ is 2p, hence $4r\sigma\neq 0$ if p>2. The Lemma is proved.

From Lemmas 8,10,13 and 14 comes Theorem 1 .

Corollary 1. Among all lens spaces, only S^1, $\mathbb{R}p^7$ and $L^1(p)$ are parallelizable.

Proof. By theorem 1, if $n\geq 2$, $p\geq 3$, then only $L^{q-1}(q)$ with q prime is a π-manifold. Since q is odd, we have

$$H^*(L^{q-1}(q),\mathbb{Z}_2) = \begin{cases} \mathbb{Z}_2, & \text{if } *=0,\ 2q-1; \\ 0, & \text{otherwise.} \end{cases}$$

Hence the Kervaire semi-characteristic number $\chi^*(L^{q-1}(q))=1$, and $L^{q-1}(q)$ is not parallelizable (cf. [10], §3). Thus, the corollary follows from the knowledge on $L^n(2)$ and the fact that any orientable 3-manifold is parallelizable.

§3 The proof of Theorem 2

From the fact that $\mathbb{R}p^n$ immerses in $\mathbb{R}^{n+2}$ iff n=1,2,3,5,6 or 7, we have immediately the following

Lemma 15. If $n\geq 4$ and p is even, then $L^n(p)$ does not immerse in $\mathbb{R}^{2n+3}$

From Mahammed [9], we have

Lemma 16. Let p be an odd prime, and $A(n,p,m)=\{i \mid 1\leq i\leq [n/2],\ \binom{n+i}{i}\not\equiv 0 \bmod p^{m+[n-2i/p-1]}\}\neq\phi$, then $L^n(p^m)$ does not immerse in $\mathbb{R}^{2n+2\ell(n,p,m)}$ where $\ell(n,p,m)= =\sup A(n,p,m)$.

Calculations by means of Lemma 16 yield

Lemma 17. The following lens spaces have no codimension 2 immersions in Euclidean spaces:

(1). $L^n(p)$ with $n\geq 4$, $n\neq p-1$ and $p-2$, and $p\geq 5$ an odd prime;

(2). $L^{p-1}(p\ )$ with m>1 and $p\geq 5$ an odd prime;

(3). $L^{p-2}(p^m)$ with m>1 and $p\geq 7$ an odd prime;

(4). $L^n(3)$ with n=6 or $n\geq 8$;

(5). $L^n(3^m)$ with m>1 and $n\geq 4$.

Now let us prove the following

Lemma 18. $L^4(3)$, $L^5(3)$ and $L^7(3)$ have no codimension 2 immersions in Euclidean spaces.

Proof. It is seen from [5] that $\widetilde{KO}(L^4(3))\cong\widetilde{KO}(L^5(3))\cong\mathbb{Z}_9$ and $\widetilde{KO}(L^7(3))\cong\mathbb{Z}_{27}$ are generated by $r\sigma$. The only elements in $\widetilde{KO}(L^n(3))$ having the form $\xi-2$ with ξ 2-plane bundles are 0 and $r\sigma$. It is obviously that $(n+1)r\sigma$ and $(n+2)r\sigma$ do not vanish in $\widetilde{KO}(L^n(3))$ for n=4,5 or 7. Thus, by Lemma 1, it is seen that, for any 2-plane bundle ξ, $\xi\oplus\tau(L^n(3))$ is not stably trivial. The lemma is proved.

Lemma 19. Let $q\geq 5$ be a prime, then $L^{q-1}(q)$ and $L^{q-2}(q)$ have codimension 2 immersions in Euclidean spaces.

Proof. The conclusion for $L^{q-1}(q)$ is obviously true by Theorem 1. Let ξ be the normal 2-plane bundle of $L^{q-2}(q)$ in $L^{q-1}(q)$, then

$$\tau(L^{q-2}(q))\oplus\xi\oplus 1 = \tau(L^{q-1}(q)\big|_{L^{q-2}(q)} \oplus 1$$

is trivial since $L^{q-1}(q)$ is a π-manifold by Theorem 1. Thus the use of Hirsch theory [11] completes the proof.

From Lemmas 17 and 18, and the fact that if p is odd, then $L^n(p)$ is covered by some $L^n(q^m)$ with q an odd prime and $m\geq 1$, we obtain

Lemma 20. Let p be odd and $n\geq 4$. If (n,p) is not the form of (q-1,q) or (q-2,q) with q a prime, then $L^n(p)$ has no codimension 2 immersions.

Now, Theorem 2 follows from lemmas 15,19 and 20.

§4. Solutions of two equations of number theory

To prove theorem 3, we have to know if there exists an orientable 2-plane bundle ξ over $L^n(p)$ such that $\xi\oplus\tau(L^n(p))$ is trivial, or equivalently $(\xi-2)+(n+1)r\sigma=0$ in $\widetilde{KO}(L^n(p))$ by Lemma 1. From Theorem 5 and lemma 7, the above is equivalent to the solvability of the equations

$$x^2+n+1 \equiv 0 \mod p \quad \text{if } p \text{ odd},$$
$$x^2+n+1 \equiv 0 \mod 2p \quad \text{if } p \text{ even},$$

for n=2 or 3. To classify codimension 2 immersions, we need also to know the number of the solutions. The reader can find the method for solving these equations in many text books on number theory.

Lemma 21. Let p be odd, then $x^2+3 \equiv 0 \mod p$ is solvable iff the standard decomposition of p has the form

$$p = 3^{\alpha}(3m_1+1)^{r_1}\cdots(3m_t+1)^{r_t}, \quad \alpha=0 \text{ or } 1.$$

Proof. It is easy to check that $x^2 \equiv -3 \mod 9$ has no solutions. Let $p\geq 5$ be a prime, then p=3m+1 or 3m+2. Now, we calculate the Legendre's symbol $(\frac{-3}{p})=(-1)^{\frac{p-1}{2}}(\frac{3}{p})$. Since both p and 3 are odd primes, we have $(\frac{3}{p})=(-1)^{\frac{p-1}{2}\cdot\frac{3-1}{2}}(\frac{p}{3})$. Thus

$$(\frac{-3}{p}) = (-1)^{p-1}(\frac{p}{3}) = \begin{cases} (-1)^{p-1}(\frac{1}{3})=(-1)^{p-1}=1, & \text{if } p=3m+1, \\ (-1)^{p-1}(\frac{-1}{3})=(-1)^{p}=-1, & \text{if } p=3m+2, \end{cases}$$

i.e., $x^2 \equiv -3 \mod p$ has solutions iff p=3m+1. Obviously, $x^2 \equiv -3 \mod 3$ has solutions, too. The proof is completed.

Lemma 22. Let p be odd, then $x^2 \equiv -4 \mod p$ has solutions iff p has the standard decomposition

$$p = (4m_1+1)^{r_1}\cdots(4m_t+1)^{r_t}.$$

Proof. Assume p to be an odd prime. Then the Legendre's symbol

$$(\frac{-4}{p})=(-1)^{\frac{p-1}{2}}(\frac{4}{p})=(-1)^{\frac{p-1}{2}}(\frac{1}{p})=(-1)^{\frac{p-1}{2}} = \begin{cases} 1, & \text{if } p=4m+1, \\ -1, & \text{if } p=4m+3, \end{cases}$$

This proves the lemma.

Lemma 23. Let p be even, then $x^2 \equiv -3 \mod 2p$ has solutions iff p has the standard decomposition

$$p = 2\cdot 3^{\alpha}(3m_1+1)^{r_1}\cdots(3m_t+1)^{r_t}, \quad \alpha=0 \text{ or } 1.$$

Proof. Suppose $p=2\beta\cdot q$ with q odd, and $\beta\geq 1$. It is easy to see that $x^2 \equiv -3 \mod 8$ has solutions, hence $x^2 \equiv -3 \mod 2^{\beta+1}$ has solutions iff $\beta=1$. This, together with lemma 21, completes the proof.

Lemma 24. Let p be even, then $x^2 \equiv -4 \mod 2p$ has solutions iff p has the standard decomposition

$$p = 2^{\alpha}\cdot(4m_1+1)^{r_1}\cdots(4m_t+1)^{r_t}, \quad \alpha=1,2.$$

Proof. Notice that $x^2 \equiv -4 \mod 2^2$ or 2^3 has solutions while $x^2 \equiv -4 \mod 2^4$ has not, thus comes the lemma.

Lemmas 21-24 together prove Theorem 3.

Now, we shall take a look at the number of the solutions. Let p be an odd prime, which is not a divisor of a, then it was proved in number theory that if $x^2 \equiv a \bmod p^m$ with $m \geq 1$ has solutions, the number of which is 2. It is easy to check that the numbers of the solutions of following equations

$$x^2 \equiv -3 \bmod 3, \quad x^2 \equiv -3 \bmod 4, \quad x^2 \equiv -4 \bmod 4, \quad x^2 \equiv -4 \bmod 8$$

are 1,2, 2 and 2 respectively. Thus, the number of the solutions of the following equations

$$x^2 \equiv -3^{\alpha} \bmod 3\,(3m_1+1)^{r_1} \cdots (3m_t+1)^{r_t}, \quad \alpha=0 \text{ or } 1,$$

$$x^2 \equiv -3 \bmod 2\cdot 2\cdot 3^{\alpha}(3m_1+1)^{r_1} \cdots (3m_t+1)^{r_t}, \quad =0 \text{ or } 1,$$

$$x^2 \equiv -4 \bmod (4m_1+1)^{r_1} \cdots (4m_t+1)^{r_t},$$

$$x^2 \equiv -4 \bmod 2\cdot 2^{\alpha}(4m_1+1)^{r_1} \cdots (4m_t+1)^{r_t}, \quad \alpha=1 \text{ or } 2$$

are $2^t, 2^{t+1}, 2^t$ and 2^{t+1} respectively, where $3m_i+1$ and $4m_j+1$ are primes. It is easily seen that neither of 0 and p is a solution to the following equations

$$x^2 \equiv -3 \bmod 2p, \quad p \text{ even},$$

$$x^2 \equiv -4 \bmod 2p, \quad p>2, \text{ even},$$

and if p=2, then both 0 and 2 are solutions of $x^2 \equiv -4 \bmod 4$. Thus, just half of the solutions of the equations

$$x^2 \equiv -3 \bmod 2p, \quad p \text{ even},$$

$$x^2 \equiv -4 \bmod 2p, \quad p \text{ even}$$

falls in to the interval [1,p]. So far, we have proved

Lemma 25. Let $p_2=2^{\alpha}\cdot 3^{\beta}(3m_1+1)^{r_1} \cdots (3m_t+1)^{r_t}$ with $\alpha,\beta=0$ or 1 and $p_3=2^{\alpha}(4m_1+1)^{r_1} \cdots (4m_t+1)^{r_t}$ with $\alpha=0,1$, or 2 be standard decompositions, then out of $r\eta^1, \cdots r\eta^{p_n}$, there are exactly 2^t that are the normal bundles of codimension 2 immersions of $L^n(p_n)$ in Euclidean spaces, where n=2 or 3.

§5. Classification of immersions

It is well known that the set of regular homotopy classes of immersions of an m-dimensional π-manifold M in $\mathbb{R}^{m+1}$, denoted by $I[M, \mathbb{R}^{m+1}]$, is in bijection with the set $[M, SO(m+1)]$ of homotopy classes of maps $M \to SO(m+1)$ (cf.[12] or [13]).

For classifying codimension 2 immersions, we may employ the method introduced in [14] by the first author. Since for lens space, $H^1(L^n(p); \mathbb{Z})=0$, we deduce from corollary 2 in [14] that the regular homotopy classes of codimension 2 immersions with normal bundles isomorphic to a fixed 2-plane bundle ν is in bijection with $[L^n(p), SO]$ if ν has orientation-reversing automorphisms and with $[L^n(p), 0]$, otherwise. And if ν has no oriention-reversing automorphisms, then by theorem 3 in [14], we have $2X(\nu)\neq 0$, where $X(\nu)$ is the Euler class of ν. $X(\nu)\neq -X(\nu)$ means that there are 2 classes of oriented 2-plane bundles under the orientation-preserving isomorphisms in the isomorphism class of ν. Therefore, we have

Lemma 26. If $L^n(p)$ has codimension 2 immersions, then $I[L^n(p), \mathbb{R}^{2n+3}]$ is in bijection with k copies of the set $[L^n(p), SO]$ and k is the cardinal of the set

$\{s/1 \leq s \leq p,\ r\eta^s \oplus \tau(L^n(p))$ is trival$\}$.

Theorem 6. Let $p \geq 3$ be a prime, then

$$I[L^{p-1}(p), \mathbb{R}^{2p}] \leftrightarrow \begin{cases} \mathbb{Z}, & \text{if } p \equiv 3 \bmod 4; \\ \mathbb{Z} \oplus \mathbb{Z}_2, & \text{if } p \equiv 1 \bmod 4. \end{cases}$$

Proof. We need only to calculate $[L^{p-1}(p), SO(2p)]$.
To do this, we use the spectral sequence given in Theorem (4.4.1) in [15, p.277].

It is well known that for the CW-decomposition

$$L^n(p) = e^0 \cup e^1 \cdots \cup e^{2n+1}$$

we have $\partial e^{2i+1}=0$, and $\partial e^{2i+2}=pe^{2i+1}$. Let X_k be the k-skeleton of $L^{p-1}(p)$, $u \in [L^{p-1}(p), e^0; SO(2p), 1]$, where 1 is the unit of $SO(2p)$, then the spectral sequence is given by

$$E_2^{s,t} = H^s(X_{s+t}, e^0; \Pi_{s+t}(SO(2p))), \quad s+t \geq 2, s \geq 1,$$

$$E_r^{s,t} = 0, \text{ for } s<1 \text{ or } t<0,$$

and $d_r : E_r^{s,t} \to E_r^{s+r,t-1}$.

Since $\Pi_i(SO(2p))=\Pi_i(SO)$, $i<2p-1$, it is easy to see that $H^i(L^{p-1}(p), e^0; \Pi_i(SO(2p))=0$, $i<2p-1$. Thus, any two maps $(L^{p-1}(p), e^0) \to (SO(2p), 1)$ are homotopic on the (2p-2)-skeleton of $(L^{p-1}(p), e^0)$. Therefore

$$[L^{p-1}(p), SO(2p)] \cong [L^{p-1}(p), e^0; SO(2p), 1]$$

$$\cong J_m(L^{p-1}(p), e^0; u) = E_{2p-1}^{2p-1,0}.$$

If $s+1<2p-1$ is odd, then $X_{s+1}=L^{s/2}(p)$ and $\Pi_{s+1}(S)(2p))=\Pi_{s+1}(S)$ is $\mathbb{Z}, \mathbb{Z}_2$ or 0 by Bott periodicity theorem, so $E_2^{s,1}=H^s(L^{s/2}(p), e^0; \Pi_{s+1}(SO(2p))$ is either $\mathbb{Z}_2$ or 0. It follows from $\partial e^{s+1}=pe^s$ that $H^s(X_{s+1}, e^0; \mathbb{Z}_2)=0$. Hence, $E_2^{s,1}=\mathbb{Z}_p$ or 0. Since

$$\Pi_{2p-1}(SO(2p)) = \begin{cases} \mathbb{Z}, & p \equiv 3 \bmod 4 \\ \mathbb{Z} \oplus \mathbb{Z}_2, & p \equiv 1 \bmod 4. \end{cases}$$

(cf.[16]), and d_r has bidegree $(r, -1)$, we have

$$\Pi_{2p-1}(SO(2p)) \cong E_0^{2p-1,0} \cong E_3^{2p-1,0} \cong \cdots \cong E_{2p-1}^{2p-1,0},$$

hence the theorem.

Remark. Except those considered in Theorem 6, the other lens spaces which are π-manifolds are $S^1, L^1(p)$ and $L^3(2)$. $I[L^3(2), \mathbb{R}^8]$ has been calculated in [13], and

$$I[L^1(p), \mathbb{R}^4] \leftrightarrow \begin{cases} \mathbb{Z} \oplus \mathbb{Z}, & p \text{ odd}, \\ \mathbb{Z}_2 \oplus \mathbb{Z} \oplus \mathbb{Z}, & p \text{ even} \end{cases}$$

follows from [17]. In this way, we have made a thorough classification of the codimension 1 immersions of lens spaces.

Theorem 7. Let $p\geq 5$ be a prime, then

$$I[L^{p-1}(p),\ \mathbb{R}^{2p+1}]\leftrightarrow\begin{cases}\mathbb{Z}_2, & \text{if } p\equiv 1 \bmod 4,\\ 0, & \text{if } p\equiv 3 \bmod 4.\end{cases}$$

$$I[L^{p-2}(p),\ \mathbb{R}^{2p-1}]\leftrightarrow \text{ 2 copies of } \mathbb{Z}.$$

Proof. The total Pontrjagin class of $L^n(p)$ is $P(L^n(p))=(1+x^2)^{n+1}$, where x is the generator of $H^2(L^n(p);\mathbb{Z})\cong\mathbb{Z}_p$. Thus, a simple calculation on Pontrjagin classes shows that among $r\eta, r\eta^2, \cdots r\eta^p$, only $r\eta^p$ is the nomoral bundles of $L^{p-1}(p)$ immersed in $\mathbb{R}^{2p+1}$, and only $r\eta$ and $r\eta^{p-1}$ are the normal bundles of $L^{p-2}(p)$ immersed in $\mathbb{R}^{2p-1}$.

Now, we use the spectral sequence of Baues to calculate $[L^n(p),SO]$. It is easy to see that $H^i(L^n(p),e^0;\ \pi_i(SO))=0$ if $i<2n+1$, since p is odd. Hence $[L^n(p),SO]\cong$ $\cong E_{2N+1}^{2n+1,0}$.

If $s\geq 2$ is even, then $X_{s+1}=L^{s/2}(p)$, and $E_2^{s,1}=H^s(X_{s+1},e^0;\pi_{s+1}(SO))=0$ or $\mathbb{Z}_p$.

For odd s, $\pi_{s+1}(SO)=0$ or $\mathbb{Z}_2$, so $E_2^{s,1}=H^s(X_{s+1},e^0;\pi_{s+1}(SO))=0$. Since $E_2^{2n+1,0}=$ $=H^{2n+1}((L^n(p),e^0;\pi_{2n+1}(SO))=\pi_{2n+1}(SO)$ is $0,\mathbb{Z}_2$ or $\mathbb{Z}$, the d_r concerning term $E_r^{2n+1,0}$ are trival. Therefore $[L^n(p),SO]\cong[L^n(p),e^0;SO,1]\cong E_{2n+1}^{2n+1,0}\cong\pi_{2n+1}(SO)$. Then the use of Lemma 26 and Bott periodicity theorem completes the proof.

Theorem 8. Let $p=2^{\alpha}3^{\beta}(3m_1+1)^{r_1}\cdots(3m_t+1)^{r_t}$, $\alpha,\beta=0$ or 1, be a standard dec decomposition of p, then $\#I[L^2(p),\mathbb{R}^7]=2^t$. Let $p=2^{\alpha}(4m_1+1)^{r_1}\cdots(4m_t+1)^{r_t}$, $\alpha=0,1,2$, be a standard decomposition of p, then $I[L^3(p),\ \mathbb{R}^9]\leftrightarrow 2^t$ copies of $\mathbb{Z}$.

Proof. Since $H^i(L^2(p);\pi_i(SO))=0$, for $i\leq 5$, it is obvious that

$$\#[L^2(p),SO]=1$$

For $L^3(p)$, we have

$$H^i(L^3(p);\pi_i(SO))\cong\begin{cases}\mathbb{Z}, & \text{if } i=7,\\ 0, & \text{ortherwise},\end{cases}$$

and that $E_2^{s,1}=H^s(X_{s+1},\pi_{s+1}(SO))$ is finite. Therefore

$$[L^3(p),\ SO]\cong E_7^{7,0}\cong E_2^{7,0}\cong\mathbb{Z},$$

and the use of lemma 25 and 26 completes the proof.

Remark. $[L^1(P),\ \mathbb{R}^5]\leftrightarrow p$ copies of $\mathbb{Z}$ can be calculated by [17] or [18], thus completing the classification of codimension 2 immersions of lens space.

References

[1] Li Bang-He, Codimension 1 and 2 immersions of Dold manifold in Euclidean space, Kexue Tongbao, 1987.

[2] Mahammed,N. Piccinini, R. and Suter, U., Some Applications of Topological K-Theory, North-Holland Publishing Company, 1980.

[3] Kobayashi, T. and Sugawara, M., $K_\wedge$-rings of Lens spaces $L^n(4)$, Hiroshima Math. J.

[4] Wu Zhende, KO-rings and J-groups of $L^n(8)$, Acta Math. Sinica 25(1982) p.49-60.

[5] Kambe, T., The structure of $K_\wedge$-rings of Lens spaces and their applications, Math. Soc. Japan. 1966, 135-146.

[6] Milnor J. W. and Stasheff J. D., Characteristic Classes, Ann. of Math. Studies No. 76, 1974.

[7] Mahammed, N., A propos de la K-theorie des espaces lenticulaires, C. R., Acad. Sci. Paris 271 (1970), 639-642.

[8] Denis Sjerve, Vector bundles over orbit manifolds, Trans. Amer. Math. Soc. 138 (1969), 97-106.

[9] Mahammed, N., K-theorie des espaces lenticulaires. C. R. Acad. Sci. Paris 272 (1971), p.1363-1365.

[10] Li Bang-He, Parallelizability of algebraic knots and canonical framings, Scientia Sinica, 27 (1984), 1164-1171.

[11] Hirsch, M., Immersions of manifolds, Trans. Amer. Math. Soc. 93 (1959), 242-276.

[12] James, I. and Thomas, E., Classifying of sections, Topology 4(1966), 351-359

[13] Li Bang-He, On immersions of m-manifolds in (m+1)-manifolds. Math. Z. 182 (1983), 311-320.

[14] Li Bang-He, On reflection of codimension 2 immersions in Euclidean spaces. Scientia Sinica (Chinese version), (1987), 793-799. (English version will be published later).

[15] Hans J. Baues, Obstruction Theory, Lecture Notes in Math. No. 628, Springer-Verlag, 1977.

[16] Kervaire, M., Some nonstable homotopy groups of Lie groups, Illinois J. of Math., 4(1960), 161-169.

[17] Wu Wen-Tsun, On the immersions of C^∞-3-manifolds in a Euclidean space, Scientia Sinica, 13(1964), 335-336.

[18] Li Bang-He, On classification of immersions of n-manifolds in (2n-1)-manifolds, Comment Math. Helv. 57(1982), 135-144.

[19] Ewing J., Moolgavkar S. and Smith L., Stable parallelizability of Lens spaces, J. Pure & Applied Algebra, 10(1977), 177-191.

Authors' address

Li Bang-He — Institute of Systems Science, Academia Sinica, Beijing,P.R.C.

Tang Zi-Zhou — Dept. of Math. Graduate School of Academia Sinica, Beijing,P.R.C.

On Third Order Nondegenerate Immersions and Maps of S^1 in R^2

Li Bang-He and Xu Tao

§1. Introduction

A C^∞ map $f: S^1 \to R^n$ is called a pth order nondegenerate map if the rank of the vectors $f'(t), f''(t), \cdots, f^{(p)}(t)$ is maximal for any $t \in S^1$. If f is also an immersion, then f is called a pth order nondegenerate immersion.

Two p^{th} order nondegenerate immersions (or maps) f and g are p^{th} order regularly homotopic iff there is a C^∞ map $F: S^1 \times I \to R^n$ such that

1) $F_0 = f, F_1 = g$;

2) F_t is a pth order nondegenerate immersion (or map) for any $t \in I$, where $F_t(x) = F(x,t)$.

By Feldman [1] and Gromov-Eliashberg [4], [5], the problem of classifying p^{th} order nondegenerate immersions and maps has been solved except for the cases of $p = n$ and $p = n+1$.

Up to now, only a few results have appeared in the left cases.

For $n = 3$ and $p = 3$, Little [8] proved that there are four classes of third order nondegenerate immersions.

For $n = 2$ and $p = 2$, Li Bang-he [6] proved that the second order nondegenerate immersions are classified by $Z - \{0\}$.

In this paper, we give two theorems to solve the problem for $n = 2$ and $p = 3$ completely.

To explain our results, we introduce first the notations $w(f)$ for a third order nondegenerate immersion $f: S^1 \to R^2 \cdot w(f)$ stands for the winding number of f, i.e., the degree of $f': S^1 \to R^2 - \{0\}$. An inflexion of f is a point of S^1 such that

$$\operatorname{rank}\{f'(t_0), f''(t_0)\} = 1;$$

i.e., the curvature of f regarded as a curve is zero at t_0. We will see that the number of the inflexion of third order nondegenerate immersion is finite and even. Thus $I(f)$ denotes the number of the inflexion pairs of f. $w(f)$ and $I(f)$ are invariant under regular homotopy of third order nondegenerate immersions. We then have

<u>Theorem 1</u>. $f \to (w(f), I(f))$ gives one-to-one correcpondence of the regular homotopy classes of third order nondegenerate immersions of S^1 in R^2 and the set $Z \times Z_+ - \{(0,0)\}$, where Z_+ stands for the nonnegative integers.

<u>Theorem 2</u>. There are exactly four regular homotopy classes of third order nondegenerate maps of S^1 in R^2. The two third order nondegenerate immersions f and g with $(w, I) \neq (\pm 1, 0)$ belong to the same classes iff

$$w(f) + I(f) \equiv w(g) + I(g) \mod 2.$$

Now we give a few examples here.

1. The algebraic curve $(x^2+y^2)^2 = x^2 - y^2$,

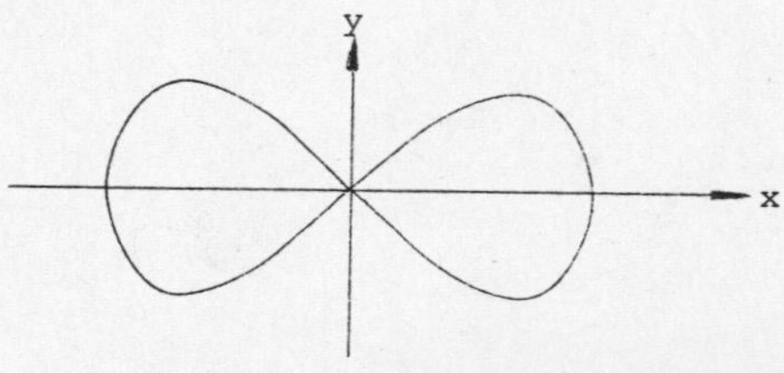

can be regarded as a third order nondegenerate immersion f with $(w(f), I(f)) = (0, 1)$.

2. $f(t) = (a + b\cos t)e^{it}$ with $a, b > 0$.

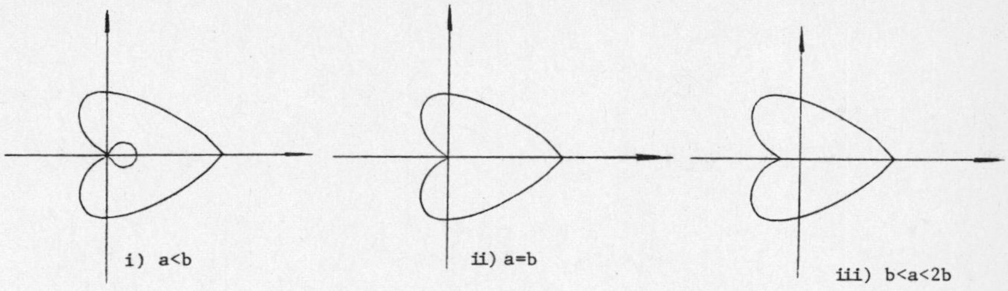

In cases (i) and (iii), f is a third order immersion with $(w(f), I(f)) = (2, 0)$ and $(1, 1)$ respectively. And in case (ii) f is not an immersion, it is a third order nondegenerate map.

Acknowledgment: The second auther would like to thank his supervisor Prof. Hu He-sheng for her encouragement.

§2. Preliminaries

At first, we need the following lemma to justify the definition of $I(f)$.

Lemma 1. For any third order nondegenerate immersion f: $S^1 \to R^2$, the number of its inflexions is finite and even.

Proof. By using the universal convering $R \to S^1$ given by $t \to e^{it}$, we may regard f as a C^∞ map $R \to R^2$ with period 2π. Then there are C^∞ real function r and θ such that

$$f'(t) = r(t)e^{i\theta(t)}, \quad r(t) > 0,$$

Thus

$$f''(t) = (r'(t) + r(t)\theta'(t)i)e^{i\theta(t)},$$

and

$$f'''(t) = [r''(t) - r(t)\theta'(t)^2 + (2r'(t)\theta'(t) + r(t)\theta''(t))i]e^{i\theta(t)}$$

imply that $\operatorname{rank}\{f'(t), f''(t)\} = 1$ iff

$$\theta'(t) = 0;$$

and if $\theta'(t) = 0$, then $\operatorname{rank}\{f'(t), f''(t), f'''(t)\} = 2$ iff

$$\theta''(t) \neq 0.$$

This shows that the inflexions of f correspond to the zeros of $\theta'(t)$ which are isolated. θ' being a c^∞ function of period 2π, the number of the zeros of θ' in $[0, 2\pi)$ is finite and even, thus completing the proof.

Let

$$P = \{p \in C^\infty(S^1, R^1) | p(s) > 0 \text{ for any } s \in S^1\},$$

and denote by $H(A)$ the set of interior points of the convex hull of a subset A of R^2. Then we have

Lemma 2. Let $h : S^1 \to S^1$ be a C^∞ map. Then there exists a C^∞ immersion $f : S^1 \to R^2$ with $f'/|f'| = h$ if and only if $0 \in H(h(S^1))$.

Proof. We shall follow the method used in Fenchel [3]. Regard h as a C^∞ map $e^{i\theta(s)}$ of R to S^1, and let

$$Q = \{q | q = \int_0^{2\pi} p(s)e^{i\theta(s)} ds, p \in P\},$$

then Q is a convex subset of R^2.

Let $m(x)$ be a C^∞ function such that

1) $0 < m(x) < 1$, if $|x| < 1$; $m(x) = 0$, otherwise;

2) $\int_{-\infty}^{\infty} m(x)dx = 1$.

For any $s_0 \in R$ and $\varepsilon \in (0, 1/2)$, let

$$m_{\varepsilon,s_0}(s) = \varepsilon + \delta m((s - s_0)/\varepsilon), \quad |s - s_0| \leq \pi$$

where $\delta > 0$ is uniquely determined by

$$\int_{s_0-\pi}^{s_0+\pi} m_{\varepsilon,s_0}(s)ds = 1,$$

thus $m_{\varepsilon,s_0}(s)$ extends to a C^∞ function on R with period 2π. Hence we may well regard $m_{\varepsilon,s_0}(s)$ as an element of P. Consequently,

$$q_{\varepsilon,s_0} = \int_0^{2\pi} m_{\varepsilon,s_0}(s)e^{i\theta(s)}ds \in Q.$$

Now

$$\begin{aligned}
|q_{\varepsilon,s_0} - e^{i\theta(s_0)}| &= \left|\int_{s_0-\pi}^{s_0+\pi} m_{\varepsilon,s_0}(s)(e^{i\theta(s)} - e^{i\theta(s_0)})ds\right| \\
&\leq \max_{|s-s_0|\leq\varepsilon} |e^{i\theta(s)} - e^{i\theta(s_0)}| \int_{|s-s_0|\leq\varepsilon} m_{\varepsilon,s_0}(s)ds \\
&\quad + 2\int_{\varepsilon\leq|s-s_0|\leq\pi} m_{\varepsilon,s_0}(s)ds \\
&\leq \max_{|s-s_0|\leq\varepsilon} |e^{i\theta(s)} - e^{i\theta(s_0)}| + 4(\pi - \varepsilon)\varepsilon \\
&\longrightarrow 0, \quad \text{as } \varepsilon \longrightarrow 0.
\end{aligned}$$

So $e^{i\theta(s_0)} \in \bar{Q}$, the closure of Q. Hence, $H(h(S^1)) \subset \bar{Q}$.

If $0 \in H(h(S^1))$, then 0 is an interior point of $\bar{Q}$, and the convexity of Q implies that $0 \in Q$, i.e., there is a $k \in P$ such that

$$0 = \int_0^{2\pi} k(s)e^{i\theta(s)}ds.$$

Let

$$f(t) = \int_0^t k(s)e^{i\theta(s)}ds,$$

then f is an immersion of S^1 in R^2 and

$$f'(t)/|f'(t)| = e^{i\theta(t)}.$$

Conversely, if $0 \notin H(h(S^1))$, then $h(S^1)$ lies in a half circle of S^1. Since $h(S^1)$ is connected,

$$\int_0^{2\pi} p(s)e^{i\theta(s)}ds \neq 0$$

for any $p \in P$. The proof is completed.

Lemma 3. Let f_0 and f_1 be two immersions of S^1 in R^2, and h_t a C^∞ homotopy of maps $S^1 \to S^1$ such that

$$h_0 = f_0'/|f_0'|, \quad h_1 = f_1'/|f_1'|$$

and

$$0 \in H(h_t(S^1)) \text{ for any } t \in [0,1].$$

Then there exists a C^∞ regular homotopy f_t of immersions S^1 in R^2 such that

$$h_t = f_t'/|f_t'| \text{ for any } t \in [0,1].$$

Proof. We can choose a C^∞ function $\theta_t(s)$ of $(t,s) \in [0,1] \times R$ to R so that h_t may be regarded as $e^{i\theta_t(s)}$. Then our problem becomes: We are to find a C^∞ homotopy p_t between $p_0 = f_0'/f_0'$ and $p_1 = f_1'/f_1'$ with $p_t \in P$ and $\int_0^{2\pi} p_t(s)e^{i\theta_t(s)}ds = 0$, for any $t \in [0,1]$.

Let $L^2[0,2\pi]$ be the Hilbert space of real square-integrable Lebesgue mesurable functions on $[0,2\pi]$ with inner product

$$\langle x,y\rangle = \int_0^{2\pi} x(s)y(s)ds.$$

It is easy to see that $\cos\theta_t(s)$ and $\sin\theta_t(s)$ span a 2-dimensional subspace A_t in $L^2[0,2\pi]$ for any $t \in [0,1]$. By using Gram-Schmidt orthogonalization, we have an orthonormal basis $(u_t(s), v_t(s))$ of A_t, and it is obvious that $u_t(s)$ and $v_t(s)$ extend to functions of s on R with period 2π so that they are C^∞ functions of $(t,s) \in [0,1] \times R$.

Since $0 \in H(h_t(S^1))$, we can choose $q_t \in P$ for any $t \in (0,1)$ so that

$$\int_0^{2\pi} q_t(s)e^{i\theta_t(s)}ds = 0.$$

We let $q_0 = f_0'/|f_0'|, q_1 = f_1'/|f_1'|$.

Let

$$q_{t,r}(s) = q_t(s) - \langle q_t, u_r\rangle u_r(s) - \langle q_t, v_r\rangle v_r(s).$$

Then

$$\langle q_{t,r}, u_r\rangle = \langle q_{t,r}, v_r\rangle = 0,$$

$q_{t,r}(s)$ can be viewed as a C^∞ function of $(r,s) \in [0,1] \times R$ with period 2π for s. Therefore, $q_{t,t} = q_t \in P$ implies that there exists a neighborhood U_t of t in $[0,1]$ such that $q_{t,r} \in P$ if $r \in U_t$. If $t \in (0,1)$, we assume $U_t \subset (0,1)$. Then there is a C^∞ partition of unit $\{\phi_i\}_{i=1}^m$ of $[0,1]$ subordinate to the open covering $\{U_t\}_{t\in[0,1]}$ such that the support of $\phi_i \subset U_{t_i}$ for some t_i.

Let

$$p_r(s) = \sum_{i=1}^m \phi_i(s) q_{t_i,r}(s),$$

then $p_r(s)$ is a C^∞ function of $(r,s) \in [0,1] \times R$ with period 2π for s. The convexity of P implies $p_r \in p$ for any r, and the assumption of $U_t \subset (0,1)$ for $t \in (0,1)$ ensures that

$$q_0 = p_0, \quad q_1 = p_1.$$

Therefore

$$f_r(t) = \int_0^t p_r(s)e^{i\theta_r(s)}ds$$

gives the regular homotopy of immersions connecting f_0 and f_1, thus completing the proof.

§3. The Proof of Theorem 1

Lemma 4. $w(f)$ and $I(f)$ are third order regular homotopy invariants.

Proof. We need only to prove the statement for $I(f)$. While this is already quite clear from the proof of Lemma 1, we leave the details to the reader.

Lemma 5. $(w(f), I(f)) \neq (0,0)$.

Proof. Let $f'(t) = r(t)e^{i\theta(t)}$. Then $w(f) = 0$ implies that $\theta(0) = \theta(2\pi)$. So there exists $t_0 \in (0, 2\pi)$ with $\theta'(t_0) = 0$. This shows that $I(f) \neq 0$, proving the lemma.

Lemma 6. For any $(n, k) \in Z \times Z_+ - \{(0,0)\}$, there exists an f with $(w(f), I(f)) = (n, k)$.

Proof. Case i), $n, k > 0$. Let

$$f(t) = (n+k)e^{\mathrm{ink}} - (n-k)e^{i(n+k)t}, \quad \varepsilon > 0,$$

then

$$\begin{aligned} f'(t) =& i(n+k)e^{int}(n - (n-\varepsilon)e^{int}), \\ f''(t) =& -(n+k)e^{int}(n^2 - (n-\varepsilon)n + k)e^{ikt}), \\ f'''(t) =& -i(n+k)e^{int}(n^3 - (n-\varepsilon)(n+k)^2 e^{ikt}). \end{aligned}$$

When ε is small enough, the map $e^{it} \to n - (n-\varepsilon)e^{ikt}$ is a one of S^1 to $R^2 - \{0\}$ with degree zero, and the map $e^{it} \to i(n+k)e^{int}$ degree n. So f is an immersion with $w(f) = n$.

Rank$\{f'(t), f''(t)\} = 1$ iff $f''(t)/f'(t)$ is a real number. This is equivalant to

$$i(n^2 - (n-\varepsilon)(n+k)e^{ikt})(n - (n-\varepsilon)e^{-ikt}) \text{ is real},$$

or

$$n^3 - (n-\varepsilon)^2(n+k) + (n^2(n-\varepsilon) - n(n-\varepsilon)(n+k)) \cos kt = 0.$$

It is easy to see that the last equation has exactly $2k$ solutions near but not equal to $t = 2m\pi/k, m = 0, 1, \cdots, k-1$, in $[0, 2\pi)$, as ε is small enough.

A simple calculation shows that $f'''(t)/f'(t)$ is real iff

$$\sin\ kt = 0.$$

If t is near but not equal to $2m\pi/k$, $\sin kt \neq 0$. So $f(t)$ is a third order nondegenerate immersion with $(w(f), I(f)) = (n, k)$ when ε is small.

Case ii), $n \neq 0, k = 0. f(t) = e^{int}$ will be the required immersion.

Case iii), $n < 0, k > 0$. Let $f(t)$ be an immersion with $(w(f), I(f)) = (-n, k)$, then its conjugacy $\bar{f}(t)$ with $(w(\bar{f}), I(\bar{f})) = (n, k)$.

Case iv), $n = 0, k > 0$. We have given an example in §1 with $(w(f), I(f)) = (0, 1)$. (Another construction of such an f will be given Lemma 12 §4). Let $g(t) = f(kt)$, then $(w(g), I(g)) = (0, k)$. The proof is completed.

Lemma 7. If $(w(f), I(f)) = (w(g), I(g))$, then f and g are third order regularly homotopic.

Proof. Assume that $w(f) = n \neq 0, I(f) = k$ and $f'(t) = r_0(t)e^{i\theta_0(t)}, g'(t) = r_1(t)e^{i\theta_1(t)}$ and let $0 \leq t_1 < t_2 < \cdots t_{2k} < 2\pi$ be the inflexion of f. Then there exists an orientation-preserving diffeomorphism h of S^1 such that $t_1, \cdots, t_{2k}$ are inflexions of $\tilde{g}(t) = g(h(t))$ with $\tilde{g}'(t) = \tilde{r}_1(t)e^{i\tilde{\theta}_1(t)}$. We can also require $\theta_0''(t_1)/\tilde{\theta}_1''(t_1) > 0$. It is easy to see that g and $\tilde{g}$ are third order regularly homotopic. So we may assume $g = \tilde{g}$.

Let

$$\theta_s(t) = s\theta_0(t) + (1-s)\theta_1(t), \quad s \in [0,1],$$

we have $\theta_s(2\pi) = \theta_s(0) + 2n\pi$. Hence

$$0 \in H\{e^{i\theta_s(t)}/t \in R\}, \quad \text{for any } s.$$

It is easy to see that, for $t \in [0, 2\pi), \theta_s'(t) = 0$ iff $t = t_j$ and $\theta_s''(t_j) \neq 0$ for $j = 1, \cdots, 2k$.

Now, assume $w(f) = 0, I(f) = k > 0$. Let $\tilde{f}(t) = h(kt)$ with $(w(h), I(h)) = (0, 1)$. By the same technique we may assume that f and $\tilde{f}$ have the same inflexions $t_1 t_2 \cdots t_{2k}$ in $[0, 2\pi)$ such that $\tilde{\theta}''(t_1)/\theta''(t_1) > 0$, where $f'(t) = r(t)e^{i\theta(t)}$ and $\tilde{f}'(t) = \tilde{r}(t)e^{i\tilde{\theta}(t)}$. We may assume that t_1 and t_j are respectively a minimum and a maximum of θ, then so are t_1 and t_j for $\tilde{\theta}$. Thus by Lemma 2,

$$\theta(t_j) - \theta(t_1) > \pi, \text{ and } \tilde{\theta}(t_j) - \tilde{\theta}(t_1) > \pi.$$

Let $\theta_s(t) = s\theta(t) + (1-s)\tilde{\theta}(t), s \in [0,1]$, then

$$\theta_s(t_j) - \theta_s(t_1) > \pi.$$

Now, by using Lemma 3, we can find a third order regular homotopy conncecting f and g or f and $\tilde{f}$, thus completing the proof.

Lemmas 4-7 together give the proof of Theorem 1.

§4. The Proof of Theorem 2.

The following lemma shows that there are at least two third order regular homotopy classes of maps.

Lemma 8. Let f_0 and f_1 be two third order nondegenerate immersions of S^1 in R^2. If they are regularly homotopic as third order nondegenerate maps, then $w(f_0)+I(f_0) \equiv w(f_1)+I(f_1)$ mod 2.

Proof. Step 1. We prove first that if f_t is a third order regular homotopy between f_0 and f_1, then f_t can be perturbed to a third order regular homotopy connecting f_0 and f_1 such that

$$f_t'(s) = 0$$

has only finite solutions for $(t,s) \in [0,1] \times [0,2\pi)$.

Let $\pi : [0,1] \times S^1 \to [0,1]$ be the projection, $T([0,1] \times S^1)$ the tangent bundle of $[0,1] \times S^1$, and E its subbundle consisting of the tangent vectors of the fibers of π. Denote by $\mathrm{Hom}(E, TR^2)$ the homomorphism bundle over $[0,1] \times S^1 \times R^2$, then its fibers are 2- dimensional and its zero-section K is 4-dimensional. Let $A = 0 \times S^1 \cup 1 \times S^1$, and

$$C_A = \{g_t \in C^\infty([0,1] \times S^1, R^2) : g_0 = f_0, g_1 = f_1\}.$$

Then A is closed in $[0,1] \times S^1$ and K cohesive in the sense of Feldman [1]. By using Theorem 5.2 in [1], we see that the subset of C_A, consisting of those g_t with g_t' transeversal to K, is dense in C_A. Since the property of being a third order regular homotopy is stable under small pertibations, we can find a $g_t \in C_A$ near f_t such that g_t' is trasnvesal to K and g_t is a third order regular homotopy. The dimension of $g_t'([0,1] \times S^1)$ being 2, the dimension of the intersection of $g_t'([0,1] \times S^1)$ and K is zero. This shows that $g_t'(s) = 0$ has only finite solutions for $(t,s) \times [0,1] \times S^1$.

Step 2. For simplicity, we assume there is only one point $(t_0, s_0) \in [0,1] \times S^1$ with $g_{t_0}'(s_0) = 0$. The general case can be similarly treated.

Since g_t' is transversal to K, the Jacobi determinant of the map $g_t'(s) : (t,s) \to R^2$ at (t_0, s_0) denoted by $|g_{t_0}''(s_0), \frac{\partial}{\partial t} g_{t_0}'(s_0)|$ does not vanish, so the degrees of g_0' and $g_1' : S^1 \to R^2 - \{0\}$ differ by ± 1.

Notice that we always use "'" to denote the derivative with repsect to s. Letting

$$h_t(s) = |g_t'(s), g_t''(s)|,$$

we have

$$h_{t_0}(s_0) = 0,\ h_{t_0}'(s_0) = |g_{t_0}'(s_0), g_{t_0}'''(s_0)| = 0,$$
$$h_{t_0}''(s_0) = |g_{t_0}''(s_0),\ g_{t_0}'''(s_0)| \neq 0.$$

Since

$$\frac{\partial}{\partial t}h_{t_0}(s_0) = |\frac{\partial}{\partial t}g'_{t_0}(s_0), g''_{t_0}(s_0)| \neq 0,$$

there exists $\xi, \eta > 0$ such that in the interval $[s_0 - \eta, s_0 + \eta]$ of s, either g_t with $t \in (t_0, t_0 + \xi]$ has 2 inflexions but g_t with $t \in (t_0 - \xi, t_0)$ does not, or g_t with $t \in (t_0, t_0 + \xi)$ has no inflexions but g_t with $t \in (t_0 - \xi, t_0)$ has two. Outside $[s_0 - \eta, s_0 + \eta]$, the number of inflexions is invariant for t. We have proved that $w(g_0) + I(g_0) \equiv w(g_1) + I(g_1) \bmod 2$, and hence the lemma.

Lemma 9. Let f be a third order nondegenerate map, then f is third order regularly homotopic to an immersion.

Proof. Suppose $f'(0) = 0$. Let $\alpha(t)$ be a C^∞ function on $[-\pi, \pi]$ with support in $[-1, 1]$ and $\alpha(0) \neq 0$. Let $0 < \varepsilon < 1$, and

$$F_u(t) = u + \alpha(\frac{t}{\varepsilon})T + f(t), \quad 0 \leq u \leq 1, \quad -\pi \leq t \leq \pi,$$

where $T = \varepsilon^3 f'''(0)$. Then

$$\begin{aligned} F''_u(t) =& 2\frac{u}{\varepsilon}\alpha'(\frac{t}{\varepsilon})T + \frac{ut}{\varepsilon^2}\alpha''(\frac{t}{\varepsilon})T + f''(t), \\ F'''_u(t) =& 3\frac{u}{\varepsilon^2}\alpha''(\frac{t}{\varepsilon})T + \frac{ut}{\varepsilon^3}\alpha'''(\frac{t}{\varepsilon})T + f'''(t). \end{aligned}$$

Since $|f''(0), f'''(0)| \neq 0$, we can choose $\varepsilon > 0$ so small that

$$|F''_u(t), F'''_u(t)| \neq 0, \quad \text{if} \quad |t| \leq \varepsilon.$$

If $\varepsilon \leq |t| \leq \pi$, however, then $F_u(t) = f(t)$, hence F_u is a third order regular homotopy. Letting

$$h(t) = |f'(t), f'''(0)|,$$

we have $h'(0) = |f''(0), f'''(0)| \neq 0$, while $h(0) = 0$. Thus we can require that ε also satisfy

$$|f'(t), f'''(0)| \neq 0, \quad \text{if} \quad 0 < |t| \leq \varepsilon.$$

Now

$$F'_1(t) = [\alpha(\frac{t}{\varepsilon}) + \frac{t}{\varepsilon}\alpha'(\frac{t}{\varepsilon})]\varepsilon^3 f'''(0) + f'(t).$$

So, if $0 < |t| \leq \varepsilon$, $F'_1(t) \neq 0$, and $F'_1(0) = \alpha(0)\varepsilon^3 f'''(0) \neq 0$. This shows that the zeros of F'_1 are less than those of f'. Since the number of the zeros of f' is finite, the proof is completed.

Below, we shall use $\underline{(n,k)}$ to denote the regular homotopy class of third order nondegenerate immersions with $(w, I) = (n, k)$ and $\underline{(n,k)} \sim \underline{(n',k')}$, the fact that they are regularly homotopic as third order nondegenerate maps.

Lemma 10. Let $n > 0$, $k > 0$. Then $\underline{(n,k)} \sim \underline{(n+k,0)}$, $\underline{(-n,k)} \sim \underline{(-n-k,0)}$.

Proof. Let

$$f_\varepsilon(t) = (n+k)e^{int} - (n-\varepsilon)e^{i(n+k)t}.$$

If $\varepsilon > 0$ is small enough, we have seen in the proof of Lemma 6 that $f_\varepsilon \in \underline{(n,k)}$. Similarly, if $\varepsilon < 0$ with $-\varepsilon$ small enough, we have $f_\varepsilon \in \underline{(n+k,0)}$. $f'_0(t) = 0$ iff $t = 2m\pi/k$, and it is obvious that $|f''(2m\pi/k), f'''(2m\pi/k)| \neq 0$. Therefore,

$$\underline{(n,k)} \sim \underline{(n+k,0)}.$$

By using the conjugacy $\bar{f}_\varepsilon$, we see that

$$\underline{(-n,k)} \sim \underline{(-n-k,0)}.$$

This proves the Lemma.

Lemma 11. If $2n > k > n > 0$, then $\underline{(n,0)} \sim \underline{(n-k,k)}$ and $\underline{(-n,0)} \sim \underline{(k-n,k)}$. If $k > 2n > 0$, then $\underline{(n,k)} \sim \underline{(n-k,0)}$ and $\underline{(-n,k)} \sim \underline{(k-n,0)}$.

Proof. Let

$$f_\varepsilon(t) = (k-n)e^{int} + (n-\varepsilon)e^{i(n-k)t},$$

then

$$\begin{aligned} f'_\varepsilon(t) =& i(k-n)e^{int}(n-(n-\varepsilon)e^{-ikt}), \\ f''_\varepsilon(t) =& -(k-n)e^{int}(n^2+(n-\varepsilon)(k-n)e^{-ikt}), \\ f'''_\varepsilon(t) =& -i(k-n)e^{int}(n^3-(n-\varepsilon)(k-n)^2e^{-ikt}). \end{aligned}$$

It is easily seen that $f'_0(t) = 0$ iff $t = 2m\pi/k, m \in Z$. If $|\varepsilon| \neq 0$ is small, then $f'_\varepsilon(t) \neq 0$, and

$$w(f_\varepsilon) = \begin{cases} n, & \text{if } \varepsilon > 0, \\ n-k, & \text{if } \varepsilon < 0. \end{cases}$$

Calculations show that when $|\varepsilon|$ is small, f_ε is a third order nondegenerate map. Now, if $\varepsilon \neq 0$, then

$f''_\varepsilon(t)/f'_\varepsilon(t)$ is real iff $n^3 - (n-\varepsilon)(k-n) = (n-\varepsilon)n(2n-k)\cos kt$. Let

$$h(\varepsilon) = \frac{n^3-(n-\varepsilon)^2(k-n)}{(n-\varepsilon)n(2n-k)},$$

then

$$h(0) = 1, \quad h'(0) \begin{cases} > 0, & \text{if } k < 2n, \\ < 0, & \text{if } k > 2n. \end{cases}$$

Hence

$$I(f_\varepsilon) = \begin{cases} 0, \text{if } \varepsilon > 0 \text{ and } k < 2n; k, \text{if } \varepsilon > 0 \text{ and } k > 2n, \\ k, \text{if } \varepsilon < 0 \text{ and } k < 2n; 0, \text{if } \varepsilon < 0 \text{ and } k > 2n. \end{cases}$$

The Lemma is proved.

Lemma 12. If f_1 is a map third order regularly homotopic to the map $f_0 : e^{it} \to e^{it}$ (or $f_0 : e^{it} \to e^{-it}$), then f_1 is second order regularly homotopic to f_0.

Proof. Let f_t be a third order regular homotopy. If $f'_t(s) = 0$ has no solutions for $(t,s) \in [0,1] \times S^1$, then by Lemma 4, f_t has no inflexion for any $t \in [0,1]$, consequently f_t provides a second order regular homotopy.

Now, suppose $f'_t(s) = 0$ has solutions among which t_0 is the smallest with $f'_{t_0}(s_0) = 0$. Since

$$\lim_{s \to s_0 \pm} f'_{t_0}(s)/f'_{t_0}(s) = \pm f''_{t_0}(s_0)/|f''_{t_0}(s_0)| \in S^1,$$

there exists $\eta > 0$ such that $f'_{t_0}(s) = 0$ has no solutions for $s \in [s_0 - \eta, s_0 + \eta]$ and

$$|f'_{t_0}(s_0 \pm \eta)/|f'_{t_0}(s \pm \eta)| \mp f''_{t_0}(s_0)/|f''_{t_0}(s_0)|| < 1.$$

Assume $f_0(e^{it}) = e^{it}$, then the picture of f_{t_0} in $[s_0 - \eta, s_0 + \eta]$ is as in Fig.1.

Let $t_1 < t_0$ be close to t_0. Then the picture of f_{t_1} near $s_0 - \eta$ and $s_0 + \eta$ is as in Fig.2. It is easily seen that we can construct a C^1 immersion $\tilde{f}_{t_1}$ such that $\tilde{f}_{t_1}(s) = f_{t_1}(s)$ if $\eta \leq |s - s_0| \leq \pi$, and $\tilde{f}'_{t_1}(s)$ rotates clockwise from $f'_{t_1}(s_0 - \eta)$ to $f'_{t_1}(s_0 + \eta)$, if $|s - s_0| \leq \eta$ (see Fig.2.). Since f_{t_1} has positive curvature, $f'_{t_1}(s)$ rotates anti- clockwise when s increases. So when s increases from $s_0 + \eta$ to $2\pi + s_0 - \eta$, the angle of f'_{t_1} increases either more than 2π or less than π. If it is less than π, then the image of $\tilde{f}'_{t_1}/|\tilde{f}'_{t_1}| : S^1 \to S^1$ is the smaller arc between

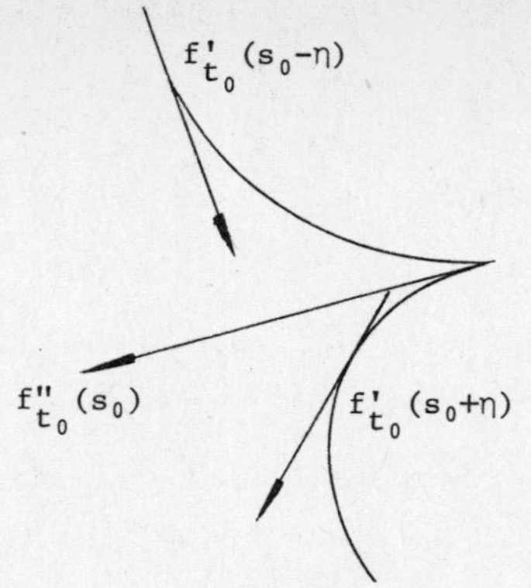

Fig.1

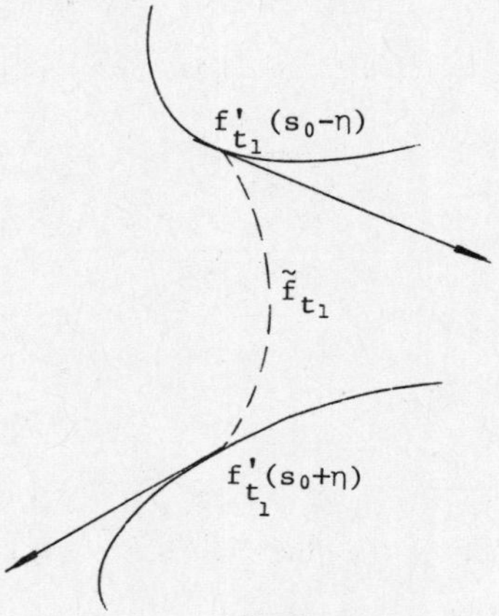

Fig.2

$f'_{t_1}(s_0-\eta)/|f'_{t_1}(s_0-\eta)|$ and $f'_{t_1}(s_0+\eta)/|f'_{t_1}(s_0+\eta)|$, contradicting to Lemma 2 (Notice that we have proved Lemma 2 for C^∞ immersions, but it works also for C^1 immersions). Therefore, $w(f_{t_1}) > 1$. But $w(f_{t_1}) = w(f_0) = 1$. This contradiction shows that $f'_t(s) = 0$ has no solutions for $(t, s) \in [0,1] \times S^1$, thus proving the lemma.

Lemma 13. $(0,1) \sim (1,2)$.

Proof. Let

$$u_\varepsilon^{(t)} = r^{-2}(2\cos t - (1+\varepsilon)\cos 2t),$$
$$v_\varepsilon^{(t)} = r^{-2}(2\sin t - (1+\varepsilon)\sin 2t),$$

where $r^2 = 4 + (1+\varepsilon)^2 - 4(1+\varepsilon)\cos t$. We have

$$(r^2)'u_\varepsilon + r^2 u'_\varepsilon = -2\sin t + 2(1+\varepsilon)\sin 2t, \tag{1}$$

$$(r^2)'u_\varepsilon + r^2 v'_\varepsilon = 2\cos t - 2(1+\varepsilon)\cos 2t, \tag{1'}$$

$$(r^2)'u_\varepsilon + 2(r^2)'u'_\varepsilon = -2\cos t + 4(1+\varepsilon)\cos 2t, \tag{2}$$

$$(r^2)''v_\varepsilon + 2(v^2)'v'_\varepsilon + r^2 v''_\varepsilon = -2\sin t + 4(1+\varepsilon)\sin 2t, \tag{2'}$$

(3) $$(r^2)'''u_\varepsilon + 3(r^2)''u_\varepsilon' + 3(v^2)'u_\varepsilon'' + r^2u_\varepsilon''' = 2\sin t - 8(1+\varepsilon)\sin 2t,$$

(3') $$(r^2)'''v_\varepsilon + 3(r^2)''u_\varepsilon' + 3(r^2)'v_\varepsilon'' = -2\cos t + 8(1+\varepsilon)\cos 2t.$$

As $|\varepsilon|$ is small, $(u_\varepsilon'(t), v_\varepsilon'(t)) = (0,0)$ has the only solution $\varepsilon = 0$, and $t = 2m\pi, m \in Z$ and $(u_0(2m\pi), v_0(2m\pi)) = (1,0)$.

Calculations show that

$$\begin{aligned} -\frac{1}{4}r^6(v_\varepsilon' u_\varepsilon'' - u_\varepsilon' v_\varepsilon'') =& 12(1+\varepsilon)^2\cos^2 t - (12(1+\varepsilon) + 9(1+\varepsilon)^3\cos t \\ &+ (1+\varepsilon)^4 + 3(1+\varepsilon)^2 + 4. \end{aligned}$$

Hence t is an inflexion of $(u_\varepsilon(t), v_\varepsilon(t))$ iff $\cos t = S_+(\varepsilon)$ or $S_-(\varepsilon)$, where

$$S_\pm = \frac{12 + 9(1+\varepsilon)^2 \pm \sqrt{72(1+\varepsilon)^2 - 15(1+\varepsilon)^4 - 48}}{24(1+\varepsilon)}.$$

Since

$$\lim_{\varepsilon\to 0} S_-(\varepsilon) = 3/4,$$

$\cos t = S_-(\varepsilon)$ has two solutions in $[0, 2\pi)$ as $|\varepsilon|$ is small. Now

$$S_+(0) = 1, \qquad S_+'(0) = 1/3 > 0.$$

So $\cos t = S_+(\varepsilon)$ has no solutions as $\varepsilon > 0$ and two solutions as $\varepsilon < 0$.

In order to see that $(u_\varepsilon, v_\varepsilon)$ is third order nondegenerate as $|\varepsilon|$ is small, we need only to see that (u_0, v_0) possesses such a property. If $\cos t = 1$, we have from (2), (2') and (3), (3') that

$$\begin{aligned} (u_0'', v_0'') &= (-2, 0), \\ (u_0''', v_0''') &= (0, 6). \end{aligned}$$

If $\cos t = 3/4$, then $\sin t = \pm\sqrt{7}/4$, and $r^2 = 2, (r^2)' = \pm\sqrt{7}, (r^2)'' = 3$, $(u_0, v_0) = (11/16, \pm\sqrt{7}/16)$, consequently

$$\begin{aligned} (u_0', v_0') &= (\mp 7\sqrt{7}/32, 13/32), \\ (u_0''', v_0''') &= (\pm 87\sqrt{7}/32, -40/32). \end{aligned}$$

Hence (u_0, v_0) is third order nondegenerate.

Let

$$f_\varepsilon(t) = e^{it}(2 - (1+\varepsilon)e^{it}), \qquad |\varepsilon| \text{ small},$$

then there is a C^∞ real function $\tilde\theta(t)$ with $\tilde\theta(0) = 0$ such that

$$2 - (1+\varepsilon)e^{it} = r(t)e^{i\tilde\theta}(t), \quad r(t) > 0.$$

Obviously

$$|\tilde\theta(t)| < \pi/2.$$

Let $\theta(t) = t + \tilde\theta(t)$, then we have $f_\varepsilon(t) = r(t)e^{i\theta(t)}$ and $(u_\varepsilon(t), v_\varepsilon(t))$ is the same as

$$g_\varepsilon(t) = \frac{1}{r(t)}e^{i\theta(t)}.$$

From $f_\varepsilon'(t) = (r'\theta t) + ir(t)\theta'(t))e^{i\theta(t)}$ and the fact that $f_\varepsilon'(t) \neq 0$ if $\varepsilon \neq 0$, it is seen that there is a C^∞ real function $\psi(t)$ with

$$r'(t) + ir(t)\theta'(t) = \rho(t)e^{i(\theta(t)+\psi(t))},$$

and

$$g_\varepsilon'(t) = \frac{\rho(t)}{r(t)^2}e^{i(\theta(t)-\psi(t)+\pi)}.$$

It has been seen in the proof of Lemma 6 and 10 that

$$w(f_\varepsilon) = \frac{1}{2\pi}(\theta(2\pi) + \psi(2\pi) - \theta(0) - \psi(0)) = \begin{cases} 2, & \text{if } \varepsilon > 0, \\ 1, & \text{if } \varepsilon < 0. \end{cases}$$

Since $\theta(2\pi) - \theta(0) = 2\pi$, we have

$$\psi(2\pi) - \psi(0) = \begin{cases} 2\pi, & \text{if } \varepsilon > 0, \\ 0, & \text{if } \varepsilon < 0. \end{cases}$$

Therefore

$$w(g_\varepsilon) = \begin{cases} 0, & \text{if } \varepsilon > 0, \\ 1, & \text{if } \varepsilon < 0. \end{cases}$$

This proves $\underline{(0,1)} \sim \underline{(1,2)}$, hence the Lemma.

Lemma 14. If $|n|, |n'| \geq 2$, then $\underline{(n,k)} \sim \underline{(n',k')}$ if $n + k \equiv n' + k' \bmod 2$.

Proof. From Lemma 13 and 10, we have

$$\underline{(0,k)} \sim \underline{(k,2k)} \sim \underline{(3k,0)}, \text{ if } k > 0.$$

Assume $r \geq 1$. Then

$$\underline{(2r,0)} \sim \underline{(2r-(2r+1), 2r+1)} = \underline{(-1, 2r+1)} \sim \underline{(-2r-2, 0)}.$$

Hence

$$\underline{(-2r,0)} \sim \underline{(2r+2,0)} \sim \underline{(-2r-4,0)}$$

and

$$\underline{(2r,0)} \sim \underline{(-2r-2,0)} \sim \underline{(2r+4,0)}.$$

Now

$$\underline{(2,0)} \sim \underline{(-4,0)} = \underline{(2-6,0)} \sim \underline{(2,6)} \sim \underline{(8,0)} \sim \underline{(4,0)}$$

and

$$\underline{(-2,0)} \sim \underline{(-4,0)},$$

so $\underline{(2m,0)} \sim \underline{(2n,0)}$ for any $m, n \in Z - \{0\}$.

We have also

$$\underline{(2r+1,0)} \sim \underline{(2r+1-(2r+2), 2r+2)} = \underline{(-1, 2r+2)} \sim \underline{(-2r-3, 0)},$$

$$\underline{(-2r-1)} \sim \underline{(2r+3,0)} \sim \underline{(-2r-5,0)},$$

$$\underline{(2r+1,0)} \sim \underline{(-2r-3,0)} \sim \underline{(2r+5,0)},$$

and

$$\underline{(3,0)} \sim \underline{(-5,0)} = \underline{(2-7,0)} \sim \underline{(2,7)} \sim \underline{(9,0)} \sim \underline{(5,0)},$$

$$\underline{(-3,0)} \sim \underline{(-5,0)},$$

hence $\underline{(2m+1,0)} \sim \underline{(2n+1,0)}$ for any $m, n \in Z - \{0,-1\}$. The Lemma is proved.

Lemma 8, together with Lemmas 9, 12, 13 and 14 proves Theorem 2.

REFERENCES

[1] E.M.Feldman, Geometry of submanifolds I, Trans. Amer. Math. Soc., 87 (1965) 185-224

[2] E.M.Feldman, Deformation of closed spase curve, J.Diff. Geom., 2 (1968) 67-75

[3] W.Fenchel, Uber Krumming und Winding geschlossener Raumkurvenn, Math. Ann., 101 (1929) 238-252

[4] M.Gromov, Partial Differential relations, Springer-Verlag 1986

[5] M.Gromov & J.Eliashberg, Removal of singularities of smooth mappings, Math. USSR Izv., 5 (1971) 615-639

[6] B.H.Li, On second order nondegenerate immersion of S^1 in R^2, Top. and its Appl., 25 (1987) 161-164

[7] T.A.Little, Nondegenerate homotopies of curves on the unit 2- sphere, J.Diff. Geom., 4 (1970) 339-348

[8] T.A.Little, Third order nondegenerate homotopies of space curves, J. Diff. Geom., 5 (1971) 503-515

[9] A.Mukherjee, Higher-order nondegenerate immersion of manifolds, Top. and its Appl., 25 (1987) 129-135

[10] W.F.Pohl, Differential geometry of higher order, Topology, 1 (1962) 169-211

[11] H.Whitney, On regular closed curves in the plane, Compos. Math., 4 (1937) 276-284

Li Bang-he
Institute of System Science
Academia Sinica
Beijiang 100080
P.R.China

Xu Tao
Institute of Mathematics
Fudan University
Shanghai, P.R.China

Complete Surfaces in H^3 with a Constant Principal Curvature

Ma Zhisheng
Sichuan normal university, Chengdu

§0. Introduction

Stoker J.J. and Massey W.S.[1,2] studied complete regular surfaces in E^3 with identically zero Gaussian curvatures. Katsuhire Shichama and Ryoichi Takagi[3] made a study of isometric immersion $M \to E^3$ with a non-zero constant principal curvature. The purpose of the present paper is to discuss complete surfaces in $H^3(-1)$ (simply H^3) with a constant principal curvature.

§1. Definitions and local formulas

An m-dimensional pseudosphere $H^m(-a^2)$ refers to the simple connected complete hypersurface with negative constant curvature $-a^2$. Just as an m-dimensional standard sphere $S(a^2)$ can be isometrically immersed into E^{m+1}, $H^m(a^2)$ can be isometrically immersed [4,5,6] into $(m+1)$-dimensional Lorentz- Minkowski space L^{m+1} as a spacelike hypersurface†. Let $(x_1, x_2, \cdots, x_m, t)$ be a Lorentz-Minkowskian orthogonal coordinates system of E^{m+1}, thus, its metric can be given by $ds^2 = \sum_{A=1}^m (dx_A)^2 - (dt)^2$. Let $\langle , \rangle$ denote Lorentz inner product. Thus the isometric immersion τ: $H^m(-a^2) \to L^{m+1}$ can be expressed by $\tau(H^m(-a^2)) = \{x \in L^{m+1}; \langle x, x\rangle = -\frac{1}{a^2}, t > 0\}$. In this paper, $a = 1$ and $H^m(-1)$ is written as H^m. We choose a local field of orthonormal frames $e_1, e_2, \cdots, e_{m+1}$ in L^{m+1}, such that

$$\begin{aligned} &\langle e_A, e_A\rangle = 1 \qquad A = 1, 2, \cdots, m. \\ &\langle e_{m+1}, e_{m+1}\rangle = -1 \end{aligned} \tag{1.1}$$

and restricted to H^m, we have

$$e_{m+1} = -x, \qquad x \in H^m. \tag{1.2}$$

Hence, $e_1, \cdots, e_m$ are tangent to H^m. Let $\omega_1, \omega_2, \cdots, \omega_{m+1}$ be the field of dual frames relative to the frame field of L^{m+1} chosen above. Restricting these forms to H^m, we have

$$\omega_{m+1} = 0, \tag{1.3}$$

consequently

$$\begin{aligned} dx &= \sum_{A=1}^m \omega_A e_A, \\ de_A &= \sum_{B=1}^m \omega_{AB} e_B - \omega_A e_{m+1}, \end{aligned} \tag{1.4}$$

where ω_{AB} are uniquely determined by structure equations of H^m [4].

$$\begin{aligned} d\omega_A &= \sum_{B=1}^m \omega_{AB} \wedge \omega_B, \qquad \omega_{AB} + \omega_{BA} = 0, \\ d\omega_{AB} &= \sum_{C=1}^m \omega_{AC} \wedge \omega_{CB} - \omega_A \wedge \omega_B. \end{aligned} \tag{1.5}$$

Throughout this paper, we let M be a two-dimensional connected, complete, orientable Riemannian manifold of class C^∞. Let $\tilde{x} : M \to H^3$ be an isometric immersion of M into a

† A hypersurface of L^{m+1} is spacelike, if its induced metric is positive definite.

3- dimensional psendosphere H^3, thus the composite map $x = \tau \circ \tilde{x} : M \to L^4$ is an isometric immersion of M into L^4, such that $\langle x, x\rangle = -1$. From now on, $m = 3$, and e_1, e_2, e_3, e_4 are chosen such that e_1 and e_2 are tangent to $x(M)$ (simply M), then restricted to M, we have

$$\omega_4 = \omega_3 = 0, \tag{1.6}$$

$$dx = \sum_{i=1}^{2} \omega_i e_i, \tag{1.7}$$

$$\omega_{A4} = -\omega_A, \qquad A = 1, 2, 3, \tag{1.8}$$

$$\omega_{i3} = \sum_{j=1}^{2} h_{ij}\omega_j, \qquad h_{ij} = h_{ji}, \qquad i, j = 1, 2. \tag{1.9}$$

The second fundamental form of M in H^3 is defined by $\sum_{i,j=1}^{2} h_{ij}\omega_i \otimes \omega_j$. A point $x \in M$ is called an umbilical point if the matrix (h_{ij}) takes the form $(h_{ij}) = \begin{pmatrix} h & 0 \\ 0 & h \end{pmatrix}$ at this point, where h is a real number. Let U denotes the set of all umbilical points on M, and $N = M - U$. If a point $x_0 \in N$, there exists a neighborhood $V \subset N$. We can take an orthonormal frame field (e_1, e_2, e_3, e_4) with respect to which (h_{ij}) takes the form $(h_{ij}) = \begin{pmatrix} h_1 & 0 \\ 0 & h_2 \end{pmatrix}, h_1 > h_2$, so that

$$\omega_{i3} = h_i\omega_i, \tag{1.10}$$

where h_1 and h_2 are principal curvatures of M, e_1 and e_2 are relative principal directions respectively. Since M is orientable, a unit normal vector field e_3 can be globally defined on M. Then we can consider h_1 and h_2 to be continuous functions on M, satisfying $h_1 \geq h_2$, and reduce the assumption that one of the principal curvature is everywhere a constant R to one of the following:

$$(i)\ R \equiv h_1 \geq h_2, \qquad (ii)\ h_1 \geq h_2 \equiv R.$$

Furthermore, we may assume that $R \geq 0$ (by replacing the unit normal vector field e_3 by $-e_3$, if necessary). Obviously, U is closed on M, hence N is open on M. If N is nonempty, then in every connected component of $N, h_1 > h_2$ and their satisfy one of the conditions (i) and (ii). Furthermore, h_1 and h_2 are differentiable on N.

§2.

Lemma 1. In lorentz-minkowski 4-dimensional space- time, the intersection S of pseudosphere H^3 with 3-flat [6] (or 3-plane, or hyperplane) $\langle y, x\rangle = -1$ (where y is a nonzero constant vector) is one of the following:

(1). If y is a spacelike vector, then the intersection is a 2- dimensional pseudosphere.

(2). If y is a timelike vector, then the intersection may be expressed as

$$\begin{cases} \langle x, x\rangle = -1 \\ t = \frac{1}{y} > 0 \end{cases} \tag{2.1}$$

where t-axis is in the direction of y.

(3). If y is a lightlike vector, then the intersection may be expressed as

$$\begin{cases} x_1^2 + x_2^2 - \frac{2t}{y} = -(1 + \frac{1}{y^2}) \\ x_3 = t - \frac{1}{y} \end{cases} \tag{2.2}$$

Proof. (1). Since y is spacelike, there is a Lorentz rotation trasformation of coordinate system [6] such that the new Lorenta orthonormal basis $\bar{e}_1, \bar{e}_2, \bar{e}_3, \bar{e}_4$ has $y = \bar{y}\bar{e}_3$, where $\bar{y}$ is a real number. Hence the intersection S is given by

$$\begin{cases} x_1^2 + x_2^2 - t^2 = -(1 + \frac{1}{\bar{y}^2}), \\ x_3 = \frac{-1}{\bar{y}}. \end{cases}$$

This equation represents a two-dimensional pseudosphere.

(2). Since y is timelike, there is a Lorentz rotation transformation of coordinate system such that the new Lorentz orthonormal basis has $y = \bar{y}e_4$, where $\bar{y}$ is a nonzero real number, hence the intersection S is given by

$$\begin{cases} \langle y, x\rangle = -1, \\ t = \frac{1}{\bar{y}}. \end{cases}$$

On the other hand, $\langle y, x\rangle = -1$ implies that x and y belong to the same funnel of the cone, and thus $t = \frac{1}{\bar{y}} > 0$.

(3). Since y is lightlike, we can choose a Lorentz rotation transformation of coordinate system such that the new basis has $y = \bar{y}(\bar{e}_3 + \bar{e}_4)$, where $\bar{y}$ is a nonzero real number, hence the intersection S is given by equations (2.2).

Proposition 1. Let M be totally umbilical, then the principal curvatures of M in H^3 are all equal to constant R, and we have in L^4:

(1). If $0 \le R < 1$, then M is a 2-dimensional pseudosphere.

(2). If $R > 1$, then M is a 2-dimensional sphere of H^3, and M is compact.

(3). If $R = 1$, then M is a 2-dimensional paraboloid†.

Proof. Since $h_1 = h_2 = h$, differentiating (1.10) exteriorly and using (1.5), (1.6) and (1.8), we get $dh \wedge \omega_1 = dh \wedge \omega_2 = 0$. Consequently $h = R = constant$. It follows from (1.4) that $de_3 = -Rdx$.

If $R = 0$, then $e_3 = constant$, and thus $M = H^3 \cap P$, where P denotes a 3-flat orthogonal to e_3 and passing through the origin. Hence, M is a 2-dimensional pseudosphere.

If $R > 0$, integrating $de_3 = -Rdx$, we obtain

$$e_3 = -R(x - y), \tag{2.3}$$

where y is an integral constant vector. By (2.3) and (1.2) we get

$$\langle y, y\rangle = \frac{1 - R^2}{R^2}, \tag{2.4}$$

hence, from (1.2), (2.3) and (2.4) we have the conclusion that, in Lorentz orthonormal coordinate system, the equations of M are given by

$$\begin{cases} \langle x, x\rangle = -1, \\ \langle y, x\rangle = -1. \end{cases} \tag{2.5}$$

If $R < 1$, y is spacelike by (2.4); if $R > 1$, y is timelike; if $R = 1$, then y is lightlike. It follows from Lemma 1 that M is a 2-dimensional pseudosphere if $0 < R < 1$, M is a 2-dimensional sphere of H^3 if $R > 1$, M is a 2-dimensional paraboloid if $R = 1$. The proof of Prop. 1 is completed.

Now, let N be a nonempty set. We first consider Case (i) $h_2 \le R$ locally.

---

† A 2-dimensional paraboloid refers to surfaces determined by equations (2.2) in L^4.

Lemma 2. For every point $x_0 \in N$, there is a neighborhood $V \subset N$ of x_0 in which there exists a local field of orthonormal frames e_1, e_2, e_3, e_4. There are differential functions u and f defined on V satisfying

$$du = \omega_1, \tag{2.6}$$

$$\omega_{12} = f\omega_2. \tag{2.7}$$

Proof. The existness of V and field of frames can be proved as before. Differentiating (1.10) exteriorly and using (1.5), we get

$$(R - h_2)\omega_{12} \wedge \omega_2 = 0.$$

Since $R - h_2 > 0$ on V, there exists C^∞ function f satisfying (2.7). Substituting (2.7) into (1.5) and using (1.6), we get $d\omega_1 = 0$. Consequently there exists C^∞ function u satisfying (2.6).

Lemma 3. With respect to the frame field e_1, e_2, e_3, e_4 on V given by lemma 2, there exixt C^∞ function g and q satisfying

$$dh_2 = (R \cdot h_2) f\omega_1 + g\omega_2 \tag{2.8}$$

$$df = (1 - Rh_2 - f^2) \cdot \omega_1 + q \cdot \omega_2. \tag{2.9}$$

Proof. Exteniorly differentiating the second equation of (1.10), and using (1.5) and (1.10), we get

$$dh_2 \wedge \omega_2 = (R - h_2) f\omega_1 \wedge \omega_2.$$

Exteriorly differentiating (2.7), and using (1.5) and (1.10), we get

$$df \wedge \omega_2 = (1 - Rh_2 - f^2)\omega_1 \wedge \omega_2.$$

The existsness of the functions g and q follows from the equations just obtained, thus proving Lemma 3.

Here after we will let V denote such a neighborhood where we always use such a frame field as above.

Proposition 2. For every point $x_0 \in N$, there exists the unique geodesic γ_{x_0} passing through x_0 and in the unique direction of the principal curvature in H^3 equal to R. Furthermore, γ_{x_0} is a part of the intersection of H^3 with a 2-flat in L^4. The equations of this intersection curve are given by (2.12)-(2.15).

proof. Since $h_2(x_0) < R$, e_1 is the unique direction of the principal curvature in H^3 equal to R at the point of V and its differential equation is $\omega_2 = 0$. Hence, for every point $x_0 \in V$, there exists the unique integral curve γ_{x_0} passing through x_0.

From (2.7), (1.10) and (1.4), we hve

$$\begin{aligned} de_1 &= (Re_3 - e_4)\omega_1 \\ de_2 &= 0, \qquad de_3 = -Rdx. \end{aligned} \tag{2.10}$$

The first equation of (2.10) implies that γ_{x_0} is a geodesic of M. The second inplies that e_2 is a constant vector along γ_{x_0}, and that γ_{x_0} lies on a 3- flat orthonormal to e_2 and passing through the origin, thus its equation is given by

$$\langle e_2, x \rangle = 0. \tag{2.11}$$

From the third it follows that, if $R = 0$, then e_3 is a constant vector along γ_{x_0}, and γ_{x_0} lies on a 3-flat:

$$\langle e_3, x \rangle = 0.$$

Hence, there exists a Lorentz rotation transformation of coordinate system such that the new basis vectors $\bar{e}_2$ and $\bar{e}_3$ are in the fixed directions of e_2 and e_3 respectively. From this and (2.11) it follows that γ_{x_0} satisfies the following equations:

$$\begin{cases} x_1^2 - t^2 = -1, & t > 0, \\ x_2 = x_3 = 0. \end{cases} \tag{2.12}$$

If $R \neq 0$, integrating $de_3 = -Rdx$ along γ_{x_0}, we get $e_3 = -R(x - y)$, where y is a nonzero constant vector. Thus, by Lem.1, Prop.1 and (2.11), we obtain the following result:

(1). If $0 < R < 1$, then γ_{x_0} satisfies

$$\begin{cases} x_1^2 - t^2 = -\frac{R^2}{1-R^2} & t > 0 \\ x_2 = 0, \quad x_3 = -\frac{R}{\sqrt{1-R^2}} \end{cases} \tag{2.13}$$

(2). If $R > 1$, then γ_{x_0} satisfies

$$\begin{cases} x_1^2 + x_3^2 = 1, \\ x_2 = 0, \quad t = R/\sqrt{-1+R^2}. \end{cases} \tag{2.14}$$

(3). If $R = 1$, then γ_{x_0} satisfies

$$\begin{cases} x_1^2 - \frac{2t}{g} = -(1 + \frac{1}{g^2}) \\ x_2 = 0, \quad x_3 = t - \frac{1}{g}. \end{cases} \tag{2.15}$$

The proof of Lem.3 is completed.

From (2.12)-(2.15), we have the following corollary.

Corollary 1. Every γ_{x_0} always lies on some 2-pseudosphere.

It is convenient to use the following terminology: An integral curve of field e_1 passing through a point $x_0 \in N$ is said to be maximal if it is not a proper subset of some integral curve passing through x_0. Hereafter, we will denote by γ_{x_0} the maximal integral curve.

By Lemmas 2 and 3, we have, along γ_{x_0}

$$\frac{dh_2}{du} = (R - h_2)f \tag{2.16}$$

$$\frac{df}{du} = (1 - Rh_2 - f^2) \tag{2.17}$$

where u is the arc length of γ_{x_0}. Since e_1 is the unique unit tangent vector field on N where the principal curvature of M in H^3 is equal to R, the quantities occurred in (3.16) and (3.17) are functions on N independent of the frame field except the sign of u and f. When we are replacing e_1 by $-e_1$, however, the sign of u and f are changed simultaneously, thus (2.16) and (2.17) are invarians.

Set $\Phi = \frac{1}{R-h_2}$, then by (2.16) and (2.17), we have along γ_{x_0}

$$\frac{d\Phi}{du} = f\Phi \tag{2.18}$$

$$\frac{d^2\Phi}{du^2} - (1 - R^2)\Phi - R = 0. \tag{2.19}$$

Proposition 3. Let γ_{x_0} be a maximal integral curve passing through the point $x_0 \in N$, then $\gamma_{x_0} \cap U = \phi$.

Proof. For several cases of R, we get the solutions of (2.19) as follows:

$$R - h_2 = 1 \,/\, (a \cosh u + b \sinh u), (R = 0), \tag{2.20}$$

$$R - h_2 = 1/\,/(a \cosh\sqrt{1-R^2}u + b \sinh\sqrt{1-R^2}u - \frac{R}{1-R^2}), \quad (0 < R < 1), \tag{2.21}$$

$$R - h_2 = 1 \,/\, (a \cos\sqrt{R^2-1}u + b \sin\sqrt{R^2-1}u + \frac{R}{R^2-1}), \quad (R > 1), \tag{2.22}$$

$$R - h_2 = 2 \,/\, (u^2 + 2au + 2b), \qquad (R = 1), \tag{2.23}$$

where a and b are integral constants.

Assume that γ_{x_0} contains a point $y_0 \in U$. Since γ_{x_0} is connected and N is open, there exists a point x_1 corresponding to u_1, such that $x_1 \in U$ and the points of γ_{x_0} with $u < u_1$ belong to N. Hence it follows from (2.20)-(2.23) and the continuity of h_2 that

$$0 = R - h_2(x_1) = \lim_{u \to u_1} [R - h_2(\gamma_{x_0}(u))] \neq 0.$$

which is a contradiction and thus concludes the proof.

Corollary 2. The equation of γ_{x_0} is one of the equations (2.12)-(2.15), and h_2 is determined by one of the equations (2.20)-(2.23).

Proof. By completeness of M and Prop.3, we conclude that γ_{x_0} is extended for arbitrarily large values of its canonical parameter (arc length) u.

On the other hand, the equations (2.12)-(2.15) satisfied by γ_{x_0} can be respectively parametrize by arc length u as follows:

$$\begin{cases} x_1 = \sinh u, \\ t = \cosh u, \\ x_2 = x_3 = 0; \end{cases} \tag{2.12'}$$

$$\begin{cases} x_1 = \frac{R}{\sqrt{1-R^2}} \sinh\sqrt{1-R^2}u, \\ t = \frac{R}{1-R^2} \cosh\sqrt{1-R^2}u, \\ x_2 = 0, \quad x_3 = -\frac{R}{\sqrt{1-R^2}}; \end{cases} \tag{2.13'}$$

$$\begin{cases} x_1 = \cos\sqrt{R^2-1}u, \\ x_3 = \sin\sqrt{R^2-1}u, \\ x_2 = 0, \quad t = \frac{R}{(R^2-1)^{\frac{1}{2}}}; \end{cases} \tag{2.14'}$$

and

$$\begin{cases} x_1 = u, \quad x_2 = 0, \\ x_3 = \frac{g}{2}(u^2 + 1 - \frac{1}{g^2}), \\ t = \frac{g}{2}(u^2 + 1 + \frac{1}{g^2}). \end{cases} \tag{2.15'}$$

Consequently (2.12)-(2.15) are the equations of γ_{x_0} and (2.20)-(2.23), the representations of h_2 along γ_{x_0} respectively.

Remark. It follows from (2.12)'-(2.15)' that if $R > 1$, then γ_{x_0} is a closed curve; if $0 \leq R \leq 1$, then γ_{x_0} is a unbounded curve in L^4.

We are now in a position to prove the following global results.

Theorem 1. Let M be a 2-dimensional connected, complete, orientable Riemannian manifold of class C^∞, $\tilde{x} : M \to H^3$ an isometric immersion with a constant principal curvature R. If $h_2 \leq h_1 \equiv R$, then the immersion is either totally umbilical or umbilically free.

proof. Assume that M is not totally umbilical, then N is a nonempty open set in M. It suffices to show that N is closed in M. Let $x_n (n = 1, 2, \cdots)$ be a sequence of points belonging to N assume x_n belongs to the same connected component N_0) such that $\lim_{n\to\infty} x_n = x_0 \in M$ and $\gamma_n = \gamma_{x_n}$. Now let us prove that $x_0 \in N$.

Assume the contrary, i.e., $x_0 \in U \cap Bd(N)$. We shall first conclude that γ_n converges to a certain geodesic γ_0 through x_0. In fact, it follows from (1.4), (1.5), Lemma 1 and Prop.3 that there exists a system of geodesic coordinates (u, v) in the entire N such that

$$N_0 : \quad \begin{cases} -\infty < u < +\infty \\ v_0 < v < v_0', \end{cases}$$

and that every curve $v = const.$ is the itnegral geodesic curve proposed in Prop.3, and that γ_n are all the curves $v = const.$ corresponding to $v_n, v_0 < v_n < v_0'$. Consequently in such coordinates, from the completeness of M, there exists a number u_0 such that (u_0, v_0) (or (u_0, v_0')) corresponds to the point x_0, and $v = v_0$ correspond to a certain curve of M through x_0. Clearly, the curve $v = v_0$ is a geodesic γ_0 of M and γ_n converges to the geodesic γ_0.

We know from (2.20)-(2.23) that in any case of R, for any curve γ_n, there exists a point y_n of γ_n such that $R - h_2(y_n) = c \neq 0$, where c is a constant independent of n. For example, if $R > 1$, then $c = \frac{R^2-1}{R} \neq 0$; if $R = 1$, then $c = \frac{1}{6}$, etc. Hence we can choose a subsequence $\{y_{kn}\}$ of $\{y_n\}$ converging to a point y_0 on γ_0. Consequently, we have $R - h_2(y_0) = \lim_{n\to\infty}[R - h_2(y_{k_n})] = c \neq 0$ by continuity of h_2. Therefore, $Y_0 \in N$. This fact together with $\lim_{n\to\infty} \gamma_n = \gamma_0$ implies that the tangent vector of γ_0 at y_0 coincides with $e_1(y_0)$, thus we have $\gamma_0 = \gamma_{y_0}$; in particular, $x_0 \in N$. This contradiction implies that N is closed on M, completing the proof of Theorem 1.

The remaining case (ii) can be treated in a similar way.

Theorem 2. let M be a 2-dimensional connected, complete orientable Riemannian manifold of class C^∞, $\tilde{x} : M \to H^3$ an isometric immersion with a constant principal curvature R. If $h_1 \geq h_2 \equiv R$, then R must satisfy $0 \leq R < 1$, and the immersion is either totally umbilical or umbilically free.

Proof. In this case, the previos discussions are also valid by exchanging the role of h_1 and the one of h_2 mutually. Assume that there is a non-umbilical point x_0 on such an M, the integral curve γ_{x_0} of the vector field e_2 must satisfy one of the equations (2.12)'-(2.15)' and h_1 is determined by one of the equations (2.20)-(2.23) in which we substitute h_1 for h_2. But from (2.22) and (2.23) we know that the inequality $R - h_1 < 0$ in the case of R satisfying $R > 1$ are not identically valid along γ_{x_0}. The proof of theorem 2 is completed.

References

[1] Stoker, J.J., Developable surfaces in the large, Comm. pure and appl. math., 14(1962), 627-635.

[2] Massey, W.S., Surfaces of Gaussian curvature zero in Euclidean spaces, Tohoku math. J., 14(1962), 73-79.

[3] Katsuhiro Shiohama and Ryoichi Takagi, A characterization of a standard torus in E^3, J. Diff. Geometry, 4(1970), 477-485.

[4] Wolf, J.A., Spaces of constant curvature, McGraw-Hill, New York, 1967.

[5] Dubrovin, B.A., Fomenko, A.T. and Novikov, S.P., Modern geometry-Methods and applications, Part I, Springer-Verlag New York Berlin Heidelberg Tokyo, 1972.

[6] Synge, J.L., Relatively: The special theory, Dublin institute for advanced studies, 1956.

EXCEPTIONAL SIMPLE LIE GROUPS AND RELATED TOPICS IN RECENT DIFFERENTIAL GEOMETRY

Shingo MURAKAMI

Contents

§0 Introduction
§1 Cayley numbers
§2 Principle of triality I
§3 Principle of triality II
§4 Exceptional group F_4
§5 Cayley projective plane
§6 Automorphisms and subgroups of F_4
§7 Spin groups and spin representations
§8 Characterzations of G_2 and Spin(7) by invariant forms

§0 Introduction

These are expository notes on exceptional simple Lie groups (except those of type E), based on my lectures delivered at the Nankai Institute of Mathematics in the spring 1987. The exceptional simple Lie groups of lower dimensions play some important roles in geometry, e.g., in the study of holonomy groups. (cf. References)- My lectures were motivated by recent works of Bryant on the construction of Riemannian manifolds whose holonomy group is G_2 or Spin(7). The results applied there on these groups should be well known among experts, but seem not so easy to approach for beginners because the litterature on the subject is generally old and dispersed. Thus I intend here to give a quick introduction to these exceptional groups to those who have basic acquirements on the theory of Lie groups. Also, basing on famous notes by Freudenthal, I show that the exceptional Lie group F_4 contains the groups G_2 and Spin(k) (k=7,8,9) as distinguished subgroups connected with an automorphism of order three.

I would express here my hearty thanks to Director S.S.Chern, Professor Yen Chih-ta, Professor Hou Zi-xin and all members of the Institute for their warm hospitality shown to me. Owing to their kind acceptance, I could enjoy my stay at the Institute and in Tianjin.

§1 Cayley numbers

Definition. A real hypercomplex system is a finite dimensional real vector space F with two structures:

1) positive definite inner product < , >
2) real bilinear multiplication $E \times E\ (x,y) \to xy \in E$ and satisfying:
3) $\exists e_0 \in E$ unit element,
4) $|xy| = |x||y|$ for $x,y \in E$ where $|\cdot| = \sqrt{\langle\cdot,\cdot\rangle}$

Examples. $\mathbb{R}$: real number field, $\mathbb{C}$: complex number field, $\mathbb{H}$: quaternion number field.

Definition. The system of Cayley numbers $\mathbb{D}$ is the 8-dimensional real vector space with the inner product < , > in which the product is defined as follows. There exists an orthonormal basis $\{e_0, e_1, \cdots, e_7\}$ such that e_0 is the unit element, $e_i^2 = -e_0$ $(1 \leqq i \leqq 7)$ and $e_i e_j = \pm e_k$ $(1 \leqq i \neq j \leqq 7)$ where e_k is the third element on the line (or circle) through e_i, e_j in the figure and ± is determined according as the direction from e_i to e_j coincides with the arrow or not.

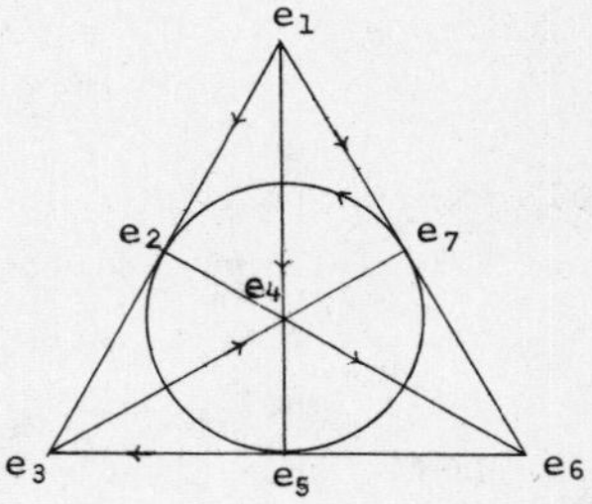

Theorem. Real hypercomplex systems are $\mathbb{R}$, $\mathbb{C}$, $\mathbb{H}$ or $\mathbb{D}$.

This theorem was first proved by Hurwicz in 1898. For proofs, see B.Eckmann[8], H.Frendenthal[9], R.Harvey and H.B.Lawson Jr. [10].

Formulas. On Cayley numbers $\mathbb{D}$, we derive the formulas as those for a hypercomplex system E. We put $\operatorname{Re} E = \mathbb{R}\cdot e_0$, $\operatorname{Im} E = (\operatorname{Re} E)^{\perp}$ where $\perp$ denotes the orthogonal complement. Therefore, an element $x \in E$ decomposes into

$$x = \operatorname{Re} x + \operatorname{Im} x$$

and $\operatorname{Re} x = \langle x, e_0\rangle e_0$. We shall often identify Re x with the real number $\langle x, e_0\rangle$. For $a \in E$, define L_a, and $R_a : E \to E$ by $L_a x = ax$, $R_a x = xa$ $(x \in E)$. Then by 4), if $|a| = 1$, L_a and R_a are orthogonal transformations of E. It follows

(1) $\langle L_a x, L_a y\rangle = |a|^2 \langle x,y\rangle$, $\langle R_a x, R_a y\rangle = |a|^2\langle x,y\rangle$.

By polarization we get

(1)' $\langle L_a x, L_b y\rangle + \langle L_b x, L_a y\rangle = 2\langle a,b\rangle\langle x,y\rangle$

Now we define for $x \in E$, $\bar{x} = \operatorname{Re} x - \operatorname{Im} x$. Then

(3) $\langle L_a x, y\rangle = \langle x, L_{\bar{a}} y\rangle$, $\langle R_a x, y\rangle = \langle x, R_{\bar{a}} y\rangle$

Proof. It is sufficient to proof this assuming $\bar{a} = -a$. Then, by (2) + (1)'

$$\langle ax, y\rangle + \langle x, ay\rangle = 2\langle e_0, a\rangle\langle x,y\rangle = 0. \quad \text{q.e.d.}$$

We see

(4) $\bar{\bar{x}} = x$

$|\bar{x}| = |x|$, $\langle x,y\rangle = \langle \bar{x}, \bar{y}\rangle$

The second follows by polarization of the first one.

(5) $\overline{xy} = \bar{y}\bar{x}$

Proof. $\langle \overline{xy}, a\rangle = \langle xy, \bar{a}\rangle = \langle e_0, \overline{yxa}\rangle = \langle a, \overline{yx}\rangle = \langle \bar{y}\bar{x}, a\rangle$ for any $a \in E$, from which (5) follows. q.e.d.

We have

(6) $\langle x,y\rangle = \langle e_0,\bar{x}y\rangle = \mathrm{Re}\ \bar{x}y = \frac{1}{2}(x\bar{y}+y\bar{x})$.

Since $\overline{x\bar{x}} = x\bar{x}$, $x\bar{x} = \mathrm{Re}\ x\bar{x}$, we get

(7) $\langle x,x\rangle = x\bar{x}$.

On associativity:

Definition. For $x,y,z \in E$, put

$[x,y,z] = (xy)z - x(yz)$ and call $[\cdot,\cdot,\cdot]$ the associator.

Lemma 1. $[x,y,z]$ on E is alternating.

Proof. Since $[x,y,z]$ is trilinear in the variable x,y,z and since $[x,y,z] = 0$ if one of x,y,z are real, i.e. belong to $\mathbb{R}e_0$, it is sufficient to show that $[x,y,z]=0$ if x,y,z are imaginary and if two of x,y,z are equal. By (1), $\langle (xw)\bar{w},y\rangle=\langle xw,yw\rangle=\langle x,y\rangle|w|^2=\langle x|w|^2,y\rangle$ for any $y\in E$, so $(xw)\bar{w}=x|w|^2=x(w\bar{w})$. Therefore $[x,w,\bar{w}]=0$. Hence $[x,w,w]=0$, and $[x,y,z]=-[x,z,y]$. In particular $[w,y,w]=-[w,w,y]$. Thus it remains to prove $[w,\bar{w},z]=0$, which follows in the same way as $[x,w,\bar{w}]=0$ q.e.d.

We get immediately two corollaries:

Corollary 1. $[x,y,z]=0$ if two of x,y,z are equal or conjugate.

Corollary 2. (a) Each nonzero element $x\in E$ has a unique left and right inverse $x=\bar{x}/|x|^2$.

(b) Given $x,y\in E$, with $x\neq 0$, the equation $xw=y$, $wx=y$ can be solved (uniquely) for w with $w=\bar{x}y/|x|^2$, and $w=y\bar{x}/|x|^2$, respectively.

From (6), $2\langle x,y\rangle = x\bar{y}+y\bar{x}$, we get $2\langle x,y\rangle w-x(\bar{y}w)-y(\bar{x}w) = [x,\bar{y},w]+[y,\bar{x},w]$. Thus by Lemma 1

(8) $x(\bar{y}w)+y(\bar{x}w)=2\langle x,y\rangle w$ and

(9) $(w\bar{y})x+(w\bar{x})y=2\langle x,y\rangle w$.

In particular, if $\langle x,y\rangle=0$,

(10) $x\bar{y} = -\bar{y}x$,

(11) $x(\bar{y})=-y(\bar{x}w)$,

(12) $w(\bar{y}x) = -(w\bar{x})y$, for any $w\in E$.

Lemma 2. (Manfang identities)

(a) $(xyx)z = x(y(xz))$ (b) $z(xyx) = ((zx)y)x$ (c) $(xy)(zx) = x(yz)x$

Proof. Consider the difference of the left and right hand sides. They vanish if any two of the variables are equal, because any two elements of E generates an associative algebra $\mathbb{R}$, $\mathbb{C}$ or $\mathbb{H}$. (Theorem of Artin). Since they are linear in y and z, we may assume $\langle y,z\rangle=0$. By repeated use of (8), we see : both sides in (a) equal $-|x|^2yz$, in (b) equal $-|x|^2zy$, and in (c) equal $-|x|^2\bar{z}\bar{y}+2\langle x,yz\rangle x$. q.e.d.

Definition. Let G_2 be the group of automorphisms of the system of Cayley numbers.

As usual let O(n) be the group of real orthogonal matrices of order n and SO(n) its subgroup consisting of matrices of determinant 1. The group O(8) acts on $\mathbb{D}$ by orthogonal transformations, $\mathbb{D}$ being identified with $\mathbb{R}^8$ by the basis $\{e_i\}$. We shall identify the group O(7) (resp. SO(7)) with the subgroup of O(8) (resp. SO(7)) consisting of elements leaving e_0 fixed.

Proposition 1. $G_2 \subset O(7)$

Proof. Let $\phi\in G_2$. Then $\phi(e_0)=e_0$. Let $(x,e_0)=0$. Then $\bar{x}=-x$ and $x^2=-\langle x,x\rangle e_0$. So $\phi(x)^2=-\langle x,x\rangle e_0$. But if $\phi(x)=\alpha e_0+y$ with $\bar{y}=-y$, $\phi(x)^2=(\alpha^2-\langle y,y\rangle)e_0+2\alpha y$. Therefore $\alpha=0$, and $\langle y,y\rangle=\langle x,x\rangle$. Therefore $\overline{\phi(x)}=-\phi(x)$, and $\langle\phi(x),\phi(x)\rangle=\langle x,x\rangle$, for any $x \in \operatorname{Im} \mathbb{D}$, Thus $\phi \in O(7)$ q.e.d.

Theorem 1. G_2 acts transitively on the 6-sphere.

$$S^6 = \{x \in \mathbb{D} \,|\, |x|=1 \ \langle x,e_0\rangle = 0\}.$$

The isotropy group at a point is isomorphic to SU(3).

Proof. By Prop. 1, G_2 acts on S^6. Let $e_1' \in S^6$, we construct $\phi \in G_2$ shch that $\phi(e_1)=e_1'$. Choose $e_2' \perp \{e_0,e_1'\}_{\mathbb{R}}^*$ and $|e_2'|=1$, and put $e_3'=e_1'e_2'$. Then $\{e_0,e_1',e_2',e_3'\}$ is an orthonormal set. Take e_4' such that $e_4' \perp \{e_0,e_1',e_2',e_3'\}_{\mathbb{R}} \langle e_4',e_4'\rangle=1$, and set $e_5'=e_1'e_4'$, $e_6'=e_5'e_3'$, $e_7'=e_6'e_1'$. Then, $\{e_0,e_1',e_2',\cdots,e_7'\}$ is an orthonormal basis of $\mathbb{D}$.

We define ϕ: $\mathbb{D} \to \mathbb{D}$ by $\phi(e_0)=e_0$ and $\phi(e_i)=e_i'$ and show $\phi \in G_2$. In fact it suffices to show $\phi(e_i)\phi(e_j)=\phi(e_ie_j)$ $(1 \leq i,\ j \leq 7)$. If $i=j$, $(e_i')^2=-e_0$ proves this. If $i \neq j$, this can be verified easily by using Lemma 1 and (8). We have thus proved that G_2 acts transitively on S_6.

The isotropy subgroups at various points of S_6 are all conjugate to each other. Therefore, to show the second part, it is sufficient to consider the isotropy subgroup H of G_2 at e_1; $H=\{\phi \in G_2;\ \phi(e_1)=e_1\}$. If $\phi \in H$, then $\phi(e_1x)=e_1\phi(x)$ for any $x \in \mathbb{D}$ and conversely. Therefore H consists of automorphisms which commute with L_{e_1}. Now L_{e_1} defines a complex structure I in the 6-dimensional real vector space $\{e_2,e_3,e_4,e_5,e_6,e_7\}_{\mathbb{R}}$. We denote the complex vector space so obtained by V. Then $\{e_2,e_4,e_6\}$ is a basis of V. An Hermitian inner product h on V can be given by $h(x,y)=\langle x,y\rangle-\sqrt{-1}\cdot\langle Ix,y\rangle$ for $x,y \in V$.

If $\phi \in G_2$, then $\phi(e_0)=e_0$, $\phi(e_1)=e_1$ and ϕ induces a unitary transformation in the Hermitian vector space (V,h). We have

(13) $\phi(e_6)=\phi(e_2)\phi(e_4)$.

Conversely if ϕ is a unitary transformation in V satisfying (13), ϕ is defined by a unique element of G_2. We shall show that the condition (13) is equivalent to that $\det(\phi|V)=1$. Put

$$\begin{cases} \phi(e_2) = \alpha_1 e_2 + \alpha_2 e_4 + \alpha_3 e_6 \\ \phi(e_4) = \beta_1 e_2 + \beta_2 e_4 + \beta_3 e_6 \\ \phi(e_6) = \gamma_1 e_2 + \gamma_2 e_4 + \gamma_3 e_6. \end{cases}$$

Now

$$\begin{aligned}((a+\sqrt{-1}b)e_4)((c+\sqrt{-1}d)e_6) &= (ae_4+be_1e_4)(ce_6+de_1e_6) = (ae_4+be_5)(ce_6-de_7) \\ &= (ac-bd)e_2-(ad+bc)e_3 = \{(ac-bd)-(ad+bc)\sqrt{-1}\}e_2.\end{aligned}$$

Therefore $\qquad (\alpha e_4)(\beta e_6) = (\bar{\alpha}\bar{\beta})e_2$ for $\alpha,\beta \in \mathbb{C}$.

Analoguely $(\alpha e_6)(\beta e_4) = -(\bar{\alpha}\bar{\beta}_2)e_2$ etc, and we get

---

* $\{\cdot,\cdot\}_{\mathbb{R}}$ is the real vector subpace spanned by { , }.

$$\phi(e_2)\phi(e_4)=\begin{vmatrix}\bar\alpha_2 & \bar\alpha_3\\ \bar\beta_2 & \bar\beta_3\end{vmatrix}e_2+\begin{vmatrix}\bar\alpha_3 & \bar\alpha_1\\ \bar\beta_3 & \bar\beta_1\end{vmatrix}e_4+\begin{vmatrix}\bar\alpha_1 & \bar\alpha_2\\ \bar\beta_1 & \bar\beta_2\end{vmatrix}e_6 .$$

If (13) holds,

$$\gamma_1=\overline{\begin{vmatrix}\alpha_2 & \alpha_3\\ \beta_2 & \beta_3\end{vmatrix}}\qquad \gamma_2=\overline{\begin{vmatrix}\alpha_3 & \alpha_1\\ \beta_3 & \beta_1\end{vmatrix}}\qquad \gamma_3=\overline{\begin{vmatrix}\alpha_1 & \alpha_2\\ \beta_1 & \beta_2\end{vmatrix}}$$

$$\det\begin{vmatrix}\alpha_1 & \alpha_2 & \alpha_3\\ \beta_1 & \beta_2 & \beta_3\\ \gamma_1 & \gamma_2 & \gamma_3\end{vmatrix}=\gamma_1\begin{vmatrix}\alpha_2 & \alpha_3\\ \beta_2 & \beta_3\end{vmatrix}+\gamma_2\begin{vmatrix}\alpha_3 & \alpha_2\\ \beta_3 & \beta_2\end{vmatrix}+\gamma_3\begin{vmatrix}\alpha_1 & \alpha_2\\ \beta_1 & \beta_2\end{vmatrix}=\gamma_1\bar\gamma_1+\gamma_2\bar\gamma_2+\gamma_3\bar\gamma_3=1$$

since $\phi|V$ is unitary. Conversely, if $\det(\phi|V)=1$, we get $h(\phi(e_6),\phi(e_2)\phi(e_4))=1$ by (14). But $|\phi(e_2)\phi(e_4)|=|\phi(e_2)||\phi(e_4)|=|e_2||e_4|=1$, and by (10),(12) $h(\phi(e_2)\phi(e_4),\phi(e_4))=h(\phi(e_2),e_0)=0$, $h(\phi(e_2)\phi(e_4),\phi(e_2))=0$, thus $\phi(e_2)\phi(e_4)=\phi(e_6)$

q.e.d.

Corollary to Theorem 1. S^6 is homeomorphic to $G_2/SU(3)$, and G_2 is connected.

§2 Principle of triality 1

2.1 Automorphisms π, κ, λ, of $\mathcal{D}_4$.

We shall identify the group SO(8) with the group of orthogonal matrices of determinant 1 acting on $\mathbb{D}=\mathbb{R}^8$. Let $\mathcal{D}_4$ be its Lie algebra; $\mathcal{D}_4$ consists of real skew-symmetric matrices and acts on $\mathbb{D}$ by skew-symmetric transformations.

Let $G_{ij}\in\mathcal{D}_4$ $(0\leqq i\neq j\leqq 7)$ be such that

$$G_{ij}e_k=\begin{cases}-e_j, & k=i,\\ e_i, & k=j,\\ 0\ , & k\neq i,j.\end{cases}\tag{1}$$

Then $\{G_{ij}|0\leqq i<j\leqq 7\}$ form a basis of $\mathcal{D}_4$ we have,

$$\begin{cases}G_{ij}+G_{ji}=0\\ [G_{ij},G_{jk}]=G_{ik}\\ [G_{ij},G_{kj}]=0 \quad \text{if } i,j,k,l \text{ are all different.}\end{cases}\tag{2}$$

Let F_{ij} $(0\leqq i\neq j\leqq 7)$ be the linear transformation of $\mathbb{D}$ defined by

$$\begin{cases}F_{i0}x=\frac{1}{2}e_ix \qquad F_{0i}=-F_{i0} & (1\leqq i\leqq 7)\\ F_{ij}x=\frac{1}{2}e_j(e_ix) & (1\leqq i,\ j\leqq 7,\ i\neq j)\end{cases}$$

Namely $\quad F_{ij}\, x = -\frac{1}{2}\, e_j(\bar{e}_i x) \qquad (1 \leqq i \neq j \leqq 7)$

Then for $i,j \geqq 1$ and $i \neq j$

$$F_{ij}\, x + F_{ji}\, x = \frac{1}{2}\{e_j(e_i x) + e_i(e_j x)\} = -\frac{1}{2}\{e_j(\bar{e}_i x) + e_i(\bar{e}_j x)\} = -\langle e_i, e_j\rangle = 0$$

by (1.8). Thus

(4) $F_{ij} + F_{ji} = 0. \qquad (0 \leqq i \neq j \leqq 7)$

Since $2\langle F_{ij}x, y\rangle = -\langle e_j(\bar{e}_i x), y\rangle = -\langle \bar{e}_i x, \bar{e}_j y\rangle = -\langle x, e_i(\bar{e}_j y)\rangle = 2\langle x, F_{ji} y\rangle$, $F_{ij} \in \mathcal{D}_4$.

We can verify easily

(5) $$\begin{cases} [F_{ij}, F_{jk}] = F_{ik} & \\ [F_{ij}, F_{kl}] = 0 & \text{if } i,j,k,l \text{ are all different.} \end{cases}$$

By (2),(4) we can define linear transformation π of $\mathcal{D}_4$. Such that

(6) $$\pi(G_{ij}) = F_{ij} \qquad (0 \leqq i \neq j \leqq 7)$$

By (2),(5) π is a Lie algebra endomorphisms of $\mathcal{D}_4$. Since $\mathcal{D}_4$ is simple and $\pi \neq 0$, we conclude that π is an automorphism of $\mathcal{D}_4$.

Since $F_{10} = \frac{1}{2} L_{e_1}$ is represented as

$$\frac{1}{2}\begin{pmatrix} 0 & -1 & & & & & & \\ 1 & 0 & & & & & & \\ & & 0 & -1 & & & & \\ & & 1 & 0 & & & & \\ & & & & 0 & -1 & & \\ & & & & 1 & 0 & & \\ & & & & & & 0 & -1 \\ & & & & & & 1 & 0 \end{pmatrix}$$

$\det F_{10} = (1/4)^4$. On the other hand $\det G_{10} = 1$. Therefore G_{10} and F_{10} are not conjugate as matrices. Thus π is an outer automorphism of $\mathcal{D}_4$.

Let γ be the conjugation of $\mathbb{D}$ and define the automorphism κ of $\mathcal{D}_4$ by (7)

$\kappa(X) = \kappa X \gamma \qquad (x \in \mathcal{D}_4)$, i.e. $\kappa(X)x = \overline{X\bar{x}} \quad (x \in \mathbb{D})$

we see $\kappa(G_{10}) = -G_{10}$. $\kappa(G_{ij}) = G_{ij} \quad (i,j \geqq 0)$. Therefore $\det \kappa = (-1)^7$. Since $Ad(SO(8))$ is a connected simple group, $Ad(SO(8))$ consists of transformations of determinant 1. Therefore κ is an outer automorphism.

2.2 Infinitesimal principle of triality.

Note that L_a and R_a belong to $\mathcal{D}_4$ if $\mathrm{Re}\, a = 0$ by(1.3). Put $T_a = L_a + R_a$. Since $[a,x,y] = [x,y,a]$ by Lemma 1.1, we get

(7) $$L_a x \cdot y + x \cdot R_a y = T_a(xy)$$

we have $\quad L_{e_i} x = e_i x = 2F_{i0} x$

$$T_{e_i} x = e_i x + x e_i = \begin{cases} -2e_0 & x = e_i \\ 2e_i & x = e_0 \\ 0 & x = e_j \quad (j \neq i) \end{cases}$$

Therefore $L_{e_i} = 2F_{i0}$, $T_{e_i} = 2G_{i0}$, and so $\pi(T_{e_i}) = L_{e_i}$. It follows

(8) $\pi(T_a)=L_a$ (Re $a=0$).

Since $\overline{a\bar{x}}=-xa$, $\overline{\bar{x}a}=-ax$ (Re $a=0$).

(9) $\kappa L_a=-R_a$, $\kappa R_a=-L_a$.

Because $\frac{1}{4}[L_{e_i},L_{e_j}]=-[F_{i0},F_{0j}]=-F_{ij}$ $(1\leq i\neq j\leq 7)$, $\{L_a | \text{Re } a=0\}$ generates the Lie algebra $\mathcal{D}_4$. Define $F'_{ij}:\mathbb{D}\to\mathbb{D}$ by

(10) $F'_{ij}x = -\frac{1}{2}(x\bar{e}_i)e_j$ $(0\leq i\neq j\leq 7)$.

Then, we can discuss exactly as above using $\{F'_{ij}\}$ instead of $\{F_{ij}\}$. Thus $\{R_a; \text{Re } a=0\}$ generate the Lie algebra $\mathcal{D}_4$, and the mapping $F_{ij}\to F'_{ij}$ defines an automorphism λ of $\mathcal{D}_4$. Since $F_{i0}=L_{e_i}$, $F'_{i0}=R_{e_i}$, we see

(11) $\lambda(L_a)=R_a$ (Re $a=0$).

Combining (7)(8)(9), we get

(12) $(L_ax)y+x((\lambda L_a)y)=(\pi^{-1}L_a)(xy)$ (Re $a=0$).

Theorem 1. (Infinitesimal principle of triality).

For any $X\in\mathcal{D}_4$, there exists $Y,Z\in\mathcal{D}_4$ such that

(13) $(Xx)y+x(Yy)=Z(xy)$ $(x,y\in\mathbb{D})$.

Moreover, Y,Z is uniquely determined by X, In more detail we have $Y=\lambda X$, $Z=\pi^{-1}X$.

Proof. Let End($\mathbb{D}$) be the Lie algebra of all linear transformation of $\mathbb{D}$. Put

$$g=\{(X,Y,Z)\in \text{End}(\mathbb{D})^3 \,|\, (X,Y,Z) \text{ satisfies (13)}\}$$

Then, it is easy to verify that g is a subalgebra of the direct sum End$(\mathbb{D})^3$. By (12),

$(La,\lambda La, \pi^{-1}La)\in g$ (Re $a=0$), and the map

$La\to(La,\lambda La,\pi^{-1}La)$

gives rise to a homomorphism of $\mathcal{D}_4$ into g, since $\{L_a\}$ generate $\mathcal{D}_4$ and λ,π^{-1} are automorphisms of $\mathcal{D}_4$, this means that $(X,\lambda X,\pi^{-1}X)$ satisfies (13) for any $X\in\mathcal{D}_4$. Putting $Y=\lambda X$, $Z=\pi^{-1}X$, we have proved the first assertion.

To show the uniqueness, it is sufficient to show $Y=Z=0$, if $X=0$. If $X=0$, we get $x(Yy)=Z(xy)$, for any $X,Y\in\mathbb{D}$. Letting $x=e_0$, we see $Y=Z$, Letting $a=Ye_0$, we have $Zx=xa$. It follows that, $x(ya)=(xy)a$ for any $x,y\in\mathbb{D}$. Then $a=\alpha e_0$ with some $\alpha\in\mathbb{R}$. Since Z is skew-symmetric, we get $a=0$, and thus $Y=Z=0$.

q.e.d. q.e.d.

By Lemma 1.1, $[a,x,y] = -[x,a,y] = [x,y,a]$

Therefore $a(xy)+x(ay)=(ax)y+(xa)y$,

$(xy)y+(xy)a=x(ay)+x(ya)$

so $(T_ax)y-x(L_ay)=L_a(xy)$, $(R_ax)y-x(T_ay)=-R_a(xy)$. By Theorem 1, we get,

(14) $$\begin{cases} -L_a = \lambda T_a & L_a = \pi^{-1}T_a \\ -T_a = \lambda R_a, & -R_a = \pi^{-1}R_a \ (\text{Re } a=0) \end{cases}$$

Combining (9),(11),(14), we get

Lemma 1. For a with Re $a=0$,

1) $\pi : L_a \to T_a \to L_a$, $\qquad \pi : R_a \to -R_a \to R_a$

2) $\lambda : L_a \to R_a \to -T_a \to L_a$

3) $\kappa : L_a \to -R_a \to L_a$, $\qquad \kappa : T_a \to -T_a \to T_a$

since $\{L_a \mid \mathrm{Re}\ a=0\}$ generate $\mathcal{D}_4$. we get

Theorem 2. In the automorphism group of $\mathcal{D}_4$, we have

$$\pi^2 = \kappa^2 = \lambda^3 = 1, \qquad \kappa\lambda^2 = \lambda\kappa = \pi$$

$\{\pi, \kappa\}$ generates a finite subgroup in $\mathrm{Aut}(\mathcal{D}_4)$ isomorphic to the symmetric group of order 3.

The formula (13) is then written as $(Xx)y+x((\lambda X)y)=(\kappa\lambda^2 X)(xy)^{\sim}$ Therefore we have

Supplement to Theorem 1. When we write (13) in the form $(Xx)y+x(Yy)=(\kappa Z)(xy)$, then (X, Y, Z) is the orbit $(X, \lambda X, \lambda^2 X)$ of the automorphism λ of order 3. In particular any one of (X, Y, Z) determines the remaining two.

Theorem 3. The Lie algebra g_2 of the automorphism group G_2 of Cayley numbers is $g_2=\{X \in \mathcal{D}_4 \mid \lambda X=\pi X=X\}=\{X \in \mathcal{D}_4 \mid \lambda X=\kappa X=X\}$; g_2 is the simple Lie algebra of type G_2.

Proof By Proposition 1.1, we know that $G_2 \subset O(7) \subset O(8)$. Therefore $g_2 \subset \mathcal{D}_4$. On the other hand, we know that the Lie algebra g_2 consists of all derivations of $\mathbb{D}$, namely of linear transformations χ of $\mathbb{D}$ such that $(Xx)y + x(Xy)=X(xy)$. By Theorem 1 and its supplement. g_2 consists of all elements $\chi \in \mathcal{D}_4$ such that $X=\lambda X=\pi X$, or equivalently such that $X=\lambda X=\kappa X$.

To prove the second half, put $X=\sum_{i\neq j}\alpha_{ij}G_{ij}$ $(\alpha_{ij}+\alpha_{ji}=0)$ for $X \in g_2$. Since $Xe_0=0$, $\alpha_{i0}=0$ for $i \geq 1$. Now $\lambda X=\sum_{i\neq j}\alpha_{ij}F_{ij}$, since $X=\lambda X$, $(\lambda X)e_0=0$, and we have $\sum_{i\neq j}\alpha_{ij}e_j e_i=0$. We get then seven relations $\alpha_{23}+\alpha_{45}+\alpha_{76}=0$ etc, as coefficients of $e_1, e_2, \cdots, e_7$ in the above relation. Therefore X is a linear combinations of the following elements

$$(7) \quad \begin{cases} \alpha G_{23} + \beta G_{45} + \gamma G_{76} \\ \alpha G_{12} + \beta G_{47} + \gamma G_{65} \\ \alpha G_{14} + \beta G_{72} + \gamma G_{36} \\ \alpha G_{31} + \beta G_{46} + \gamma G_{57} \\ \alpha G_{51} + \beta G_{62} + \gamma G_{73} \\ \alpha G_{71} + \beta G_{42} + \gamma G_{53} \\ \alpha G_{61} + \beta G_{25} + \gamma G_{34} \end{cases}$$

where $\alpha+\beta+\gamma=0$. These elements belongs to g_2. In fact, each one of these, X, is such that $Xe_0=(\lambda X)e_0=0$. Therefore $\kappa X=\kappa\lambda X=X$ and so $\kappa X=\lambda X=X$. It follows that $\dim g_2=14$, we see easily $\{G_{23}+G_{45}+G_{76} \mid \alpha+\beta+\gamma=0\}$ is a maximal abelian subalgebra of dim 2 of g_2. The group G_2 is compact. Thus, as the Lie algebra of compact group of dim 14, and of rank 2, g_2 is simple of type G_2.

q.e.d.

Remark: (7) determines explicit forms of matrices belonging to g_2, namely g_2 consists of matrices $A=(a_{ij})$ where a_{ij} $(1\leq i,\ j\leq 7)$ satisfy $a_{ij}+a_{ji}=0$ and equations corresponding to (7); e.g., $a_{23}+a_{45}+a_{76}=0$ etc.

§3 Principle of triality II

The conjugation in $\mathbb{D}$ defines an automorphism, denoted also κ, of the group $O(8)$ which induces the automorphism κ on the Lie algebra $\mathcal{D}_4$. By definition, $(\kappa\theta)(x)=\overline{\theta(\overline{x})}$ for $\theta\in SO(8)$ and κ defines an outer automorphism of $SO(8)$.

Theorem 1 (Principle of triality)

For each $\theta_1\in SO(8)$, there exist $\theta_2,\theta_3\in SO(8)$, such that $\theta_1(x)\theta_2(y)=\theta_3(xy)$ for $x,y\in\mathbb{D}$. Moreover, if θ_2', θ_3' satisfy the same condition as θ_2, θ_3, then $\theta_2'=\varepsilon\theta_2$, $\theta_3'=\varepsilon\theta_3$ with $\varepsilon=\pm 1$.

Proof: Since $SO(8)$ is a compact connected Lie group, we may write $\theta_1=\exp X$ with $X\in\mathcal{D}_4$, let Y and Z be elements in $\mathcal{D}_4$ determined by the infinitesimal principle of triality for X namely, we have $(Xx)y+x(Yy)=Z(xy)$ for any $x,y\in\mathbb{D}$. Then

$$Z^2(xy) = (X^2x)y+2(Xx)(Yy)+x(Y^2y)$$

and, by induction, we see

$$Z^p(xy)=\sum_{i+j=p}\frac{p!}{i!j!}(X^ix)(Y^jy) \qquad (p\geq 1)$$

Using this, we can easily prove

$$(\exp Z)(xy) = \sum_{p=0}^{\infty}\frac{1}{p!}Z^p(xy) = \sum_{p=0}^{\infty}\sum_{i+j=p}\frac{1}{i!j!}(X^ix)(Y^jy)$$
$$=(\exp X)(x)\cdot(\exp Y)(y)$$

For $\theta_1=\exp X$, $\theta_2=\exp Y$ and $\theta_3=\exp Z$ satisfy the requirement. To show the uniqueness of (θ_1,θ_2) up to sign, assume first $\theta_1=1$, then $x(\theta_2y)=\theta_3(xy)$. Putting $x=e_0$, we see $\theta_2=\theta_3$. Put $\theta_2e_0=a$. Then $\theta_3(x)=xa$. And we have $x(ya)=(xy)a$. This implies $a=\alpha e_0$ with $\alpha\in\mathbb{R}$, Since $\theta_3\in SO(8)$, $a=\pm 1$. Therefore we have have $\theta_1=\theta_2=\pm 1$. Suppose in general $\theta_1(x)\theta_2(y)=\theta_3(xy)$ and $\theta_1(x)\theta_2'y)=\theta_3'(xy)$. Then

$$\theta_3(\theta_1^{-1}(x)\theta_2^{-1}(y))=\theta_1\ \theta_1^{-1}(x))\theta_2\ \theta_2^{-1}(y)=xy,\quad \theta_1^{-1}(x)\theta_2^{-1}(y)=\theta_3^{-1}(xy).$$

Applying θ_3' we get $(\theta_3'\theta_3^{-1})(xy)=\theta_1'(\theta_1^{-1}(x))\theta_2'(\theta_1^{-1}(y))=\theta_1(\theta_2'\theta_1^{-1}(y))$.

By the result proved above for the case $\theta_1=1$, if follows

$$\theta_3'\theta_3^{-1} = \theta_2'\theta_2^{-1}=\pm 1$$

which proves the theorem.

Lemma 1 Let θ_1, θ_2, $\theta_3\in O(8)$ and assume that

$$\theta_1(x)\theta_2(y) = (\kappa\theta_3)(xy)$$

for any $x,y\in\mathbb{D}$. Then

i) $\theta_2(x)\theta_3(y)=(\kappa\theta_1)(xy)$, $\theta_3(x)\theta_1(y)=(\kappa\theta_2)(xy)$ $(x,y\in\mathbb{D})$

ii) $\theta_1,\theta_2,\theta_3$ belong to SO(8).

Proof. (i) The assumption is : $\theta_1(x)\theta_2(y)=\overline{\theta_3(xy)}$ $(x,y\in\mathbb{D})$.
Then $\overline{\theta_2(y)}\,\overline{\theta_1(x)}=\theta_3(\overline{xy})$, multiplying $\theta_2(y)/|y|^2$ to both side. Since $|\theta_2(y)|^2=|y|^2$, we get

$$\overline{\theta_1(x)} = \theta_2(y)\theta_3(\overline{xy})/|y|^2 \quad \text{(we may assume } y\neq 0)$$

Put x=yz. Then it follows $\overline{\theta_1(\overline{yz})}=\theta_2(y)\theta_3(\bar{y}(yz)/|y|^2=\theta_2(y)\theta_3(z)$. This proves $\theta_2(x)\theta_3(y)=(\kappa\theta_1)(xy)$ $(x,y\in\mathbb{D})$.
The second formula follows then obviously.

(ii) Assuming that at least one of $\theta_1,\theta_2,\theta_3$ does not belong to SO(8), we shall derive a contradiction. By (i), we may assume $\theta_1\notin SO(8)$. By Theorem 1, for any $\tau_1\in SO(8)$, we find $\tau_2,\tau_3\in SO(8)$ such that

$$\tau_1(\theta_1(x))\tau_2(\theta_2(y)) = (\kappa\tau_3)(\kappa\theta_3(xy)),$$

$$(\tau_1\theta_1)(x)(\tau_2\theta_2)(y) = (\kappa(\tau_3\theta_3))(xy).$$

Let $\gamma\in O(8)$ be the conjugation of $\mathbb{D}$; $\gamma x=\bar{x}$. There exists $\tau_1\in SO(8)$ such that $\tau_1\theta_1=\gamma$. In this case, put $\xi_2=\tau_2\theta_2$ and $\xi_3=\tau_3\theta_3$. Then, we get

(*) $\qquad \bar{x}\xi_2(y) = (\kappa\xi_3)(xy) \quad (x,y\in\mathbb{D})$

Putting $x=e_0$ or $y=e_0$, we see $\xi_2=\kappa\xi_3$, $\bar{x}\xi_2(e_0)=\kappa\xi_3(x)$ and $\xi_2(x)=\bar{x}\cdot a$, where $a=\xi_2(e_0)\neq 0$. Therefore, by (*), $\bar{x}(\bar{y}a)=(\overline{xy})a=(\bar{y}\cdot\bar{x})a$,

(**) $\qquad x(ya) = (yx)a, \quad (x,y\in\mathbb{D})$.

Since $a(ya)=(ay)a$, it follows $(ay)a=(ya)a$ and $ay=ya$ $(y\in\mathbb{D})$. Combining $|a|=|e_0|=1$, it follows $a=\pm e_0$. Then by (**) $xy=yx$ for any $x,y\in\mathbb{D}$, which is a contradiction.

q.e.d.

Subgroups of SO(8).

As in §1, we shall use the identification $SO(7)=\{\theta\in SO(8)\,|\,\theta(e_0)=e_0\}$.

Proposition 1. An element $\theta\in SO(8)$ belongs to SO(7) if and only if there exists $\tilde{\theta}\in SO(8)$ such that

(1) $\qquad \theta(x)\tilde{\theta}(y)=\tilde{\theta}(xy)$ for any $x,y\in\mathbb{D}$.

Proof. Let θ_2 and θ_3 be elements associated to $\theta=\theta_1$ by Theorem 1. Namely $\theta(x)\theta_2(y)=\theta_3(xy)$ $(x,y\in\mathbb{D})$. Then $\theta(e_0)\theta_2(y)=\theta_3(y)$ $(y\in\mathbb{D})$. Therefore $\theta(e_0)=e_0$ if and only if $\theta_2=\theta_3$, q.e.d.

q.e.d.

Analogously, we get

Proposition 1'. An element $\theta\in SO(8)$ belongs to SO(7) if and only if there exists $\tilde{\theta}'\in SO(8)$ such that

(2) $\qquad \tilde{\theta}'(x)\theta(y)=\tilde{\theta}'(xy)$ for any $x,y\in\mathbb{D}$

Proof. Taking conjugate of (1) $\overline{\tilde{\theta}(y)}\,\overline{\theta(x)}=\overline{\tilde{\theta}(xy)}$, and so

$$(\kappa\tilde{\theta})(\bar{y})(\kappa\theta)(\bar{x}) = \kappa\tilde{\theta}(\overline{xy})$$

Since $\kappa\theta=\theta$ for $\theta\in SO(7)$, we have $(\kappa\tilde{\theta})(x)\theta(y)=(\kappa\tilde{\theta})(xy)$. Thus $\tilde{\theta}'=\kappa\tilde{\theta}$ satisfies (2).

then $\kappa\tilde{\theta}'$ satisfies (1) and θ belongs to SO(7).

Definition. Put $\text{Spin}^+(7) = \{\tilde{\theta} \mid (1) \text{ holds for some } \theta \in SO(7)\}$,

$\text{Spin}^-(7) = \{\tilde{\theta}' \mid (2) \text{ holds for some } \theta \in SO(7)\}$.

We see easily that $\text{Spin}^{\pm}(7)$ are closed subgroups of SO(8). By the proof of Prop.1', we see $\text{Spin}^-(7) = \kappa(\text{Spin}^+(7))$ or

(3) $$\text{Spin}^-(7) = \gamma \text{Spin}^+(7)\gamma^{-1}.$$

Proposition 2. i) The action of $\text{Spin}^{\pm}(7)$ on S^7 as subgroups of SO(8) is transitive on S^7, and the isotropy subgroup is G_2, $S^7 = \text{Spin}^{\pm}(7)/G_2$,

ii) Define the maps $p^{\pm}: \text{Spin}^{\pm}(7) \to SO(7)$ by (1) and (2) in an obvious way. Then the groups $\text{Spin}^{\pm}(7)$ are double covering group of SO(7) by these maps.

Proof. It is easy to check that $p^+: \tilde{\theta} \to \theta$ is a homomorphism of $\text{Spin}^+(7)$ onto SO(7). The kernel is $\{\pm 1\}$ by Theorem 1. Therefore $\text{Spin}^+(7)$ is locally isomorphic to SO(7), and so, $\dim \text{Spin}^+(7) = 21$. On the other hand, if $\theta \in G_2$, then $\tilde{\theta} = \theta$. Therefore $G_2 \subset \subset \text{Spin}^+(7)$ and $\tilde{\theta}(e_0) = e_0$, i.e., $G_2 \subset \text{Spin}^+(7) \cap SO(7)$. Conversely, if $\theta \in \text{Spin}^+(7) \cap SO(7)$ then we see $\tilde{\theta} = \theta$ by (1), and $\theta \in G_2$. Thus $G_2 = \text{Spin}^+(7) \cap SO(7)$, which means that the isotropy subgroup at e_0 of $\text{Spin}^+(7)$ is G_2. Then we have a natural imbedding of $\text{Spin}^+(7)/G_2$ into S^7. But $\dim \text{Spin}^+(7)/G_2 = 21-14 = 7$. Thus, $S^7 = \text{Spin}^+(7)/G_2$ and $\text{Spin}^+(7)$ acts transitively on S^7.

Since S^7 and G_2 are connected $\text{Spin}^+(7)$ is connected. Therefore $p^+: \tilde{\theta} \to \theta$, is a double covering map.

Similar argument holds for $\text{Spin}^-(7)$. q.e.d.

Proposition 3. $\text{Spin}^+(7) \cap \text{Spin}^-(7) = G_2$.

Proof. Obviously $G_2 \subset \text{Spin}^+(7) \cap \text{Spin}^-(7)$. Let $\tilde{\theta} \in \text{Spin}^+(7) \cap \text{Spin}^-(7)$. Then there are $\theta_1, \theta_2 \in SO(7)$ such that $\theta_1(x)\tilde{\theta}(y) = \tilde{\theta}(xy) = \tilde{\theta}(x)\theta_2(y)$. Putting x or $y = e_0$, we see $\theta_1 = \tilde{\theta} = \theta_2$ and so $\tilde{\theta}$ belong to G_2. q.e.d.

Now, the left multiplication L_{e_1} defines a complex structure on the 8-dimensional real vector space $\mathbb{D}$. Put

$$\tilde{H} = \{\tilde{\theta} \in SO(8) \mid \tilde{\theta} L_{e_1} = L_{e_1}\tilde{\theta}\},$$

then $\tilde{H} \cong U(4)$.

Proof. Let E be the complex vector space $\mathbb{D}$ endowed, with the complex structure L_{e_1} Define a Hermitian inner product in E, by $((x,y)) = \langle x,y\rangle - \sqrt{-1}\langle L_{e_1}x, y\rangle$ $(x,y \in \mathbb{D})$. Then any element of H is a unitary transformation of E, and conversely. q.e.d.

Let H be the commutator subgroup of $\tilde{H}$. Since $\tilde{H} \cong U(4)$, we see

$$H \cong SU(4) \text{ and } \dim H = 15.$$

Proposition 4. We have $H = (\tilde{H} \cap \text{Spin}^+(7))^\circ$, the superscript $\circ$ denoting the identity component. The map $p^+: \text{Spin}^+(7) \to SO(7)$ induces covering map $H \to SO(6)$ where $SO(6) = \{\theta \in SO(7) \mid \theta(e_1) = e_1\}$.

Proof. Suppose $\theta \in SO(6)$. Then there exists $\tilde{\theta} \in \text{Spin}^+(7)$, such that $\theta(x)\tilde{\theta}(y) = \tilde{\theta}(xy)$. Put $x = e_1$ then $e_1\tilde{\theta}(y) = \tilde{\theta}(e_1 y)$, which shows $\tilde{\theta} \in \tilde{H} \cap \text{Spin}^+(7)$. Conversely, if $\tilde{\theta} \in \tilde{H} \cap \text{Spin}^+(7)$, then $\theta = p^+(\tilde{\theta})$ satisfies $\theta(e_1)\tilde{\theta}(y) = \tilde{\theta}(e_1 y) = e_1\tilde{\theta}(y)$, and $\theta(e_1) = e_1$, which means $\theta \in SO(6)$.

Therefore the map p^+ induces a homomorphism from $H \cap \text{Spin}^+(7)$ onto SO(6), which is obviously locally isomorphic. Therefore $\tilde{H} \cap \mathrm{Spin}^+(7)$ is a Lie group whose dimension is dim $SO(6)=15$ and whose Lie algebra is simple. Since H is the commutor subgroup of $\tilde{H}$, H contains the connected component of the identify $(\tilde{H} \cap \mathrm{Spin}^+(7))^\circ$, this being equal to its own commutator group. Comparing the dimension, we conclude $H = (\tilde{H} \cap \mathrm{Spin}^+(7))^\circ$.

Remark. This proves also that $\mathrm{Spin}^+(7)$ acts transitively on S^7, because H does so. It follows that the group $\mathrm{Spin}^+(7)$ acts irreducibly on the vector space $\mathbb{D}$.

Definition. We put $\mathrm{Spin}(8)=\{\theta=(\theta_1,\theta_2,\theta_3)\,|\,\theta_1(x)\theta_2(y)=(\kappa\theta_3)(xy)\}\subset SO(8)\times SO(8)\times SO(8)$. This is a closed subgroup of $SO(8)\times SO(8)\times SO(8)$. The Lie algebra Lie (Spin(8)) is Lie $(\mathrm{Spin}(8))=\{(\lambda X,\lambda^2 X,X)\,|\,X\in \mathfrak{D}_4\}$. This follows easily from supplement to Theorem 2.1, and the proof of Theorem 1.

The projections of $SO(8)\times SO(8)\times SO(8)$ on its factors define homomorphisms Δ^+,Δ^-, $\Delta_\circ$ of Spin(8) to SO(8), which are onto by Theorem 1. By definition, we have $\Delta^+(\theta)=\theta_1$, $\Delta^-(\theta)=\theta_2$, $\Delta_\circ(\theta)=\theta_3$ for $\theta=(\theta_1,\theta_2,\theta_3)\in\mathrm{Spin}(8)$. Then $(\Delta^+)^{-1}(SO(7))=$ $=\{(\theta,\tilde{\theta},\kappa\tilde{\theta})\,|\,\tilde{\theta}\in\mathrm{Spin}^+(7)\}$, $(\Delta^-)^-(SO(7))=\{(\tilde{\theta},\theta,\kappa\tilde{\theta})\,|\,\tilde{\theta}\in\mathrm{Spin}^-(7)\}$. The restrictions of Δ^- and Δ^+ define the isomorphism

(4) $$(\Delta^\pm)^{-1}(SO(7))\cong\mathrm{Spin}^\pm(7).$$

We shall consider $\mathrm{Spin}^\pm(7)\subset\mathrm{Spin}(8)$ by these isomorphisms.

The group Spin(8) acts on the sphere S^7 transitively through homomorphism Δ^+. The isotropy subgroup at the point $e_\circ$ is $(\Delta^+)^{-1}(SO(7))=\mathrm{Spin}^+(7)$. Thus $\mathrm{Spin}(8)/$ $/\mathrm{Spin}^+(7)=S^7$. Since $\mathrm{Spin}^+(7)$ is connected, it follows that Spin(8) is connected. By Theorem 1, the kernel of Δ^+ is of order 2. Therefore, Δ^+ gives the double covering Spin(8) of SO(8). By Lemma 1, the same is true for Δ^- and $\Delta_\circ$.

Since SO(8) acts transitively on S^7, $\Delta^\pm$ gives rise to irreducible representations of Spin(8). Moreover, we have

$$\mathrm{Ker}\Delta^+ = \{(1,\pm1,\pm1)\},\qquad \mathrm{Ker}\Delta^- = \{(\pm1,1,\pm1)\}$$

and $\mathrm{Ker}\Delta^+\cap\mathrm{Ker}\Delta^- = \{(1,1,1)\}$. Therefore Δ^+ and Δ^- are inequivalent representations of Spin(8), and their sum $\Delta^+\oplus\Delta^-$ is a faithful representation of Spin(8).

Let $(\theta_1,\theta_2,\theta_3)\in\mathrm{Spin}(8)$. By definition, we have $\theta_1(x)\theta_2(y) = (\kappa\theta_3)(xy)$ for any $x,y\in\mathbb{D}$. Then, $\overline{\theta_2(y)}\,\overline{\theta_1(x)}=\theta_3(\bar{y}\bar{x})$. Replacing x and y by $\bar{y}$ and $\bar{x}$, respectively, we have $(\kappa\theta_2)(x)(\kappa\theta_1)(y) = \theta_3(xy)$. Thus $(\kappa\theta_2,\ \kappa\theta_1,\ \kappa\theta_3)\in\mathrm{Spin}(8)$.

Together with Lemma 1. We may now put

Definition: We define the automorphisms κ, λ, π of the group Spin(8) as follows:

$$\kappa(\theta_1,\theta_2,\theta_3) = (\kappa\theta_2,\kappa\theta_1,\kappa\theta_3),$$

$$\lambda(\theta_1,\theta_2,\theta_3) = (\theta_2,\theta_3,\theta_1),$$

$$\pi(\theta_1,\theta_2,\theta_3) = (\kappa\theta_1,\kappa\theta_3,\kappa\theta_2).$$

We have

(5) $$\Delta^+ = \Delta_\circ \circ \lambda, \qquad \Delta^- = \Delta_\circ \circ \lambda^2.$$

Now we study the Lie algebra Lie (Spin(8)) of the Lie group Spin(8), The Lie algebra of the group $SO(8) \times SO(8) \times SO(8)$ is the direct Sum $\mathcal{D}_4^3 = \mathcal{D}_4 + \mathcal{D}_4 + \mathcal{D}_4$, and an element $(x,y,z) \in \mathcal{D}_4^3$ belongs to Lie (Spin(8)) if and only if,

$$\text{expt } (X,Y,Z) = (\text{expt } X, \text{ expt } Y, \text{ expt } Z)$$

belongs to Spin(8) for all $t \in \mathbb{R}$. Therefore

$$((\text{expt } X)x)((\text{expt } Y)y) = (\kappa \text{expt } Y)(xy)$$

For any $x,y \in \mathbb{D}$ and $t \in \mathbb{R}$, By differentiating at $t = 0$, we have

$$(Xx)y + x(Yy) = (\kappa(z))(xy), \qquad (x,y \in \mathbb{D}).$$

Conversely, $(X,Y,Z) \in \mathcal{D}_4^3$ satisfying this condition belongs to Lie (Spin(8)) (cf. Proof of Theorem 1). By the supplement to Theorem 2.1, we conclude that Lie (Spin (8)) consists of $(X,Y,Z) \in \mathcal{D}_4^3$, which represent cyclic orbit of the automorphism λ of $\mathcal{D}_4$.

Since Δ^+, Δ^-, $\Delta_\circ$ are locally isomorphic homomorphism of Spin(8) onto SO(8), their differentials $d\Delta^+$, $d\Delta^-$, $d\Delta_\circ$ are isomorphisms of Lie (Spin(8)) onto $\mathcal{D}_4$. We shall identify Lie (Spin(8)) with $\mathcal{D}_4$ by the isomorphism $d\Delta_\circ$. This means that $X \in \mathcal{D}_4$ is identified with $(\lambda X, \lambda^2 X, X) \in$ Lie (Spin(8)); $X = (\lambda X, \lambda^2 X, X)$. We see that the automorphisms κ, λ, π of the group Spin(8) defined above induce, under this identification, the automorphisms κ, λ, π of $\mathcal{D}_4$ defined in §2 (6),(8),(11), repectively. If follows from (5) that the differentials $d\Delta^\pm$: Lie (Spin(8)) $\rightarrow \mathcal{D}_4$ are represented by

(6) $$d\Delta^+ = \lambda, \qquad d\Delta^- = \lambda^2.$$

Finally, the restrictions of $\Delta^\pm$ to $\text{Spin}^\pm(7)$ $(=(\Delta^\pm)^{-1}(SO(7)))$ are covering homomorphisms $p^\pm$: $\text{Spin}^\pm(7) \rightarrow SO(7)$. When we identify Lie $(\text{Spin}^\pm(7))$ with Lie (SO(7)) by the differentials $dp^\pm$, the injective map $\Delta^\pm$: $\text{Spin}^\pm(7) \rightarrow SO(8)$ have λ and λ^2 as their differentials. In fact $Y \in$ Lie (SO(7)) is identified with $(Y, \lambda Y, \lambda^2 Y) \in$ Lie $(\text{Spin}^+(7)) \subset$ Lie (Spin(8)) by dp^+ and this is mapped to λY by the differential $d\Delta^-$.

§4. Exceptional group F_4

In this and next sections we follow mostly Frendenthal [9].

4.1 Jordan algebra $\mathfrak{J}$

Definition. Let $\mathfrak{J}$ be the set of hermitian 3×3 matrices with components in $\mathbb{D}$, equipped with the product

$$X \circ Y = \frac{1}{2}(XY + YX).$$

$\mathfrak{J}$ is then a real Jordan algebra. A real Jordan algebra is, by definition, a finite-dimensional real vector space with multiplication $\circ$ which is commutative and such that

$$X^2\circ(X\circ Y) = X\circ(X^2\circ Y)$$

for any two elements X, Y, where $X^2=X\circ X$.

A general element $X \in \mathfrak{J}$ is of the form

$$(1) \qquad X = \begin{pmatrix} \xi_1 & a_3 & \bar{a}_2 \\ \bar{a}_3 & \xi_2 & a_1 \\ a_2 & \bar{a}_1 & \xi_3 \end{pmatrix} \qquad \text{where } \xi_1,\xi_2,\xi_3 \in \mathbb{R}, \quad a_1,a_2,a_3 \in \mathbb{D}.$$

We express this X by $X=X(\xi,a)=X(\xi_1,\xi_2,\xi_3,a_1,a_2,a_3)$. The real vector space $\mathfrak{J}$ is spanned by the following elements:

$$E_1 = \begin{pmatrix} 1 & & \\ & 0 & \\ & & 0 \end{pmatrix}, \qquad E_2 = \begin{pmatrix} 0 & & \\ & 1 & \\ & & 0 \end{pmatrix}, \qquad E_3 = \begin{pmatrix} 0 & & \\ & 0 & \\ & & 1 \end{pmatrix},$$

$$F_1(a) = \begin{pmatrix} 0 & & \\ & 0 & a \\ & \bar{a} & 0 \end{pmatrix}, \qquad F_2(a) = \begin{pmatrix} 0 & & \bar{a} \\ & 0 & \\ a & & 0 \end{pmatrix}, \qquad F_3(a) = \begin{pmatrix} 0 & a & \\ \bar{a} & 0 & \\ & & 0 \end{pmatrix},$$

where $a \in \mathbb{D}$ and components not written are zero. We get

$$(2) \qquad \begin{cases} E_i \circ E_i = E_i & E_i \circ E_j = 0 \quad (i\neq j) \\ E_i \circ F_i(a) = 0 & 2E_i \circ F_j(a) = F_j(a) \quad (i\neq j) \\ F_i(a) \circ F_i(b) = a,b\,(E_{i+1} + E_{i+2}) & \\ 2F_i(a) \circ F_{i+1}(6) = F_{i+2}(\overline{ab}) & \end{cases}$$

where $(i,i+1,i+2)$ is counted modulo 3.

For $X,Y \in \mathfrak{J}$, let trX be the sum of diagonal elements of X and put

(3) $\qquad (X, Y) = \operatorname{tr}(X \circ Y)$

Then (X,Y) is a real bilinear form on $\mathfrak{J}$, and

$$(4) \qquad \begin{cases} (X, Y) = (Y, X) \\ (X, X) \geqq 0, \quad \text{and} \quad (X, X) = 0 \Longleftrightarrow X = 0. \end{cases}$$

In fact, the first is obvious, and, if X is of the form (1)

$$(X, X) = \operatorname{tr}(X \circ X) = \xi_1^2+\xi_2^2+\xi_3^2+|a_1|^2+|a_2|^2+|a_3|^2 \geqq 0.$$

Definition. Let F_4 be the group of the automorphisms of $\mathfrak{J}$. This a closed subgroup of the real linear transformation group of $\mathfrak{J}$. As we shall see later, F_4 is a subgroup of orthogonal transformations in $\mathfrak{J}$, and so F_4 is a compact Lie group.

Theorem 1. Let N be the subgroup of F_4 consisting of elements α such that

$$\alpha(E_i) = E_i, \qquad (i=1,2,3).$$

Then $\qquad N \cong \mathrm{Spin}\,(8).$

Proof. Let $\alpha \in N$. Then by (2) $E_i \circ \alpha(F_i(a)) = 0$,

$$2E_j \circ \alpha(F_i(a)) = \alpha(F_i(a)) \qquad (i \neq j).$$

Therefore we may put $\alpha(F_i(a))=F_i(\theta_i(a))$ $(i=1,2,3)$. Then θ_i is a real linear transformation of $\mathbb{D}$.

Again by (2), we get $F_i(\theta_i(a)) \circ F_i(\theta_i(b))=\langle a,b\rangle(E_{i+1}+E_{i+2})$ and so $\langle\theta_i(a),\theta_i(b)\rangle=\langle a,b\rangle$. Therefore $\theta_i \in O(8)$, From the last equality of (2) follows

$$F_i(2a)\circ F_{i+1}(2b) = F_{i+2}(2\overline{ab}).$$

Applying α, we get

$$F_i(2\theta_i(a)) \circ F_{i+1}(2\theta_{i+1}(b)) = F_{i+2}(2\theta_{i+2}(\overline{ab}))$$

and so

$$\theta_i(a)\theta_{i+1})(b) = \overline{\theta_{i+2}(\overline{ab})}$$

Then, by Lemma 3.1, $\theta_1,\theta_2,\theta_3$ belong to $SO(8)$, and by the definition of Spin(8), $(\theta_1,\theta_2,\theta_3) \in \mathrm{Spin}(8)$.

Conversely, if $(\theta_1,\theta_2,\theta_3) \in \mathrm{Spin}(8)$, the map α of $\mathfrak{J}$ into itself defined by

$$\alpha(x)=\begin{pmatrix} \xi_1 & \theta_3(a_3) & \overline{\theta_2(a_2)} \\ \overline{\theta_3(a_3)} & \xi_2 & \theta_1(a_1) \\ \theta_2(a_2) & \overline{\theta_1(a_1)} & \xi_3 \end{pmatrix}$$

is an automorphism of $\mathfrak{J}$. By the correspondence α to $(\theta_1,\theta_2,\theta_3)$, the subgroup N is isomorphic to Spin(8). q.e.d.

4.2 The Lie algebra of F_4.

Let m_n be the real vector space formed by n×n matrices with components in $\mathbb{D}$. For $A \in m_n$, Put $A^*={}^t\bar{A}$ and

$$m_n^+ = \{A \in m_n \mid A^*=A\}, \quad m_n^-=\{A \in m_n \mid A^*=-A\}.$$

Thus $\mathfrak{J}=m_3^+$, Since $A=\frac{1}{2}(A+A^*)-\frac{1}{2}(A-A^*)$,

$m_n=m_n^+ +m_n^-$, $m_n^+ \cap m_n^- = \{0\}$. For $X\cdot Y \in m_n$, we have $(XY)^*=Y^*X^*$.

Define $(X,Y) = \mathrm{Re}\,\mathrm{tr}(XY)$.

Remark that, for $a,b \in \mathbb{D}$, we have $\mathrm{Re}\, ab=\langle e_0,ab\rangle=\langle\bar{a},b\rangle=\langle\overline{ab},e_0\rangle=\langle\overline{ba},e_0\rangle=\mathrm{Re}\,\overline{ba}=\mathrm{Re}\, ba$.

Therefore, for $X=(x_{ij}) \in m_n$, we have

$$\mathrm{Re}\,\mathrm{tr}(XY) = \mathrm{Re}(\sum_{ij} x_{ij}y_{ji}) = \mathrm{Re}(\sum_{ij} y_{ji}x_{ij})=\mathrm{Re}\,\mathrm{tr}(YX)$$

and so

(5) $$(X,Y) = (Y,X)$$

we see

(6) $$(X,X^*)=\sum_{ij} |x_{ij}|^2 > 0, \quad \text{for } X \neq 0$$

(7) $$(X,Y) = \mathrm{tr}(\tfrac{1}{2}(XY+Y^*X^*))$$

Proof.
$$(X,Y) = \text{Re } tr(XY) = \frac{1}{2}\{tr(XY) + tr(\overline{XY})\}$$
$$= \frac{1}{2}\{tr(XY) + tr(\bar{Y}\bar{X})\} = \frac{1}{2}\{tr(XY) + tr(Y^*X^*)\}$$
$$= \frac{1}{2}\, tr(XY + Y^*X^*)$$

In partivular, if both X, Y belong to $m_n^{\pm}$,

$$(X,Y) = tr[\tfrac{1}{2}(XY + YX)]. \quad (8)$$

We see therefore that in case n=3 (X,Y) coincides with that defined by (3) on $\mathfrak{J}=m_3^+$.

If $X \in m_n^+$, $Y \in m_n^-$, then $(X,Y)=-(Y,X)$ by (7), and, combining with (5), $(X,Y)=0$. Therefore m_n^+ and m_n^- are orthogonal w.r.t.$(\cdot,\cdot)$, and, by (6), $(\cdot,\cdot)$ is positive- and negative-definite on m_n^+ and m_n^- respectively.

Let $X=(x_{ij})$, $Y=(y_{ij})$, $Z=(z_{ij}) \in m_n$. Then, by Lemma 1.1,

$$z_{ij}(x_{jk}y_{ki}) + (x_{jk}y_{ki})z_{ij} = (z_{ij}x_{jk})y_{ki} + x_{jk}(y_{ki}z_{ij}).$$

Therefore $\quad tr(Z(XY)) + tr((XY)Z) = tr((ZX)Y) + tr(X(YZ))$.

Taking real part, we get $\quad 2(XY,Z) = (ZX,Y) + (X,YZ)$.

Therefore, we have also $\quad 2(YZ,X) = (XY,Z) + (Y,ZX)$.

From these two equalities, we get

$$(XY,Z) = (X,YZ) = (Y,ZX). \quad (9)$$

The second equality follows: $(ZX,Y) = (Z,XY) = (X,YZ)$. Denoting $[X,Y] = XY - YX$, we get then

$$([Z,X],Y) + (X,[Z,Y]) = 0. \quad (10)$$

We see easily

$$\begin{cases} [m_n^+, m_n^+] \subset m_n^+ \\ [m_n^-, m_n^-] \subset m_n^+ \\ [m_n^+, m_n^-] \subset m_n^+. \end{cases} \quad (11)$$

Now, <u>we assume n=3</u>. Thus $m_3^+=\mathfrak{J}$. Define the trilinear form (X,Y,Z) on $\mathfrak{J}$ by $(X,Y,Z)=(X\circ Y,Z)$.

Then by (9),
$$(X,Y,Z)=(X\circ Y,Z)=\frac{1}{2}\{(XY,Z)+(YX,Z)\}$$
$$=\frac{1}{2}\{(X,YZ)+(X,ZY)\}$$
$$=(X,Y\circ Z)$$

Therefore, (X,Y,Z) is symmetric in (X,Y) and (Y,Z) and so is a symmetric trilinear form on $\mathfrak{J}$.

<u>Lemma 1</u>. A linear transformation α which preserues (X,Y) and (X,Y,Z) is an

automorphism of $\mathfrak{J}$ and so belongs to F_4.

Proof $(\alpha(X\circ Y),Z) - (\alpha(X)\circ\alpha(Y),Z) = (X\circ Y,\alpha^{-1}(Z)) - (\alpha(X),\alpha(Y),Z)$

$= (X,Y,\alpha^{-1}(Z)) - (X,Y,\alpha^{-1}(Z)) = 0.$

Therefore $\alpha(X\circ Y) = \alpha(X) \circ \alpha(Y)$. q.e.d.

We shall denote by $\mathfrak{F}_4$ the Lie algebra of the automorphism group F_4 of the algebra $\mathfrak{J}$. $\mathfrak{F}_4$ consists of all derivations of $\mathfrak{J}$.

Lemma 2 Any linear transformation D of $\mathfrak{J}$ satisfying the following conditions belongs to $\mathfrak{F}_4$. $(D(X),Y) + (X,D(Y)) = 0$,

$$(D(X),Y,Z) + (X,D(Y),Z) + (X,Y,D(Z)) = 0$$

for any $X, Y, Z \in \mathfrak{J}$.

Proof. If D satisfyies these conditions, then

$$(D(X)\circ Y + X\circ D(Y) - D(X\circ Y),Z) = (D(X),Y,Z) + (X,D(Y),Z) - (D(X\circ Y),Z) = (D(X),Y,Z) + (X,D(Y),Z) + (X,Y,D(Z)) = 0$$

for any $Z \in \mathfrak{J}$. Therefore D is a derivation of $\mathfrak{J}$ and belongs to $\mathfrak{F}_4$. q.e.d.

Remark. By Lemmas 1 and 2, $\mathfrak{F}_4$ coincides with the Lie algebra of the group F_4 of linear transformation of $\mathfrak{J}$ preserving (X,Y) and (X,Y,Z). By (11), any $A \in m_3^-$ defines a linear transformation $\tilde{A}$ of $\mathfrak{J}$ by

(12) $$\tilde{A}(X) = [A,X].$$

Put $k = \{A \in m_3^- | \mathrm{tr}A = 0\}$. We shall prove $\tilde{A} \in \mathfrak{F}_4$ for $A \in k$. We show first

(13) $$X(XX) - X(XX) = a1_3 \quad (X \in \mathfrak{J})$$

where $a \in \mathbb{D}$ is such that Re a=0.

Proof. The (i,1) - component of the left-side of (13) is

$$\sum_{j,k=1}^{3} \{x_{ij}(x_{jk}x_{k1}) - (x_{ij}x_{jk})x_{k1}\}.$$

Note that x_{pp} (p=1,2,3) are real and that $\bar{x}_{pq}=x_{qp}$ $(p\neq q)$. By corollary 1 of Lemma 1.1, all terms in this expression is zero except when i,j,k are all different and $k\neq 1$, $j\neq 1$ and hence i=1. Thus the (1.1) - component is

$$\{x_{12}(x_{23}x_{31}) - (x_{12}x_{23})x_{31}\} + \{x_{13}(x_{32}x_{21}) - (x_{13}x_{32})x_{21}\}$$
$$= -[x_{12},x_{23},x_{31}] - [x_{13},x_{32},x_{21}].$$

Similary we see that the (i,2) and (j,3) - components are non-zero only if i=2 and j=3. By Lemma 1.1, all the diagonal elements of X(XX)-(XX)X are equal to an element $a \in \mathbb{D}$. Since $XX \in m_3^+$, $X(XX) - (XX)X \in m_3^-$ and so Re a=0, This proves (13).

Lemma 3. For $A \in k$, the endomorphism $\tilde{A}$ defined by (12) belongs to $\mathfrak{J}_4$.

Proof. By lemma 2, it is sufficient to show

$$(\tilde{A}(X),Y) + (X,\tilde{A}(Y)) = 0,$$

$$(\tilde{A}(X),Y,Z) + (X,\tilde{A}(Y),Z) + (X,Y,\tilde{A}(Z)) = 0$$

for $X,Y,Z \in \mathfrak{J}$. The first equality holds by (10). To prove the second, let $A \in k$, and so $\mathrm{tr}A=0$. Then by (9) and (13)

$$\begin{aligned}(\tilde{A}(X),XX) &= (AX,XX) - (XA,XX) = (A,X(XX) - (XX)X) = (A,a1_3)\\ &= \mathrm{Re}\ \mathrm{tr}(Aa) = \mathrm{Re}((\mathrm{tr}A)a) = 0.\end{aligned}$$

The left-hand side of the second equation to be proved is obviously symmetric in (X,Y,Z), and vanishes for $X=Y=Z$. Lemma 3 follows then from these facts. q.e.d.

Lemma 4. The subalgebra of $\mathfrak{F}_4$ consisting of the derivations D such that $D(E_i) = 0$ $(i=1,2,3)$ is canonically isomorphic to $\mathfrak{D}_4$. In more detail, D is written as

$$D(X) = \begin{pmatrix} 0 & \delta_3(a_3) & \overline{\delta_2(a_2)} \\ \overline{\delta_3(a_3)} & 0 & \delta_1(a_1) \\ \delta_2(a_2) & \overline{\delta_1(a_1)} & 0 \end{pmatrix} \qquad \text{for } X=X(\xi,a)$$

where $\delta_i \in \mathfrak{D}_4$ and $(\delta_1,\delta_2,\delta_3)$ are related by the infinitesimal principle of triality, namely, $\delta_i(a)\cdot b + a\cdot\delta_{i+1}(b) = \overline{\delta_{i+2}(\overline{ab})}$ $(i=1,2,3)$ $(a,b \in \mathbb{D})$.

Proof is the infinitesimal version of the proof of Theorem 1.
Let $D \in \mathfrak{F}_4$ annihilate E_i $(i=1,2,3)$. By (2), if follows that

$$E_i \circ D(F_i(a)) = 0,$$
$$2E_j \circ D(F_i(a)) = D(F_j(a)).$$

Therefore, we may put

$$D(F_i(a)) = F_i(\delta_i(a))$$

$\delta_i(a)$ is linear in a, Moreover $F_i(a) \circ F_i(a) \equiv 0 \mod E_1,E_2,E_3$. So applying D,

$$F_i(\delta_i(a)) \circ F_i(b) + F_i(a) \circ F_i(\delta_i(b)) = 0,$$

$$\delta_i(a)\cdot\bar{b} + b\cdot\overline{\delta_i(a)} + a\cdot\overline{\delta_i(b)} + \delta_i(b)\bar{a} = 0,$$

$$\langle\delta_i(a),b\rangle + \langle a,\delta_i(b)\rangle = 0$$

and hence $\delta_i \in \mathfrak{D}_4$. Also from

$$F_i(2a) \circ F_{i+1}(2b) = F_{i+2}(2\overline{ab})$$

we get $F_i(2\delta_i(a))\circ F_{i+1}(2b) + F_i(2a)\circ F_{i+1}(2\delta_{i+1}(b)) = F_{i+2}(2\delta_{i+2}(\overline{ab}))$

$$F_{i+2}(2\overline{\delta_i(a)\cdot b}) + F_{i+2}(\overline{2a\cdot\delta_{i+1}(b)}) = F_{i+2}(2\delta_{i+1}(\overline{ab}))$$

and $\delta_i(a)\cdot b + a\cdot\delta_{i+1}(b) = \overline{\delta_{i+2}(\overline{ab})}$ $(a,b \in \mathbb{D})$.

Therefore D detemines $(\delta_1,\delta_2,\delta_3)$ which is uniquely determined by δ_1 in view of the infinitesimal principle of triality (Theorem 1.1). onversely, such a triple $(\delta_1,\delta_2,\delta_3)$ defines a derivation D such that $D(E_i)=0$ $(1\leqq i\leqq 3)$. By the correspondance $D\leftrightarrow\delta_1$, the subalgebra is isomorphic to $\mathfrak{D}_4$, q.e.d.

We shall regard $\mathfrak{D}_4$ as contained in $\mathfrak{F}_4$ by the isomorphism of this lemma.

Theorem 2 A derivation Δ of $\mathfrak{J}$ is uniquely determined as $\Delta=D+\tilde{A}_\circ$ where $A_\circ \in k$ whose diagonal elements are all zero and $D\in\mathfrak{D}_4$.

Proof. Since $E_i\circ E_i=E_i$, $E_i\circ E_j=0$ $(i\neq j)$,

we have

$$2\Delta(E_i)\circ E_i = \Delta(E_i),$$

$$\Delta(E_i)\circ E_j + E_i\circ\Delta(E_j) = 0 \qquad (i\neq j)$$

From these, we see that $\Delta(E_i)$ $(i=1,2,3)$ are of the form:

$$\Delta(E_1) = \begin{pmatrix} 0 & -a_3 & \bar{a}_2 \\ -\bar{a}_3 & 0 & 0 \\ a_2 & 0 & 0 \end{pmatrix},\quad \Delta(E_2) = \begin{pmatrix} 0 & a_3 & 0 \\ \bar{a}_3 & 0 & a_1 \\ 0 & \bar{a}_1 & 0 \end{pmatrix},\quad \Delta(E_3) = \begin{pmatrix} 0 & 0 & \bar{a}_2 \\ 0 & 0 & -a_1 \\ a_2 & -\bar{a}_1 & 0 \end{pmatrix}.$$

Put $A = \begin{pmatrix} 0 & a_3 & \bar{a}_2 \\ -\bar{a}_3 & 0 & a_1 \\ -a_2 & -\bar{a}_1 & 0 \end{pmatrix} \in k.$

Then $\tilde{A}(E_i) = [A,E_i] = \Delta(E_i)$ $(i=1,2,3)$

We know $\tilde{A}\in\mathfrak{F}_4$ by lemma 3, and $\tilde{A}-\Delta\in\mathfrak{D}_4$. Thus $\Delta = D+\tilde{A}$. Since the coefficient of A is determined by $A(E_i)=\Delta(E_i)$ $(i=1,2,3)$, this decomposition is unique. q.e.d.

Corollary 1. The Lie algebra $\mathfrak{F}_4$ is generated by the subspace $\{\tilde{A}\,|\,A\in k\}$.

Proof. By Theorem, it is sufficient to show that the subalgebra $\mathfrak{D}_4$ is contained in the subalgebra generated by $\{\tilde{A}\,|\,A\in k\}$.
As a Lie algebra of linear transformations of $\mathbb{D}$, we know that $\mathfrak{D}_4$ is generated by $\{L_b\,|\,\mathrm{Re}\, b=0\}$ (Cf. §2.2). Therefore, those elements D in $\mathfrak{F}_4$ for which $\delta_1=L_b$ $(\mathrm{Re}\, b=0)$ generates $\mathfrak{D}_4$ in $\mathfrak{F}_4$. Put

$$A_b = \begin{pmatrix} -b & & \\ & b & \\ & & 0 \end{pmatrix}.$$

Then, $A_b\in k$ since $\bar{b}=-b$, and

$$\tilde{A}_b(F_1(a)) = [A_b,F_1(a)] = F_1(ba) = F_1(\delta_1(a)),$$

$$\tilde{A}_b(F_2(a)) = [A_b,F_2(a)] = F_2(ab) = F_2(\delta_2(a)),$$

$$\tilde{A}_b(F_3(a)) = [A_b,F_3(a)] = F_3(-ba-ab) = F_3(\overline{T_b\bar{a}}) = F_3(\overline{\delta_3(\bar{a})}),$$

$$\tilde{A}_b(E_i) = [A_b,E_i] = 0 \qquad (i=1,2,3)$$

where we used $L_b a+R_b a=T_b a$ so that (L_b,R_b,T_b) is $(\delta_1,\delta_2,\delta_3)$. Therefore $\{\tilde{A}_b \mid \mathrm{Re}\ b = 0\}$ generates the subalgebra $\mathfrak{D}_4$. q.e.d.

Corollary 2. The Lie algebra $\mathfrak{F}_4$ consists of all linear endomorphisms D such that

$$(D(X),Y) + (X,D(Y)) = 0$$
$$(D(X),Y,Z) + (X,D(Y),Z) + (X,Y,D(Z)) = 0$$

for any $X,Y,Z \in \mathfrak{J}$.

Proof. The linear transformations D satisfying the conditions form a Lie algebra, say g. By lemma 2, $g \subset \mathfrak{F}_4$. By the proff of Lemma 3, we see $\{\tilde{A} \mid A \in k\}$ is contained in g. By corollary 1, we conclude $g = \mathfrak{F}_4$. q.e.d.

Remark. By this Corollary, the automorpzism group F_4 of $\mathfrak{J}$ and the group of linear transformations of $\mathfrak{J}$ preserving (X,Y) and (X,Y,Z) have the same identify component. We shall see later these two groups coincide (cf. Lemma 1).

In the remaining part of this section, we show that $\mathfrak{F}_4$ is a simple Lie algebra of type F_4.

Let $D \in \mathfrak{D}_4 \subset \mathfrak{F}_4$. Then for $X=(x_{ij}) \in \mathfrak{J}$, we may write

$$D(X) = (\delta_{ij}x_{ij}) = \begin{pmatrix} \alpha_{11}x_{11} & \alpha_{12}x_{12} & \alpha_{13}x_{13} \\ \alpha_{21}x_{21} & \alpha_{22}x_{22} & \alpha_{23}x_{23} \\ \alpha_{31}x_{31} & \alpha_{32}x_{32} & \alpha_{33}x_{33} \end{pmatrix} \tag{15}$$

where $\delta_{ii}=0$ $(i=1,2,3)$ and $(\delta_{23},\delta_{31},\delta_{12})=(\delta_1,\delta_2,\delta_3)$ are elements in $\mathfrak{D}_4$ related by the principle of triality and $\delta_{ji}=\kappa\delta_{ij}$ (see (14)). Then for $\{i,j,k\}=\{1,2,3\}$.

$$(\delta_{ij}a)b + a(\delta_{jk}b) = \delta_{ik}(ab) \qquad (a,b \in \mathbb{D}). \tag{16}$$

Proof. When (i,j,k) is an even permutation of (1,2,3), $(\delta_{ij},\delta_{jk},\delta_{ki})$ is an even permutation of $(\delta_1,\delta_2,\delta_3)$ and (16) holds by the triality. When (i,j,k) is an odd permutation of (1,2,3), $(\delta_{ij},\delta_{jk},\delta_{ki})$ is an even permutation of $(\delta_{32},\delta_{21},\delta_{13}) = (\kappa\delta_{23},\kappa\delta_{12},\kappa\delta_{31})=(\kappa\delta_1,\kappa\delta_3,\kappa\delta_2)$. Since $\lambda\kappa=\kappa\lambda^2$, and $(\delta_1,\delta_2,\delta_3)$ is an cyclic orbit of λ, we see that $(\kappa\delta_1,\kappa\delta_3,\kappa\delta_2)$ is a cyclic orbit of λ, therefore (16) follows also from the triality. This proves (16).

Now, put $m_3^r=\{X \in m_3 \mid$ diagonal elements of X are real$\}$ and define $X\circ Y=\frac{1}{2}(XY+Y^*X^*)$, $X,Y \in m_3^r$. Then, m_3^r is a distributive algebra, containing $\mathfrak{J}=m_3$. For $X \in m_3^r$, we define D(X) by (15). Then

$$D(X\circ Y) = D(X)\circ Y + X\circ D(Y) \tag{17}$$

for $X,Y \in m_3^r$.

Proof. Let $X=(x_{ij})$, $Y=(y_{ij})$. Then

$$(\delta_{ij}x_{ij})y_{jk} + x_{ij}(\delta_{jk}y_{jk}) = \delta_{ik}(x_{ij}y_{jk})$$

for any $i,j,k=1,2,3$, $i\neq k$. In fact, if $\{i,j,k\}=\{1,2,3\}$, this holds by (16). If $i=j\neq k$, $\delta_{ii}=0$ and $x_{ii}\in\mathbb{R}$, so the above holds; analogously for the case $i\neq j=k$.

Therefore, the matrix $Z=D(X)\cdot Y+X\cdot D(Y)-D(XY)$ is a diagonal matrix whose diagonal elements are

$$\sum_j\{(\delta_{ij}x_{ij})y_{ji} + x_{ij}(\delta_{ji}y_{ji})\} \quad (i=1,2,3).$$

Since $\overline{\delta_{ji}x_{ji}}=(\kappa\delta_{ji})(\bar{x}_{ji})=\delta_{ij}(\bar{x}_{ji})$, $D(X)^*=D(X^*)$. Therefore

$$D(X)\circ Y + X\circ D(Y) - D(X\circ Y) = \frac{1}{2}\{D(X)\cdot Y + X\cdot D(Y) - D(XY) + Y^*D(X)^* + D(Y)^*X^* - D(Y^*X^*)\} = \frac{1}{2}(Z + Z^*).$$

Now the diagonal elements of $\frac{1}{2}(Z+Z^*)$ are equal to

$$\mathrm{Re}\sum_j \{(\delta_{ij}x_{ij})y_{ji} + x_{ij}(\delta_{ji}y_{ji})\} = \sum_j \{\langle\delta_{ij}x_{ij},\bar{y}_{ji}\rangle + \langle x_{ij},\delta_{ij}\bar{y}_{ji}\rangle\} = 0$$

since $\delta_{ij}=\kappa\delta_{ji}$, and $\delta_{ij}\in\mathcal{D}_4$, Thus $\frac{1}{2}(Z+Z^*)=0$, which proves (17).

Let $A\in m_3^-$ whose diagonal elements are all zero. Then $A\in k\cap m_3^r$. Since $A^*=-A$, $[A,X]=(AX+X^*A^*)=2A\circ X$ for $X\in\mathcal{J}$. We see $D(A)\in m_3^-$, and

$$[D(A),X] = 2D(A)\circ X$$
$$[A,D(x)] = 2A\circ D(X)$$

and, using (17), we get.

$D[A,X]=2D(A\circ X)=2D(A)\circ X+2A\circ D(X)=[D(A),X]+[A,D(X)]$, $X\in\mathcal{J}$. Therefore, in the Lie algebra $\mathcal{F}_4$ of all derivatives of $\mathcal{J}$, we have

$$[D,\tilde{A}] = \widetilde{D(A)} \tag{18}$$

for $D\in\mathcal{D}_4$ and $A\in m_3^-$ with zero diagonal elements.

We shall represent $\mathcal{D}_4$ as the Lie algebra of 8×8 real skew-symmetric matrices. Then

$$H = \{G_{04},G_{15},G_{26},G_{37}\}_{\mathbb{R}} = \left\langle \left(\begin{array}{c|c} 0 & \begin{matrix} -\alpha_1 & & & \\ & -\alpha_2 & & \\ & & -\alpha_3 & \\ & & & -\alpha_4 \end{matrix} \\ \hline \begin{matrix} \alpha_1 & & & \\ & \alpha_2 & & \\ & & \alpha_3 & \\ & & & \alpha_4 \end{matrix} & 0 \end{array} \right) \right\rangle$$

is a Cartan subalgebra of $\mathcal{D}_4$, where $\{\cdots\}_{\mathbb{R}}$ denotes the real span of $\{\cdots\}$. By direct calculation, we see that

$2F_{04}=G_{04}-G_{15}-G_{26}-G_{37}$ etc. are represented formally as

$$(2F_{04},\ 2F_{15},\ 2F_{26},\ 2F_{37}) = (G_{04},\ G_{15},\ G_{26},\ G_{37})\begin{pmatrix} 1 & -1 & -1 & -1 \\ -1 & 1 & -1 & -1 \\ -1 & -1 & 1 & -1 \\ -1 & -1 & -1 & 1 \end{pmatrix}.$$

Therefore, H is invariant under the automorphism $\pi : G_{ij} \to F_{ij}$ of $\mathfrak{F}_4$. Also, κ maps G_{0i} to $-G_{0i}$, and G_{ij} to G_{ij} $(i,j>0)$, and so H is invariant under κ. Therefore $\lambda=\pi\kappa$ leaves stable H and induces the transformation expressed by the orthogonal matrix

$$(19) \qquad \frac{1}{2}\begin{pmatrix} -1 & -1 & -1 & -1 \\ 1 & 1 & -1 & -1 \\ 1 & -1 & 1 & -1 \\ 1 & -1 & -1 & 1 \end{pmatrix}.$$

We prove that H is a Cartan subalgebra of $\mathfrak{F}_4$. By Theorem 2, we have the vector space direct sum decomposition

$$\mathfrak{F}_4 = \mathfrak{D}_4 + a$$

where $a=\{\tilde{A} \mid A \in \bar{m}_3$ whose diagonals are zero$\}$. Now every $A \in \bar{m}_3$ with zero diagonal elements is a unique sum of elements of the form

$$A_1(a)=\begin{pmatrix} 0 & & \\ & 0 & a \\ & -\bar{a} & 0 \end{pmatrix}, \qquad A_2(a)=\begin{pmatrix} 0 & & -\bar{a} \\ & 0 & \\ a & & 0 \end{pmatrix},$$

$$A_3(a)=\begin{pmatrix} 0 & a & \\ -\bar{a} & 0 & \\ & & 0 \end{pmatrix} \qquad (a \in \mathbb{D}).$$

An element $D \in H$ is written uniquely as

$$D = \alpha_1 G_{04} + \alpha_2 G_{15} + \alpha_3 G_{26} + \alpha_4 G_{37}$$

$$= \left(\begin{array}{cccc|cccc} & & & & -\alpha_1 & & & \\ & 0 & & & & -\alpha_2 & & \\ & & & & & & -\alpha_3 & \\ & & & & & & & -\alpha_4 \\ \hline \alpha_1 & & & & & & & \\ & \alpha_2 & & & & 0 & & \\ & & \alpha_3 & & & & & \\ & & & \alpha_4 & & & & \end{array}\right)$$

which we write $D(\alpha)=D(\alpha_1,\alpha_2,\alpha_3,\alpha_4)$. We have $\lambda D=D(\lambda\alpha)$, $\lambda^2 D=D(\lambda^2\alpha)$. By lemma 4, $\mathfrak{D}_4$ is identified with a subalgebra of $\mathfrak{F}_4$ by assigning to $D \in \mathfrak{D}_4$ the derivation (14) such that $\delta_1=D$, $\delta_2=\lambda D$. $\delta_3=\lambda^2 D$. In particular, $D(\alpha)$ is identified with the derivation (14), where

$$\delta_1=D(\alpha), \qquad \delta_2=D(\lambda\alpha), \qquad \delta_3=D(\lambda^2\alpha)$$

By (18), we get $[D(\alpha),\ \tilde{A}_1(a)] = D(\alpha)\widetilde{(A_1(a))} = A_1\widetilde{(D(\alpha)a)}$,

$$[D(\alpha),\ \tilde{A}_2(a)] = D(\alpha)\widetilde{(A_2(a))} = A_2\widetilde{(D(\lambda(\alpha))a)},$$

$$[D(\alpha), \tilde{A}_3(a)] = D(\alpha)(A_3(a)) = \widetilde{A_3(D(\lambda^2\alpha)a)}.$$

We conclude therefore : Under the adjoint action of H in $\mathfrak{F}_4$, the subspace a is stable; as H-module a is the sum of three 8-dimensional H-submodules $\{A_i(a)\}$(i=1,2,3) and, each being identified with $\mathbb{D}$ in an obvious way, $D(a) \in H$ acts by $D(a)$, $D(\lambda a)$ and $D(\lambda^2(a))$. In follow that H is a maximal abelian subalgebra, and so a Cartan subalgebra of $\mathfrak{F}_4$ (which is the Lie algebra of the compact group F_4).

Moreover, we can verify the roots of $\mathfrak{F}_4$ with respect to the Cartan subalgebra . Since $H \subset \mathfrak{D}_4 \subset \mathfrak{F}_4$ all roots of $\mathfrak{D}_4$ are roots of $\mathfrak{F}_4$. The adjoint action of H on the subspace $\{\widetilde{A_1(a)}\} \subset a$ is given $D(a)$, and therefore the roots $\pm\sqrt{-1}\alpha_i$ $(1 \leqq i \leqq 4)$ appear here. The actions of H on the two other subspaces of a yield the roots $\pm\sqrt{-1}\alpha_i \circ \lambda$, $\pm\sqrt{-1}\alpha_i \circ \lambda_i^2$. Applying (19), we see that there are of the form $\frac{\sqrt{-1}}{2}(\pm\alpha_1 \pm\alpha_2 \pm\alpha_3 \pm\alpha_4)$ where $\pm$ run over all possible combinations.

To sum up, we have thus proved

Theorem 3. The Lie algebra $\mathfrak{F}_4$ is a compact Lie algebra of rank 4 with roots $\pm\sqrt{-1}(\alpha_i \pm \alpha_j)$, $(i \neq j)$, $\pm\sqrt{-1}\,\alpha_i$, $\frac{\sqrt{-1}}{2}(\pm\alpha_1 \pm\alpha_2 \pm\alpha_3 \pm\alpha_4)$ $(1 \leqq i \leqq j \leqq 4)$. Therefore, $\mathfrak{F}_4$ is a simple Lie algebra of type F_4.

Corollary 1. The subalgebra of $\mathfrak{F}_4$ consisting of derivations Δ of $\mathfrak{J}$ such that $\Delta(E_1)=0$ is isomorphic to the simple Lie algebra of type B_4.

Proof If $\Delta(E_1)=0$, then in the decomposition $\Delta = D + \tilde{A}$ ($A \in m_3^-$ with zero diagonal elements) we have $\tilde{A}(E_1)=0$. Since $\tilde{A}(E_1)=[A,E_1]$, it follow that $A=A_1(a)$ for some $a \in \mathbb{D}$.

Conversely, $A=A_1(a)$ satisfies $\tilde{A}(E_1)=\{0\}$ therefore

$$\{\Delta \mid \Delta(E_1)=0\} = \{D+A_1(a) \mid D \in \mathfrak{D}_4,\ a \in \mathbb{D}\}.$$

By the observation above, we see that this is a compact Lie algebra of rank 4 with roots

$$\sqrt{-1}(\pm\alpha_i \pm\alpha_j), \qquad \pm\sqrt{-1}\alpha_i,$$

which is a simple Lie algebra of type B_4. q.e.d.

Gorollary 2. The subalgebra of $\mathfrak{F}_4$ consisting of derivations Δ of $\mathfrak{J}$ such that $\Delta(X)=0$ for real matrix $X \in \mathfrak{J}$ is isomorphic to the Lie algebra g_2.

Proof If $\Delta(x)=0$ for real X, we have $\Delta(E_i)=0$ (i=1,2,3) and so $\Delta = D \in \mathfrak{D}_4$. Also, we see from (14) that D corresponds to $(\delta_1,\delta_2,\delta_3)$ such that $\delta_1 e_0 = \delta_2 e_0 = \delta_3 e_0 = 0$, therefore $\kappa\delta_i = \delta_i$ (i=1,2,3). Then $\pi\delta_i = \lambda\kappa\delta_i = \lambda\delta_i = \delta_{i+1}$, $\pi\delta_i = \kappa\lambda^2\delta_i = \kappa\delta_{i+2} = \delta_{i+2}$ and $\delta_{i+1} = \delta_{i+2}$. Therefore $\delta_1 = \delta_2 = \delta_3 = D$. Conversely such a D defines a derivation Δ of $\mathfrak{J}$ satisfying the condition. Therefore, the subalgebra is isomorphic to g_2 by Theorem 1.3. q.e.d.

§5. Cayley projective plane

Let F_4' be the group of real linear transformations α of the exceptional Jordan algebra $\mathfrak{J}$ which preserves $(\cdot,\cdot)$ and $(\cdot,\cdot,\cdot)$, i.e., such that

$$(1)\qquad \begin{cases} (\alpha(X),\ \alpha(Y)) = (X,Y) \\ (\alpha(X),\ \alpha(Y),\ \alpha(Z)) = (X,Y,Z) \end{cases}$$

for any $X.Y.Z \in \mathfrak{J}$. Recall that

$$(X,Y) = \mathrm{tr}(X \circ Y),$$
$$(X,Y,Z) = (X \circ Y,\ Z) = \mathrm{tr}((X \circ Y) \circ Z),$$

and these are symmetric forms. By Lemma 4.1, $F_4' \subset F_4$ we shall show $F_4'=F_4$. First, we see

$$(2)\qquad F_4' = \{\alpha \in F_4 \mid \mathrm{tr}\alpha(X) = \mathrm{tr}X (X \in \mathfrak{J})\}$$

In fact, if $\alpha \in F_4'$, $\alpha \in F_4$ and so $\mathrm{tr}\alpha(X) = (\alpha(X),E) = (\alpha(X),\alpha(E))=(X,E)=\mathrm{tr}\,X$. Conversely, if $\alpha \in F_4$ and $\mathrm{tr}\alpha(X)=\mathrm{tr}X$, then $(\alpha(X),\alpha(Y))=\mathrm{tr}(\alpha(X)\circ\alpha(Y))=\mathrm{tr}\alpha(X\circ Y)= =\mathrm{tr}(X\circ Y)=(X,Y)$ and similarly $(\alpha(X),\alpha(Y),\alpha(Z))=(X,Y,Z)$. This proves (2).

The Lie algebra of F_4' consists of real linear transformation Δ of $\mathfrak{J}$ such that

$$(\Delta(X),Y)+(X,\Delta(Y))=0,$$
$$(\Delta(X),Y,Z)+(X,\Delta(Y),Z)+(X,Y,\Delta(Z))=0$$

for any $X,Y,Z \in \mathfrak{J}$. By Lemma 4.2, this coincides with the Lie algebra $\mathfrak{F}_4$ of the group F_4. Therefore, we get

$$(3)\qquad F_4^\circ = (F_4')^\circ$$

where $\circ$ denotes the connected component of the identity of the gruop.

Since (,) is positive definite on $\mathfrak{J}$, the group F_4 is a closed subgroup of orthogonal linear transformations of $\mathfrak{J}$. In particular F_4' is compact. By (3), F_4° is also compact.

Lemma 1. Any element $X \in \mathfrak{J}$ is transformed by an automorphism α belonging to F_4° into a diagonal form :

$$(4)\qquad \alpha(X) = \begin{pmatrix} \xi_1 & & \\ & \xi_2 & \\ & & \xi_3 \end{pmatrix},$$

Moreover, ξ_1,ξ_2,ξ_3 are determined by X uniquely up to order,

Proof. Consider the orbit $F_4^\circ(X)=\{\alpha(X) \mid \alpha \in F_4^\circ\}$ of X in $\mathfrak{J}$. Since F_4° is compact, this orbit is compact. Write a general element X as

$$X = \begin{pmatrix} \xi_1 & x_3 & \bar{x}_2 \\ \bar{x}_3 & \xi_2 & x_1 \\ x_2 & \bar{x}_1 & \xi_3 \end{pmatrix}.$$

Let X_0 be an element of $F_4^\circ(X)$ at which the function $\xi_1^2+\xi_2^2+\xi_3^2$ takes its maxinum value on $F_4^\circ(X)$. We prove X_0 is of a diagonal form.

Suppose x_1-component $x_1^{(0)} \neq 0$ for X_0. Let

$$A = \begin{pmatrix} 0 & & \\ & 0 & a \\ & -\bar{a} & 0 \end{pmatrix} .$$

Then $\tilde{A} \in \mathfrak{F}_4$ and determines the one-parameter subgroup $\exp t\tilde{A}$ in F_4°.

Put
$$X_t = (\exp t\tilde{A})X_0$$

The components of X_t are functions of t with value in $\mathbb{D}$, and we see

(*)
$$\frac{dX_t}{dt} = [A,X_t](= \tilde{A}(X_t))$$

Since
$$[A,X_t] = \begin{pmatrix} 0 & 0 & 0 \\ ax_2 & a\bar{x}_1 & a\xi_3 \\ -\bar{a}\bar{x}_3 & -\bar{a}\xi_2 & -\bar{a}x_1 \end{pmatrix} - \begin{pmatrix} 0 & -\bar{x}_2\bar{a} & x_3 a \\ 0 & -x_1\bar{a} & \xi_2 a \\ 0 & -\xi_3\bar{a} & \bar{x}_1 a \end{pmatrix}$$
$$= \begin{pmatrix} 0 & \bar{x}_2\bar{a} & -x_3 a \\ ax_2 & a\bar{x}_1 + x_1\bar{a} & -\xi_2 a \\ -\bar{a}x_3 & -\bar{a}\xi_2 & -(\bar{a}x_1+\bar{x}_1 a) \end{pmatrix},$$

the components of X_t satisfy, as functions of t, the following differential equations

$$\begin{cases} \dfrac{d\xi_1}{dt} = 0 \\[2mm] \dfrac{d\xi_2}{dt} = a\bar{x}_1 + x_1\bar{a} = 2<a,x_1> \\[2mm] \dfrac{d\xi_3}{dt} = -(a\bar{x}_1 + \bar{x}_1 a) = -2<a,x_1> . \end{cases}$$

Then $\frac{d}{dt}(\xi_1^2+\xi_2^2+\xi_3^2)=2\{\xi_1\frac{d\xi_1}{dt} + \xi_2\frac{d\xi_1}{dt} + \xi_3\frac{d\xi_3}{dt}\}=2<a,x_1>(\xi_2-\xi_3)$.

Put $a=x_1^{(0)}\neq 0$. Since $\xi_1^2+\xi_2^2+\xi_3^2$ is maximal at X_0, it follows that $\xi_2-\xi_3=0$ at X_0 .

Then
$$\frac{d^2}{dt^2}(\xi_1^2+\xi_2^2+\xi_3^2)\Big|_{t=0} = 2\frac{d<a,x_1>}{dt}\Big|_{t=0}(\xi_2-\xi_3)\Big|_{t=0}+2<a,a>(\frac{d\xi_2}{dt} - \frac{d\xi_3}{dt})\Big|_{t=0}=$$
$$=4<a,a>^2 > 0$$

Thus $\xi_1^2+\xi_2^2+\xi_3^2$ takes its strict minimum at t=0 on the curve $\{X_t\}$, which is a contradiction. Analogously, we see $x_2^{(0)}=x_3^{(0)}=0$.

We see that ξ_1,ξ_2,ξ_3 in (4) is uniquely determined by X. Indeed, this follows from the following equalities obtained by (1) and (2)

$$\begin{cases} \xi_1+\xi_2+\xi_3 = tr(\alpha(X)) = tr(X) \\ \xi_1^2+\xi_2^2+\xi_3^2 = tr(\alpha(X)\circ\alpha(X)) = (\alpha(X),\alpha(X)) = (X,X) \\ \xi_1^3+\xi_2^3+\xi_3^3 = tr((\alpha(X)\circ\alpha(X))\circ\alpha(X)) = (\alpha(X),\alpha(X),\alpha(X)) = (X,X,X). \end{cases}$$

q.e.d.

Corollary. If we define $X^n=X^{n-1}\circ X$ inductively. Then $X^p\circ X^q=X^{p+q}$.

Proof. Obvious for the case X diagonal, so also for general X by Theorem. q.e.d.

Definition. An element $X \in \mathfrak{J}$ is said to be irreducible idempotent if it satisfies the following conditions:

(a) $X^2 = X \neq 0$

(b) If $X = X_1 + X_2$, $X_1^2 = X_1$, $X_2^2 = X_2$, and $X_1 \circ X_2 = 0$, then either X_1 or X_2 is zero.

The set of all irreducible idempotents of $\mathfrak{J}$ is called the Cayley projective plane and will be denoted by Π.

Lemma 2 the following conditions for $x \in \mathfrak{J}$ are equivalent.

(i) $X \in \Pi$,

(ii) $X^2 = X$, $\quad tr(X) = 1$,

(iii) $tr(X) = (X,X) = (X,X,X) = 1$,

(iv) $X = \alpha(E_i)$ for some $i=1,2,3$ and $\alpha \in F_4$,

(v) $X = \alpha(E_1)$ for some $\alpha \in F_4$.

Proof. (i) $\Rightarrow$ (ii). By Lemma 1, there is $\alpha \in F_4^\circ$ such that $\alpha(X)$ is of the diagonal form (4).

Since Π is stable under the action of F_4 on $\mathfrak{J}$, $\alpha(X) \in \Pi$. Therefore $\xi_i^2=\xi_i$ and therefore $\xi_i=0$ or 1 $(1 \leqq i \leqq 3)$. But $\alpha(X) \in \Pi$ implies that exactly one of ξ_i is 1. Thus $tr(X)=tr(\alpha(X))=1$ by (2).

(ii) $\Rightarrow$ (iii), $\quad (X,X) = tr(X \circ X) = tr(X) = 1$,

$(X,X,X) = (X \circ X, X) = (X,X) = 1.$

(iii) $\Rightarrow$ (iv). Apply Lemma 1 to obtain $\alpha \in F_4^\circ$ for which (4) holds. Then, by (2) and (3)

$$\xi_1 + \xi_2 + \xi_3 = \xi_1^2 + \xi_2^2 + \xi_3^2 = \xi_1^3 + \xi_2^3 + \xi_3^3 = 1.$$

Therefore $(\xi\ ,\xi\ ,\xi\ )$ is equal to (1,0,0) up to order. This prove (iv). (iv) $\Rightarrow$ (v). It suffices to show that E_i is transformed to E_1 by an automorphism of $\mathfrak{J}$.

Let $\alpha_t \in F_4^\circ$ be

$$\alpha_t(X) = \begin{pmatrix} \cos t & -\sin t & 0 \\ \sin t & \cos t & 0 \\ 0 & 0 & 1 \end{pmatrix} X \begin{pmatrix} \cos t, & -\sin t & 0 \\ \sin t & \cos t & 0 \\ 0 & 0 & 1 \end{pmatrix}^{-1}$$

which is well-defined as the matrices used are real matrices. Then $\alpha_{\frac{\pi}{2}}(E_1)=E_2$.

Analogously, we can find $\alpha \in F_4^\circ$ such that $\alpha(E_1)=E_3$.

(v)$\Rightarrow$(i). Since $\alpha \in F_4$, it suffices to show that E_1 is an irreducible idempotent. Suppose $E_1=X_1+X_2$, with $X_1^2=X_1$, $X_2^2=X_2$ and $X_1 \circ X_2=0$. Then $X_1 \circ E_1=X_1 \circ X_1=X_1$. Therefore

$$\frac{1}{2}\begin{pmatrix} 2\xi_1 & x_3 & \bar{x}_2 \\ \bar{x}_3 & 0 & 0 \\ x_2 & 0 & 0 \end{pmatrix} = \begin{pmatrix} \xi_1 & x_3 & \bar{x}_2 \\ \bar{x}_3 & \xi_2 & x_1 \\ x_2 & \bar{x}_1 & \xi_3 \end{pmatrix} = X_1$$

and so $x_1=x_2=x_3=\xi_2=\xi_3=0$. Namely $X_1=\xi_1E_1$. Then $X_2=(1-\xi_1)E_1$. Since $X_2\circ X_2=0$, we have $\xi_1(1-\xi_1)=0$ and $\xi_1=1$ or 0. Then either X_1 or X_2 is zero. q.e.d.

From the proof of this Lemma, we get

Corollary. the group F_4° acts transitively on Π. In particular, Π is a connected set

Theorem 1. $F_4 = F_4'$.

Proof. Remark first that $tr(\alpha(E_i))=1$ for $\alpha \in F_4$ by Lemma 2. Let $\alpha \in F_4$ and $X \in \mathfrak{J}$. By Lemma 1, we can find $\beta \in F_4^\circ$ such that $\beta(x)=\sum_{i=1}^{3}\xi_iE_i$, Then by (2) and (3), $trX=tr\beta(X)=\sum_{i=1}^{3}\xi_i trE_i=\sum_{i=1}^{3}\xi_i$. We see then

$$\alpha(X) = (\alpha\beta^{-1})(\beta(X)) = (\alpha\beta^{-1})(\Sigma\xi_iE_i) = \sum_{i=1}^{n}\xi_i(\alpha\beta^{-1})(E_i).$$

By the remarks above, we have $tr\alpha(X)=\sum_i\xi_i tr(\alpha\beta^{-1})(E_i)=\sum_i\xi_i=trX$.

This being true for any $X \in \mathfrak{J}$, we see $\alpha \in F_4$ by (2). Since $F_4' \subset F_4$, this proves Theorem.

Since F_4' is a closed subgroup of the orthogonal transformations of $\mathfrak{J}$, we get

Corollary. The group F_4 is a compact Lie group.

The following theorem is due to Matsushima [11].

Theorem 2. For i=1,2,3, put

$$H_i = \{\alpha \in F_4 | \alpha(E_i)=E_i\}.$$

Then H_i is isomorphic to the double covering group Spin(9) of SO(9).

Proof. We consider the case i=1; the other cases are similar. The Lie algebra of H_1 consists of these elements $\Delta \in \mathfrak{F}_4$ such that $\Delta(E_1)=0$. By corollary 1 to Theorem 4.3, this is a simple Lie algebra of type B_4. Therefore H_1 is locally isomorphic to SO(9) .

Put
$$\Pi_1 = \{X \in \Pi | E_1\circ X = 0\}$$

An element $X \in \mathfrak{J}$ satisfying $E_1\circ X=0$ is of the form

$$X = \begin{pmatrix} 0 & 0 & 0 \\ 0 & \xi_1 & x_1 \\ 0 & \bar{x}_1 & \xi_2 \end{pmatrix} \tag{5}$$

and, by Lemma 2, this X belongs to Π if and only if

$$\begin{cases} \xi_1 + \xi_2 = 1 \\ \xi_2 = \xi_2^2 + \bar{x}_1x_1, \quad \xi_1 = \xi_1^2 + x_1\bar{x}_1 \end{cases}$$

or equivalently, if and only if

$$\begin{cases} \xi_1 + \xi_2 = 1 \\ \xi_1^2 + \xi_2^2 + 2Rex_1\bar{x}_1 = 1. \end{cases}$$

Let $\|X\|$ be the norm in $\mathfrak{J}$ define by $(\cdot,\cdot)$. Then

$$\|X\|^2 = \xi_1^2 + \xi_2^2 + 2Rex_1\bar{x}_1$$

for X in (5). Therefore, in the 10-dimensional real euclidean space $V=\{(\xi_1,\xi_2,x_1)\}$ formed by X of the form (5), Π_1 is the intersection of the unit sphere and the hyperplane defined by $\xi_1+\xi_2=1$.

Suppose now $\alpha \in H_1$. Then, if $X \in \Pi$, $E_1 \circ \alpha(X)=\alpha(E_1 \circ X)=0$, and so α leaves stable the space V and $\Pi_1=\Pi \cap V$. Therefore α defines an element $p(\alpha)$ of $O(10)$ which leaves the intersection Π_1 of the unit sphere with the hyperplane $\xi_1+\xi_2=1$. We can verify easily α leaves stable the vector subspace $\xi_1+\xi_2=0$ in V, and so $p(\alpha)$ belongs to a subgroup isomorphic to $O(9)$ in $O(10)$. We have thus obtained a homemorphism $p : H_1 \to O(9)$.

The subgroup $N = \{\alpha \in F_4 \mid \alpha(E_i) = E_i \quad (i=1,2,3)\}$ is clearly contained in H_1. By the proof of Theorem 4.1, $\alpha \in N$ acts on Π_1 by

$$\alpha(x) = \begin{pmatrix} 0 & 0 & 0 \\ 0 & \xi & \theta_1(x_1) \\ 0 & \overline{\theta_1(x_1)} & \xi_3 \end{pmatrix}$$

and $\alpha \to \theta_1$ is a double covering map of N to $SO(8)$. Let Γ be the kernel of the homomorphism $p: H_1 \to O(9)$. Then α leaves fixed both E_2 and $E_3 \in \Pi_1$, therefore α belongs to N. Thus, the kernel Γ coincides with the kernel of the map p restricted to N.

Since Γ is discrete, we see that $p: H_1 \to O(9)$ is locally isomorphic. In particular, we see $p(H_1^\circ) = SO(9)$ where H_1° is the connected component of the identity. Because $N \cong Spin(8)$ (Theorem 4.1) is connected and that $\Gamma \subset N \subset H_1^\circ$, we conclude that $p^{-1}(SO(9))=H_1^\circ$ and $p: H_1^\circ \to SO(9)$ is a double covering homomorphism. To complete the proof, it is sufficient to show $H_1=H_1^\circ$, suppose $H_1 \neq H_1^\circ$. Then $\{e\} \neq H_1/H_\circ^\circ=H_1/p^{-1}(SO(9)) \cong p(H_1)/SO(9) \subset O(9)/SO(9)$. Since $O(9)/SO(9)$ is a group of order 2, we see then $p(H_1)=O(9)$. We may then find $\alpha \in H_1$ such that $\alpha \bar{\in} H_1^\circ$ and $p(\alpha)$ maps

$$\begin{pmatrix} \xi_2 & x_1 \\ \bar{x}_1 & \xi_3 \end{pmatrix} \longrightarrow \begin{pmatrix} \xi_3 & x_1 \\ \bar{x}_1 & \xi_2 \end{pmatrix}$$

in V, since this transformation belongs to $O(10)$ but not to $SO(10)$ and leaves Π_1 stable. Then

$$\alpha(E_2) = E_3, \qquad \alpha(E_3) = E_2 .$$

We can verify easily that $SO(9)$ acts transitively on Π_1, as it does on the unit sphere S^8. Therefore, there exists $\beta \in H_1^\circ$ such that $\beta(E_2)=E_3$. Then $\beta(E_3) \in \Pi_1$ and $E_3 \circ \beta(E_3)=\beta(E_2 \circ E_3)=0$. Therefore, $\beta(E_3)=E_2$. In this case $\alpha\beta^{-1}(E_i)=E_i$ $(i=1,2,3)$ and $\alpha^{-1}\beta \in N \subset H_1^\circ$. This implies $\alpha \in H_1^\circ\beta=H^\circ$, which is a contradiction.

q.e.d.

Theorem 3. Π can be identified with F_4/H_1 where $H_1 \cong Spin(9)$. The group F_4 is connected.

Proof. By Corollary to Lemma 2, the group F_4 acts transitively on Π and its isotropy subgroup at E_1 is H_1. Thus we have the identification $\Pi = F_4/H_1$. By Theorem 3, $H_1 \cong Spin(9)$ and this is connected. On the other hand, Π is connected by Corollary to Lemma 2. It follows that F_4 is connected. q.e.d.

§6 Automorphisms and subgroups of F_4

Let

$$A = \begin{pmatrix} 0 & 1 & 0 \\ 1 & 0 & 0 \\ 0 & 0 & 1 \end{pmatrix}$$

then, AXA^{-1} is well-defined for $X \in \mathfrak{J}$, because A is a real matrix, and the map $\alpha_0 : X \to AXA^{-1}$ is an automorphism of the algebra $\mathfrak{J}$, i.e., $\alpha_0 \in F_4$. We see

(1) $$\alpha_0 \begin{pmatrix} \xi_1 & x_3 & \bar{x}_2 \\ \bar{x}_3 & \xi_2 & x_1 \\ x_2 & \bar{x}_1 & \xi_3 \end{pmatrix} = \begin{pmatrix} \xi_2 & \bar{x}_3 & x_1 \\ x_3 & \xi_1 & \bar{x}_2 \\ \bar{x}_1 & x_2 & \xi_3 \end{pmatrix}$$

Let K_{12} be the inner automorphism of F_4 defined by α_0. Since $A^2 = 1_3$, $\alpha_0^2 = id$ and

(2) $\quad K_{12}^2 = 1 \quad$ in $\quad Aut(F_4)$.

We see $\alpha_0(E_1) = E_2$, $\alpha_0(E_2) = E_1$. As the group F_4 acts transitively on Π with the isotropy group H_i at E_i, it follows that

$$\alpha_0^{-1} H_1 \alpha_0 = H_2, \qquad \alpha_0^{-1} H_2 \alpha_0 = H_1.$$

This can be expressed as

(3) $\quad K_{12}(H_1) = H_2, \qquad K_{12}(H_2) = H_1$

We define the automorphisms K_{23} and K_{31} in a similyr way.

Put $\Lambda = K_{31} K_{23}$.

This is the inner automorphism of F_4 defined by the following element $\beta_0 \in F_4$

(4) $$\beta_0 \begin{pmatrix} \xi_1 & x_3 & \bar{x}_2 \\ \bar{x}_3 & \xi_2 & x_1 \\ x_2 & \bar{x}_1 & \xi_3 \end{pmatrix} = \begin{pmatrix} \xi_2 & x_1 & \bar{x}_3 \\ \bar{x}_1 & \xi_3 & x_2 \\ x_3 & \bar{x}_2 & \xi_1 \end{pmatrix}.$$

Because $\beta_0^3 = id$ we get

(5) $\quad \Lambda^3 = 1.$

We may check easily $\quad K_{12}\Lambda^2 = \Lambda K_{12}$.

In the following, we put $\quad K = K_{12}$.

Then in the group of automorphisms $Aut(F_4)$ of the group F_4, $\{K, \Lambda\}$ generates finite subgroup S_3 isomorphic to the symmetric group of order 3.

We shall consider the effect of S_3 on subgroups of F_4.

Since $\alpha_0(E_3) = E_3$, $\alpha_0 \alpha \alpha_0^{-1}(E_3) = E_3$ for $\alpha \in H_3$. Therefore

(6) $\quad K(H_3) = H_3$

The restriction $K_{H_3}=K|_{H_3}$ is the inner automorphism of H_3 defined by $\alpha_0 \in H_3$ so K_{H_3} leaves fixed every element of the center of H_3, and in particular induces an automorphism of $p(H_3)=SO(9)$. Now, since $\alpha_0(E_i)=E_j$, $\beta_0(E_i)=E_{i-1}$, we see

(7) $$K(N) = N, \qquad \Lambda(N)=N$$

where $N=\{\alpha \in F_4 \mid \alpha(E_i) = E_i \ (i=1,2,3)\}$. By Theorem 4.1, $N \cong Spin(8)$, and the element $\alpha \in N$ corresponding to $\theta=(\theta_1,\theta_2,\theta_3) \in Spin(8)$ is given by

(8) $$\alpha \begin{pmatrix} \xi_1 & x_3 & \bar{x}_2 \\ \bar{x}_3 & \xi_2 & x_1 \\ x_2 & \bar{x}_1 & \xi_3 \end{pmatrix} = \begin{pmatrix} \xi_1 & \theta_3(x_3) & \overline{\theta_2(x_2)} \\ \overline{\theta_3(x_3)} & \xi_2 & \theta_1(x_1) \\ \theta_2(x_2) & \overline{\theta_1(x_1)} & \xi_3 \end{pmatrix}$$

which will be denoted by $\alpha(\theta_1,\theta_2,\theta_3)$.

Let K_N and Λ_N be the restrictions of K and Λ to N. Then by (1) and (8), we see

(9) $K_N(\alpha(\theta_1,\theta_2,\theta_3))=\alpha_0^{-1}\alpha(\theta_1,\theta_2,\theta_3)\alpha_0=\alpha(\kappa\theta_2,\kappa\theta_1,\kappa\theta_3)$.

The canonical projection $\Delta_0 : Spin(8) \to SO(8)$ given in §3 is

$$\Delta_0(\alpha(\theta_1,\theta_2,\theta_3)) = \theta_3$$

Let Γ be the kernel of Δ_0. Then $\Gamma=\{\alpha(\pm 1,\pm 1,1)\}$ which is invariant under K_N. Therefore K_N induces an automorphism of SO(8) which is equal to κ as seen from (9). Also we have

(10) $$\Lambda_N\alpha(\theta_1,\theta_2,\theta_3) = \beta_0\alpha(\theta_1,\theta_2,\theta_3)\beta_0^{-1} = \alpha(\theta_2,\theta_3,\theta_1)$$

Because Λ_N does not leave Γ stable, Λ_N does not induce an automorphism of SO(8).

We make the identification $SO(7)=\{\theta \in SO(8) \mid \theta(e_0)=e_0\}$. Then $SO(7)=\{\theta \in SO(8) \mid \kappa(\theta)=\theta\}$ Define $M_1= \{(\theta_1,\theta_2,\theta_3) \in N \mid \theta_1 \in SO(7)\}$. Under the canonical isomorphism $N \cong Spin(8)$, M_1 corresponds to the subgroup $(\Delta^+)^{-1}(SO(7))=$ $=Spin^+(7)$. Similarly, put

$$M_2 = \{\alpha(\theta_1,\theta_2,\theta_3) \in N \mid \theta_2 \in SO(7)\}.$$

Then $M_2 \cong Spin^-(7)$ canonically. Define also

$$M_3 = \{\alpha(\theta_1,\theta_2,\theta_3) \in N \mid \theta_3 \in SO(7)\}.$$

Then we see

(11) $$\begin{cases} K_N(M_1) = M_2, & K_N(M_2) = M_1, & K_N(M_3) = M_3, \\ \Lambda_N(M_1) = M_3, & \Lambda_N(M_2)_1= M_1, & \Lambda_N(M_3) = M_2. \end{cases}$$

we see $\Gamma \cap M_1=\Gamma \cap M_2=\{1\}$, Therefore $\Delta_\circ$ maps M_1,M_2 in SO(8) injectively and we have

(12) $$\begin{cases} \Delta_\circ(M_1) = \kappa(Spin^+(7)) = Spin^-(7) \\ \Delta_\circ(M_2) = \kappa(Spin^-(7)) = Spin^+(7). \end{cases}$$

By definition $N=H_1 \cap H_2 \cap H_3$ and we see $N=H_i \cap H_{i+1}$ (i=1,2,3).

Also, put $\bar{G}_2=M_1 \cap M_2 \cap M_3$. Then $\bar{G}_2=\{\alpha(\theta_1,\theta_2,\theta_3) \left| \begin{array}{l}\kappa\theta_i=\theta_i \\ \kappa\theta_i=\theta_{i+1}\end{array} \right. (i=1,2,3)\ \} =$

$=\{\alpha(\theta,\theta,\theta)\,|\,\theta(a)\theta(b)=\theta(ab)\quad (a,b \in \mathbb{D})\}$ and $\bar{G}_2$ is isomorphic to the group G_2.
One can also see

$$\bar{G}_2 = M_i \cap M_{i+1} \quad (i=1,2,3). \qquad \bar{G}_2 = \{\alpha \in N\,|\,\Lambda_N\alpha=\alpha\}$$

We have thus obtained the following inclusions with types of subgroups

$$\begin{array}{ccccccc} F_4 \supset H_i \supset & \bigcap_{i=1}^{3} H_i = N & \supset & M_i \supset & \bigcap_{i=1}^{3} M_i = \bar{G}_2 \\ \cong & \cong & & \cong & \cong \\ \mathrm{Spin}(9) & \mathrm{Spin}(8) & & \mathrm{Spin}(7) & G_2 \end{array}$$

Together with (11) and (12), this shows that the group F_4 is a good group to interpret the principle of triality and which contains various interesting groups.

§7 Spin group and spin representations

We summarize definitions and fundamental properties concerning spin groups Spin(n). Cf. [1],[7], [13].

Definition. Let V be a real vector space of dimension n endowed with a positive definite inner product $<\cdot, \cdot>$. Put

$$q(v) = -<v,v>$$

Let $T(V)=\sum_{i=0} \otimes^i V$ be the tensor algebra over V, and I(q) be the ideal of T(V) generated by $\{v\otimes v-q(v)\,|\,v \in V\}$. The quotient algebra $C(V,q)=T(v)/I$ is called the Clifford algebra associated to (V,q).

Let $V=\mathbb{R}^n$ and $<\ ,\ >$ be the standard inner product such that $\{e_1,\cdots,e_n\}$ is an orthonormal basis where $e_i=(0,\cdots 0,1,0\cdots 0)$.

Then we write C_n for $C(\mathbb{R}^n,q)$. The following basic properties are well-known.

A) V is imbedded in C_n and $\{e_{i_1}\cdots e_{i_k}\,|\,i_1<\cdots<i_k\ ,\ 0\leq k\leq n\}$ is a basis of C_n, so $\dim C_n=2^n$.

B) The endomorphisms α,β in T(E) defined by

$$\alpha(v_1 \otimes \cdots \otimes v_k) = (-1)^k v_1 \otimes \cdots \otimes v_k,$$
$$\beta(v_1 \otimes \cdots \otimes v_k) = v_k \otimes \cdots \otimes v_1$$

induce involutions α,β in C_n: α is an automorphism and β is an anti-automorphism of C_n. Define C_n^0, C_n^1 to be the 1- and -1 eigenspace of α respectively. Then $C_n=C_n^0 + C_n^1$ and C_n^0 is an subalgebra.

C) Algebraic structure of C_n, we have $e_i^2=-1$. $e_ie_j+e_je_i=0$ $(1\leq i\neq j\leq n)$. The center of C_n is $\mathbb{R}\cdot 1$ (n even) or $\mathbb{R}\cdot 1+\mathbb{R}e_N$ (n odd) where $e_N=e_1\cdots e_n$. If n is even or if n is odd and $j=\sqrt{(-1)^{n(n+1)/2}} \bar{\in} \mathbb{R}$ then C_n is simple. Otherwise, C_n is the sum of two ideals generated by $u=\frac{1}{2}(1+je_N)$ and $v=\frac{1}{2}(1-je_N)$.

D) We have algebra isomorphism $C_n^0 \cong C_{n-1}$ defined by the following map $\psi: C_{n-1}\to C_n^0$
For $x \in C_{n-1}$, put $x=a+b$, $a \in C_{n-1}^0$, $b \in C_{n-1}^1$, then $\psi(x)=a+be_n$.

E) $C_1 \cong \mathbb{C}$ (Complex number field),
$C_2 \cong \mathbb{H}$ (Quaternion numbers field).

Definition. Let C_n^x be the multiplicative group of invertible elements in C_n. Put $\Gamma_n=\{x \in C_n^x \mid \alpha(x)vx^{-1} \in \mathbb{R}^n,\ \text{for any } v \in \mathbb{R}^n\}$. Then Γ_n is a subgroup of C_n^x called Clifford group. The map $\rho:\Gamma \to GL(\mathbb{R}^n)$ defined by $\rho(x)v=\alpha(x)vx^{-1}$ is called twisted adjoint representation of Γ_n.

The subgroup Γ_n is stable under α, β. Put $\bar{x}=\alpha(\beta(x))=\beta(\alpha(x))$ and define $N(x)=x\bar{x}$ $(x \in C_n)$. Then N defines a multiplicative homomorphism $N: \Gamma_n \to \mathbb{R}^x$. Put Pin(n)= =kernel of $N: \Gamma_n \to \mathbb{R}^x$. Then we have the exact sequence

$$\{1\} \to \mathbb{Z}_2 \to \mathrm{Pin}(n) \overset{\rho}{\to} O(n) \to \{1\}.$$

We define the spinor group by $\mathrm{Spin}(n)=\{x \in \mathrm{Pin}(n) \mid \rho(x) \in SO(n)\}$. The following facts are known.

F) $\mathrm{Spin}(n)=\mathrm{Pin}(n) \cap C^0$ and we have the exact sequence

$$1 \to \mathbb{Z}_2 \to \mathrm{Spin}(n) \overset{\rho}{\to} SO(n) \to 1.$$

The center of Spin(n) is $\{\pm 1\}$ if n is odd, and if n is even

$$\{\pm 1, \pm e_N\} \cong \mathbb{Z}_2 \times \mathbb{Z}_2 \quad (n \equiv 0\ (4)) \quad \text{or} \quad \mathbb{Z}_4 \quad (n \not\equiv 0\ (4))$$

G) C_n^x has a natural structure of Lie group (by means of regular representation of C_n) with Lie algebra C_n (in which $[x,y]=xy-yx$). As a closed subgroup of C_n^x, Spin(n) is a Lie group to which corresponds the subalgebra m of C_n spanned by $\{e_ie_j \mid i<j\}$. In fact we have $[e_ie_j, e_ie_\ell]=2e_je_\ell$, $[e_ie_\ell, e_ke_\ell]=2e_ie_k$ and $[e_ie_j, e_ke_\ell]=0$ for other (i,j,k,ℓ).

Moreover, the representation ρ: Spin(n) $\to$ SO(n) induces the isomorphism $d\rho: m_2 \to \mathcal{SO}(n)$ defined as follows. The adjoint action X of the Lie algebra C_n restricted to m leaves stable $\mathbb{R}_n \subset C_n$, and $d\rho(x)=X(x) \mid \mathbb{R}^n$. In fact,

$$X(e_ie_j)e_k = \begin{cases} 0 & k \neq i,j, \\ 2e_j & k=i, \\ -2e_i & k=j. \end{cases}$$

(1) $d\rho(e_ie_j)=2g_{ji}$ $(1 \leq i,\ j \leq n)$ in the notation (2.1).

H) Any algebra representation $\sigma: C_n^+ \to \mathrm{End}(W)$ (W real or complex vector space) gives rise to a representation of Spin(n). The differential of this representation $\sigma|_{\mathrm{Spin}(n)}$ is equal to $\sigma|_{m_2}$.

I) Spin(n) is a simply connected compact Lie group.

J) Let $C_n^{\mathbb{C}}$ be the complexification of C_n. Then, if n=2r, $C_n^{\mathbb{C}}$ has unique irreducible representation of degree 2^r; if n=2r+1, $C_n^{\mathbb{C}}$ has only two non-equivalent irreducible representations of degree 2^r.

Since $\mathrm{Spin}(n) \subset C_n^0 \cong C_{n-1} \subset C_{n-1}^{\mathbb{C}}$ any irreducible representation of $C_n^{\mathbb{C}}$ defines a representation of Spin(n). In this way, we get: For n=2r, two complex irreducible representations Δ^+, Δ^- of degree 2^{r-1} called half spin representation of Spin(n)

For $n=2r+1$, one complex irreducible representation Δ of degree 2^r called spin representation of $Spin(n)$.

Spin representation of Spin(8) and Spin(7)

We shall prove that the representation $\Delta^{\pm}$ of Spin(8) and its restrictions to Spin(7) constructed in §3 are essentially spin representations.

The spin representations being obtained from the relation $Spin(8) \subset C_8^0 \cong C_7$, we begin with exhibiting representations of C_7 where we identify $\mathbb{R}^7 = \mathrm{Im}\,\mathbb{D}$.

By the fact C), C_7 has center $\mathbb{R}1 + \mathbb{R}e_N$ and C_7 is the sum of two simple ideals generated by $1+e_N$ and $1-e_N$.

Let $\mathbb{R}^8 = \mathbb{D}$ and $\{e_0, e_1, \cdots, e_7\}$ be the basis of $\mathbb{D}$ introduced in §1. $\mathbb{R}^7 = \mathrm{Im}\,\mathbb{D}$ has the basis $\{e_1, \cdots e_7\}$. Denoting by L_v and R_v the left and right multiplication by an element $v \in \mathbb{D}$, if $v \in \mathbb{R}^7 = \mathrm{Im}\mathbb{D}$, we have $(L_{-v})^2 x = (-v)(-vx) = \bar{v}(-vx) = -\|v\|^2 x$, and $(R_{-v})^2 x = -\|v\|^2 x \quad (x \in \mathbb{R}^8)$.

Therefore by the universal property of C_7, we get representations Δ^+, Δ^- of C_7 on $\mathrm{End}(\mathbb{R}^8)$ such that $\Delta^+(v) = L_{-v}$, $\Delta^-(v) = R_{-v}$ $(v \in \mathbb{R}^n)$. We have $\dim C_7 = 2^7$, $\dim \mathrm{End}(\mathbb{R}^8) = 64 = 2^6$. Since $\Delta^{\pm}$ are non-trivial, the kernel of $\Delta^{\pm}$ is a proper ideal of C_7 which is of dimension 2^6. Then $\Delta^{\pm}$ are surjective and real irreducible representations of C_7. Moreover

$$\Delta^{\pm}(e_N)e_0 = \begin{Bmatrix} (-1)^7 e_1(e_2(e_3(\cdots e_6|e_7e_0))) \\ (-1)^7 ((((e_0e_7)e_6\cdots)e_3)e_2)e_1 \end{Bmatrix} = \pm e_0$$

Since the ideals of C_7 are generated by $1+e_N$ and $1-e_N$, Δ^+ and Δ^- are not equivalent, and the map

$$\Delta^+ \oplus \Delta^- : C^7 \to \mathrm{End}(\mathbb{R}^8) \oplus \mathrm{End}(\mathbb{R}^8)$$

is an algebra isomorphism.

It follows from J) that $\Delta^{\pm}$ are also irreducible as complex representations.

By definition, $\Delta^{\pm}$ give rise to half spin representations of Spin(8), denoted also by $\Delta^{\pm}$, through $Spin(8) \subset C_8^0 \cong C_7$.

We shall study differentials of these spin representations $\Delta^{\pm}$.

The isomorphism $C_7 \cong C_8^0$ recalled in D) is the map $a+b \to a+be_0$ $(a \in C_7^0,\ b \in C_7^1)$
The Lie algebra m of Spin(8) is spanned by

(2) $\qquad e_ie_0 \quad (1 \le i \le 7), \qquad e_ie_j \quad (1 \le i < j \le 7),$

which are mapped by the above isomorphism to

(3) $\qquad e_i \quad (1 \le i \le 7), \qquad e_ie_j \quad (1 \le i < j \le 7).$

By the fact H), the differentials $d\Delta^{\pm}$ of the representations $\Delta^{\pm}: Spin(8) \to \to \mathrm{End}(\mathbb{R}^8)$ is the restriction of $\Delta^{\pm}$ to the Lie algebra m. Comparing (2) and (3), for $1 \le i,\ j \le 7$ we get

$$(5) \quad \begin{cases} d\Delta^+(e_ie_0) = \Delta^+(e_i) = L_{-e_i} = 2F_{0i}, \\ d\Delta^+(e_ie_j) = \Delta^+(e_ie_j) = L_{-e_i}L_{-e_j} = 2F_{ji}, \end{cases}$$

(see (2.3) for the definition of F_{ij}). Similarly

(6) $$\begin{cases} d\Delta^-(e_i e_0) = \Delta^-(e_i) = R_{-e_i} = 2F'_{ji} \\ d\Delta^-(e_i e_j) = \Delta^-(e_i e_j) = R_{-e_i}R_{-e_j} = F'_{ji} \end{cases}$$

(see (2.10) for F'_{ij})

When we identify m with the Lie algebra $\mathfrak{D}_4$ by the canonical isomorphism $d\rho$: $m \to \mathfrak{D}_4$, we get from (1),(5),(6)

$$d\Delta^+(G_{ij}) = F_{ij}, \qquad d\Delta^-(G_{ij}) = F'_{ij}.$$

It follows that $\Delta^{\pm}(\mathrm{Spin}(8)) \subset SO(8)$ and

(6) $$d\Delta^+ = \Pi = \lambda\kappa, \qquad d\Delta^- = \lambda\Pi = \lambda^2\kappa.$$

in the notation of §2 (p.12, p.16).

Comparing with (3.5)(p.35) we see that the representations $\Delta^{\pm}$ con tructed in §3 are half-spin representations of Spin(8) when they are composed with the automorphism κ of Spin(8).

In general, it is known that the two inequivalent representations $\Delta^{\pm}$ of C_{2r+1} of degree 2^r restrict to equivalent irredecible representations of degree 2^r of C_{2r} ($\cong C^0_{2r+1}$) Therefore, we see that the spin representation of Spin(7), say Δ, is obtained from $\Delta^{\pm}$: Spin(8) $\to$ End($\mathbb{R}^8$) by restriction.

As remarked at the end of §3, the inclusions $\Delta^{\pm}$: $\mathrm{Spin}^{\pm}(7) \to SO(8)$ have λ and λ^2 as their differentials. Note that the automorphism κ of Spin(8) goes down to κ on SO(8) which maps $\mathrm{Spin}^+(7)$ to $\mathrm{Spin}^-(7)$ by an conjugation of O(8). Combining these observations with (6), we conclude that:

The identity representations of $\mathrm{Spin}^{\pm}(7)$ into SO(8) are equivalent to the spin representation of the group Spin(7),

§8 Characterzations of G_2 and Spin(7) by invariant forms.

8.1. The group G_2

In this section, we put $V = \mathrm{Im}\mathbb{O}$, Then V has $\{e_1, \cdots, e_7\}$ as an orthonormal basis, and the group SO(7) acts canonically on V. The group G_2 may be identified with a subgroup of SO(7).

We define the trilinerr form ϕ on V by

(1) $$\phi(x,y,z) = \langle x, yz \rangle$$

where $x,y,z \in V$.

Lemma 1. ϕ is a G_2-invariant alternating form on V.

Proof. Let $g \in G_2$. Then it is easy to see $\phi(gx, gy, gz) = \phi(x, y, z)$, which shows ϕ is G_2-invariant. We have by (1,1)

$$\phi(x,x,z) = \langle x,xz \rangle = |x|^2 \langle e_0, z \rangle = 0 \qquad \text{for } x,z \in V.$$

Analogously $\phi(x,y,y)=0$ for $x,y \in V$. Therefore, ϕ is an alternating form. q.e.d.

Let $\{\omega^1, \cdots, \omega^7\}$ be the system of 1-forms on V such that $\omega^i(e_j) = \delta_{ij}$ $(1 \leq i, j \leq 7)$. We write $\omega^{ijk} = \omega^i \wedge \omega^j \wedge \omega^k$. Then, by the multiplication table for $\{e_1, \cdots e_7\}$, we get easily the following expression of ϕ.

(2) $\phi = \omega^{123}+\omega^{145}-\omega^{167}+\omega^{246}+\omega^{257}+\omega^{347}-\omega^{356}$. We put

(3) $\omega = \omega^1 \wedge \omega^2 \wedge \cdots \wedge \omega^7$

Theorem 1. $G_2 = \{g \in GL(V) \mid g^*\phi = \phi\}$.

Proof. Recall that $g^*\phi$ is the form given by

$$(g^*\phi)(x,y,z) = \phi(g^{-1}x,\ g^{-1}y,\ g^{-1}z) \qquad \text{for } x,y,z \in V.$$

By Lemma 1, $g^*\phi=\phi$ for $g \in G_2$. Conversely, suppose $g \in GL(V)$ be such that $g^*\phi=\phi$.

First, we show $g \in SO(7)$. Define the bilinear form $b: V\times V \to \Lambda^7 V$ by

$$b(x,y) = \frac{1}{6}\,(i(x)\phi \wedge i(y)\phi \wedge \phi)$$

where $i(x)$ is the interior product by $x \in V$; namely $i(x)\phi$ is the bilinear form defined by $i(x)\phi(y,z)=\phi(x,y,z)$. Then we see easily by (2) and (3).

(4) $$b(x,y) = \langle x,y\rangle\omega \qquad (x,y \in V).$$

On the other hand, since $g^*\phi=\phi$ we get easily

(5) $$i(gx)\phi = g^*(i(x)\phi)$$

It follows then $b(gx,gy) = \frac{1}{6}\,\{g^*(i(x)\phi) \wedge g^*(i(y)\phi) \wedge g^*\phi\} = (g^*b)(x,y)$

Since $g^*\omega = (\det g)^{-1}\omega$, this and (4) imply

(6) $$\langle gx,gy\rangle = (\det g)^{-1}\langle x,y\rangle.$$

Now, for any $x_1,\cdots,x_7 \in V$ we have

$$\det(\langle x_i,x_j\rangle = (\omega(x_1,\cdots,x_7))^2.$$

Putting $x_i=g^{-1}e_i$ $(1 \leqq i \leqq 7)$, we get $\det(\langle g^{-1}e_i,g^{-1}e_j\rangle) = \omega(g^{-1}e_1,\cdots,g^{-1}e_7)^2 =$

$$=(\det g)^{-2}\omega(e_1,\cdots,e_7) = (\det g)^{-2}.$$

On the other hand, it follows from (6)

$$\det(\langle g^{-1}e_i,\ g^{-1}e_j\rangle) = \det(\det g\langle e_i,e_j\rangle) = (\det g)^7.$$

Therefore, we have $(\det g)^9=1$ and $\det g=1$ Then, by (6), we conclude that $g \in SO(7)$.

We now prove $g \in G_2$. Let $P: V\times V \to V$ be $P(x,y)=\mathrm{Im}\, xy$. Then, we may write $\phi(x,y,z)=\langle x,P(y,z)\rangle$. Since $g^*\phi=\phi$ and $g \in SO(7)$, it follows

$$\langle x,P(gx,gy)\rangle = \langle g^{-1}x,\ P(y,z)\rangle = \langle x,gP(y,z)\rangle$$

and therefore $P(gx,gy)=gP(y,z)$ for any $y,z \in V$. Now for two elements $ae_0+x,\ be_0+y \in \mathbb{D}$ with $a,b \in \mathbb{R}$, $x,y \in V$, we have $(ae_0+x)(be_0+y)=(ab-\langle x,y\rangle)e_0+(ay+bx+P(x,y))$. Then, g being extended to an orthogonal transformation of $\mathbb{D}$ by $g(e_0)=e_0$, we see $(g(ae_0+x))(g(be_0+y))=(ae_0+gx)(be_0+gy)=(ab-\langle gx,gy\rangle)+(a(gy)+b(gx)+P(gx,gy))=g((ae_0+x)(be_0+y))$.

Therefore, we have proved $g \in G_2$. q.e.d.

We define a 4-linear form ψ on the space of Cayley numbers $\mathbb{D}$ by $\psi(x,y,z,w)=\frac{1}{2}\langle x,[y,z,w]\rangle$ for $x,y,z,w \in \mathbb{D}$, where $[\cdot,\cdot,\cdot]$ denotes the associator (defined in §1).

Lemma 2. ψ is a G_2-invariant alternating form on $\mathbb{D}$.

Proof. G_2-invariance of ψ is easily verified. We have

$$2\psi(x,x,z,w)=\langle x,[x,z,w]\rangle=\langle x,(xz)w\rangle-\langle x,x(zw)\rangle=\langle x\bar{w},xz\rangle-\langle \bar{x}x,zw\rangle = |x|^2(\langle \bar{w},z\rangle-\langle \bar{w},z\rangle)=0.$$

Since the associator $[\cdot,\cdot,\cdot]$ is alternating (Lemma 1.1), it follows that ψ is an alternating form.

Let $\omega^0,\omega^1,\cdots,\omega^7$ be 1-forms on $\mathbb{D}$ such that $\omega^i(e_j)=\delta_{ij}$ $(0\leq i,j\leq 7)$, and write $\omega^{ijk\ell}=\omega^i \wedge \omega^j \wedge \omega^k \wedge \omega^\ell$. Then, $\omega(e_i,e_j,e_k,e_\ell)=0$ if $[e_j,e_k,e_\ell]=0$, and we get the following expression

(7) $$\psi = \omega^{4567}+\omega^{2367}-\omega^{2145}+\omega^{1357}+\omega^{1346}+\omega^{1256}-\omega^{1247}.$$

Extending the 3-form ϕ on V onto $\mathbb{D}$ in an obvious way, i.e., by the formula (2), we get $\psi=*(\omega^0 \wedge \phi)$ where $*$ is the $*$-opertor acting on forms on the vector space $\mathbb{D}$ with the inner product $\langle\cdot,\cdot\rangle$ and with the orientation such that $\omega^0 \wedge \omega^1 \wedge \cdots \wedge \omega^7$ is a positive 8-form.

8.2 The group Spin(7)

In this section, Spin(7) denotes the group $Spin^-(7)$ in the notation in 8.3. Thus, Spin(7) consists of $g \in SO(8)$ for which there exists $\chi_g \in SO(7)$ such that $g(x)\chi_g(y)=g(xy)$ for any $x,y \in \mathbb{D}$. The map $g \to \chi_g$ is a covering homomorphism χ: Spin(7) $\to$ SO(7) with kernel $\{\pm 1\}$.

Proposition 1. The group Spin(7) coincides with the subgroup of SO(8) generated by the right multiplications R_u where u runs over the unit sphere $S^6=\{u \in \operatorname{Im}\mathbb{D} \mid |u|=1\}$.

Proof. We use the Manfang's identity: $z(xyx)=((zx)y)x$ (Lemma 1.2.b). Replacing z by xu and x by u, we get $(xu)(uyu)=((xu)u)y)u$. Suppose $u \in S^6$. Then, $u^{-1}=\bar{u}=-u$. And the map $\chi_u: \mathbb{D} \to \mathbb{D}$ defined by $\chi_u(y)=\bar{u}yu$. belongs to SO(7). The above formula implies

$$(R_u x)(\chi_u(y)) = R_u(xy).$$

Therefore $R_u \in$ Spin(7) for $u \in S^6$. and $\chi_{R_u}=\chi_u$.

Let G be the subgroup of SO(8) generated by $\{R_u | u \in S^6\}$. Since $\chi_u(v)=-v$, if v is orthogonal to u in $\operatorname{Im}\mathbb{D}$ and $\chi_u(u)=u$, $-\chi_u$ is the reflection of $\operatorname{Im}\mathbb{D}$ with respect to the hyperplane orthogonal to u. The group O(7) is then generated by $\{-\chi_u | u \in S^6\}$, and therefore SO(7) is generated by $\{\chi_u | u \in S^6\}$. It follows that $\chi(G)=SO(7)$. On the other hand, $R_{e_1}^2=-1$, and so G contains the kernel of χ. So, we conclude G=Spin(7) q.e.d.

Now, following Harvey-Lawson [10], we define the triple cross-product $x\times y\times z$ of three elements $x,y,z \in \mathbb{D}$ by the formula

$$x\times y\times z =\frac{1}{2}(x(\bar{y}z)-z(\bar{y}x)).$$

And, define 4-linear form Φ on $\mathbb{D}$ by

$$\Phi(x,y,z,w)=\langle x,y\times z\times w\rangle \qquad \text{for } x,y,z,w \in \mathbb{D}.$$

Lemma 3. Φ is an alternating form on $\mathbb{D}$ and we have

$$\Phi = \omega^0 \wedge \phi + \psi.$$

Proof. See [10] P.120.

From this Lemma and (2), (7), we get the following expression

(9) $\Phi = \omega^{0123}-\omega^{0167}-\omega^{0527}-\omega^{0563}-\omega^{0154}-\omega^{0264}-\omega^{0374}-\omega^{4567}-\omega^{4523}-\omega^{4163}-$

$-\omega^{4127}-\omega^{2367}-\omega^{1357}-\omega^{1256}$

Theorem 2 $\quad \mathrm{Spin}(7) = \{g \in GL(\mathbb{D})\ g^*\Phi = \Phi\}$

Proof. Let G be the group on the right-hand side. We refer to [10] p.122. Prop. 1.6 for the proof of the fact $\mathrm{Spin}(7) \subset G$, in which one applies Proposition 1.

Spppose $g \in G$ and assume $g(e_0) = \lambda e_0$ with some $\lambda>0$. We shall prove $g \in G_2$. For any form α on $\mathbb{D}$, we see easily $i(gx)(g^*\alpha)=g^*(i(x)\alpha)$. In particular, for $x=e_0$ we have

$$\lambda i(e_0)(g^*\alpha) = g^*\ (i(e_0)\alpha)$$

By (8), we have $\phi=i(e_0)\Phi$ since $i(e_0)\phi=i(e_0)\psi=0$. Therefore, since $g^*\Phi=\Phi$, we get $g^*\phi=\lambda i(e_0)(g^*\Phi)=\lambda\phi$. Namely

$$\phi(g^{-1}x,\ g^{-1}y,\ g^{-1}z) = \lambda\phi(x,y,z)$$

for any $x,y,z \in \mathbb{D}$. By identifying $V=\mathbb{R}^7$ with the quotient space $\mathbb{D}/\mathbb{R}e_0$, $h=\lambda^{-1/3}g^{-1}$ induces an action of G_2 on V by Theorem 1. The form ψ may be considered as a form on this qoutient V and invariant under the action of G_2. Therefore $h^*\psi=\psi$ and $\psi(g^{-1}x,\ g^{-1}y,\ g^{-1}z,\ g^{-1}w)=\lambda^{4/3}\psi(x,y,z,w)$ and $g^*\psi=\lambda^{4/3}\psi$. On the other hand, put $g^*\omega^0=\sum_{i=o}^{7} a_i\omega^i$. Then $a_i=(g^*\omega^0)(e_i)=\omega^0(g^{-1}e_i)$, and so $a_0=1/\lambda$. Then, putting $\gamma=\sum_{i=1}^{7} a_i\omega^i$

$$g^*\Phi = g^*\omega^0\ \Lambda\ g^*\phi + g^*\psi = \frac{1}{\lambda}\ \omega^0\ \Lambda\ (\lambda\phi)+\gamma\ \Lambda\ \lambda\phi + g^*\psi$$

$$= \omega^0\ \Lambda\ \psi + \gamma\ \Lambda\ \lambda\phi + \lambda^{4/3}\psi.$$

Since $g^*\Phi=\Phi$, we get $\psi = \gamma\,\Lambda\,\lambda\phi+\lambda^{4/3}\psi$

Now by (2) $\gamma\,\Lambda\,\lambda\phi$ is a linear combination of $\omega^{ijk\ell}$ such that some three of $\{e_i, e_j, e_k, e_\ell\}$ are associative. while by (7), ψ is a linear combination of $\omega^{ijk\ell}$ such that any three of $\{e_i,e_j,e_k,e_\ell\}$ are not associative. It follows that $\gamma\ \Lambda\ \lambda\phi=0$ and $\lambda^{4/3}=1$. If follows again by (2) that $\gamma=0$ and $\lambda=1$. Thus $ge_0=e_0$ and $g(\mathrm{Im}\ \mathbb{D})=\mathrm{Im}\ \mathbb{D}$. Since $g^*\phi=\phi$, we see $g \in G_2$ by Theorem 1.

Finally let g be an arbitrary element of G. Since Spin(7) acts transitively on S^7 and $-1 \in \mathrm{Spin}(7)$, we can find $g_1 \in \mathrm{Spin}(7)$ such that $g_1^{-1}g(e_0)=\lambda e_0$ with $\lambda>0$. As G contains Spin(7), $g_1^{-1}\ g \in G$. By what we have proved above, $g_1^{-1}g$ belongs then to G_2 which is contained in Spin(7). Therefore $g \in \mathrm{Spin}(7)$. Thus $G \subset \mathrm{Spin}(7)$. We have thus proved G=Spin(7). q.e.d.

Proofs of Theorems 1 and 2 are due to Bryant [5].

We cite another theorem of Bryant without proof.

Theorem 3. (1) On $\mathbb{R}^7=\mathrm{Im}\ \mathbb{D}$, any G_2 invariant alternating form is a linear combination of $1,\phi,\psi$ and ω.

(2) On $\mathbb{R}^8 = \mathbb{D}$, any Spin(7)-invariant alternating form is a linear combination of $1, \Phi$, and $\omega^0 \wedge \omega$.

Remark. The characterizations of the groups G_2 and Spin(7) are essentially applied by Bryant [5] to construct examples of Riemannian manifolds whose holonomy group is G_2 or Spin(7) (Cf. Besse [3] Chap 10 and Appendix B). The examples in [5] being not complete as Riemannian manifolds, Bryant [6] and Salamon [12] construct complete Riemannian manifolds with the same type of holonomy group.

END.

References

[1] M.F.Atiyah, R.Bott and S.Sapiro: Clifford modules, Topology 3(Supplement) (1964), 3-38.

[2] M.Berger: Sur les groupes d'holonomie des variétés à connexions affines et des variétés riemanniennes, Bull. Math. Soc. France 83(1955), 279-330.

[3] A.Besse: Einstein manifolds, Springer, 1987.

[4] A. Borel: Le plan projectif des octaves et les sphères comme espaces homogenès, C.R. Paris 230(1950), 1878-1380.

[5] R.L. Bryant: Metrics with exceptional holonomy, (to appear).

[6] R.L. Bryant: Note on the construction of metrics with holonomy G_2 or Spin(7), (to appear).

[7] C.Chevalley: Theory of Lie groups, Princeton, 1946.

[8] B.Eckmann: Gruppentheoretischer Beweis des Satzes von Hurewitz-Radon über die Koposition quadratischer Formen, Commentarii Math. Helv. 15(1942/43), 358-366.

[9] H.Freudenthal: Oktaven, Ausnahmegruppen und Oktaven-geometrie, Math. Inst. der Rijsuniversiteit te Utrecht, 1951.

[10] R.Harvey and H.B. Lawson: Calibrated geometry, Acta Math. 148(1982), 47-157.

[11] Y. Matsushima: Some remarks on the exceptional simple Lie groups F_4, Nagoya Math. J. 4(1954), 83-88.

[12] S.M. Salamon: Self-duality and exceptional geometry, Topology and its applications, 1-8, Baku, 1987.

[13] H.Wu: The Bochner technique in differential geometry, (to appear in the series "Mathematical Reports").

[14] Yen Chih-tà: Sur les polynomes de Poincaré des groupes exceptionels, C.R. Paris 228(1949), 628-630.

Shingo MURAKAMI
Department of Mathematics
Osaka University
Toyonaka 560, Japan

A Remark on the Isoparametric Polynomials of Degree 6

Peng Chia-Kuei
Graduate School of University of Science and Technology of China
Beijing, China

Hou Zixin
Department of Mathematics, Nankai University
Tianjin, China

There has been an extensive study of isoparametric hypersurfaces in spheres and mighty advances have been made by many authors. In this respect, we refer to the expository work [1] for the results and references.

In 1976 H.Ozeki and M.Takeuchi [2] gave the explicit forms of isoparametric polynomials representing the homogeneous isoparametric hypersurfaces in spheres. For the case of degree 6, however, their forms are rather conceptional than computable.

Our present short paper is intended to give more explicit forms of the isoparametric polynomials with degree 6 by using the matrix expression of exceptional simple Lie algebra $\mathfrak{g}_2$.

We hope this will be useful in understanding the isoparametric hypersurfaces with 6 distinct principal curvatures.

First of all, we notice that complex Lie algebra $\mathfrak{g}_2^{\mathbb{C}}$ has the following matrix expression:

$$\mathfrak{g}_2^{\mathbb{C}} = \left\{ \begin{bmatrix} 0 & \sqrt{2}a & \sqrt{2}b & \sqrt{2}c & \sqrt{2}u & \sqrt{2}v & \sqrt{2}w \\ -\sqrt{2}u & x & l & m & o & -c & +b \\ -\sqrt{2}v & r & y & n & c & o & -a \\ -\sqrt{2}w & p & q & z & -b & a & 0 \\ -\sqrt{2}a & o & -w & v & -x & -r & -p \\ -\sqrt{2}b & w & o & -u & -l & -y & -q \\ -\sqrt{2}c & -v & u & o & -m & -n & -z \end{bmatrix} \right\} \tag{1}$$

where all the letters stand for complex numbers and $x+y+z=0$.

Its compact real form $\mathfrak{g}_2$ consists of all the following skew- Hermitian matrices:

$$\begin{bmatrix} 0 & \sqrt{2}z_1 & \sqrt{2}z_2 & \sqrt{2}z_3 & \sqrt{2}\bar{z}_1 & \sqrt{2}\bar{z}_2 & \sqrt{2}\bar{z}_3 \\ -\sqrt{2}\bar{z}_1 & \sqrt{-1}x_1 & z_4 & z_5 & o & -z_3 & z_2 \\ -\sqrt{2}\bar{z}_2 & -\bar{z}_4 & \sqrt{-1}x_2 & z_6 & z_3 & o & -z_1 \\ -\sqrt{2}\bar{z}_3 & -\bar{z}_5 & -\bar{z}_6 & \sqrt{-1}x_3 & -z_2 & z_1 & o \\ -\sqrt{2}z_1 & o & -\bar{z}_3 & \bar{z}_2 & -\sqrt{-1}x_1 & \bar{z}_4 & \bar{z}_5 \\ -\sqrt{2}z_2 & \bar{z}_3 & o & -\bar{z}_1 & -z_4 & -\sqrt{-1}x_2 & \bar{z}_6 \\ -\sqrt{2}z_3 & -\bar{z}_2 & \bar{z}_1 & o & -z_5 & -z_6 & -\sqrt{-1}x_3 \end{bmatrix} \tag{2}$$

where $z_1, \cdots, z_6$ are complex numbers, x_1, x_2, x_3 real and

$$x_1 + x_2 + x_3 = 0.$$

For the sake of convenience, we can rewrite the matrix in $\mathfrak{g}_2$ as follows:

$$Q = \begin{bmatrix} O & Z & \bar{Z} \\ -\bar{Z}' & A & B \\ -Z' & \bar{B} & \bar{A} \end{bmatrix} \tag{2'}$$

where $Z = \frac{1}{\sqrt{3}}(z_1, z_2, z_3)$; z_1, z_2, z_3 are complex numbers;

$$A = \begin{bmatrix} \sqrt{-1}x_1 & \frac{1}{\sqrt{2}}z_4 & \frac{1}{\sqrt{2}}z_5 \\ -\frac{1}{\sqrt{2}}\bar{z}_4 & \sqrt{-1}x_2 & \frac{1}{\sqrt{2}}z_6 \\ -\frac{1}{\sqrt{2}}\bar{z}_5 & -\frac{1}{\sqrt{2}}\bar{z}_6 & \sqrt{-1}x_3 \end{bmatrix}$$

(here z_4, z_5, z_6 are complex numbers, x_1, x_2, x_3 real and $x_1 + x_2 + x_3 = 0$);

$$B = \frac{1}{\sqrt{6}} \begin{bmatrix} 0 & -z_3 & z_2 \\ z_3 & 0 & -z_1 \\ -z_2 & z_1 & 0 \end{bmatrix}.$$

The inner product on $\mathfrak{g}_2$ is

$$\langle Q_1, Q_2 \rangle = -\frac{1}{2} tr Q_1 Q_2 \quad \text{(where } Q_1 \text{ and } Q_2 \in \mathfrak{g}_2\text{)},$$

which turns out to be standard Euclidean metric.

(1) The first case corresponding to the symmetric space $G_2/SU(2) \times SU(2)$:

Considering Cartan involution $\theta(Q) = \overline{Q}$, we have Cartan decomposition of $\mathfrak{g}_2$ as follows

$$\mathfrak{g}_2 = k + \sqrt{-1}\, p$$

k consists of all the following real skew symmetric matrices:

$$K = \begin{bmatrix} 0 & u & u \\ -u' & U & V \\ -u' & V & U \end{bmatrix}$$

where $u = \frac{1}{\sqrt{3}}(u_1, u_2, u_3)$,

$$U = \frac{1}{\sqrt{2}} \begin{bmatrix} 0 & u_4 & u_5 \\ -u_4 & 0 & u_6 \\ -u_5 & -u_6 & 0 \end{bmatrix},$$

$$V = \frac{1}{\sqrt{6}} \begin{bmatrix} 0 & -u_3 & u_2 \\ u_3 & 0 & -u_1 \\ -u_2 & u_1 & 0 \end{bmatrix}$$

(here $u_i (i = 1, \cdots, 6)$ are real numbers);

p consists of all the following symmetric matrices:

$$P = \begin{bmatrix} 0 & Y & -Y \\ Y' & T & S \\ -Y' & -S & -T \end{bmatrix}$$

where $Y = \frac{1}{\sqrt{3}}(y_1, y_2, y_3)$,

$$T = \begin{bmatrix} x_1 & \frac{1}{\sqrt{2}} y_4 & \frac{1}{\sqrt{2}} y_5 \\ \frac{1}{\sqrt{2}} y_4 & x_2 & \frac{1}{\sqrt{2}} y_6 \\ \frac{1}{\sqrt{2}} y_5 & \frac{1}{\sqrt{2}} y_6 & x_3 \end{bmatrix},$$

$$S = \begin{bmatrix} 0 & -y_3 & y_2 \\ y_3 & 0 & -y_1 \\ -y_2 & y_1 & 0 \end{bmatrix}$$

in which $x_i (i = 1, 2, 3)$ and $y_j (j = 1, \cdots, 6)$ are real numbers. The isoparametric polynomial is $f = 18 tr P^6 - \frac{5}{4}(tr P^2)^3$.

To make the computation simpler, let us consider the following transformation:

$$P \longmapsto J P J' = \tilde{P}$$

where

$$J = \frac{1}{\sqrt{2}} \begin{bmatrix} \sqrt{2} & 0 & 0 \\ 0 & I & I \\ 0 & -I & I \end{bmatrix} \in SO(7)$$

in which I is a 3×3 identity matrix.

It is easy to see that

$$\tilde{P} = - \begin{bmatrix} 0 & 0 & \sqrt{2}Y \\ 0 & 0 & T-S \\ \sqrt{2}Y' & T+S & 0 \end{bmatrix},$$

consequently

$$\tilde{P}^2 = \begin{bmatrix} 2YY' & \sqrt{2}Y(T+S) & 0 \\ \sqrt{2}(T-S)Y' & (T-S)(T+S) & 0 \\ 0 & 0 & 2Y'Y + (T+S)(T-S) \end{bmatrix} = \begin{bmatrix} P_1 & 0 \\ 0 & P_2 \end{bmatrix}.$$

Hence

$$trP^6 = tr\tilde{P}^6 = trP_1^3 + trP_2^3.$$

We notice, however, that $trP_1^3 = trP_2^3$. In fact, considering the matrix

$$W = \begin{bmatrix} 0 & \sqrt{2}Y \\ 0 & T-S \end{bmatrix}$$

we have $WW' = P_1$ and $W'W = \begin{bmatrix} 0 & 0 \\ 0 & P_2 \end{bmatrix}$. The statement follows evidently.

To sum up, the isoparametric polynomial can be rewritten as

$$f = 36trP_2^3 - 10\langle P, P\rangle^3$$

where $P_2 = 2Y'Y + (T+S)(T-S)$ which is a 3×3 symmetric matrix

(2) The second case corresponding to symmetric space $G_2 \times G_2/G_2$:

The isoparametric polynomial is still as follows

$$f = 18trQ^6 - \frac{5}{4}(trQ^2)^3$$

where Q is the same as (2).

The corresponding representation of $\mathfrak{g}_2$ being irreducible, the isoparametric polynomial in this case is much more complicated than the first one. However, we notice that the representative matrices Q have symmetry between the blocks and $ZB = 0$. Hence the expression of the polynomial can be reduced to the computation of the traces of some 3×3 matrices, Further calculation is omitted here.

References

[1] Cecil, T.E. and Ryan, P.J. **Tight and Taut Immersions of Mamifolds** Pitman Advanced Publishing Program 1985.

[2] Ozeki, H. and Takeuchi, M. **On some types of isoparametric hypersurfaces in spheres I, II.** Tohoku Math. J. **27** (1975) 515- 559, **28** (1976) 7-55.

[3] Wan, Z.X. **Lie algebra** Pergammon Press, New York, 1975.

On The Holomorphic Maps From Riemann Surfaces to Grassmannians

Shen Chun-li

Dept. of Math., Fudan Univ.,
Shanghai, China

Abstract

With the recent results of Atiyah-Bott [2] about the Yang-Mills connections over the Riemann surface, and those of Narasimha- Seshadri [6] and Donaldson [3] about the stable holomorphic vector bundles, we have proved the following:

Theorem Let M be a compact Riemann surface with genus $g (g \geq 2)$, $Hol_d(M, G_r(N))$ the set of all full, indecomposable holomorphic maps with degree d from M to $G_r(N)$, (see §2 for the detailed definitions of the degree and full property). Then we have

$$\dim Hol_d(M, G_r(N)) = N(d + r - rg) + r^2(g - 1),$$

if $d \geq r(r-1)(3g-2) + 2rg$,

§1. Introduction

It is very interesting to search for whether there exist holomorphic maps between two complex manifolds and how many of those are between them. There are very few results about these problems, in which mathematical physists are more interested. In the non-linear σ-model theory one wants to study the harmonic maps from compact Riemann surface M with genus g into complex projective space $\mathbb{C}P^N$ or complex Grassmannian $G_r(N) = G_r(\mathbb{C}^N)$. The holomorphic map from M to $\mathbb{C}P^N$ or $G_r(N)$ is a harmonic one with minimal energy. According to Elles-Wood [8], J.Ramanatham [9] and Chern-Wolfson [7], any harmonic map from $\mathbb{C}P^1$ to $\mathbb{C}P^N$ or complex Grassmannian can be obtained from a holomorphic one from $\mathbb{C}P^1$ to $\mathbb{C}P^N$ or complex Grassmannian by a given way. So the investigation of the holomorphic maps from M to $\mathbb{C}P^N$ or $G_r(N)$ is very important not only in the theory of complex geometry but also in that of harmonic maps. Recently, Killingback [5] calculated the dimension of the space of holomorphic maps from M to the $\mathbb{C}P^N$, and noted that the corresponding problem for the holomorphic maps from M to the $G_r(N)$ has not been solved yet. Here in the paper, we have claculated the dimension of the space of holomorphic maps from M to the $G_r(N)$, and proved the following theorem.

Theorem Let M be a compact Riemann surface with genus $g (g \geq 2)$, $Hol_d(M, G_r(N))$ the set of all full, indecomposable holomorphic maps with degree d from M to $G_r(N)$, (see §2 for the detailed definitions of the degree and full property). Then if $d \geq r(r-1)(3g-2) + 2rg$, we have

$$\dim Hol_d(M, G_r(N)) = N(d + r - rg) + r^2(g - 1).$$

The main idea of the proof is to reduce the study of $Hol_d(m, G_r(N))$ to that of the holomorphic vector bundles over M with rank r. With the recent results of Atiyah-Bott [2] about the Yang-Mills connections over the Riemann surface, and those of Narasimma-Seshadri [6] and Donaldson [3] about the stable holomorphic vector bundles, we can prove the theorem stated above.

§2. The relation between the holomorphic map and the holomorphic vector bundle.

Let M be a compact Riemann surface with genus g, $G_r(N)$ complex Grassmannian, namely, all complex r-planes in $\mathbb{C}^N$, and

$$G_r(N) = \frac{U(N)}{U(r) \times U(N-r)}.$$

1. Let $f : M \to G_r(N)$ be a differentiable map and assume $U \to G_r(N)$ to be a universal r-plane bundle over $G_r(N), U^* \to G_r(N)$ the dual bundle of U. From the theory of fibre bundle, we have a commutative diagram

$$\begin{array}{ccc} f^*U^* & \longrightarrow & U^* \\ \downarrow & & \downarrow \\ M & \xrightarrow{f} & G_r(N) \end{array}$$

where pull-back bundle f^*U^* is a complex differentiable vector bundle over M with rank $r, c_1(f^*U^*) = f^*c_1(U^*)$, and c_1 represents the first Chern class of complex vector bundles. On the other hand, for any complex differentiable bundle F over M with rank r, let $\deg F = \int_M c_1(F)$, then

$$\deg(f^*U^*) = \int_M f^*c_1(U^*).$$

Definition 1 We call

$$\deg f = \deg(f^*U^*)$$

the degree of the map $f : M \to G_r(N)$.

Conversely, for any complex differentiable vector bundle E over M with rank r, let $\Gamma(E)$ be the set of all differentiable sections of E. If there is an N-dimensional subspace V of $\Gamma(E)$ without base point (i.e., for any $x \in M$, the fibre E_x of E at x can be spanned by $\{\varphi(x) | \varphi \in V\}$.), then by a standard way, we can find a differentiable map $f : M \to G_r(N)$ such that $E = f^*(U^*)$, and the commutative diagram

$$\begin{array}{ccc} E = f^*U^* & \longrightarrow & U^* \\ \downarrow & & \downarrow \\ M & \xrightarrow{f} & G_r(N) \end{array}$$

holds, and $\deg f = \deg E$, (see [10]), and the equivalent complex vector bundles E corresponds to the homotopic maps, and vice versa. Consequently, the set of all equivalent classes of complex differentiable vector bundles over M with rank r is just that of all homotopic classes of differentiable maps from M to Grassmannian $G_r(N)$. Therefore, the problem to determine how many non-equivalent complex differentiable vector bundles with rank r over a compact Riemann surface with genus g reduces to how to compute the homotopic class

$$[M, G_r(N)]_*.$$

Since the first homology group of $G_r(N)$ is zero, its first homotopic group is also zero, thus $G_r(N)$ is simply connected. Consequently

$$[M, G_r(N)]_* \simeq H^2(M, \pi_2(G_r(N))) \simeq \pi_2(G_r(N)).$$

When $N - r \geq 2$, it follows from

$$\pi_2(G_r(N)) \simeq \pi_1(U(r)) \simeq \mathbf{Z},$$

that

$$[M, G_r(N)]_* = \mathbf{Z}.$$

This implies that the complex differentiable vector bundle E of rank r over the compact Riemann surface M with genus g can be characterized by the degree of this vector bundle.

2. Using homogeneous coordinates we can express the element x of $G_r(N)$, i.e., express x by a $r \times N$ matrix

$$x = \begin{pmatrix} x_1^1 & \cdots & x_N^1 \\ & \cdots & \\ x_1^r & \cdots & x_N^r \end{pmatrix}$$

with rank r. Taking $\varphi \in GL(N, \mathbb{C})$, we define the action of φ on $G_r(N)$ as

$$\varphi(x) \stackrel{\text{def}}{=} \begin{pmatrix} x_1^1 & \cdots & x_N^1 \\ & \cdots & \\ x_1^r & \cdots & x_N^r \end{pmatrix} \begin{pmatrix} \varphi_{11} & \cdots & \varphi_{1N} \\ \vdots & & \vdots \\ \varphi_{N1} & \cdots & \varphi_{NN} \end{pmatrix}.$$

Obviously,

$$\begin{aligned} \varphi \ : \ G_r(N) &\longrightarrow G_r(N) \\ x &\longmapsto \varphi(x) \end{aligned}$$

is a holomorphic transformation. We also call φ a projective automorphism of $G_r(N)$. All projective automorphisms of $G_r(N)$ form a group, known as the group of projective automorphisms and denoted by $PGL(N, \mathbb{C})$. Since when φ is the scalar matric, $\varphi(x) = x$ in $G_r(N)$, i.e., φ is fixed at all points of $G_r(N)$, we have

$$\dim PGL(N, \mathbb{C}) = \dim GL(N, \mathbb{C}) - 1 = N^2 - 1.$$

The dual bundle $U^* \to G_r(N)$ of the universal r-plane bundle over $G_r(N)$ has global sections. For example, when

$$x = \begin{pmatrix} x_1^1 & \cdots & x_N^1 \\ \vdots & & \vdots \\ x_1^r & \cdots & x_N^r \end{pmatrix},$$

let

$$\begin{aligned} \eta_p : \ G_r(N) &\longrightarrow U \\ x &\longmapsto \eta_p(x) = \begin{pmatrix} x_p^1 \\ \vdots \\ x_p^r \end{pmatrix} \end{aligned}$$

for $p = 1, \cdots, N$.

Definition 2 We call the above sections $\eta_1, \cdots, \eta_N$ the standard global sections of U^*.

When $N' < N, G_r(N')$ can be naturally embedded into $G_r(N)$, i.e., inclusion map i : $G_r(N') \hookrightarrow G_r(N)$ is an embedding. Therefore, we usually consider $G_r(N')$ to be a submanifold of $G_r(N)$, and omit the notation i.

3. Further more, if $f : M \to G_r(N)$ is a holomorphic map, then $E = f^*U^*$ in the commutative diagram in §2.1 is a holomorphic vector bundle over M with rank r. Obviously, if $f : M \to G_r(N)$ is a holomorphic map, then for every $\varphi \in PGL(N, \mathbb{C})$, $\varphi \circ f : M \to G_r(N)$ remains a holomorphic map.

Definition 3 For a holomorphic map $f : M \to G_r(N)$, if there is an integer $N' < N$, such that

$$\varphi \cdot f : M \longrightarrow G_r(N') \subset G_r(N),$$

where φ is a projective automorphism in $G_r(N)$, then we say f is non-full, or otherwise, f is full.

Now we denote by $\Gamma(U^*)$ and $\Gamma(E)$ the sets of all holomorphic maps of holomorphic vector bundles U^* and E respectively. We have the following

Proposition 1 Suppose a holomorphic map $f : M \to G_r(N)$ is full, the corresponding commutative diagram is

$$\begin{array}{ccc} E = \; f^*U^* & \longrightarrow & U^* \\ \big\downarrow & & \big\downarrow \\ M & \xrightarrow{\;f\;} & G_r(N) \end{array}$$

Let $\eta_1, \cdots, \eta_N$ be the standard global sections of U^*, denote $T = span(\eta_1, \cdots, \eta_N) \subset \Gamma(U^*)$, and let $V = f^*T = span(f^*\eta_1, \cdots, f^*\eta_N) \subset \Gamma(E)$. Then

(1) V has no base point,

(2) $\dim V = N$.

Proof

(1) Since $U^*_{f(x)}$ has been full-spanned by $T_{f(x)}$ for every point x in M, $E_x = (f^*U^*)_x = U^*_{f(x)}$ is full- spanned by V_x. Consequently, V has no base point.

(2) As $V = f^*T, \dim V \leq \dim T = N$. If $\dim V = N' < N$, it would reach a contradiction. Without loss of generality, we may well assume the first N' sections $f^*\eta_1, \cdots, f^*\eta_{N'}$ to be a base of V, thus $V = span(f^*\eta_1, \cdots, f^*\eta_{N'})$. Then by the commutative diagram

$$\begin{array}{ccc} E = f^*U = \; f'^*(U^*) & \longrightarrow & U^* \\ \big\downarrow & & \big\downarrow \\ M & \xrightarrow{\;f'\;} & G_r(N') \subset G_r(N) \end{array}$$

in §2.1, we can construct a holomorphic map

$$f' : M \longrightarrow G_r(N'),$$

and for every point $x \in M$, we have

$$\begin{aligned} f'(x) =& ((f^*\eta_1)(x), \cdots, (f^*\eta_{N'})(x)) \\ \overset{i}{\hookrightarrow}& ((f^*\eta_1)(x), \cdots, (f^*\eta_{N'})(x), 0, \cdots, 0). \end{aligned}$$

On the other hand, if we let

$$f^*\eta_\alpha = A^p_\alpha(f^*\eta_p),$$

where $\alpha = N' + 1, \cdots, N, p = 1, \cdots, N'$, then

$$f(x) = ((f^*\eta_1)(x), \cdots, (f^*\eta_{N'})(x),$$
$$A^p_{N'+1}(f^*\eta_p)(x), \cdots, A^p_N(f^*\eta_p)(x)).$$

And we further assume that

$$\varphi = \begin{pmatrix} 1 & & 0 & A^1_{N'+1} & \cdots & A^1_N \\ & \ddots & \vdots & & & \vdots \\ 0 & & 1 & A^{N'}_{N'+1} & \cdots & A^{N'}_N \\ & & & 1 & & 0 \\ & 0 & & & \ddots & \\ & & & 0 & & 1 \end{pmatrix} \in GL(N, \mathbb{C}),$$

so $f(x) = \varphi \circ (i \circ f')$, then

$$i \circ f' = \varphi^{-1} \circ f : M \longrightarrow G_r(N') \hookrightarrow G_r(N),$$

hence we get to the contradiction with the full properties.

QED.

Definition 4 Suppose we have two holomorphic maps

$$f_1: \quad M \longrightarrow G_{r_1}(N) \subset G_r(N),$$
$$f_2: \quad M \longrightarrow G_{r_2}(N) \subset G_r(N),$$

(here $r = r_1 + r_2$), hence for any $x \in M, f_1(x)$ and $f_2(x)$ are the subspaces in $\mathbb{C}^N$ with dimensions r_1 and r_2 respectively. If

$$f_1(x) \cap f_2(x) = \{0\},$$

then we can construct a new holomorphic map

$$\begin{aligned} f: \quad M &\longrightarrow G_r(N) \\ x &\longmapsto f_1(x) \oplus f_2(x). \end{aligned}$$

We call such an $f : M \to G_r(N)$ a decomposable holomorphic map, and denote it by $f = f_1 \oplus f_2$. Otherwise, f is indecomposable.

Proposition 2 $f : M \to G_r(N)$ is a decomposable holomorphic map iff $E = f^*U^*$ is a decomposable holomorphic vector bundle over M.

Proof

If f is decomposable, then using

$$f(x) = f_1(x) \oplus f_2(x),$$

we have

$$U_{f(x)} = U_{1,f_1(x)} \oplus U_{2,f_2(x)},$$

where $U_i \to G_{r_i}(N)$ is the universal r_i-plane bundle over $G_{r_i}(N), (i = 1, 2)$. Therefore

$$U^*_{f(x)} = U^*_{1,f_1(x)} \oplus U^*_{2,f_2(x)},$$

where U^*_i is the dual bundle of $U_i, (i = 1, 2)$. Let $E_1 = f_1^*U_1^*, E_2 = f_2^*U_2^*$, then

$$E_x = E_{1x} \oplus E_{2x}.$$

Hence E is decomposable.

Coversely, if E is decomposable, then $E = E_1 \oplus E_2$, where E_1 and E_2 are the holomorphic vector bundles over M with ranks r_1 and r_2 respectively. Following the same notations as in Prop.1, we have

$$f^*\eta_j = (f^*\eta_j)_1 + (f^*\eta_j)_2, \quad j = 1, \cdots, N,$$

where $(f^*\eta_j)_i$ is the projective part of $f^*\eta_j$ to E_i.

For any $x \in M$,

$$(f^*\eta_1)(x), \cdots\cdots, (f^*\eta_N)(x)$$

span $E_x = U^*_{f(x)}$, thus

$$(f^*\eta_1)_1(x), \cdots\cdots, (f^*\eta_N)_1(x)$$

span E_{1x}, and

$$(f^*\eta_1)_2(x), \cdots\cdots, (f^*\eta_N)_2(x)$$

span E_{2x}. Then we can construct a holomorphic map

$$f_i : M \longrightarrow G_{r_i}(N)$$

by

$$f_i(x) = ((f^*\eta_1)_i(x), \cdots\cdots, (f^*\eta_N)_i(x)).$$

Obviously,

$$f_1(x) \cap f_2(x) = \{0\},$$

hence $f = f_1 \oplus f_2$.

QED.

4. Denote

$$A = Hol_d(M, G_r(N)) = \left\{ f : M \to G_r(N) \;\middle|\; \begin{array}{l} f \text{ is a full, indecomposable} \\ \text{holomorphic map with degree } d \end{array} \right\}.$$

We may introduce an equivalent relation $\sim$ in A as follows:

For $f, g \in A$, we say $f \sim g$ if there is an element of the group of projective automorphisms $PGL(N, \mathbb{C})$ in $G_r(N)$, such that

$$f = \varphi \circ g.$$

Denote the space of equivalent classes in A by

$$\tilde{A} = A/\sim,$$

and write

$$B = \left\{ (E, V) \;\middle|\; \begin{array}{c} E \to M \text{ is an indecomposable holomorphic vector bundle over} \\ M \text{with rank } r \text{ and } \deg E = d, \\ V \subset \Gamma(E) \text{is an} N - \text{dimensional subspace} \\ \text{without base point} \end{array} \right\}.$$

For $f \in A$, using the above method in §2.1 we can construct an element of B. Denote this corresponding by

$$\Phi : A \longrightarrow B,$$

and the image element by $\Phi(f)$. Conversely, for every element (E, V) in B, we can still use the method in §2.1 to construct an element f(a holomorphic map) in A, denoted by $\Psi(E, V) \in A$.

This element is not uniquely determined, only up to an equivalent relation $\sim$. Hence we obtain a correspondence

$$\Psi : B \longrightarrow \tilde{A}.$$

Proposition 3 $\Phi : \tilde{A} \longrightarrow B$ is bijective.

Proof It needs only to check

$$\Psi \circ \Phi = id|_{\tilde{A}}$$
$$\Phi \circ \Psi = id|_{B},$$

by a straight-forward computation.

QED.

Denote

$\mathbf{m}(r,d) = \{$ indecomposable holomorphic vector bundle E with rank $r| \deg E = d\}$, then we have

Proposition 4

$$\dim Hol_d(M, G_r(N)) = \dim \mathbf{m}(r,d) + \dim G_N(\dim \Gamma(E)) + \dim(PGL(N,\mathbb{C})).$$

Proof From Proposition 3, we know that

$$\dim A = \dim \left\{ (E,V) \middle| \begin{array}{c} E \in \mathbf{m}(r,d) \\ V \subset \Gamma(E) \text{ is an} N - \text{dimensional space} \\ \text{without base point} \end{array} \right\}.$$

But

$$\left\{ (E,V) \middle| \begin{array}{c} E \in \mathbf{m}(r,d) \\ V \subset \Gamma(E) \text{ is an} N - \text{dimensional subspace} \\ \text{which has base point} \end{array} \right\}$$

forms a proper closed subset of set

$$\left\{ (E,V) \middle| \begin{array}{c} E \in \mathbf{m}(r,d) \\ V \subset \Gamma(E) \text{ is an} N - \text{dimensional subspace} \end{array} \right\}$$

in Zariski topology. Hence in computing $\dim \tilde{A}$, we can omit the condition that V has no base point. Furthermore, as V is an element of Grassmannian $G_N(\dim \Gamma(E))$, we have

$$\dim \tilde{A} = \dim \mathbf{m}(r,d) + \dim G_N(\dim \Gamma(E)).$$

Since $\tilde{A} = A/\sim$, if two elements of A can be transformed by an element of $PGL(N,\mathbb{C})$, then they correspond to the same element of $\tilde{A}$. Hence

$$\begin{aligned} \dim A &= \dim \tilde{A} + \dim PGL(N,\mathbb{C}) \\ &= \dim \mathbf{m}(r,d) + \dim G_N(\dim \Gamma(E)) + \dim PGL(N,\mathbb{C}). \end{aligned}$$

QED.

§3. A theorem and its proof.

In this section, we are going to prove the following:

<u>Theorem</u> Let M be a compact Riemann surface with genus g, and $Hol_d(M, G_r(N))$ the set of all full, indecomposable holomorphic maps from M to $G_r(N)$ with degree d. Then we have

$$\dim Hol_d(M, G_r(N)) = N(d + r - rg) + r^2(g-1)$$

if $d \geq r(r-1)(3g-2) + 2rg$.

<u>Proof</u> From Proposition 4 in §2, we know that if the dimensions of three spaces can be computed, then we can prove this theorem. The proof will be given in several steps:

1. Computation of $\dim G_N(\dim \Gamma(E))$.

First, let us compute $\dim \Gamma(E)$. Since $\Gamma(E) = H^\circ(M, \mathrm{O}(E))$, where $H^\circ(M, \mathrm{O}(E))$ is the 0-th cohomology group of M with coefficients in the sheaf $\mathrm{O}(E)$, then

$$\dim \Gamma(E) = \dim H^\circ(M, \mathrm{O}(E)).$$

Considering the elliptic complex

$$0 \longrightarrow \wedge^\circ(M) \otimes E \overset{\bar\partial}{\longrightarrow} \wedge^{0,1}(M) \otimes E \longrightarrow 0,$$

and using the Riemann-Roch-Hirzebruch index theorem, we have

$$\begin{aligned}&\dim H^\circ(M, \mathrm{O}(E)) - \dim H^1(M, \mathrm{O}(E))\\ &= \left(e^{c_1(M)/2}(e^{\delta_1} + \cdots + e^{\delta_r}) \frac{\frac{c_1(M)}{2}}{\sinh \frac{c_1(M)}{2}} \right)[M],\end{aligned}$$

where $c_1(M)$ is the first Chern class of M and $\delta_1, \cdots, \delta_r$, Chern roots of E, i.e.,

$$\begin{aligned}&1 + c_1(E)x + c_2(E)x^2 + \cdots + c_r(E)x^r\\ =&(1 + \delta_1 x) \cdots (1 + \delta_r x),\end{aligned}$$

hence

$$c_1(E) = \delta_1 + \cdots + \delta_r.$$

Therefore

$$\begin{aligned}&\dim H^\circ(M, \mathrm{O}(E)) - \dim H^1(M, \mathrm{O}(E))\\ =&(1 + \frac{c_1(M)}{2})((1+\delta_1) + \cdots + (1+\delta_r))[M]\\ =&(1 + \frac{c_1(M)}{2})(r + c_1(E))[M]\\ =&(\frac{c_1(M)}{2} r + c_1(E))[M]\\ =&d + r/2 \cdot \chi(M)\\ =&d + r(1-g).\end{aligned}$$

On the other hand, under the assumptions of this theorem, if we can deduce that

$$H^1(M, \mathrm{O}(E)) = 0,$$

then by using the Riemann-Roch-Hirzebruch index theorem, we can calculate

$$\dim \Gamma(E) = \dim H^\circ(M, \mathrm{O}(E)).$$

Let L be the canonical line bundle over M, and

$$E(n) = E \otimes \underbrace{L \otimes \cdots \otimes L}_{n \text{ times}}.$$

From Atiyah [1], if E is indecomposable and

$$n \geq N(g,r,d) = -\frac{d}{r} + (r-1)(3g-2) + 2g,$$

then $E(n)$ is an ample vector bundle, that is to say, $E(n)$ satisfies the following conditions:

(i) $E(n)$ has sufficiently many sections (to be exact, for every point x of M, the restricted map

$$\Gamma(E(n)) \longrightarrow E(n)_x$$

is surjective).

(ii) $H^p(M, \mathrm{O}(E(n))) = 0$, when $p > 0$. Hence E is itself an ample vector bundle if $d \geq r(r-1)(3g-2)+2rg$. Consequently, $H^1(M, \mathrm{O}(E)) = 0$, and $\Gamma(E)$ has no base point. Therefore we get

$$\dim \Gamma(E) = \dim H^{\circ}(M, \mathrm{O}(E)) = d + r(1-g).$$

2. $\dim PGL(N, \mathbb{C}) = N^2 - 1$.

3. Now we will compute $\dim \mathbf{m}(r,d)$.

For every holomorphic vector bundle F over M, let

$$\mu(F) = \frac{\deg F}{rank\ F}.$$

If for every proper holomorphic subbundle D in the holomorphic vector bundle E with rank r over M, it holds that

$$\mu(D) < \mu(E) \quad (\mu(D) \leq \mu(E)),$$

then we say that E is a stable bundle (semi-stable bundle).

Let $N(r,d)$ be the set of all stable holomorphic vector bundles with rank r and degree d. Since $N(r,d)$ is an open set in $\mathbf{m}(r,d)$, in order to compute the dimension of $\mathbf{m}(r,d)$, it needs only to compute $\dim N(r,d)$. Let J_d be the Jacobi variety of Riemann surface M, it describes the set of all holomorphic line bundle over M with degree d, and

$$\dim_{\mathbb{C}} J_d = g.$$

In [2], using the theory of Yang-Mills connections, Atiyah-Bott proved that

$$N(r,d) \longrightarrow J_d$$

is a fibre bundle whose fibre at $L \in J_d$ is

$$N_0(r,d) = \{E \in N(r,d) | \det E = L\}.$$

According to Narasimha-seshadri [6] and Donaldson [3], we know that

$$\dim N_0(r,d) = (r^2-1)(g-1)$$

when $g \geq 2$, consequently

$$\begin{aligned}\dim N(r,d) &= \dim N_0(r,d) + \dim J_d\\ &= (r^2-1)(g-1) + g\\ &= r^2(g-1) + 1.\end{aligned}$$

Hence, we conclude that

$$\dim \mathbf{m}(r,d) = r^2(g-1)+1$$

4. Since

$$\begin{aligned}
&\dim Hol_d(M, G_r(N)) \\
&= \dim \mathbf{m}(r,d) + \dim G_N(\dim \Gamma(E)) + \dim(PGL(N,\mathbb{C})) \\
&= r^2(g-1) + 1 + N(d + r(1-g) - N) + N^2 - 1 \\
&= N(d + r - rg) + r^2(g-1),
\end{aligned}$$

the theorem of this paper is now proved.

References

[1] M.F.Atiyah, Vector bundles over an elliptic curve, Proc. Lond. Math. Soc., 7, 414-452.

[2] M.F.Atiyah and R.Bott, The Yang-Mills equations over Riemann surfaces, Phil. Trans. R. Soc. Lond. A 308, 523-615 (1982).

[3] S.K.Donaldson, A New proof of a theorem of Narasimhan and Seshadri, Jour. Diff. Geom., 18 (1983), 269-277.

[4] P.Griffiths and J.Harris, Principles of algebraic geometry, New York, Interscience, 1978.

[5] T.P.Killingback, Non-linear σ-models on compact Riemann surfaces, Commun. Math. Phys. 100, 481-494 (1985).

[6] M.S.Narasimhan and C.S.Seshadri, Stable and unitary vector bundles on a compact Riemann surface, Ann. Math., 82, 540-567.

[7] S.S.Chern and J.G.Wolfson, Harmonic maps of the two-spheres into a complex Grassman manifold II, preprint.

[8] J.Eells and J.Wood, Harmonic maps from surfaces to complex projective space, Advances in Math., to appear.

[9] J.Ramanathan, Harmonic maps from S^2 to $G_{2,4}$, Jour. Diff. Geom., 19 (1984) 207-219.

[10] R.O.Wells, Differential analysis on complex manifolds, Prentice-Hall, Inc., 1973.

STABILITY OF TOTALLY REAL MINIMAL SUBMANIFOLDS

Shen Yibing

(Hangzhou University)

Introduction

Among all submanifolds of a Kaehler manifold there are two important typical classes which are called Kaehler (holomorphic) submanifolds and totally real submanifolds, respectively. As is well known, every Kaehler submanifold is minimal and moreover has the minimizing volume among all homologous competitors with the same boundary (possibly empty). In other words, all of Kaehler submanifolds are stably minimal. It is natural thereby to study the stability of totally real minimal submanifolds.

In this paper, we will give an intrinsic necessary condition for a Riemannian manifold to be isometrically immersed into a complex space form as a totally real minimal submanifold. This is an upper bound estimation of the scalar curvature for the conformal metrics of such a submanifold (Theorem 1). Then, we apply it to the investigation of the stability of totally real minimal submanifolds in a complex space form. Concretely, given such a minimal submanifold M^n, we will find a simple condition such that if a simply-connected domain $\mathcal{D}\subset M^n$ satisfies such a condition, then $\mathcal{D}$ is stable (Theorem 2 and 3). Analogous problems for minimal submanifolds in a real space form have been studied by Barbosa-Do Carmo ([1],[7]), Mori ([8]) and the author ([10]).

The author would like to thank Professor S.S.Chern and the Nankai Institute of Mathematics for their hospitality.

1. Preliminaries

A complete Kaehler manifold of constant holomorphic sectional curvature is called a complex space form. We denote by $CF^n(\tilde{c})$ a simply-connected complex space form of complex dimension n with constant holomorphic sectional curvature $\tilde{c}$, which is endowed with the Fubini-Study metric for $\tilde{c}>0$, or the Bergman metric for $\tilde{c}<0$, and or the usual Hermitian metric for $\tilde{c}=0$, respectively. Let (M^n,g) be an n-dimensional totally real minimal submanifold in $CF^n(\tilde{c})$ with induced Riemannian metric g. Choose a local field of orthonormal frames $\{e_i,\ e_{i^*}=\tilde{J}e_i\}$ in $CF^n(\tilde{c})$, where $\tilde{J}$ stands for the complex structure of $CF^n(\tilde{c})$, in such a way that, restricted to M^n, the vectors $\{e_i\}$ are tangent to M^n. Throughout this paper we assume $n\geq 2$ and use the following convention on the range of indices:

$$A,B,C,\cdots = 1,2,\cdots,n,\ 1^*,2^*,\cdots,n^*; \qquad i,j,k,\cdots = 1,2,\cdots,n.$$

Then, the curvature tensor $\tilde{R}_{ABCE}$ of $CF^n(\tilde{c})$ can be represented as

$$\tilde{R}_{ABCE} = \frac{1}{4}\tilde{c}(\delta_{AC}\delta_{BE} - \delta_{AE}\delta_{BC} + \tilde{J}_{AC}\tilde{J}_{BE} - \tilde{J}_{AE}\tilde{J}_{BC} + 2\tilde{J}_{AB}\tilde{J}_{CE}). \tag{1.1}$$

Let $\{w_i,w_{i^*}\}$ be the field of dual frames with respect to $\{e_i,e_{i^*}\}$. Then, the second fundamental form σ of M^n can be written as (cf.[4])

$$\sigma = \sum_k \sigma_{k^*}, \qquad \sigma_{k^*} = \sum_{i,j} h_{ij}^{k^*}\, w_i\otimes w_j\otimes e_{k^*}, \tag{1.2}$$

where $h_{ij}^{k^*}=h_{ji}^{k^*}=h_{ik}^{j^*}=h_{jk}^{i^*}$ are the coefficients of the second fundamental form σ_{k^*} of M^n in the normal direction e_{k^*}. The Gauss-Codazzi equations of M^n are

$$\tilde{R}_{ijkl} = \frac{1}{4}\tilde{c}(\delta_{ik}\delta_{jl} - \delta_{il}\delta_{jk}) + \sum_m (h^{m*}_{ik}h^{m*}_{jl} - h^{m*}_{il}h^{m*}_{jk}), \tag{1.3}$$

$$h^{m*}_{ijk} = h^{m*}_{jij}, \tag{1.4}$$

where R_{ijkl} is the curvature tensor of (M^n,g) and h^{m*}_{ijk} is the covariant derivative of h^{m*}_{ij}.

By the minimality, it follows from (1.3) that the scalar curvature ρ of (M^n,g) is

$$\rho = \frac{1}{4}n(n-1)\tilde{c} - \|\sigma\|^2, \tag{1.5}$$

where $\|\sigma\|^2 = \sum_{i,j,k}(h^{k*}_{ij})^2$ is the length square of σ. Let Δ denote the Laplacian of (M^n,g). It was shown in [4] that

$$\frac{1}{2}\Delta(\|\sigma\|^2) \geq \|\nabla\sigma\|^2 - (2-\frac{1}{n})\|\sigma\|^4 + \frac{1}{4}(n+1)\tilde{c}\|\sigma\|^2, \tag{1.6}$$

where $\|\nabla\sigma\|^2 = \sum_{i,j,k,m}(h^{m*}_{ijk})^2$. Moreover, in a way similar to Proposition 1 of [10], it is not hard to prove that

$$\|\nabla\sigma\|^2 \geq \frac{n+2}{4n}\frac{\|\nabla(\|\sigma\|^2)\|^2}{\|\sigma\|^2} \tag{1.7}$$

provided that M^n is not totally geodesic.

Now let $\mathcal{D} \subset M^n$ be a simply-connected domain with compact closure $\bar{\mathcal{D}}$ and piecewise smooth boundary $\partial\mathcal{D}$. The minimality means that $\mathcal{D}$ is the critical point for the volume functional of the induced metric g for all variations of $\mathcal{D}$ keeping $\partial\mathcal{D}$ fixed. Let $V = \sum_k v_{k*}e_{k*}$ be a normal vector field on $\bar{\mathcal{D}}$ which vanishes on $\partial\mathcal{D}$. Then the formula of second variation along V is ([6])

$$(V,V) = -\int_{\mathcal{D}}\{\sum_k v_{k*}(\Delta v_{k*} + \sum_{i,j} R_{k*ij*i}v_{j*} + \sum_{i,j,l} h^{k*}_{il}h^{j*}_{il}v_{j*})\}dv_g, \tag{1.8}$$

where dv_g stands for the volume element of M^n in the metric g. Introducing (1.1) into (1.8), we have

$$I(V,V) = -\int_{\mathcal{D}}\{\sum_k v_{k*}(\Delta v_{k*} + \frac{1}{4}(n+3)\tilde{c}v_{k*} + \sum_{i,j,l} h^{k*}_{il}h^{j*}_{il}v_{j*})\}dv_g. \tag{1.9}$$

we say that $\mathcal{D}$ is stable if I(V,V) is nonnegative for all variation vector fields V with $V|_{\partial\mathcal{D}}=0$. A normal vector field V is called a Jacobi field if it satisfies the following elliptic linear system ([11])

$$\Delta v_{k*} + \frac{1}{4}(n+3)\tilde{c}v_{k*} + \sum_{i,j,l} h^{k*}_{il}h^{j*}_{il}v_{j*} = 0. \tag{1.10}$$

Assume that $\mathcal{D} \subset M^n$ is not stable. Then by Smale's version of the Morse index theorem (cf. [11] and [12]), there exists a domain $\mathcal{D}' \subseteq \mathcal{D}$ and a Jacobi field V on $\bar{\mathcal{D}}'$ vanishing on $\partial\mathcal{D}'$. Since $CF^n(\tilde{c})$ is analytic, it follows that totally real minimal submanifolds in $CF^n(\tilde{c})$ are analytic and hence, as solutions of the system (1.10) of elliptic differential equations, Jacobi fields are analytic. Then, without loss of generality, we may assume that $\partial\mathcal{D}'$ is the "first" conjugate boundary ([11]), that is, the only zeros of V in $\mathcal{D}'$ are isolated points $P_1,\cdots P_\nu$. Away from these points, we can choose an adapted frames $\{e_i, e_{i*}\}$ such that $V=ue_{1*}$, $u>0$ on $\mathcal{D}'/\{P_1,\cdots,P_\nu\}$ and $u|_{\partial\mathcal{D}'}=0$. With such a choice, the e_{1*}-component v_{1*} of (1.10) becomes

$$\Delta u + \frac{1}{4}(n+3)\tilde{c}u + \|\sigma_{1*}\|^2 u = 0. \tag{1.11}$$

From (1.2) and (1.5) it follows that

$$\|\sigma_{1*}\|^2 = \|\sigma\|^2 - \sum_{k\neq 1}\|\sigma_{k*}\|^2 = \frac{1}{4}\, n(n-1)\tilde{c} - \rho - \sum_{k\neq 1}\|\sigma_{k*}\|^2,$$

which together with (1.11) yields

$$u\Delta u + \{\frac{1}{4}(n^2+3)\tilde{c} - \rho\}u^2 \geqslant 0 \tag{1.12}$$

on $\mathcal{D}'/\{P_1,\cdots,P_\nu\}$.

By Stokes' theorem and the fact that $u|_{\partial\mathcal{D}'}=0$, we get from (1.12) that

$$\int_{\mathcal{D}'}\{\frac{1}{4}(n^2+3)\tilde{c}-\rho\}u^2 dv_g \geqslant \int_{\mathcal{D}'}|\nabla u|^2 dv_g. \tag{1.13}$$

It will be the fundamental inequality used in our process below.

2. Scalar Curvature

Let (M^n,g) be an n-dimensional totally real minimal submanifold in $CF(\tilde{c})$. Set

$$\phi = \frac{1}{4}n(n-1)\tilde{c}a - \rho, \tag{2.1}$$

where ρ is the scalar curvature for the induced metric g and a is a real number satisfying

$$\left.\begin{array}{ll} a \geqslant 1+n/(4n^2 - 7n + 2) & \text{for } \tilde{c}>0; \\ a \leqslant 1-(n+2)/(8n^2 - 13n + 4) & \text{for } \tilde{c}<0. \end{array}\right\} \tag{2.2}$$

Substituting (1.5) into (2.1), we get

$$\phi = \|\sigma\|^2 + \frac{1}{4}n(n-1)(a-1)\tilde{c}, \tag{2.3}$$

which together with (2.2) implies that $\phi>0$ on M^n when either $\tilde{c}>0$ or $\tilde{c}<0$. Moreover, in the case that $\tilde{c}=0$, it follows from (2.3) that $\phi>0$ if M^n is not totally geodesic. Thus, we can introduce a conformal metric $\tilde{g}=\phi g$ in M^n. For the scalar curvature $\tilde{\rho}$ of $\tilde{g}$ we have the following estimate.

THEOREM 1. Let (M^n,g) be a totally real minimal submanifold in $CF^n(\tilde{c})$. Denote by ρ and $\tilde{\rho}$ the scalar curvatures for the induced metric g and the conformal metric $\tilde{g}=\{\frac{1}{4}n(n-1)a\tilde{c}-\rho\}g$, respectively, where a is a real number such that $a \geqslant 1+n/(4n^2-7n+2)$ for $\tilde{c}>0$ and $a \leqslant 1-(n+2)/(8n^2-13n+4)$ for $\tilde{c}<0$. Then we have

$$\tilde{\rho} \leqslant \frac{2}{n}(n-1)(2n-1) - 1. \tag{2.4}$$

Proof. It is well known that the relation between ρ and $\tilde{\rho}$ is ([3])

$$\phi\tilde{\rho} = \rho-(n-1)\Delta\log\phi - \frac{1}{4}(n-1)(n-2)|\nabla\log\phi|^2 \tag{2.5}$$

for the conformal change $\tilde{g}=\phi g$.

Substituting (2.1) into (2.5), we have

$$-4\phi\tilde{\rho} = 4\phi-n(n-1)ac+4(n-1)\frac{\Delta\phi}{\phi} + (n-1)(n-6)|\nabla\log\phi|^2. \tag{2.6}$$

From (2.3) it follows that

$$|\nabla\phi|^2 = |\nabla(\|\sigma\|^2)|^2, \qquad \Delta\phi = \Delta(\|\sigma\|^2). \tag{2.7}$$

Introducing (1.6) and (2.7) into (2.6), we get

$$-4\phi\tilde{\rho} \geqslant 4\phi - n(n-1)a\tilde{c} - 8(n-1)(2-\frac{1}{n})\frac{\|\sigma\|^4}{\phi} + 2(n^2-1)\tilde{c}\frac{\|\sigma\|^2}{\phi}$$

$$+8(n-1)\frac{\|\nabla\sigma\|^2}{\phi} + (n-1)(n-6)|\nabla\log\phi|^2 . \qquad (2.8)$$

Since (2.3) together with (2.2) implies that $\phi \geqslant \|\sigma\|^2$, one can easily see from (1.7) and (2.7) that

$$\|\nabla\sigma\|^2 \geqslant \frac{n+2}{4n}\frac{|\nabla\phi|^2}{\phi} ,$$

which implies

$$8(n-1)\frac{\|\nabla\sigma\|^2}{\phi} + (n-1)(n-6)|\nabla\log\phi|^2 \geqslant \frac{1}{n}(n-1)(n-2)^2|\nabla\log\phi|^2 \geqslant 0. \qquad (2.9)$$

Thus, it follows from (2.8) and (2.9) that

$$-4\phi^2\tilde{\rho} \geqslant 4\phi^2 - n(n-1)a\tilde{c}\phi - 8(n-1)(2-\frac{1}{n})\|\sigma\|^4 + 2(n^2-1)\tilde{c}\|\sigma\|^2. \qquad (2.10)$$

If $\tilde{c} \geqslant 0$, then (2.10) together with (2.3) gives rise to

$$-4\phi^2\tilde{\rho} \geqslant 4\phi^2 - n(n-1)a\tilde{c}\phi - 8(n-1)(2-\frac{1}{n})\phi\|\sigma\|^2$$

$$+2n(n-1)\tilde{c}\|\sigma\|^2\{(n-1)(2-\frac{1}{n})(a-1)+\frac{1}{n}+1\}. \qquad (2.11)$$

The condition (2.2) for $\tilde{c}>0$ means that the last term in the right hand side of (2.11) is nonnegative. Thus, (2.11) and (2.3) yield

$$-\tilde{\rho} \geqslant 1-2(n-1)(2-\frac{1}{n}) + \frac{1}{4}n(n-1)\tilde{c}\{2(n-1)(2-\frac{1}{n})(a-1)-a\}/\phi. \qquad (2.12)$$

Clearly, by (2.2) for $\tilde{c}>0$, the last term in (2.12) is also nonnegative. Hence, (2.4) is proved for $\tilde{c} \geqq 0$.

We now consider the case of $\tilde{c}<0$. Put

$$b = \frac{1}{4}n(n-1)(a-1)\tilde{c}. \qquad (2.13)$$

Then,

$$\phi = \|\sigma\|^2 + b. \qquad (2.14)$$

Substituting (2.14) into (2.10), we get

$$-4\phi^2\tilde{\rho} \geqslant 4\phi^2\{1-2(n-1)(2-\frac{1}{n})\} + (n-1)L(\phi), \qquad (2.15)$$

where

$$L(\phi) = \{4b(8-\frac{4}{n}-\frac{1}{n-1}) + (n+2)\tilde{c}\}\phi - 2b\{4b(2-\frac{1}{n})+(n+1)\tilde{c}\}, \qquad (2.16)$$

which is a linear function of ϕ. Moreover, by (2.14), its domain of definition is $[b,+\infty)$.

By the condition (2.2) for $\tilde{c}<0$, it is easy to see from (2.16) and (2.13) that

$$L'(\Phi) = 4b(8-\frac{4}{n}-\frac{1}{n-1})+(n+2)\tilde{c}$$

$$= \{[4(n-1)(2n-1)-n](a-1)+n+2\}\tilde{c} \geqslant 0.$$

Thus, in order to prove that $L(\phi) \geqslant 0$ for $\phi \geqslant b$, it suffices to show

$$L(b) = \{[2(2n-1)(n-1)/n-1](a-1)-1\}\cdot nb\tilde{c} \geqslant 0.$$

But, noting that $\tilde{c}<0$, this is quite clear since $a<1$ and $n \geqq 2$. Thus, (2.4) now follows directly from (2.15) and the fact that $L(\phi) \geqq 0$

Hence, Theorem 1 is proved completely.

REMARK. Theorem 1 may be regarded as an intrinsic necessary condition for a Riemannian manifold (M^n,g) to be immersed totally really and minimally into $CF^n(\tilde{c})$.

It is easy to see that $(n+3)/n(n-1) \geqslant n/(4n^2-7n+2)$. So, we may take $a=1+(n+3)/$ $/n(n-1)$ for $\tilde{c} \geqslant 0$ in (2.1). Thus, we have the following

Corollary 1.1. Under the same hypothesis as in Theorem 1, if $\tilde{c} \geqslant 0$, then the scalar curvature $\tilde{\rho}$ of the conformal metric $\tilde{g}=\{\frac{1}{4}(n^2+3)\tilde{c}-\rho\}g$ satisfies (2.4).

By taking $a=1-(n+2)/(8n^2-13n+4)$ for $\tilde{c}<0$ in (2.1), we have

Corollary 1.2. Under the same hypothesisias in Theorem 1, if $\tilde{c}<0$, then the scalar curvature $\tilde{\rho}$ of the conformal metric $\tilde{g}=\{n(n-1)(1-\frac{n+2}{8n^2-13n+4})\tilde{c}-\rho\}g$ satisfies (2.4).

3. Stability

We first consider the case that n=2, i.e., the totally real minimal furface (M^2,g) in $CF^2(\tilde{c})$. Let K be the Gauss curvature of g. Then (1.13) becomes

$$\int_{\mathcal{D}} (\tfrac{7}{4}\tilde{c} - 2K)u^2 dv_g \geqslant \int_{\mathcal{D}} |\nabla u|^2 dv_g . \tag{3.1}$$

THEOREM 2. Let (M^2,g) be a totally real minimal surface in $CF^2(\tilde{c})$. Set

$$\tilde{c} = \max\{\tfrac{7}{4}\tilde{c}, \tfrac{3}{10}\tilde{c}\}. \tag{3.2}$$

Assume that $\mathcal{D} \subset M^2$ is a simply-connected domain such that

$$\int_{\mathcal{D}} (\hat{c} - 2K)dv_g < 8\pi/3, \tag{3.3}$$

where K is the Gauss curvature of (M^2,g). Then, $\mathcal{D}$ is stable.

Proof. We use reductio ad absurdum. Suppose that $\mathcal{D}$ is not stable. Then, there would exist a domain $\mathcal{D}' \subseteq \mathcal{D}$ such that (3.1) holds, from which and (3.2) it follows that

$$\int_{\mathcal{D}'} (\hat{c} - 2K)u^2 dv_g \geqslant \int_{\mathcal{D}'} |\nabla u|^2 dv_g . \tag{3.4}$$

Introducing the conformal metric $\tilde{g}=(\hat{c}-2K)g$ on M^2, (3.4) becomes

$$\int_{\mathcal{D}'} u^2 dv_{\tilde{g}} \geqslant \int_{\mathcal{D}'} |\nabla u|^2 dv_{\tilde{g}}, \tag{3.5}$$

where $dv_{\tilde{g}}$ stamds for the area element of M^2 in the conformal metric $\tilde{g}$. By the max-min principle, (3.5) means that

$$1 \geqslant \frac{\int_{\mathcal{D}'} |\nabla u|^2 dv_{\tilde{g}}}{\int_{\mathcal{D}'} u^2 dv_{\tilde{g}}} \geqslant \tilde{\lambda}_1(\mathcal{D}') \tag{3.6}$$

where $\tilde{\lambda}_1(\mathcal{D}')$ is the first Dirichlet eigenvalue of the Laplacian on $\bar{\mathcal{D}}'$ in the metric $\tilde{g}$.

On the other hand, by Theorem 1 and its corollaries, the Gauss curvature $\tilde{K}$ of $\tilde{g}$ satisfies $\tilde{K} \leqslant 1$. Then, by Proposition (3.3) and Corollary (3.20) of [1], we have

$$\tilde{\lambda}_1(\mathcal{D}') \geqslant \lambda_1(\Omega) \tag{3.7}$$

where $\tilde{\lambda}_1(\Omega)$ is the first Dirichlet eigenvalue of a geodesic disk Ω in a standard unit sphere $S^2(1)$ such that $\mathrm{Area}(\Omega)=\tilde{A}(\mathcal{D}')$, the area of $\mathcal{D}'$ in the metric $\tilde{g}$. But, under the hypothesis (3.3), we have

$$\mathrm{Area}(\Omega) = \int_{\mathcal{D}'} dv_{\tilde{g}} \leqslant \int_{\mathcal{D}'} (\hat{c}-2K)\,dv_g < 8\pi/3. \tag{3.8}$$

By Proposition (3.10) of [1], for the disk Ω in $S^2(1)$, $\lambda_1(\Omega)$ can be estimated as follows. If $\mathrm{Area}(\Omega) < 2\pi$, then $\lambda_1(\Omega) \geqslant 4\pi/\mathrm{Area}(\Omega) > 2$. If $2\pi \leqslant \mathrm{Area}(\Omega) \leqslant 4\pi$, then (3.8) implies that

$$\lambda_1(\Omega) \geqslant 2\left(\frac{4\pi}{\mathrm{Area}(\Omega)} - 1\right) > 1.$$

This fact together with (3.6) and (3.7) leads to a contradiction. Hence, Theorem 2 is proved.

We now consider the case that $n \geqslant 3$ and denote by w_n the n-dimensional volume of the unit sphere $S^n(1) \subset R^{n+1}$.

<u>THEOREM 3</u>. Let (M^n,g) be a toatlly real minimal submanifold in $CF^n(\tilde{c})$ with $n \geqslant 3$ and $\tilde{c} \geqslant 0$. Assume that the scalar curvature ρ of (M^n,g) is bounded from below by k. Set $b=\frac{1}{2}n(n-1)\tilde{c}-2k$. If $\mathcal{D} \subset M^m$ is a geodesic disk such that

$$\mathrm{Vol}(\mathcal{D}) \leqslant 2^{n-1} \cdot \left\{\frac{n\tilde{c}+2b}{(\tilde{c}+b)[(n+3)\tilde{c}+2b]}\right\}^{n/2} \cdot w_n \tag{3.9}$$

then $\mathcal{D}$ is stable.

Proof. From (1.5) and the assumption that $\rho \geqslant k$ we have

$$0 \leqslant \|\sigma\|^2 \leqslant \frac{1}{4}n(n-1)\tilde{c}-k = \frac{1}{2}b. \tag{3.10}$$

Suppose now that $\mathcal{D}$ is not stable. Then, there would exist a domain $\mathcal{D}' \subseteq \mathcal{D}$ such that (1.13) holds, from shich it follows that

$$\int_{\mathcal{D}'}\{\tfrac{1}{4}(n^2+3)\tilde{c}-k\}u^2 dv \geqslant \int_{\mathcal{D}'}\{\tfrac{1}{4}(n^2+3)\tilde{c}-\rho\}u^2 dv \geqslant \int_{\mathcal{D}'}|\nabla u|^2 dv. \tag{3.11}$$

By the max-min principle, (3.11) means that

$$\frac{1}{4}(n^2+3)\tilde{c}-k \geqslant \frac{\int_{\mathcal{D}'}|\nabla u|^2 dv}{\int_{\mathcal{D}'} u^2 dv} \geqslant \lambda_1(\mathcal{D}'), \tag{3.12}$$

where $\lambda_1(\mathcal{D}')$ is the first Dirichlet eigenvalue of the Laplacian on $\overline{\mathcal{D}}'$. Since $\mathcal{D}' \subseteq \mathcal{D}$, (3.12) together with (3.10) implies[2]

$$\frac{1}{4}\{(n+3)c+2b\} \geqslant \lambda_1(\mathcal{D}') \geqslant \lambda_1(\mathcal{D}), \tag{3.13}$$

where $\lambda_1(\mathcal{D})$ stands for the first eigenvalue of the Laplacian on the geodesic disk $\mathcal{D}$.

On the other hand, by (1.3) and (3.10), for $i \neq j$ we have

$$R_{ijij} = \frac{1}{4}\tilde{c} + \sum_m h_{ii}^{m*}h_{jj}^{m*} - \sum_m (h_{ij}^{m*})^2$$
$$\leqslant \frac{1}{4}\tilde{c} + \frac{1}{2}\sum_m\{(h_{ii}^{m*})^2 + (h_{jj}^{m*})^2\}$$
$$\leqslant \frac{1}{4}\tilde{c} + \frac{1}{2}\|\sigma\|^2 \leqslant \frac{1}{4}(\tilde{c} + b) \leqslant K,$$

where

$$K = (\tilde{c}+b)[(n+3)\tilde{c}+2b]/4(n\tilde{c}+2b). \tag{3.14}$$

This means that the sectional curvature of (M^n,g) is bounded from above by the constant $K>0$. By Cheng's eigenvalue comparison theorem[5], it follows that

$$\lambda_1(\mathcal{D}) \geqslant \tilde{\lambda}_1(\Omega), \tag{3.15}$$

where Ω is a geodesic disk in a standard n-sphere $S^n(1/\sqrt{K})$ of curvature K such that $\mathrm{Vol}(\mathcal{D})=\mathrm{Vol}(\Omega)$. By the condition (3.9) together with (3.14), one can see that

$$\mathrm{Vol}(\Omega) = \mathrm{Vol}(\mathcal{D}) \leqslant \frac{1}{2} w_n\cdot(1/\sqrt{K})^n,$$

which implies that Ω is contained in a closed hemisphere of $S^n(1/\sqrt{K})$, of which the first Dirichlet eigenvalue is equal to nK. Hence, combining (3.13) with (3.14) and (3.15), we get

$$\frac{1}{4}\{(n+3)\tilde{c}+2b\} \geqslant \tilde{\lambda}_1(\Omega) \geqslant nK = \frac{1}{4}\{(n+3)\tilde{c}+2b\}\frac{n(\tilde{c}+b)}{n\tilde{c}+2b},$$

i.e.,

$$1 \geqslant \frac{n(\tilde{c}+b)}{n\tilde{c}+2b} > 1$$

because of $n\geqslant3$. This contradiction proves our theorem.

<u>THEOREM 4</u>. Let (M^n,g) be a totally real minimal submanifold in $CF^n(\tilde{c})$ with $n\geqslant3$. Denote by ρ the scalar curvature of (M^n,g) and let

$$\hat{c} = \max\{(n^2+3)\tilde{c},\ n(n-1)(1-\frac{n+2}{8n^2-13n+4})\tilde{c}\}. \tag{3.16}$$

For a simply-connected domain $\mathcal{D}\subset M^n$, there is a constant $c_1(n)>0$ depening only on n such that if

$$\{\int_{\mathcal{D}}(\frac{1}{4}\hat{c}-\rho)^n dv\}^{1/n} < c_1(n), \tag{3.17}$$

then $\mathcal{D}$ is stable.

Proof. The idea of the proof is analogous to that of J.Spruck [15]. If $\mathcal{D}$ is not stable, then there would exist a domain $\mathcal{D}'\subseteq\mathcal{D}$ such that (1.13) holds, from which together with (2.1), (2.3) and (3.16) it follows that

$$\int_{\mathcal{D}'}(\frac{1}{4}\hat{c}-\rho)u^2 dv \geqslant \int_{\mathcal{D}'}|\nabla u|^2 dv. \tag{3.18}$$

Applying the Sobolev inequality of Hoffman-Spruck ([14]) to u, we can obtain

$$c_1(n)\left(\int_{\mathcal{D}'}u^{\frac{2n}{n-2}}dv\right)^{\frac{n-2}{2n}} \leqslant \left(\int_{\mathcal{D}'}|\nabla u|^2 dv\right)^{1/2}, \tag{3.19}$$

where $c_1(n)$ is a positive constant depending only on n. Then, from (3.18) and (3.19) we have

$$c_1(n)\left(\int_{\mathscr{D}'} u^{\frac{2n}{n-2}} dv\right)^{\frac{n-2}{2n}} \leq \left\{\int_{\mathscr{D}'} \left(\frac{1}{4}\hat{c}-\rho\right)u^2 dv\right\}^{1/2} . \tag{3.20}$$

By Hölder's inequality

$$\int_{\mathscr{D}'} \left(\frac{1}{4}\hat{c}-\rho\right)u^2 dv \leq \left\{\int_{\mathscr{D}'} \left(\frac{1}{4}\hat{c}-\rho\right)^n dv\right\}^{\frac{2}{n}} \cdot \left\{\int_{\mathscr{D}'} u^{\frac{2n}{n-2}} dv\right\}^{\frac{n-2}{n}} .$$

(3.20) becomes

$$c_1(n)\left(\int_{\mathscr{D}'} u^{\frac{2n}{n-2}} dv\right)^{\frac{n-2}{2n}} \leq \left\{\int_{\mathscr{D}'} \left(\frac{1}{4}\hat{c}-\rho\right)^n dv\right\}^{\frac{1}{n}} \cdot \left(\int_{\mathscr{D}'} u^{\frac{2n}{n-2}} dv\right)^{\frac{n-2}{2n}} .$$

Hence,

$$\left\{\int_{\mathscr{D}} \left(\frac{1}{4}\hat{c}-\rho\right)^n dv\right\}^{1/n} \geq \left\{\int_{\mathscr{D}'} \left(\frac{1}{4}\hat{c}-\rho\right)^n dv\right\}^{1/n} \geq c_1(n)$$

contradicting the assumption (3.17). Thus, Theorem 4 is proved.

References

[1] Barbosa,J.L. & Do Carmo,M., "Stability of minimal surfaces and eugenvalues of Laplacian", Math. Z., 173(1980), 13-28.

[2] Chavel,I., "Lowest eigenvalue inequalities", Proc. Symp. Pure Math. A. M. S., 36(1980), 79-89.

[3] Chen,B.Y., "Geometry of submanifolds", Mar. Dek. Inc., 1973.

[4] Chen,B.Y. & Ogiue,., "On totally real submanifolds", Trans. A. M. S., 193 (1974), 257-266.

[5] Cheng,S.T., "Eigenfunctions and eigenvalues of Laplacian", Proc. Symp. Pure Math. A. M. S., Part II, 27(1975), 185-193.

[6] Chern,S.S., "Minimal submanifolds in a Riemannian manifold", Min. Lect. Notes, Univ. of Kansas, 1968.

[7] Do Carmo,M., "Stability of minimal submanifolds", Springer Notes 838, 129-139.

[8] Mori,H., "Notes on the stability of minimal submanifolds of Riemannian manifolds", Yokahama Math. J., 25(1977), 9-15.

[9] Osserman,R., "The isoperimetric inequality", Bull. A. M. S., 84(1978), 1182-1238.

[10] Shen,Y.B., "Curvature estimate and stability for minimal submanifolds", Scientia Sinica, 30A(1987), 9:917-926.

[11] Simons,J., "Minimal varieties in Riemannian manifolds", Ann. of Math., 88(1968), 62-105.

[12] Smale,S., "On the Morse index theorem", J. Math. Mech., 14(1965), 1049-1056.

[13] Yau,S.T., "Seminar on diff. geom. ", Ann. of Math. Studies 102, 1982, 1-71.

[14] Hoffman,D. & Spruck,J., "Sobolev and isoperimetric inequalities for Riemannian submanifolds", Comm. Pure & Appl. Math., 27(1974), 715-727.

[15] Spruck,J., "Remarks on the stability of minimal submanifolds of R^n", Math. Z., 144(1975), 169-174.

Dirichlet problems and the Laplacian in affine hypersurface theory.

Udo SIMON

There is a recent interest in affine differential geometry in the last decades, especially in global problems. This paper contains uniqueness results on compact hypersurfaces in equiaffine differential geometry.

In section 1 we present a short introduction to the equiaffine differential geometry of regular hypersurfaces (basic geometric quantities, structure equations, integrability conditions, main local theorem) using the invariant calculus. From section 2 on we restrict to locally strongly convex hypersurfaces. It is our aim to present different methods for proving uniqueness results, especially for compact hypersurfaces with boundary. Often such uniqueness problems lead to the solution of eigenvalue- or Dirichlet-problems. Most of our results are new, at least we give new or modified proofs.

A typical example of such a geometric problem is R. Schneider's conjecture ((SCHN) § 6) which after nearly 2 decades was solved by A. Schwenk(SCHW-1):

Theorem (Schneider-Schwenk). Let $x: M \longrightarrow \mathcal{A}$ be a hypersurface immersion of an oriented, connected n-dimensional differentiable manifold into real affine space, $n \geq 2$. Assume $x(M)$ to be a locally strongly convex affine sphere with closed boundary $x(\partial M)$. Assume that $x(\partial M)$ is the intersection of $x(M)$ with a hyperplane and at the same time a shadow boundary with respect to parallel light. Then $x(M)$ is a half ellipsoid.

Plane shadow lines were considered the first time by Brunn: An ovaloid in $\mathcal{A}_3$ is an ellipsoid iff each shadow line with respect to parallel light is a plane curve. - Later Blaschke realized that this - the coincidence of two properties for curves - characterizes quadrics locally ((B-2) p. 215-222). It was Schneider's idea to investigate whether a compact surface is a quadric if the boundary curve has this two properties. A. Schwenk transfered the geometric boundary conditions into two Dirichletproblems; she could finally conclude that both solutions must coincide, which led to quadrics via the structure equations. Using this method she proved a series of related results characterizing quadrics within the class of affine spheres with boundaries. In (SCHW-S), (K-S) we could prove similar characterizations within other classes of hypersurfaces using new integral formulas.

In this paper we will derive further integral formulas and finally demonstrate a third method, based on estimates for the first eigenvalue λ_1 for compact Riemannian manifolds with boundary (K), (R). Analogously to investigations for submanifolds in different spaces (for a survey cf. (S-W-1), (S-W-2)) we derive estimates for λ_1 on compact hypersurfaces in affine space.

As one application of the different methods we are able to give 3 different proofs for the Schneider-Schwenk-Theorem. The following table of contents shows which types of hypersurfaces are investigated in this paper.

1. Review of the structure equations for hypersurfaces.
2. Boundary conditions.
3. Integral formulas.
4. Estimates for the first eigenvalue of the Laplacian.
5. Affine hyperspheres with boundary.
6. Complete hypersurfaces with constant mean curvature.
7. Equiaffine Einstein spaces.
8. References.

Some of the results in this paper are based on discussions which I had with V. Oliker at Emory University Atlanta in spring 1986. The first version of this paper is part of my lectures notes on "Affine Differential Geometry" which I wrote for the participants of a course on this topic at the NIM in March 1987. This final version is an extension of some sections of these lecture notes; it was written during a stay at UC Berkeley in August/September 1987. I would like to thank all three institutions, and especially the geometers there, for their invitations and their great hospitality.

1. Review of the structure equations for hypersurfaces

In this section we give a review of the structure equations which can serve as a brief introduction to the classical equiaffine theory of hypersurfaces (cf. e.g. (B-1), (SCHI)), using the invariant calculus.

Let $\mathcal{A}$ denote a real affine space of dim $\mathcal{A} = n + 1$, $n \geqslant 2$, V the corresponding $\mathbb{R}$-vectorspace, V* it's dual, and $<\,,\,>: V^* \times V \longrightarrow \mathbb{R}$ the standard scalar product. We fix a nontrivial determinantform <u>Det</u> in the 1-dimensional vectorspace of such forms; <u>Det</u> induces a volume-structure on $\mathcal{A}$. We consider the geometry with respect to the unimodular (volumepreserving) group of transformations on $\mathcal{A}$. Finally, the structure of the ambient space $\mathcal{A}$ includes the standard flat connection ∂ (this notation allows to denote partial derivation as usual).

For an n-dimensional connected and oriented differentiable manifold we consider a hypersurface-immersion

$$x: M \longrightarrow \mathcal{A}.$$

For convenience we fix an origin $O \in \mathcal{A}$ and denote by x also the position vector of x(M) with respect to $O \in \mathcal{A}$.

We identify the tangent hyperplane $T_{x(p)}x(M)$ at x(p) with an n-dimensional subspace of V; via duality, to this subspace there corresponds a 1-dimensional subspace of V*, the <u>conormalspace</u>, thus defining a line-bundle over M, the <u>conormal-bundle</u> C(M). A nowhere vanishing differentiable section $X: M \longrightarrow C(M)$ is called a <u>conormalfield</u>.

In the following we denote tangent vectorfields over M by v,w and correspondingly by $v_1, \ldots, v_n$ a local frame field. We restrict to so-called <u>regular</u> immersions which are characterized by the condition

(1.1) $\quad \underline{\mathrm{Det}}^* \, (X, dX(v_1), \ldots, dX(v_n)) \neq 0,$

where <u>Det</u>* is the form dual to <u>Det</u>, X is an arbitrary conormal-field and $v_1, \ldots, v_n$ an arbitrary framefield. Obviously (1.1) does not depend on the choice of X (neither of that of <u>Det</u>, nor the framefield), but only on the immersion x.

For a regular immersion there exist - up to the sign - exactly one conormalfield which is invariant under the unimodular group ((CA), (N)). For locally strongly convex hypersurfaces which we will consider from section 2 on, we fix the sign of X(p) in $p \in M$ s.t. for a suitable neighborhood U of p and any point $q \in U$, $q \neq p$, we have

(1.2) $\quad \langle X(p), x(q) - x(p) \rangle > 0.$

The regularity condition allows to interprete X as a hypersurface-immersion $X: M \longrightarrow V^*$ s.t. X(p) is transversal to X(M) at each point $p \in M$. X(M) is called the <u>conormal-indicatrix</u>.

Furthermore, because of the regularity, the solution y of the system (i = 1, ..., n)

(1.3) $\quad \langle X, y \rangle = 1, \ \langle dX(v_i), y \rangle = 0$

is uniquely determined. y is called the equiaffine normal vectorfield of x(M). y is transversal to x(M). The induced mapping $y : M \longrightarrow V$ is called the <u>normal-indicatrix</u>, but y must not be an immersion. The mappings X und y are analoga to the Euclidean Gauss map.

From x, X and y we derive the structure equations analogously to the Euclidean hypersurface theory

(1.4) $dy(v) = dx(-B(v))$ equiaffine Weingarten equation,

(1.5) $\partial_v dx(w) = dx({}^1\nabla_v w) + G(v,w)y$
(1.6) $\partial_v dX(w) = dX({}^2\nabla_v w) - \hat{B}(v,w)X$ } equiaffine Gauß equations.

(1.4) - (1.6) form a linear system of p.d.e. We are going to list the properties of the geometric quantities ${}^1\nabla$, ${}^2\nabla$, B, G, $\hat{B}$ which are invariant under the unimodular group.

${}^1\nabla$, ${}^2\nabla$ are two affine connections on M induced by the immersions x, X resp.; the symmetric (0.2)-tensorfields G, $\hat{B}$ fulfil the relations

(1.7) $G(v,w) = \langle X, \partial_v dx(w)\rangle = -\langle dX(v), dx(w)\rangle$,

(1.8) $\hat{B}(v,w) = -\langle \partial_v dX(w), y\rangle = \langle dX(v), dy(w)\rangle$.

G is regular (positive definite) iff x is regular (locally strongly convex). (1.4) is an immediate consequence of (1.3), and the induced (1.1)-tensorfield B is called the equiaffine Weingarten field (equiaffine shape operator). B is selfadjoint with respect to G and fulfils

(1.9) $\hat{B}(v,w) = G(B(v),w) = G(v,B(w))$.

The mean connection

(1.10) $\nabla := \frac{1}{2}({}^1\nabla + {}^2\nabla)$

is the Levi-Civita connection of G considered as a Pseudo-Riemannian metric on M; G is called the Berwald-Blaschke metric.

(1.11) $A: = \frac{1}{2}({}^1\nabla - {}^2\nabla)$

and G give rise to the definition of the so-called cubic-form tensorfield $\hat{A}$

(1.12) $\hat{A}(v,w,z): = G(A(v,w),z)$,

which is totally symmetric. The relation

(1.13) $\operatorname{trace} A(z, \ ) = 0$

is known as apolarity condition. The simplest invariant to derive from G and A is the Pick invariant J

(1.14) $n(n-1)J: = ||A||^2 = ||\hat{A}||^2 = \operatorname{trace}_G \alpha$.

Here, for a tensorfield C we denote by $||C||^2$ the square of the norm induced from G. α is defined by

(1.15) $\alpha(v,w) = \sum_{i=1}^{n} G^{-1}(A(v,v_i), A(w,v_i))$,

where $G(v_i, v_j) = \delta_{ij}$. Maschke's theorem states that x(M) lies on a quadric iff J = 0 (or equivalently $\hat{A} = 0 \Leftrightarrow {}^1\nabla = {}^2\nabla \Leftrightarrow \alpha = 0$) on M.

Because of (1.4) and (1.9) the equiaffine shape operator B is the adequate quantity to define equiaffine curvature functions for x(M): The real eigenvalues $k_1 \leq k_2 \leq \cdots \leq$

k_n of B are called the equiaffine principal curvatures, and their normed elementary symmetric functions H_r

(1.16) $\binom{n}{r} H_r := \sum_{i_1<\ldots<i_r}^{1\ldots n} k_{i_1} \ldots k_{i_r}$ are the equiaffine curvature functions;

$H := H_1 = \frac{1}{n}$ trace B is the equiaffine mean curvature. We have the identity

(1.17) $n\,||\hat{B}-HG||^2 = \sum_{i<j} (k_i - k_j)^2 =: S.$

$S(p) = 0$ characterizes an equiaffine umbilic $p \in M$, and $S \equiv 0$ on M characterizes an equiaffine sphere (which is proper for $H \neq 0$ and improper for $H = 0$ on M).

For the equiaffine Gauss map y we get as an immediate consequence of (1.5) and (1.13)

(1.18) $y = \frac{1}{n} \Delta x,$

where Δ denotes the Laplace-Beltrami operator with respect to G and ∇. From (1.5) y is an immersion iff B is regular. In this case the hypersurface classically was normalized centroaffinely (with $-y$ as transversal vectorfield to $y(M)$); then $\hat{B}$ becomes the centroaffine metric of this regular hypersurface. Assume M furthermore to be closed, then $\hat{B}$ must be (positive) definite on M (cf. (S-3) § III.2), i.e. $y(M)$ must be a hyperovaloid. Thus the regularity of $\hat{B}$ has an interesting geometric interpretation for global problems.

The integrability conditions of the system (1.4) - (1.6) are given by (1.19) - (1.20):

(1.19) $\nabla_v \hat{B}(w_1, w_2) - \nabla_{w_1} \hat{B}(v, w_2) = \hat{B}(A(w_1, w_2), v) - \hat{B}(A(v, w_2), w_1)$

which are the equiaffine Codazzi-equations. If R denotes the curvature tensor of G and

$$\hat{R}(w_1, v_1, w_2, v_2) := G(R(w_1, v_1) w_2, v_2),$$

then the Gauss-integrability conditions read:

(1.20) $\hat{R}(w_1, v_1, w_2, v_2) = G(A(v_1, v_2), A(w_1, w_2))$
$- G(A(w_1, v_2), A(v_1, w_2)) + \frac{1}{2}\{ \hat{B}(w_1, v_2) G(v_1, w_2) - \hat{B}(v_1, v_2) G(w_1, w_2)$
$+ \hat{B}(w_1, w_2) G(v_1, v_2) - \hat{B}(v_1, w_2) G(w_1, v_2) \}.$

As a consequence we get the following expression for the Ricci-tensor Ric (we choose the sign of Ric s.t. it is positive definite on ellipsoids):

(1.21) $\mathrm{Ric} = \alpha + \frac{n}{2} HG + \frac{1}{2}(n-2)\hat{B}.$

Denoting by $\varkappa$ the normed scalar curvature $n(n-1)\varkappa = \mathrm{trace}_G \mathrm{Ric}$, the equiaffine Theorema egregium relates three basic invariants

(1.22) $\varkappa = J + H.$

Some of the integrability conditions get a more convenient form if one uses ${}^1\nabla$ or ${}^2\nabla$ for covariant differentiation ((S-3) p.50, (S-4)); especially we get

(1.23) $\quad (n-1)\hat{B} = {}^2\mathrm{Ric},$

where ${}^2\mathrm{Ric}$ is the <u>Ricci-tensor</u> of ${}^2\nabla$. This formula together with (1.9) - (1.12) and the regularity of G show that G and $\hat{A}$ can be used as fundamental geometric quantities to describe the geometry of the hypersurface immersion. Namely, G and $\hat{A}$ allow to determine all coefficients in the linear system (1.4) - (1.6) of p.d.e. Integration of this system gives the classical main local theorem for hypersurfaces in $\mathcal{A}$ with respect to the unimodular group:

<u>Fundamental Theorem</u>. <u>A regular immersion</u> $x: M \longrightarrow \mathcal{A}$ <u>is uniquely determined by G and $\hat{A}$ up to equiaffine equivalences of $\mathcal{A}$. Moreover, if a regular symmetric (0.2)-tensorfield G and a totally symmetric (0.3)-tensorfield $\hat{A}$ are given s.t. (1.12), (1.13) and the integrability conditions (1.19), (1.20) of the linear p.d.e. system (1.4) - (1.6) are fulfilled, then there exists a regular immersion x s.t. G is the equiaffine metric and $\hat{A}$ the cubic form tensor</u>.

For other versions of this theorem cf. (S-4).-

<u>Remark</u>. In sections 2 - 7 we restrict to locally strongly convex hypersurfaces; then G is positive definite from (1.7) and the orientation of X defined in (1.2).

2. Boundary conditions.

In this section for later purpose we describe different types of boundary conditions by properties of appropriate functions on M.

<u>(2.1) Plane intersections.</u> Let $\mathfrak{h}$ be an affine hyperplane in $\mathcal{A}$ described by $\langle L, z-z_1\rangle = 0$, where $z_1 \in \mathfrak{h}$ and $0 \neq L \in V^*$. If x is a hypersurface immersion, the function $h := \langle L, x-z_1\rangle : M \to \mathbb{R}$ has the property: $x(p) \in \mathfrak{h} \cap x(M)$ iff $h(p) = 0$. Using ${}^1\nabla = \nabla + A$, the structure equations (1.5) and (1.18) imply the following relation for the ∇-<u>covariant Hessian</u>

$$(2.1.1) \qquad (\mathrm{Hess}\, h)(v,w) = \hat{A}(v,w,\mathrm{grad}\, h) + \langle L,y\rangle G(v,w) = \hat{A}(v,w,\mathrm{grad}\, h) + \tfrac{1}{n}\Delta h G(v,w).$$

In case x is a proper affine sphere with center x_o we get $H = \mathrm{const}$ and

$$(2.1.2) \qquad y = -H(x-x_o) \text{ and } \Delta h = n\langle L,y\rangle = -nHh + nH\langle L, x_o - z_1\rangle;$$

thus, h is an eigenfunction corresponding to the eigenvalue $\lambda = nH$.

In case x is an improper affine sphere we have

(2.1.3) $\Delta h = n\langle L,y\rangle = \text{const.}$

2.1.4 Lemma. Let x be an immersion. Then the zero-set of grad h is nowhere dense on M. Proof. Assume there exists an open nonempty set $U \subset M$ s.t. grad h = 0 on U; $\langle L,dx(v)\rangle = 0$ for all tangent vectorfields v on U implies $L = \sigma X$ on U where $\sigma \in C^{\infty}(U)$ is nowhere vanishing. This contradicts the regularity of $x|U$.∎

2.2 p-shadow boundaries. Let x be a hypersurface-immersion and $0 \neq b \in V$ the direction of parallel light in $\mathcal{A}$. Then the set $\{q \in M \mid \langle X(q),b\rangle = 0\}$ is called the p-shadow boundary with respect to parallel light of direction $b \in V$. Thus this set is described by the zeroes of the function $f: M \to \mathbb{R}$, $f(q) := \langle X(q),b\rangle$. From the structure equations (1.6) we immediately get

(2.2.1) $(\text{Hess} f)(v,w) = -\hat{A}(v,w,\text{grad} f) - \hat{B}(v,w)f$,

(2.2.2) $(\Delta + nH)f = 0$.

This last equation is a Schrödinger-type equation with operator $\Delta + nH$ (Laplacian plus a potential).

2.2.3 Lemma. (i) Assume $(\text{grad} f)(q) = 0$ for $q \in M$. Then there exists $0 \neq \sigma \in \mathbb{R}$ s.t. $b = \sigma y(q)$.

(ii) $\{q \in M \mid f(q) = 0\} \cap \{p \in M \mid \text{grad} f(p) = 0\} = \emptyset$.

(iii) Assume there exists an open nonempty set $U \subset M$ s.t. grad f = 0 on U. Then x(U) lies on an improper affine sphere.

(iv) The zeroes of f are nowhere dense on M.

Proof. (i) Assume $0 = (\text{grad} f)(q)$, then $\langle dX(v),b\rangle = 0$ for $v \in T_qM$. As at the same time $\langle dX(v),y\rangle = 0$ and rank dX = n, $b \neq 0$, there exists $0 \neq \sigma \in \mathbb{R}$ s.t. $b = \sigma y(q)$.

(ii) Assume $(\text{grad} f)(q) = 0$ and $f(q)=0$ at $q \in M$. Then $\langle dX(v),b\rangle = 0$ for $v \in T_qM$ and $\langle X,b\rangle = 0$, which implies b = 0 because of the regularity of x. But $b \neq 0$.

(iii) From (i) we get $b = \sigma y$ on U where $\sigma \in C^{\infty}(M)$, i.e. the equiaffine normal field is parallel to b on U.

(iv) f = 0 on an open set U would imply grad f = 0 on U; therefore from (ii) we get $U = \emptyset$.∎

2.3 c-shadow boundaries. Let $z_o \in \mathcal{A}$, $z_o \notin x(M)$, be a center of central light. Then the set $\{q \in M \mid \langle X(q), z_o - x(q)\rangle = 0\}$ for the immersion x is called the c-shadow boundary with respect to the center z_o (here as before z_o denotes at the same time the position vector with respect to $O \in \mathcal{A}$). This boundary is described by the zeroes of the affine support function $\rho(z_o)(q) := \langle X(q), z_o - x(q)\rangle$. Again from the structure equations (1.6) we get

(2.3.1) $(\operatorname{Hess} \rho(z_o))(v,w) = -\hat{A}(v,w,\operatorname{grad}(z_o)) - \hat{B}(v,w)f + G(v,w)$,

(2.3.2) $(\Delta + nH)\rho(z_o) = n$.

2.3.3 Lemma. (i) Let $z_o \notin x(M)$ and $q \in M$ be a zero of $\operatorname{grad}\rho(z_o)$. Then there exists $0 \neq \sigma \in \mathbb{R}$ s.t. $y(q) = \sigma(z_o - x(q))$.
(ii) Assume $\operatorname{grad}\rho(z_o) = 0$ on an open connected nonempty set $U \subset M$. Then $x(U)$ lies on a proper affine hypersphere with center z_o.
Proof. (i) Cf. (2.2.3.i). (ii) From (i) and (1.4) we get σ=const on U, i.e. z_o is the center of a proper affine sphere $x(U)$.■

2.4. Totally geodesic boundaries. Let $\tilde{M}$ be a totally geodesic submanifold of the Riemannian manifold (M,G). Then the vector-valued second fundamental form $\widetilde{II}$ vanishes identically. For any $F \in C^\infty(M)$ and all tangentvectorfields $\tilde{v},\tilde{w}$ of the submanifold $\tilde{M}$ this implies

(2.4.1) $(\operatorname{Hess} F)(\tilde{v},\tilde{w}) - (\widetilde{\operatorname{Hess}}(F|\tilde{M}))(\tilde{v},\tilde{w}) = -dF(\widetilde{II}(\tilde{v},\tilde{w})) = 0$,

i.e. the two Hessians coincide on a totally geodesic submanifold ((R) p. 462-63). Later we will apply this to a compact manifold M with boundary $\tilde{M} = \partial M$.

2.5. Dirichlet problems. Boundary conditions like (2.2) lead to a Dirichlet problem on compact manifolds

(2.5.1) $\Delta f + nHf = 0 \qquad f \mid \partial M = 0$.

As an immediate consequence of Green's integral formula we get

$$\max_{p \in M} H(p) > 0$$

on manifolds with such boundaries.

3. Integral formulas

The integral formulas of this section are one of our tools to prove uniqueness results in sections 5 - 7.

3.1. Lemma. Let x be a hyperovaloid and $f := \langle X,b \rangle$, $b \in V$, a function on M. Then

$$\int_M f\,do = 0,$$

where do denotes the volume-element of (M,G).
Proof. The proof is similar to the proof of the relative Minkowski formulas in ((SI-1) § 4).■

3.2. Proposition. Let M be compact with boundary ∂M and x a hypersurface-immersion. Let ν denote the normal vectorfield along ∂M with respect to M and with the induced orientation, and $d\omega$ the volume element of ∂M.
(a) The function $f := \langle X,b \rangle$ for $b \in V$ fulfils the integral

formula

$$- n \int_{\partial M} \hat{A}(\mathrm{grad} f, \mathrm{grad} f, \nu)\, d\omega + \frac{1}{2} \int_{\partial M} f (HG - \hat{B})(\mathrm{grad} f, \nu)\, d\omega$$
$$= \int_M \{ n \| \mathrm{Hess} f \|^2 - (\Delta f)^2 \} do + \int_M (n \alpha(\mathrm{grad} f, \mathrm{grad} f) + \frac{1}{2}(n-2) S f^2) do$$
$$- \frac{1}{2} n(n-1) \int_M f G(\mathrm{grad} f, \mathrm{grad} H) do.$$

(b) The same integral formula as in (a) holds for the affine support function $\rho(z_o) := \langle X, z_o - x \rangle$, $z_o \in V$, instead of f

(c) Let $h := \langle L, x - z_1 \rangle$ for $L \in V^*$, $z_1 \in V$. Then

$$\int_{\partial M} (\mathrm{Hess} h)(\mathrm{grad} h, \nu)\, d\omega - \frac{n+2}{2n} \int_{\partial M} G(h \,\mathrm{grad} \Delta h + \Delta h \,\mathrm{grad} h, \nu)\, d\omega$$
$$- \frac{1}{2}(n+2) \int_{\partial M} h \hat{B}(\mathrm{grad} h, \nu)\, d\omega = 2 \int_M \alpha(\mathrm{grad} h, \mathrm{grad} h) do$$
$$+ \frac{1}{2} \int_M \{ n H \| \mathrm{grad} h \|^2 - (\Delta h)^2 \} do.$$

<u>3.2.1. Remark.</u> Note that the above integrands of the following type are non-negative: $\{ n \| \mathrm{Hess} f \|^2 - (\Delta f)^2 \} \geq 0$ (Newton's inequality); $\alpha(\mathrm{grad} f, \mathrm{grad} f) \geq 0$ (by definition of α); $S \geq 0$ (by definition).

<u>Proof of Theorem 3.2.</u> (a) and (b) cf. (SCHW-S).

(c) To apply the local Bochner-Lichnerowicz formula (cf. (SCHW-S) formula (2.1)) we need the calculations in the steps I-III of our proof.

Step I. Formula (2.1.1) implies

(Ia) $- \frac{1}{n} \mathrm{Hess}(\Delta h)(v,w) = - (\mathrm{Hess} \langle L, y \rangle)(v,w)$

$$= \frac{1}{n} \hat{B}(v,w) + (\nabla \hat{B})(v,w,\mathrm{grad} h) + \hat{A}(B(v), w, \mathrm{grad} h)$$
$$+ \hat{A}(v, B(w), \mathrm{grad} h) - \hat{A}(v, w, B(\mathrm{grad} h)),$$

(Ib) $G(\mathrm{grad} h, \mathrm{grad} \Delta h) = - n \hat{B}(\mathrm{grad} h, \mathrm{grad} h)$.

Step II. Let $v_1, \ldots, v_n$ denote a local orthonormal frame field on (M,G). Then

$$\hat{B}(\mathrm{grad} h, \mathrm{grad} h) = \mathrm{div}\,(h B(\mathrm{grad} h)) - h (\nabla B)(\mathrm{grad} h)$$
$$- h \sum_{i,j} \hat{B}(v_i, v_j)(\mathrm{Hess} h)(v_i, v_j);$$

the Codazzi-equation (1.19) and Step I.a give the following integral formula

$$\int_M \hat{B}(\mathrm{grad} h, \mathrm{grad} h) do - \frac{1}{n} \int_M h \Delta \Delta h do = \int_{\partial M} h \hat{B}(\mathrm{grad} h, \nu)\, d\omega.$$

Step III. Green's theorem, (Ib) and (II) give the following integral formula

$$\int_{\partial M} G(h \,\mathrm{grad} \Delta h + \Delta h \,\mathrm{grad} h, \nu)\, d\omega$$
$$= 2 \int_M G(\mathrm{grad} h, \mathrm{grad} \Delta h) do + \int_M (h \Delta \Delta h + (\Delta h)^2) do$$
$$= \int_M ((\Delta h)^2 - \hat{B}(\mathrm{grad} h, \mathrm{grad} h)) do - n \int_{\partial M} h \hat{B}(\mathrm{grad} h, \nu)\, d\omega.$$

Step IV. The local Bochner-Lichnerowicz formula reads

$\frac{1}{2}\Delta\|\mathrm{grad}h\|^2 = \|\mathrm{Hess}h\|^2 + G(\mathrm{grad}h,\mathrm{grad}\Delta h) + \mathrm{Ric}(\mathrm{grad}h,\mathrm{grad}h).$

We obey (1.21), and as a consequence of (2.1.1)

$$\alpha(\mathrm{grad}\,h,\mathrm{grad}\,h) = \|\mathrm{Hess}\,h\|^2 - \frac{1}{n}(\Delta h)^2,$$

and integrate:

$$\int_{\partial M}(\mathrm{Hess}\,h)(\mathrm{grad}\,h,\nu)d\omega = \frac{1}{2}\int_M \Delta\|\mathrm{grad}\,h\|^2 do$$
$$= 2\int_M \alpha(\mathrm{grad}\,h,\mathrm{grad}\,h)do + \frac{1}{n}\int_M(\Delta h)^2 do$$
$$+ \frac{n}{2}\int_M H\|\mathrm{grad}\,h\|^2 do - \frac{1}{2}(n+2)\int_M \hat{B}(\mathrm{grad}h,\mathrm{grad}h)do.$$

III and IV together prove the assertion.■

<u>3.3. Proposition.</u> (K-S) Let $b\in V$ and x induce an equiaffine Einstein metric on M with constant scalar curvature $\varkappa$, $n \geq 2$. Then $f := \langle X,b\rangle$ fulfils the following integral formula

$$\int_{\partial M}(-\hat{A}(\mathrm{grad}f,\mathrm{grad}f,\nu) - (n-1)JfG(\mathrm{grad}f,\nu) + f(HG-\hat{B})(\mathrm{grad}f,\nu))d\omega$$
$$= \int_M\left\{\|\mathrm{Hess}f\|^2 - \frac{1}{n}(\Delta f)^2\right\}do + n(n-1)\int_M JHf^2 do.$$

The proof is a consequence of (3.2).■

<u>4. Estimates for the first eigenvalue of the Laplacian.</u>

(S-2), (S-W-1), (S-W-2) contain surveys about the first eigenvalue λ_1 for submanifolds of different spaces. So far there are only few results on $\lambda_1 = \lambda_1(M,G)$ in affine differential geometry (cf. (K-S)). Off course, results from Riemannian geometry as Theorem A in (B-K-S-S) can be applied to the affine differential geometry, giving lower and upper bounds for $\lambda_1(M,G)$ on ovaloids; similarly, (K), (R) together with (2.2.2) give lower estimates for λ_1 on hypersurfaces in $\mathcal{A}$ under special boundary conditions.

<u>4.1. Proposition.</u> Let x be a hyperovaloid in $\mathcal{A}$. Then $\lambda_1 = \lambda_1(M,G)$ fulfils

(4.1.1) $0 < \lambda_1 \leq n \max H$;

equality holds iff x(M) is an ellipsoid.

Proof. $f := \langle X,b\rangle$ fulfils (2.2.2) and (3.1). Green's theorem and the Raleigh minimum principle ((CH), p. 16) give

$$n \max H \int_M f^2 do \geq n\int_M Hf^2 do = -\int_M f\Delta f do = \int_M \|\mathrm{grad}f\|^2 do \geq \lambda_1 \int_M f^2 do.$$

As (2.2.3.iv) implies $\int_M f^2 do > 0$, (4.1.1) is proved.-

Equality in (4.1.1) gives $\max H = H$ in (4.1.2). It is well known that $H = \mathrm{const}$ on a hyperovaloid implies that x(M) is an ellipsoid (for references cf. (S-3) § III.5.5; we will give another proof of this in section 6).

<u>4.2 Corollary.</u> (CHE) . There are no hyperovaloids with $H \leq 0$.

It is a well known consequence of the Bochner-Lichnerowicz-formula ((B-G-M) p. 179-180) that $\mathrm{Ric} \geq (n-1)k^*G$

for $0 < k^* \in \mathbb{R}$ on a Riemannian manifold (M,G) implies

$$\lambda_1(M,G) \geq nk^*.$$

Equality holds iff (M,G) is isometrically diffeomorphic to a canonical sphere $S^n(k^*)$. Estimating the Ricci-curvature from (1.21) we get the following result.

4.3. Proposition. Let x be a hyperovaloid with B regular. Then

$$\lambda_1(M,G) \geq \frac{n}{n-1} \min_{q\in M} \left(\frac{n}{2} H(q) + \frac{1}{2}(n-2)k_1(q) \right) \geq n \min_{q\in M} k_1(q) > 0.$$

Equality $\lambda_1 = \frac{n}{n-1} \min(\ldots)$ holds iff $x(M)$ is an ellipsoid.

Proof. From section 1 we know that $\hat{B}$ must be positive definite, therefore $\min k_1 > 0$ on M.-

Equality implies that $\min k_1 = H = \text{const}$ on M, which gives the assertion as above.■

We can prove other estimates for λ_1 for the Dirichlet-problem under special boundary conditions, using Kasue's Lemma 1.1 in (K).

4.4. Proposition. Let M be compact with boundary ∂M. Assume one of the following conditions (i) - (iii) to be fulfilled.

(i) x is a graph, i.e. there exists $b \in V$ s.t. $\langle X,b\rangle \neq 0$ on $\mathring{M}$ (so $\langle X,b\rangle > 0$ by the proper choice of the orientation of b) and $\langle X,b\rangle = 0$ on ∂M.

(ii) $x(M)$ is convex and $x(\partial M)$ is a p-shadow boundary with respect to parallel light $b \in V$.

(iii) $x(M)$ is convex and $x(\partial M)$ is a c-shadow boundary with respect to the center z_o, and $H > 0$ on M.

Then $n \max H \geq \lambda_1 \geq n \min H$.

Equality on the left implies that $H = \text{const}$ on M. Equality on the right implies that $H = \text{const}$ and f is a first eigenfunction of the Laplacian.

Proof. We consider $f := \langle X,b\rangle$ for (i) and (ii), and $f := \rho(z_o) - \frac{1}{H}$ in case (iii). In all cases f (or $-f$ resp.) fulfils $\Delta f + nHf = 0$, and $f = 0$ on ∂M. The left-hand inequality can be proved analogously to (4.1), as $f|\partial M = 0$ (cf. (CHA) p.16). The right-hand inequality follows from Kasue's Lemma, as f has one sign on $\mathring{M}$ (thus we can choose f to be positive) and

$$0 = \Delta f + nHf \geq \Delta f + n(\min H)f.■$$

4.4.1. Corollary. Let M be compact with boundary and $H = \text{const}$. Then each of the conditions (i) - (iii) in (4.4) implies $0 < \lambda_1 = nH$ and $f = \langle X,b\rangle$ for $b \in V$ is a first eigenfunction of the Laplacian.

5. Affine hyperspheres with boundary.

Considering different types of boundary conditions, A. Schwenk proved the following result in her thesis ((SCHW-1) p. 48, (SCHW-2) p. 300).

5.1. Proposition. Let x be a compact proper affine hypersphere with closed totally geodesic boundary ∂M. Then (i) and (ii) are equivalent:

(i) $x(\partial M)$ is a plane intersection (cf. (2.1)).

(ii) $x(\partial M)$ is a p-shadow boundary (2.2).

5.1.1. Remark. A. Schwenk assumed that the hyperplane in (5.1.i) contains the center of the sphere. This assumption is not needed.-

The following Proposition extends the above equivalence in the sense that on proper affine spheres any two of the boundary conditions imply the third one.

5.2. Proposition. Let x be a proper affine hypersphere with center x_o and closed boundary ∂M. Assume that (i) $x(\partial M)$ is a plane intersection, and (ii) $x(\partial M)$ is a p-shadow boundary. Then ∂M is totally geodesic in M.

Proof. I. We choose x_o as origin in $\mathcal{A}$ and describe the boundary conditions by appropriate functions h,f as in (2.1), (2.2), which both are eigenfunctions corresponding to the eigenvalue $\lambda = nH$, vanishing on ∂M. Define $\bar{h}: M \to \mathbb{R}$ by $\bar{h}(q) = h(q) + c$, $q \in M$, where $-c := \langle L, x_o - z_1 \rangle$. Thus $\bar{h}|\partial M = c$, and $\Delta\bar{h} + nH\bar{h} = 0$.

II. We prove $c = 0$ repeating verbatim the proof in (SCHW-1) p. 38 (cf. also (SCHW-2) p. 301, Nr. 2); therefore the center $x_o \in \mathfrak{h}$, and $\bar{h} = h = \langle L, x - x_o \rangle$.

III. f and h are solutions of the same Dirichletproblem

(5.2.1) $\Delta f + nHf = 0$ and $f|\partial M = 0$.

As the space of solutions of this problem is one-dimensional, by a proper choice of $L \in V^*$ we have $f = h$. But then Hessf = Hessh, and on proper affine spheres $B = HG$, $\Delta h = -nHh$, so comparing (2.1.1) and (2.2.1) we get

(5.2.2) $\hat{A}(v, w\ \mathrm{grad}h) = 0$ and $(\mathrm{Hess}h)(v,w) + HhG(v,w) = 0$.

IV. $h|\partial M = 0$ and (5.2.2) give $(\mathrm{Hess}h)|\partial M = 0$; furthermore $h|\partial M = 0$ implies $\widetilde{\mathrm{Hess}}(h|\partial M) = 0$ (we use the notation from (2.4)). Therefore $dh(\widetilde{II}(\tilde{v},\tilde{w})) = 0$ from (2.4.1). But $G(\mathrm{grad}h, \nu) \neq 0$ along ∂M (the boundary is a level-submanifold of h), so $\widetilde{II} = 0$, i.e. ∂M is totally geodesic.∎

The following result was conjectured by R. Schneider ((SCHN) p. 399) and proved by A. Schwenk ((SCHW-1) p. 37, Satz 13; p. 52, Korollar 2; cf. also (SCHW-2) p. 300, Satz 1). We give different proofs for the Schneider-Schwenk result.

5.3. Theorem. Let x be a proper affine sphere with closed boundary ∂M. Assume that two of the following three boundary conditions are fulfilled.

(i) $x(\partial M)$ is a plane intersection;

(ii) $x(\partial M)$ is a p-shadow boundary;

(iii) $\partial M \subset M$ is totally geodesic.

Then $x(M)$ is a half ellipsoid.

Proof. I. Assume (i) and (iii) are fulfilled. We use the integral formula (3.2.c) and (2.1), (2.4), $0 \in \mathfrak{h}$.

Then $h|\partial M=0$ and furthermore $(\text{Hess}\, h)\,|\,\partial M = 0$ because of (2.4.1) and (iii); therefore all boundary-integrals vanish. Furthermore $H = \text{const}$ and $\Delta h + nHh = 0$ (2.1.2), so from Green's theorem

$$\int_M \{nH\,||\text{grad}\,h\,||^2 - (\Delta h)^2\}\,do = -\int_M \Delta h(nHh + \Delta h)do = 0.$$

α is semi-positive definite, so (3.2.c) gives $\alpha(\text{grad}\,h,\text{grad}\,h) = 0$. (2.1.4) implies $\alpha = 0$, i.e. $x(M)$ lies on a quadric. $\Delta h + nHh = 0$ and $h|\partial M = 0$ give $H > 0$, so $x(M)$ is an ellipsoid. A plane section of an ellipsoid being totally geodesic must contain the center.

III. Assume (ii) and (iii) are fulfilled. We use (2.2), (2.4) and the integral-formula (3.2.a): Again $f|\partial M = 0$ implies $(\text{Hess}\, f)\,|\,\partial M = 0$ as in (I), and (2.2.1) gives $\hat{A}(\tilde{v},\tilde{w},\text{grad}\, f) = 0$ on ∂M for $\tilde{v},\tilde{w}$ tangent to ∂M. Therefore all boundary-integrals in (3.2.a) vanish; this and (3.2.1) lead to $\alpha(\text{grad}\, f,\text{grad}\, f) = 0$. Using Lemma (2.2.3), we conclude as in I.

III. (The Schneider conjecture).

1. Proof. Use Proposition (5.2) and I.

2. Proof. Consider f and h as in (2.1), (2.2). As in the proof of Proposition (5.2), steps I - III, we arrive at $f = h$, $\text{Hess}\, f = \text{Hess}\, h$, and $\hat{A}(v,w,\text{grad}\, h) = 0$ (cf. (5.2.2)). But then $\alpha(\text{grad}\, h,\text{grad}\, h) = 0$ from (1.15), and $\alpha = 0$ from Lemma (2.1.4). Proceed again as at the end of I. ∎

5.3.1. Remark. Another proof for II (and III resp.) comes from a result of Reilly ((R) Theorem 4). For a proper affine sphere $\text{Ric} \geqslant (n-1)HG$ (use (1.21), $\hat{B} = HG$, α semi-positive definite); as ∂M is totally geodesic, Reilly's result implies $\lambda_1 \geqslant nH$. On the other hand nH is an eigenvalue (2.2.2), so $\lambda_1 = nH$. Reilly's discussion of equality states that $M \cup \partial M$ is isometrically diffeomorphic to a Euclidean half-sphere of curvature $0 < \varkappa = H$; comparison with the affine Theorema egregium (1.22) gives $J = 0$. The rest is again obvious. ∎

Considering proper affine spheres with plane c-shadow boundaries A. Schwenk ((SCHW-1) p. 41-46 ; (SCHW-2) p. 306) could prove some results only in dimension 2, giving different proofs for $H > 0$ and $H < 0$, resp. In the following we present a simplified proof, considering two Dirichlet-problems as in (SCHW-1).

5.4. Theorem. Let x be a proper compact affine hypersphere with center x_o and with closed boundary ∂M. Assume that $x(M)$ is a graph and that

(i) $x(\partial M)$ is a plane intersection,

(ii) $x(\partial M)$ is a c-shadow boundary with z_o as center of light.

Then $x(M)$ lies on a quadric.- Furthermore, z_o is contained in the following line: There is a unique point $p_o \in M$ s.t. the tangentplane at $x(p_o)$ is parallel to the plane $\mathfrak{h}$, and z_o lies on the line uniquely determined by x_o and $x(p_o)$. If $H > 0$, z_o lies not between x_o and $x(p_o)$; if $H < 0$, z_o does.

Proof. We choose x_o as origin and consider the functions h and $\rho(z_o)$ from (2.1) and (2.3), resp. As $\langle L, z_1 \rangle \neq 0$, we choose L s.t. $\langle L, x \rangle | \partial M = \langle L, z_1 \rangle = -\frac{1}{H}$. Again from (2.1) and (2.3)

$$\Delta h + nHh = n \text{ and } h|\partial M = 0$$

$$\Delta \rho(z_o) + nH\rho(z_o) = n \text{ and } \rho(z_o)|\partial M = 0.$$

From Schneider's theorem ((SCHN-1) p. 385, Satz 3.4) both functions coincide: $h = \rho(z_o)$; the comparison of the Hessians finally gives $\alpha = 0$ as in (5.3), i.e. x is a quadric. The rest is elementary. ■

For the following theorem of A. Schwenk ((SCHW-1) p. 46, Satz 16) we give a modified shorter proof.

5.5 Theorem. Let x be a compact improper affine hypersphere with closed boundary ∂M. Assume that

(i) $x(\partial M)$ is a plane intersection,

(ii) $x(\partial M)$ is a c-shadow boundary with z_o as center of light.

Then $x(M)$ lies on an elliptic paraboloid.

Proof. Choose the origin to lie in the boundary plane and use (2.1), (2.3); choose $L \in V^*$ s.t. $\langle L, y \rangle = 1$ on M. From (2.1) and (2.3)

$$\Delta h = n \quad \text{ and } h|\partial M = 0,$$

$$\Delta \rho(z_o) = n \text{ and } \rho(z_o)|\partial M = 0.$$

Considering the difference-function, the maximumprinciple implies $h - \rho(z_o) = 0$ on M. We proceed as above to prove $\alpha = 0$, i.e. $x(M)$ lies on an elliptic paraboloid (as $H = 0$). ■

6. Complete hypersurfaces with constant mean curvature.

In Lemma 6.1 we give a minor extension of a result of R. Schneider ((SCHN-1) p. 403, Hilfssatz); this is the basis for a first step to classify the complete hypersurfaces of constant equiaffine mean curvature. - We refer to (SCHW-S) for results on compact hypersurfaces with boundary and constant mean curvature. Li (L) proved various uniqueness-results using curvature functions.

6.1. Lemma. Let (M,G) be complete and assume there exists $\delta > 0$ s.t. for x

(6.1.1) $nHG + (n-2)\hat{B} \geqslant \delta G$.

Then M is closed.

Proof. From (6.1.1) we immediately get a lower bound for the Ricci-curvature. It is a standard-argument that M must be closed; (M). ■

6.2. Remark. If $\hat{B}$ is semi-positive definite and there exists $0 < \delta^* \in \mathbb{R}$ s.t. $H \geqslant \delta^*$, then (6.1.1) is fulfilled. From section 1 the condition on $\hat{B}$ can be interpreted geometrically.

6.3. Theorem. Let (M,G) be complete and x be an immersion of constant equiaffine mean curvature.

(i) ((SI-3) Satz III. 5.6.1). If $n = 2$ and $H > 0$, then x(M) lies on an ellipsoid.

(ii) If $n \geqslant 2$ and $\hat{B}$ is positive definite, then x(M) lies on an ellipsoid.

Proof. M is closed (6.1 - 6.2). From (3.2.a) we get $\alpha(\operatorname{grad} f, \operatorname{grad} f) = 0$ and as before $\alpha = 0$. ■

6.4. Theorem. Let M be compact with boundary and x(M) be convex. Assume H = const and $\mathrm{Ric} \geqslant (n-1)HG$. Then the following two boundary conditions

(i) $x(\partial M)$ is p-shadow boundary,

(ii) ∂M is totally geodesic

imply that x(M) is a half ellipsoid.

Proof. From (4.4.1) we get $\lambda_1 = nH > 0$. As in (5.3.1) we apply now Reilly's Theorem 4 in (R). ■

7. Equiaffine Einstein spaces.

In this section we investigate equiaffine Einstein-spaces and similar types of manifolds. Theorem (7.1) and (7.3) are minor generalizations of Satz A and Satz C, resp., in (K-S).

7.1. Theorem. Let x be a hypersurface of constant equiaffine scalar curvature $\varkappa$ and $\mathrm{Ric} \geqslant (n-1)\varkappa G$. Furthermore assume

(M,G) to be complete. Then x(M) is an ellipsoid.

Proof. From the affine Theorema egregium (1.22), (4.2) and $Ric \geq (n-1)\varkappa G$ we get $n\varkappa \geq n \max H \geq \lambda_1 \geq n\varkappa$, i.e. $\lambda_1 = n\varkappa = n \max H = n H$. (4.1) gives the assertion.■

As a corollary we get a proof of a conjecture of Blaschke (B-2) p. 223-228; cf. also (SCHN) Satz 4.6.

7.2. Corollary. Let x be an equiaffine complete surface in $\mathcal{A}_3$ with $0<\varkappa=$const. Then x(M) lies on an ellipsoid.

7.3. Theorem. Let M be compact with boundary and (M,G) an Einstein space with $\varkappa =$ const $=: c$ and $n \geq 2$. If (i) $J|\partial M = 0$ and (ii) $x(\partial M)$ is a p-shadow boundary, then x(M) lies on a half ellipsoid.

Proof. (3.3) and the assumptions imply $JH = 0$. But $0 < \max H \leq \varkappa = c = J+H$, therefore $J(J-c) = JH = 0$. The zeroes of $(J-c)$ are nowhere dense (otherwise $H \equiv 0$, which is impossible), so $J=0$ on M, i.e. x describes a quadric, which is an ellipsoid because of $H > 0$. As in section 5 the boundary $x(\partial M)$ is a plane which contains the center.■

REFERENCES

(B-K-S-S) K. Benko, M. Kothe, K.D. Semmler, U. Simon: Eigenvalues of the Laplacian and curvature. Colloquium Math. 42, 19-31 (1978).

(B-G-M) M. Berger, P. Gauduchon, E. Mazet: Le spectre d'une variété riemannienne. Lecture Notes Math. 194. Springer (1971).

(B-1) W. Blaschke: Vorlesungen über Differentialgeometrie. Bd. II. Affine Differentialgeometrie. Springer (1923).

(B-2) W. Blaschke: Gesammelte Werke. Vol. 4. Eds.: W. Burau, S.S. Chern, K. Leichtweiß, H.R. Müller, L.A. Santalo, U. Simon, K. Strubecker. Thales Verlag Essen (1985).

(CA) E. Calabi: Hypersurfaces with maximal affinely invariant area. Amer. J. Math. 104, 91-126 (1982).

(CHA) I. Chavel: Eigenvalues in Riemannian Geometry. Academic Press (1984).

(CHE) S.S. Chern: Affine minimal hypersurfaces. Proc. Jap. - U.S. Semin. Tokyo 1977, 17-30 (1978).

(K) A. Kasue: On a lower bound for the first eigenvalue of the Laplace operator on a Riemannian manifold. Ann. scient. Ec. Norm. Sup., 4^{e} ser., 17, 31-44 (1984).

(K-S) M. Kozlowski, U. Simon: Hyperflächen mit äquiaffiner Einsteinmetrik. Festschrift E. Mohr zum 75. Geb., TU Berlin 1985, 179-190.

(L) A.-M, Li: Uniqueness theorems in affine differential geometry. Results in Mathematics 1988. To appear.

(M) S.B. Myers: Riemannian manifolds with positive mean curvature. Duke Math. J. 8, 401-404 (1941).

(N) K. Nomizu: On completeness in affine differential geometry. Geometriae Dedicata 20, 43-49 (1986).

(R) R.C. Reilly: Applications of the Hessian operator in a Riemannian manifold. Indiana Univ. Math. J. 26, 459-472 (1977).

(SCHI) P.A. and A.P. Schirokow: Affine Differentialgeometrie. Leipzig, Teubner 1962 (Zbl. 106.147; russ. Original Zbl. 85.367).

(SCHN) R. Schneider: Zur affinen Differentialgeometrie im Großen. I. Math. Z.101, 375-406 (1967).

(SCHW-1) A. Schwenk: Eigenwertprobleme des Laplace-Operators und Anwendungen auf Untermannigfaltigkeiten. Dissertation TU Berlin 1984.

(SCHW-2) A. Schwenk: Affinsphären mit ebenen Schattengrenzen. In: D. Ferus, R.B. Gardner, S. Helgason, U. Simon (Eds.): Global Differential Geometry and Global Analysis 1984. Lecture Notes Math. 1156. Springer, 296-315 (1985).

(SCHW-S) A. Schwenk, U. Simon: Hypersurfaces with constant equiaffine mean curvature. Arch. Math. 46, 85-90 (1986).

(S-1) U. Simon: Minkowskische Integralformeln und ihre Anwendungen in der Differentialgeometrie im Großen. Math. Annalen 173, 307-321 (1967).

(S-2) U. Simon: Hypersurfaces in equiaffine differential geometry and eigenvalue problems. Proc. Conf. Diff. Geom. Appl. Nové Město na Morave (ČSSR) 1983. Part I, 127-136 (1984).

(S-3) U. Simon: Zur Entwicklung der affinen Differentialgeometrie nach Blaschke. In: (B-2), 35-88 (1985).

(S-4) U. Simon: The fundamental theorem in affine hypersurface theory. Geometriae Dedicata. To appear.

(S-W-1) U. Simon, H. Wissner: Geometry of the Laplace Operator: Proceedings Conference Algebra and Geometry. Kuwait University. Alden Press Oxford, 171-191 (1981).

(S-W-2) U. Simon, H. Wissner: Geometrische Aspekte des Laplace Operators. Jahrbuch Überblicke Math. 73-92 (1982).

TECHNISCHE UNIVERSITÄt BERLIN

A class of symmetric functions and Chern classes of projective varieties†

Hsin-sheng Tai
Institute of Mathematics
Academia Sinica
Beijing, China

Introduction.

Let $\varphi : X \to P^N$ be a smooth projective variety of dimension n in the complex projective space P^N, T be the holomorphic tangent bundle of X, and $H = \varphi^* O_{P^N}(1)$ be the line bundle on X corresponding to a hyperplane section. The imbedding φ defines a Gauss map from X into the Grassmannian G_{n+1}^{N+1}. Let E be the pull back of the universal bundle of the Grassmannian by the Gauss map. Then E is nothing but the dual of the jet bundle $J_1(H)$. T and E are related by the exact sequence of Nakano and Serre ([1]):

$$0 \to 1 \to E \otimes H \to T \to 0, \tag{1}$$

where 1 stands for the trivial line bundle.

The aim of this paper is to exploit some information about Chern classes of T and E from (1).

Suppose $c(E) = \prod_{i=0}^{n}(1 + x_i)$ is a formal factorization of the total Chern class of E. Then $c(T) = \prod_{i=0}^{n}(1 + x_i + h)$, where $h = c_1(H)$. The idea goes as follows. Firstly, we determine the structure of the class of symmetric polynomials which are invariant under the change of indeterminates $x_i \mapsto x_i + h, 0 \le i \le n$. Secondly, corresponding to each polynomial of this class, the characteristic classes of T and E are the same. It turns out that, for a *complete intersection*, the Chern classes of E have the simple form:

$$c_k(E) = (-1)^k w_k(b) h^k, \tag{2}$$

where $w_k(b) = \sum b_{i_1} b_{i_2} \cdots b_{i_k}, i_1 \le i_2 \le \cdots \le i_k$, is the k-th homogeneous product sum of Wronski symmetric function in $(b_1, \cdots, b_r) = (a_1 - 1, \cdots, a_r - 1)$, and $a_1, \cdots, a_r$ are the degrees of the defining hypersurfaces of the complete intersection. Thus, for a complete intersection, we can get a number of interesting formulas of Chern classes this way. For example, the linear combination of Chern numbers corresponding to

$$nS_n - (n-1)S_{n-1}\sigma_1 + \cdots + (-1)^{n-1} S_1 \sigma_{n-1} + (-1)^n \sigma_n, \tag{3}$$

where σ_j is the j-th elementary symmetric function and S_{n-j} is the $(n-j)$-th power sum in $x_0, \cdots, x_n$, is equal to

$$\left(-\sum_{j=0}^{n-1}(n-j)S_{n-j}(b)w_j(b) + w_n(b)\right)a_1 \cdots a_r \le 0, \tag{4}$$

---

†Research partially by the Chinese Science Foundation and the Royal Swedish A cademy of Sciences

with a_i and $w_j(b)$ described as above and $S_{n-j}(b)$ is the $(n-j)$-th power sum in $a_1 - 1, \cdots, a_r - 1$. In particular, we have the inequalities

$$
\begin{aligned}
3c_2 &\geq c_1^2, && \text{if } n = 2,\\
4c_2c_1 &\geq 8c_3 + c_1^3, && \text{if } n = 3,\\
15c_4 + 5c_2c_1^2 &\geq 6c_3c_1 + 4c_2^2 + c_1^4, && \text{if } n = 4,
\end{aligned} \tag{5}
$$

and n is the dimension of the complete intersection. The equalities hold if and only if the complete intersection is a projective subspace.

The paper is organized as follows.

The structure of the symmetric polynomials which are invariant under the change of indeterminates is completely determined in §1. A multiplicative basis for this class of polynomials is given. The rather short section §2 deals with a nonsingular projective variety in general. Then the results of the first two sections are applied to a complete intersection in §3. The final section §4 contains two classes of examples which are defined intrinsically. The quotients of the unit ball satisfy a large number of equalities of Chern numbers corresponding to polynomials described in §1. And the quotients of the polydisk serve as counter-examples to the validity of some results in §3 for more general varieties of odd dimensions.

An earlier draft of this paper has existed for some time. It contains most of the results in §§1,2, and 4 in less precise form. Results in §3 were proved only for hypersurfaces. Wu Wen- tsün got interested and extended inequalities in (5) to hypersurfaces with arbitrary singularities within the framework of his theory of Chern classes of projective varieties [11]. We would like to thank him for his interest in our work which stimulated us to work out §3 for complete intersections.

1. Symmetric functions in the differences.

Symmetric functions of $n+1$ indeterminates $x_0, \cdots, x_n$ over the rational numbers form a subalgebra $\sum \subset Q[x_0, \cdots, x_n]$. $\sum$ has a natural grading $\sum = \oplus_{k \geq 0} \sum^k$, where $\sum^k$ consists of the homogeneous polynomials of degree k, together with the zero polynomial. The fundamental theorem of symmetric functions asserts that $\sum$ is a polynomial algebra on $n+1$ algebraically independent generators

$$\sum = Q[\sigma_1, \cdots, \sigma_{n+1}] \tag{1.0}$$

where σ_k denotes k-th elementary symmetric function.

We are interested in a subalgebra of symmetric polyunomials which are invariant under the change of indeterminates $x_i \mapsto x_i + h, 0 \leq i \leq n$. We call them the symmetric functions in the differences $x_i - x_j$ for reasons explained below.

Consider the following differential operator on symmetric functions:

$$D = (n+1)\frac{\partial}{\partial \sigma_1} + n\sigma_1 \frac{\partial}{\partial \sigma_2} + \cdots + \sigma_n \frac{\partial}{\partial \sigma_{n+1}}. \tag{1.1}$$

In terms of a new basis, it takes the form

$$D = \frac{\partial}{\partial a_1} + 2a_1 \frac{\partial}{\partial a_2} + \cdots + (n+1)a_n \frac{\partial}{\partial a_{n+1}}, \tag{1.2}$$

where a_k is related to σ_k by

$$\sigma_k = \binom{n+1}{k} a_k. \tag{1.3}$$

D is a natural derivation on the algebra of symmetric functions. It differentiates a_k just like powers in ordinary differentiation:

$$Da_k = ka_{k-1}, \tag{1.4}$$

and

$$D^p a_k = \frac{k!}{(k-p)!} a_{k-p}, \tag{1.5}$$

for $1 \leq p \leq k \leq n+1$.

THEOREM 1 *A symmetric function f is invariant under the change of indeterminates $x_i \mapsto x_i + h, 0 \leq i \leq n$, if and only if it satisfies a formal partial differential equation*

$$(n+1)\frac{\partial f}{\partial \sigma_1} + n\sigma_1 \frac{\partial f}{\partial \sigma_2} + \cdots + \sigma_n \frac{\partial f}{\partial \sigma_{n+1}} = 0.$$

In other words, the set of all symmetric functions in the differences is nothing but $\ker D$.

PROOF: By looking at the Taylor expansion of $f(x_0 + h, \cdots, x_n + h)$ at $h = 0$, it suffices to prove

$$\frac{d^p}{dh^p} f(x_0 + h, \cdots, x_n + h)|_{h=0} = D^p f. \tag{1.6}$$

for $p \geq 1$.

It is not hard to derive

$$a_k(x_0 + h, \cdots, x_n + h) = \sum_{j=0}^{k} \binom{k}{j} a_{k-j} h^j. \tag{1.7}$$

Applying to symmetric functions in $x_0 + h, \cdots, x_n + h$; the chain rule implies

$$\frac{d}{dh} = \sum_{j=1}^{k} \binom{k}{j} j a_{k-j} h^{j-1} \frac{\partial}{\partial a_k}. \tag{1.8}$$

Differentiate (1.7) p times with respect to h. We obtain from (1.5)

$$\frac{d^p}{dh^p} a_k(x_0 + h, \cdots, x_n + h)|_{h=0} = D^p a_k. \tag{1.9}$$

Since D and $\frac{d}{dh}$ are derivations and f is a polynomial in $a_1, \cdots, a_{n+1}$, (1.6) follows from (1.9). The theorem is proved.

REMARK: Theorem 1 is inspired by an exercise in [10] (§26, ex.7), where it is stated as a necessary condition satisfied by the discriminant.

Let $\sum_D^k = Ker D \cap \sum^k$, the set of polynomials of degree k in the differences. We want to take a look of $\sum_D^k$, for $k \leq 4$.

It is clear from (1.1) that $\sum_D^1 = \{0\}$. We make an agreement that $\sum_D^0 = \sum^0 = Q$.

Both $\sum_D^2$ and $\sum_D^3$ are of dimension 1 over Q. The following are respectively their bases with integral coefficients.

$$2(n+1)\sigma_2 - n\sigma_1^2, \tag{1.10}$$

$$3(n+1)^2\sigma_3 - 3(n^2-1)\sigma_2\sigma_1 + n(n-1)\sigma_1^3. \tag{1.11}$$

While $\dim_Q \sum_D^4 = 2$. A set of basis with integral coefficients are the square of (1.2) and

$$2n(n+1)\sigma_4 - 2n(n-2)\sigma_3\sigma_1 + (n-1)(n-2)\sigma_2^2. \tag{1.12}$$

This can be derived as follows. Let

$$\lambda_0\sigma_4 + \lambda_1\sigma_3\sigma_1 + \lambda_2\sigma_2\sigma_1^2 + \lambda_3\sigma_2^2 + \lambda_4\sigma_1^4$$

be an element of $\sum_D^4$. Applying the differential operation (1.1), we arrive at a set of linear equations.

$$\begin{cases} (n+1)\lambda_1 + (n-2)\lambda_0 = 0, \\ 2(n+1)\lambda_2 + 2n\lambda_3 + (n-1)\lambda_1 = 0, \\ 4(n+1)\lambda_4 + n\lambda_2 = 0. \end{cases} \tag{1.13}$$

It is easy to see that $\dim \sum_D^4 = 2$. Setting $\lambda_0 = \lambda_1 = 0$, we find that $\lambda_2 = -4n(n+1), \lambda_3 = 4(n+1)^2$, and $\lambda_4 = n^2$ are a set of integral solution which gives the square of (1.10). While letting $\lambda_2 = \lambda_4 = 0$, we are led to $\lambda_0 = 2n(n+1), \lambda_1 = -2n(n-2)$, and $\lambda_3 = (n-1)(n-2)$ which give (1.12).

(1.10) and (1.11) can be derived in a similar way.

There is a general way of producing an element of $\sum_D^k$, for $2 \leq k \leq n$. Let

$$\tau_k = \sum_{0\leq i\leq n} \Big(\sum_{0\leq j_1<\cdots<j_k\leq n} \prod_{1\leq r\leq k} (x_i - x_{j_r}) \Big). \tag{1.14}$$

It is straightforward to derive

$$\tau_k = \sum_{j=0}^{k} (-1)^j \binom{n+1-j}{k-j} S_{k-j}\sigma_j, \tag{1.15}$$

where S_i is the i-th power sum. Note that $S_0 = n+1$ and $\sigma_0 = 1$.

Then $\frac{-2}{n-1}\tau_2$ is (1.10) and $\frac{6}{(n-2)}\tau_3$ is (1.11). However, τ_4 is quite different from (1.12), e.g. when $n = 4$,

$$\tau_4 = \sigma_1^4 - 5\sigma_2\sigma_1^2 + 6\sigma_3\sigma_1 + 4\sigma_2^2 - 15\sigma_4. \tag{1.16}$$

Equation (1.15) can be put in a different form by the Newton formulas. For example, the Newton formula

$$S_n - S_{n-1}\sigma_1 + \cdots + (-1)^{n-1}S_1\sigma_{n-1} + (-1)^n n\sigma_n = 0$$

reduces (1.15) for $k = n$ to the form

$$\tau_n = nS_n - (n-1)S_{n-1}\sigma_1 + \cdots + (-1)^{n-1}S_1\sigma_{n-1} + (-1)^n\sigma_n. \tag{1.17}$$

However, it is more convenient to write (1.15) for $2 \leq k \leq n$.

We will write a partition $I = (i_1, \cdots, i_r)$ of a nonnegative integer k in decreasing order $i_1 \geq i_2 \geq \cdots \geq i_r$. The weight of the partition $|I| = i_1 + \cdots + i_r = k$. The following notation will be used:

$$\sigma_I = \sigma_{i_1} \cdots \sigma_{i_r}.$$

Let $p(k)$ denote the number of partitions of k. If $k > 1$, by just putting an one in the end, then every partition of $k-1$ corresponds to a partition of k containing 1.

The following theorem gives the structure of the quotient algebra

$$\sum_D^{(n)} = \bigoplus_{0 \leq k \leq n} \sum_D^k.$$

THEOREM 2.
i) $\dim \sum_D^k = p(k) - p(k-1)$, *the number of partition without ones for* $2 \leq k \leq n$.
ii) $\{\tau_k\}_{2 \leq k \leq n}$, *is a set of generators of the quotient algebra* $\sum_D^{(n)}$.

PROOF: $\dim \sum^k = p(k)$ and $\{\sigma_I\}$, $|I| = k$, is a basis of the vector space $\sum^k$ over Q for $1 \leq k \leq n$. Let D be the differential operator (1.1). We claim that

$$D : \sum^k \to \sum^{k-1}, \quad 2 \leq k \leq n,$$

are surjective. Thanks to the fundamental theorem (1.0), we can proceed on the total degree in σ_i's. $\sigma_{k-1} \in \sum^{k-1}$ is the only element of the basis with σ- degree 1, and $D(\frac{1}{n-k+2}\sigma_k) = \sigma_{k-1}$. Extend linearly, we see that every element of σ-degree 1 is in the image of D. Suppose the claim is true for elements of σ-degrees less than d. Let σ_I be of σ- degree d, where $I = (i_1, \cdots, i_d)$ is a partition of $k-1$. Then $I_1\,(i_2, \cdots, i_d)$ is a partition of $k - i_1 - 1$ and $\sigma_{i_1+1}D(\sigma_{I_1})$ is of σ-degree $d-1$. Hence by induction assumption there is an $f \in \sum^k$ such that $D(f) = \sigma_{i_1+1}D(\sigma_{I_1})$. Whence $D((\frac{1}{n-i_1+1}(\sigma_{i_1+1}\sigma_{I_1} - f)) = \sigma_I$. Extend linearly, the claim holds for all elements of σ-degree d. Since $\sum_D^k = \ker D$, we have $\dim \sum_D^k = p(k) - p(k-1)$. This proves i). For each partition J without ones, we

can form τ_J. Therefore $\{\tau_J\}, |J| = k$ and J contains no ones, form an additive basis of $\sum_D^k$. Hence $\{\tau_k\}$, $2 \le k \le n$, form a multiplicative basis of $\sum_D^{(n)}$.
COROLLARY. *$\{\sigma_1, \tau_2, \cdots, \tau_n, \sigma_{n+1}\}$ is a multiplicative basis of $\sum$.*

2. Chern classes of projective varieties.

Let X be a nonsingular algebraic variety of dimension n imbedded in the complex projective space P^N of dimension N. Let T be the holomorphic tangent bundle of X, H be the line bundle on X corresponding to a hyperplane section, and G_{n+1}^{N+1} denote the Grassmannian of n-dimensional subspaces of P^N. Then the imbedding of X in P^N defines a Gauss map of X into G_{n+1}^{N+1} by assigning to each point p the tangent projective n-space to X at p. Let E be the $(n+1)$-dimensional vector bundle over X induced by the Gauss map. Then ([1], p. 199) the following exact sequence of Nakano and Serre

$$0 \to 1 \to E \otimes H \to T \to 0 \tag{2.1}$$

implies

$$c_k(T) = c_k(E \otimes H), \qquad 1 \le k \le n. \tag{2.2}$$

Let

$$c(E) = \prod_{0 \le i \le n} (1 + x_i) \tag{2.3}$$

be a formal factorization of Chern classes of E. Then by the axioms of the Chern classes ([7], p. 64)

$$c(E \otimes H) = \prod_{0 \le i \le n} (1 + x_i + h), \tag{2.4}$$

where $h = c_1(H)$ denote the hyperplane class of X.

Let $f \in \sum_D^{(n)}$ be expressed as a polynomial in terms of elementary symmetric functions. It follows from (2.4) that, when σ_k is replaced by c_k, f satisfies

$$f(c_k(T)) = f(c_k(E)). \tag{2.5}$$

In other words, the characteristic classes corresponding to symmetric polynomials in the differences of T and E are the same.
REMARK 1. (2.2) and (2.4) imply the following classical formula of Todd:

$$c_k(T) = \sum_{j=0}^{k} \binom{n+1-j}{k-j} c_j(E) h^{k-j}. \tag{2.6}$$

By eliminating the mixed terms with $j > 0$, we came across the fact (2.5) for $3\sigma_2 - \sigma_1$ when $n = 2$ and $8\sigma_3 - 4\sigma_2\sigma_1 + \sigma_1^3$ when $n = 3$. We found out later that Van de Ven [9] has already noticed the fact for $n = 2$. §1 gives a complete solution of such polynomials.

2. $c(E)$ and $c(E\otimes H)$ both have $(n+1)$ Chern roots. However $c(T)$ has only n Chern roots, say, $y_1,\cdots,y_n$. But nevertheless $c_k(T)=c_k(E\otimes H)$, and Newton's formulas assure us that

$$S_k(y_1,\cdots,y_n)=S_k(x_0+h,\cdots,x_n+h),\quad 1\le k\le n. \tag{2.7}$$

We will need this fact in computing the characteristic corresponding to τ_n (1.17) for some examples in §4.

3. Complete intersections.

We follow the notation of Hirzebruch ([5], [7]).

Consider r nonsingular hypersurfaces $F^{(a_1)},\cdots,F^{(a_r)}$ of degrees $a_1,\cdots,a_r$ in the complex projective space P^{n+r}. The intersection $V_n^{(a_1,\cdots,a_r)}=F^{(a_1)}\cap\cdots\cap F^{(a_r)}$ is a complete intersection of dimension n if the hypersurfaces $F^{(a_1)},\cdots,F^{(a_r)}$ are in general position. Our goal is to calculate the characteristic classes, corresponding to symmetric polynomials $\tau_k, 2\le k\le n$, of a complete intersection.

We need some elementary lemmas which are combinatorial in nature.

LEMMA 1.

$$\binom{r+k-1}{k}\binom{k}{j}=\binom{r+j-1}{j}\binom{r+k-1}{k-j}$$

PROOF: Straightforward.

LEMMA 2. *Let $b_i=a_i-1, 1\le i\le r$, and let $w_k(a)$ and $w_k(b)$ denote the k-th homogeneous product sum or Wronski symmetric polynomial in $a_1,\cdots,a_r$, and in $b_1,\cdots,b_r$ respectively. Then*

$$w_k(b)=\sum_{j=0}^{k}(-1)^{k-j}\binom{r+k-j}{k-j}w_j(a).$$

PROOF: $w_k(b)=\sum b_{i_1}b_{i_2}\cdots b_{i_k}, i_1\le i_2\le\cdots\le i_k$. Expand $\binom{r+k-1}{k}$ terms of $w_k(b)$ into monomials in $a_1,\cdots,a_r$. Among these monomials, there are $\binom{r+k-1}{k}\binom{k}{j}$ terms of degree j. Since each monomial in $w_j(a), 1\le j\le k$, has equal chance to come out of the expansion of $w_k(b)$ and have the sign $(-1)^{k-j}$, and $w_j(a)$ has $\binom{r+j-1}{j}$ terms, it follow from Lemma 1 that the coefficient of $w_j(a)$ in $w_k(b)$ is $(-1)^{k-j}\binom{r+k-1}{k-j}$.

Q.E.D.

LEMMA 3.

$$\binom{m+q+1}{\ell}=\sum_{i=0}^{\ell}\binom{m-i+1}{\ell-i}\binom{q+i-1}{i}.$$

PROOF: For simplicity, let us call a set containing N elements an N-set. Count the number $\binom{m+q+1}{\ell}$ of ℓ-subsets Y in an $(m+q+1)$-set S in the following way.

Choose an $(m+1)$-subset S_0 and a sequence of elements $\{s_1, \cdots, s_\ell\}$ from S_0. Let $S_i, 1 \le i \le \ell$, be the $(m+1-i)$-subset of S_0 which excludes the first i elements in the sequence. Split the class of ℓ-subsets Y of S into two subclasses depending on whether $Y \subset S_0$ or $Y \cap S_0$ has less than ℓ elements. The first subclass has $\binom{m+1}{\ell}$ elements. Split the complement into two more subclasses, depending on $Y \subset S_1$ (call this the second subclass) or $Y \cap S_1$ containing less than $\ell - 1$ elements. The second subclass has $\binom{m}{\ell-1}\binom{q}{1}$ elements, because those containing s_1 should be in the first subclass. Continuing this process, we get $\ell + 1$ mutually exclusive subclasses. The j-th subclass has $\binom{m-j+2}{\ell-j+1}\binom{q+j-2}{j-1}$ elements. The formula now follows by the rule of sum.

Let T_V and E_V denote the holomorphic tangent bundle and the induced bundle via the Gauss map of a complete intersection $V_n^{(a_1,\cdots,a_r)}$. It is well known that ([5], [7])

$$c(T_V) = (1+h)^{n+r+1}(1+a_1h)^{-1}\cdots(1+a_rh)^{-1}. \tag{3.1}$$

It is more convenient to write

$$\prod_{\ell=1}^{r}(1+a_\ell h)^{-1} = \sum_j (-1)^j w_j(a)h^j.$$

Hence

$$c(T_V) = \left(\sum_{i=0}^{n+r+1}\binom{n+r+1}{i}h^i\right)\left(\sum_j(-1)^j w_j(a)h^j\right).$$

In particular,

$$c_k(T_V) = \left(\sum_{j=0}^{k}(-1)^j\binom{n+r+1}{k-j}w_j(a)\right)h^k. \tag{3.2}$$

Comparing (3.2) with (2.6), one readily sees that

$$\begin{aligned} c_1(E_V) &= (r - w_1(a))h \\ &= \left(-\sum_{\ell=1}^{r}(a_\ell - 1)\right)h \\ &= -w_1(b)h, \end{aligned}$$

and

$$\begin{aligned} c_2(E_V) &= \left(\binom{r+1}{2} - (r+1)w_1(a) + w_2(a)\right)h^2 \\ &= \left(\sum_\ell (a_\ell-1)^2 + \sum_{\ell<m}(a_\ell-1)(a_m-1)\right)h^2 \\ &= w_2(b)h^2. \end{aligned}$$

THEOREM 3.

$$c_k(E_V) = (-1)^k w_k(b) h^k, \tag{3.3}$$

where $w_k(b)$ is the k-th Wronski symmetric polynomial in $a_1 - 1, \cdots, a_r - 1$.
PROOF: Proceed by induction. Since

$$\begin{aligned} c_k(T_V) &= \sum_{j=0}^{k} \binom{n+1-j}{k-j} c_j(E_V) h^{k-j} \\ &= \left(\sum_{j=0}^{k} (-1)^j \binom{n+r+1}{k-j} w_j(a) \right) h^k, \end{aligned}$$

the induction assumption and Lemma 2 imply

$$\begin{aligned} c_k(E_V) =& (\sum_{j=0}^{k} (-1)^j \binom{n+r+1}{k-j} w_j(a) \\ &- \sum_{p=0}^{k-1} \binom{n+1-p}{k-p} (\sum_{j=0}^{p} (-1)^j \binom{r+p-1}{p-j} w_j(a)))h^k \\ =& (-1)^k w_k(a) h^k + (\sum_{j=0}^{k-1} (-1)^j (\binom{n+r+1}{k-j} \\ &- \sum_{i=0}^{k-j-1} \binom{n+1-j-i}{k-j-i} \binom{r+j+i-1}{i}) w_j(a)) h^k, \end{aligned}$$

where $i = p - j$. Apply Lemma 3 with $m = n + 1 - j, q = r + j$ and $\ell = k - j$,

$$\begin{aligned} c_k(E_V) =& (-1)^k \left(\sum_{j=0}^{k} (-1)^{k-j} \binom{r+k-1}{k-j} w_j(a) \right) h^k \\ =& (-1)^k w_k(b) h^k \end{aligned}$$

by Lemma 2. Q.E.D.
REMARK 1. $c_k(E_V)$ depends only on the integers $a_1 - 1, \cdots, a_r - 1$, and not on $F^{(a_1)}, \cdots, F^{(a_r)}$, and n.

2. We can now put

$$c_k(T_V) = \left(\sum (-1)^j \binom{n+1-j}{k-j} w_j(b) \right) h^k, \tag{3.4}$$

which depends only on $a_1 - 1, \cdots, a_r - 1$, and n.

Let

$$F(t) = \prod_{i=0}^{n} (1 - x_i t) = \sum_{i=0}^{n} (-1)^i \sigma_i(x) t^i \pmod{t^{n+1}},$$

and

$$G(t) = 1/\prod_{i=1}^{r}(1 - b_i t) = \sum_{i=0}^{n} w_i(b)t^i \pmod{t^{n+1}},$$

we have, following the usual way of deriving Newton's formulas,

$$tF'(t)/F(t) = -\sum_{k=1}^{n} S_k(x)t^k \pmod{t^{n+1}}$$

and

$$tG'(t)/G(t) = \sum_{k=1}^{n} S_k(b)t^k \pmod{t^{n+1}}.$$

Clear the denominators and replace t by ht in the second equation, then

$$\begin{aligned}(3.5)\qquad &\sum_{j=1}^{n}(-1)^j j\sigma_j(x)t^j \\ &= -\left(\sum_{i=0}^{n}(-1)^i\sigma_i(x)t^i\right)\left(\sum_{k=1}^{n} S_k(x)t^k\right) \pmod{t^{n+1}},\end{aligned}$$

$$\begin{aligned}(3.6)\qquad &\sum_{j=1}^{n} jw_j(b)h^j t^j \\ &= \left(\sum_{i=0}^{n} w_i(b)h^i t^i\right)\left(\sum_{k=1}^{n} S_k(b)h^k t^k\right) \pmod{t^{n+1}}.\end{aligned}$$

Thanks to Theorem 3, $c_j(E_V) = \sigma_j(x) = (-1)^j w_j(b)h^j$. It follows immediately, by comparing (3.5) and (3.6), that

THEOREM 4.

$$(3.7)\qquad S_k(x) = -S_k(b)h^k, \quad 1 \le k \le n.$$

THEOREM 5.

$$(3.8)\qquad \tau_k(x) = -\beta_k(b)h^k, \quad 2 \le k \le n,$$

where

$$\begin{aligned}(3.9)\qquad \beta_k(b) = &\sum_{j=0}^{k-2}\left(\binom{n+1-j}{k-j} - 1\right) S_{k-j}(b)w_j(b) \\ &+ (n+1-k)(S_1(b)w_{k-1}(b) - w_k(b))\end{aligned}$$

is a nonnegative integer. It vanishes if and only if $b_1 = \cdots = b_r = 0$, *or equivalently* $a_1 = \cdots = a_r = 1$.

PROOF: Substracting Newton's formula

$$\sum_{j=0}^{k-1}(-1)^j S_{k-j}(x)\sigma_j(x) + (-1)^k k\sigma_k(x) = 0$$

from (1.15), and noticing that $S_0(x) = n+1$, we obtain

$$\tau_k(x) = \sum_{j=0}^{k-1}(-1)^j \left(\binom{n+1-j}{k-j} - 1\right) S_{k-j}(x)\sigma_j(x) + (-1)^k(n+1-k)\sigma_k(x).$$

Substituting (3.3) and (3.7) to the right hand side, we have

$$\begin{aligned}\tau_k(x) &= -\left(\sum_{j=0}^{k-1}\left(\binom{n+1-j}{k-j} - 1\right) S_{k-j}(B)w_j(b) - (n+1-k)w_k(b)\right) h^k \\ &= -\beta_k(b)h^k.\end{aligned}$$

Since $b_1, \cdots, b_r$ are nonnegative integers, it is not hard to see that $S_1(b)w_{k-1}(b) \geq w_k(b)$, and obviously $\binom{n+1-j}{k-j} > 1$, for $0 \leq j < k$. Therefore $\beta_k(b)$ is a sum of nonnegative integers which vanishes if and only if each term does. The first term is a positive multiple of $S_k(b)$ and it vanishes if and only if $b_1 = \cdots = b_r = 0$. Q.E.D.

Thanks to (2.5), we have now

COROLLARY 2. *The characteristic class corresponding to τ_k of a complete intersection $V_n^{(a_1,\cdots,a_r)}$ has the expression*

$$\begin{aligned}\tau_k(c_1,\cdots,c_k)(T_V) &= \tau_k(c_1,\cdots,c_k)(E_V) \\ &= -\beta_k(b)h^k,\end{aligned} \tag{3.10}$$

where $\beta_k(b)$ is defined by (3.9). It vanishes if and only if $a_1 = \cdots = a_r = 1$.

Let $I = (i_1, \cdots, i_p)$ be a partition of n containing no ones, i.e. $i_1 \geq i_2 \geq \cdots \geq i_p \geq 2$, and $\sum_{a=1}^{p} i_a = n$. We can form

$$\tau_I(x) = \tau_{i_1}(x)\cdots\tau_{i_p}(x) = (-1)^p \beta_{i_1}(b)\cdots\beta_{i_p}(b)h^n,$$

and evaluate the corresponding characteristic class over the fundamental class [V]. The result is the following formula of the Chern numbers of $V_n^{(a_1,\cdots,a_r)}$.

THEOREM 6. *For a complete intersection $V_n^{(a_1,\cdots,a_r)}$*

$$(-1)^p\tau_I(c_1,\cdots,c_n) = \beta_{i_1}(b)\cdots\beta_{i_p}(b)a_1\cdots a_r \geq 0. \tag{3.11}$$

In particular,

$$-\tau_n(c_1,\cdots,c_n) = \beta_n(b)a_1\cdots a_r \geq 0.$$

The Chern number vanishes if and only if $a_1 = \cdots = a_r = 1$.

To give explicit inequalities for complete intersections of dimensions 2, 3, and 4, we deduce from (1.10), (1.11), and (1.16) the following

COROLLARY 3. *Let $V_n^{(a_1,\cdots,a_r)}$ be a complete intersection, then the Chern numbers satisfying*

$$\begin{aligned} 3c_2 &\geq c_1^2, && \text{if } n=2,\\ 4c_2c_1 &\geq 8c_3 + c_1^3, && \text{if } n=3, \qquad (3.12)\\ 15c_4 + 5c_2c_1 &\geq 6c_3c_1 + 4c_2^2 + c_1^4, && \text{if } n=4, \end{aligned}$$

equalities hold if and only if V is a projective subspace.

REMARK: For a nonsingular hypersurface of degree d in $\mathbf{P}^{n+1}$, β_n has the simple form $\beta_n = \frac{1}{2}(n-1)(n+2)(d-1)^n$. This can be deduced easily from (1.17).

4. Quotients of the unit ball and the polydisk.

The induced bundle E is trivial over $\mathbf{P}^n$. Then the exact sequence (2.1) takes the form ([3], p. 409)

$$0 \to 1 \to H^{\oplus(n+1)} \to T_{\mathbf{P}^n} \to 0. \tag{4.1}$$

It implies

$$c(T_{\mathbf{P}^n}) = (1+h)^{n+1}. \tag{4.2}$$

Whence

$$f(c_1,\cdots,c_n)(\mathbf{P}^n) = 0, \tag{4.3}$$

for all $f \in \sum_D^n$, because of the factorization.

The following examples are usually called the examples of Borel and Hirzebruch ([8]).

A. Quotients of the unit ball.

Let B_n denote the unit ball $\{x \in \mathbf{C}^n | \sum_{1\leq x\leq n} |x_i|^2 < 1\}$. It carries a complete Kähler metric of a negative constant holomorphic curvature and is a bounded symmetric domain $U(1,n)/U(1)\times U(n)$. Suppose a group G acts freely on B_n and B_n/G is compact. Then $X = B_n/G$ is a nonsingular algebraic variety of general type [2]. The compact dual of B_n is $U(n+1)/U(1)\times U(n) = \mathbf{P}^n$ [4]. By Hirzebruch's Proportionality theorem ([6], [7])

$$c_I(X) = \lambda c_I(\mathbf{P}^n) \tag{4.4}$$

for all Chern numbers, where I is a partition of n, and the constant λ of proportion is the volume of X in the nature metric. Therefore (4.3) also holds for X. Hence there is an equality of Chern numbers of X for each polynomial in $\sum_D^n$.

$$f(c_1,\cdots,c_n)(X) = 0, \quad f \in \sum_D^n. \tag{4.5}$$

Some examples are

$$\begin{aligned} 3c_2 &= c_1^2, & \dim X &= 2,\\ 8c_3 + c_1^3 &= 4c_2c_1, & \dim X &= 3,\\ 15c_4 + 5c_2c_1^2 &= 6c_3c_1 + 4c_2^2 + c_1^4, & \dim X &= 4. \end{aligned}$$

Since X is covered holomorphically by the unit ball and its nature metric has negative constant holomorphic curvature, the first Chern form is proportional to the Kähler form. We also have the equalities

$$\begin{aligned} 2(n+1)c_2c_1^{n-2} &= nc_1^n,\\ 3(n+1)^2c_3c_1^{n-3} + n(n-1)c_1^n &= 3(n^2-1)c_2c_1^{n-2}. \end{aligned} \tag{4.6}$$

In general for each polynomial $\tau_k, 2 \leq k \leq n$, there is an equality of Chern numbers of X of the following form

$$\tau_k(c_1, \cdots, c_k)c_1^{n-k}(X) = 0. \tag{4.7}$$

In fact (4.7) holds for every polynomial in $\sum_D^k, 2 \leq k \leq n$. By Theorem 2, they are all contained in those produced by τ_k.

Now, let $X = \mathbf{P}^1 \times \cdots \times \mathbf{P}^1$ be the Segre variety of n copies of the projective line. Let $h_i = c_1(pr_i^* O_{\mathbf{P}^1}(1))$, then $c(T_X) = \sum_{i=1}^n (1 + 2h_i)$. In other words, $c(T_X)$ has $2h_1, \cdots, 2h_n$ as its Chern roots. Keeping remark 2 of §2 in mind, we may take τ_n as a symmetric polynomial in n indeterminates in computing the characteristic class corresponding to τ_n intrinsically. To compute the Chern number, we can ignore terms containing higher powers $h_i^p, p > 1, 1 \leq i \leq n$. Hence there are only the last two terms in (1.17) to be considered: $(-1)^{n-1}S_1\sigma_{n-1} + (-1)^n\sigma_n$. One readily sees that

$$\tau_n(c_1, \cdots, c_n)(\mathbf{P}^1 \times \cdots \times \mathbf{P}^1) = (-1)^{n-1}2^n(n-1). \tag{4.8}$$

B. Quotients of the polydisk.

Let $D_n = \{x \in \mathbf{C}^n | \, |x_i|^2 < 1, 1 \leq i \leq n\}$. Its compact dual is $\mathbf{P}^1 \times \cdots \times \mathbf{P}^1$. Suppose G is a discontinuous group of transformations of D_n acting without fixed points and with compact quotient D_n/G. By Hirzebruch's Proportionality Principle again,

$$\tau_n(c_1, \cdots, c_n)(D_n/G) = (-1)^{n-1}\lambda 2^n(n-1), \tag{4.9}$$

where λ is a positive constant. These quotients are varienties of general type.

We see that, for odd n, $\mathbf{P}^1 \times \cdots \times \mathbf{P}^1$ and the quotients D_n/G serve as counterexamples to the validity of Theorem 6 for more general variety. Similarly,

$$\tau_n(c_1, \cdots, c_n)(\mathbf{P}^{n-1} \times \mathbf{P}^1) = (-1)^{n-1}n^2(n-1), \tag{4.10}$$

we can produce another set of quotients which are varieties of general type.

We have not been able to find counterexamples for even n. For dimension 4, the only other choice of a bounded symmetric domain is $B_2 \times B_2$. However, its compact dual $\mathbf{P}^2 \times \mathbf{P}^2$ has the Chern number -9 corresponding to τ_4.

As a consequence to his solution of Calabi conjecture, Yau [12] proves the inequality

$$(-1)^n 2(n+1)c_2c_1^{n-2} \geq (-1)^n nc_1^n$$

for an n-dimensional variety with ample canonical bundle, (in particular, $3c_2 \geq c_1^2$ for a surface), the equality holds if and only if the variety is covered holomorphically by a unit ball. The proof works for any Kähler-Einstein manifold. We would like to mention the coincidence of coefficients in this inequality and the polynomial (1.10).

1. M.F.Atiyah, *Complex analytic connections in fibre bundles*, Trans. Amer. Math. Soc. 85 (1957), 181-207.

2. A. Borel, *Les functions automorphes de plusieur variables complexes*, Bull. Soc. Math. France 80 (1952), 167-182.

3. P.A. Griffiths and J. Harris, "Principles of Algebraic Geometry," John Wiley, 1978.

4. S. Helgason, "Differential Geometry and Symmetric Spaces," Academic Press, 1962.

5. F. Hirzebruch, *Der Satz von Riemann-Roch in Faisceau- Theoretischer Formulierung, einige Anwendungen und offene Fragen*, Proc. of the Int. Congress of Math. Amsterdam (1954), 457-473.

6. F. Hirzebruch, *Automorphe Formen und der Satz von Riemann-Roch*, Symp. Int. de Topologica Algebraica, Mexico (1958), 129-144.

7. F. Hirzebruch, "Topological Methods in Algebraic Geometry," Springer, 1966.

8. Y. Miyaoka, *On the Chern numbers of surfaces of general type*, Invent. Math. 42 (1977), 225-237.

9. A. Van de Ven, *On the Chern numbers of surfaces of general type*, Invent. Math. 36 (1976), 285-293.

10. B.L. van der Waerden, "Modern Algebra," Vol. 1, Frederick Ungar, 1949.

11. W.-T. Wu, *On Chern numbers of algebraic varieties with arbitrary singularities*, to appear in Acta Mathematica Sinica.

12. S.-T. Yau, *On Calabi's conjecture and some new results in algebraic geometry*, Proc. Nat. Acad. Sci. USA 74 (1977), 1798-1799.

ESSENTIAL INVARIANT CIRCLES OF SURFACE AUTOMORPHISMS OF FINITE ORDER

Shicheng Wang

UCLA and Peking University

Introduction and Preparation

Professor R.D.Edwards asked when an orientation preserving periodic map on a closed oriented surface has an invariant essential circle. This note contains two theorems. The first one gives the complete answer to the prime order case of his question. The second one deals with the general case. For a given surface, we will determine the maximum order under which a periodic map might have an invariant essential circle.

Theorem 1: If f is an orientation preserving periodic map of prime order p on an oriented closed surface of genus $g, g \geq 2$, then f has an invariant essential circle if and only if $p \leq g+1$.

Theorem 2: Let $O(g)$ be the integer associated with oriented closed surface of genus g, $g \geq 2$, denoted by F_g, such that for any orientation preserving periodic map f of order p on F_g, $p > O(g)$ implies that f has no invariant essential circle and there is f' of order $O(g)$ on F_g which has an invariant essential circle. Then $O(g) = 2g+2$, if g is even; $O(g) = 2g-2$ if g is odd and $g > 3$; $O(3) = 6$.

Remark 1: Lemma 7 and Lemma 8 which will be used in the proof of theorem 2 have some independent interest. Lemma 7 studies the invariant nonseparating circle of periodic map. The maximum order under which a periodic map might have an invariant nonseparating circle is found there. The example we give in lemma 8 is unique in the following sense: every periodic map of maximum order is conjugate to a power of our example. The proof is just to use the technique in the proof of lemma 5 and Hurwitz classification theorem about branched coverings.

Remark 2: More information is provided by the following facts: There is a map of order 3570 on F_{5339} without invariant circle (therefore without invariant essential circle.) Here the order < the genus. There is a map of order 3 on F_g without invariant nonseparating circle, for any genus g.(Outline of the proof: First choose a set of suitable numbers satisfying Riemann-Hurwitz formula, then use the technique in the end of the proof of lemma 5, finally use lemma 2,3. Hint: $3570 = 2\times 3\times 5\times 7\times 17$)

Remark 3: When $g=0,1$, the answer to Edwards' question is complete. If $g=0$, there is no essential circle. If $g=1$, every periodic map which is isotopic to the identity has an invariant essential circle. There are only four kinds of orientation preserving periodic maps which are not isotopic to the identity. Their orders are 2,3,4,6, respectively. The map with order 2 has an invariant essential circle. The map with order 3,4 or 6 has no invariant essential circle. (See [He])

Remark 4: It is interesting to compare our theorems with the result of W.H.Meeks which claim that if f is an orientation preserving map of order $p^r q^s$, where p,q are prime numbers, then f has an invariant circle. One corollary of the comparison is: If f is an orientation preserving periodic map of prime order p on the oriented closed surface of genus $g, g\geqq 2$, then $p\leqq 2g+1$. (See[W])

I would like to thank Professor R.D.Edwards for his many helps(especially for his bring my attention to the branched covering and the Riemann-Hurwitz formula). The example in lemma 8 was motivated by Professor R.C.Kirby's example which he used to teach his students how F_1 branch covers over S^2. The proof of lemma 5 and lemma 6 were greatly simplified by Y.Q.Wu and Q.Zhou. Some suggestion and comment from Professor B.J.Jiang and Y.Q.Wu make this note less uglily. I would also like to thank them.

Now we give some definitions, notations and basic facts of the geometry and topology of surface which will be used in the proof of the theorems.

1) Let F_g denote the oriented closed surface of genus $g, g\geqq 2$. f denotes an orientation preserving periodic map on F_g. p always denotes the order of f. F_1 denotes the tours.

2) Given F_g and f as in 1), there exists a hyperbolic metric on F_g such that f becomes an isometry.

3) f determines a group action $\langle f\rangle$ on F_g. The quotient space induced by this group action, denoted by Q, is a hyperbolic orbifold. The quotient map $q\colon F_g \to Q$ is a cyclic branched covering of degree p. The hyperbolic orbifold Q is determined by its genus and branch points $S=(x_1, \cdots, x_l)$ which are the image of singular points on the up stair under the quotient map q. Let $\bar{g}$ be the genus of Q, $\nu_1, \cdots, \nu_l$ be the indices of the branch points. Then we have the so called Riemann-Hurwitz formula:

$$\frac{2g-2}{p} = 2\bar{g} - 2 + \sum_{i=1}^{l} (1-\frac{1}{\nu_i}).$$

The cyclic branched covering $q\colon F_g \to Q$ is represented by its well-known Hurwitz homomorphism $h\colon \pi_1(Q-S) \to Z_p \subset S_p$. Here Z_p is the cyclic group of order p and S_p is the permutation group of p letters. h is surjective to Z_p.

Furthermore $\nu_1, \cdots, \nu_l$ and p satisfy two conditions. D condition: $\nu_i | p$, i.e.

ν_i is a divisor of p. LCM condition: If $\bar{g}=0$, then $l.c.m.\ (\nu_1,\cdots,\nu_i^{\vee},\cdots,\nu_l)=p$ for all i, here $\nu_i^{\vee}$ denotes the omission of ν_i. $l.c.m.$ denotes the least common multiple. (See [H]).

The maximum order of f on F_g is $4g+2$. (See [H])

Among the maps of finite order on F_1 which are not isotopic to identity, the maximum order is 6. (See [He]).

4) A circle C on F_g is a simple closed curve on F_g. C is essential if it does not bound a disk on F_g. C is an invariant circle of f, if $f(C)=C$. If an invariant circle C of f is essential, then we call C an invariant essential circle of f. C is nonseparating if F_g-C is connected.

Let C be an invariant circle of f. The hyperbolic metric is given as in 2). If C is essential, then there is a unique geodesic circle isotopic to C and this geodesic circle must be invariant under the isometry f. If C is not essential, then C is an equidistance curve to a fixed point of f. So we may assume that C is smooth.

Proof of theorem 1

We first prove several elementary lemmas.

Lemma 1: If $p>2$, C is an invariant circle of f on F_g, r is the order of the restriction of f on C, then $r=p$ and C does not pass any singular point.

Proof: Choose any point $y\in C$ and any short geodesic are A perpendicular to C at y. f^r fixes every point of C. As an orientation preserving isometry, the only choice for f^r is to fix every point on A; therefore f^r fixes an orthogonal frame on F_g. Hence f^r is the identity of F_g, i.e, $r=p$.

Let $x\in C$ be a singular point, then $f^w(x)=x$ for some $0<w<p$. We already knew that f^w is not identity on C. By Lefschetz fixed point theorem, f^w must be orientation reversing on C. Hence f itself is orientation reversing on C. It implies that the restriction of f on C has order 2. By the fact we proved in the last paragraph it implies $p=2$. It is a contradiction.

Lemma 2 (See [M]): Let $h:\pi_1(Q-S)\to Z_p$ be the representation of the branched covering induced by $q:F_g\to Q$ and $p>2$. If C is an invariant circle of f, then $\bar{C}=q(C)$ is a circle in $Q-S$ and $h(\bar{C})$ is a generator of Z_p. Conversely, if $\bar{C}$ is a circle in $Q-S$ and $h(\bar{C})$ is a generator of Z_p, then $q^{-1}(\bar{C})$ is an invariant circle of f.

Proof: By lemma 1, C dose not pass singular point, cyclic group $\langle f\rangle$ acts freely on C, so $q(C)$ is a circle in $Q-S$. Again by lemmal, C is p-fold covering of $\bar{C}$; therefore $h(\bar{C})$ must be a generator of Z_p.

On the other hand, let $\bar{C}$ be a circle in Q-S. $q^{-1}(\bar{C})$ is a union of circles which is invariant under f. If $h(\bar{C})$ is a generator of Z_p, then $q^{-1}(\bar{C})$ is just a circle.

Lemma 3: Let $p>2$. If C is an invariant circle of f, then C separates F_g if and only if $q(C)$ separates Q.

Proof: By above two lemmas, we have $C=q^{-1}(q(C))$. Then the lemma becomes obvious.

Lemma 4: As in lemma 2, C is an invariant circle of f and Q is a 2-sphere S^2, $\bar{C}$ is a circle in S^2 and $h(\bar{C})$ is a generator of Z_p. Then C is essential if and only if each component of S^2-$\bar{C}$ contains more then one branch point.

Proof: Fact 1: Let D and $\bar{D}$ be disks and $q:D \to \bar{D}$ be a cyclic branched covering of degree p. Let $x_1,\cdots,x_k$ be the branch points with order $\nu_1,\cdots,\nu_k,\nu_i \geqq 2$ on the downstairs. By counting Eular number, we have

$$\left(\sum_{i=1}^{k}\frac{p}{\nu_i}\right) - 1 = p(k-1).$$

Thus we have

$$k\frac{p}{2} > \left(\sum_{i=1}^{k}\frac{p}{\nu_i}\right) - 1 = p(k-1).$$

Hence we get

$$\frac{kp}{2} < p \; .$$

It holds only if $k=1$. This implies that there is only one branch point on the downstairs disk.

Fact 2: Let q: $M \to \bar{D}$ be a cyclic branched covering with only one branch point x on the downstairs disk. Then the Eular number of $\bar{D}$-x is 0. As a cover of $\bar{D}$-x, M-q^{-1}(x) must have Eular number 0. Hence M must be a disk.

Now, if C is not essential, then C bounds a dick D. This D cyclic branch covers one component of S^2-$\bar{C}$. By fact 1, there is only one branch point on this component of S^2-$\bar{C}$.

Conversely, if one component $\bar{D}$ of S^2-$\bar{C}$ contains only one branch point, then one component of F_g-C($\bar{C}$ separates S^2 implies C separates F_g by lemma 3) cyclic branch covers $\bar{D}$. By fact 2, this component is a disk. Hence C is not essential.

Proof of theorem 1: In the proof of this theorem, p, the order of f, is always prime. First we assume $p>2$.

Because p is prime, every branch point has order p. Remember $\bar{g}$ is the genus of Q. The Riemann-Hurwitz formula has the form

$$\frac{2g-2}{p} = 2\bar{g}-2+\sum_{i=1}^{l}(1-\frac{1}{p}) .$$

Let $h:\pi_1(Q-S)=\{a_i,b_i,c_j; i=1,\cdots,\bar{g}; j=1,\cdots,l \mid \prod_{i=1}^{\bar{g}}[a_i,b_i]\prod_{j=1}^{l}c_j=1\} \to Z_p$. Here a_i,b_i are standard generators of Q. c_j is represented by a loop C_j around the branch point x_j. In the coming context, we will use C_j to represent the circle, its class in the first homology H_1 and its class in π_1. The homomorphism $h:\pi_1(Q-S)\to Z_p$ which represents cyclic branched covering $q:F_g\to Q$ will send C_j to a generator of Z_p. (Every nonzero element of Z_p is a generator).

Case a): $\bar{g}\geq 1$. $h:\pi_1(Q-S)\to Z_p$ induces a surjective homomorphism $H_1(Q-S)\to Z_p$. If branch set S is empty, then there is a nonseparating circle $\bar{C}$ on $Q-S$ which is sent to a generator of Z_p by h. If S is not empty, we take any nonseparating circle $\dot{C}$. If $h(\dot{C})$ is a generator of Z_p, set $\bar{C}=\dot{C}$. Otherwise set $\bar{C}=\dot{C}\#C_1$. We have $h(\bar{C})= h(\dot{C})+h(C_1)=h(C_1)$ is a generator of Z_p. (See picture 1). By Lemma 2, $q^{-1}(\bar{C})=C$ is an invariant circle of f. By lemma 3, C is nonseparating. Hence in case 1, f has an invariant essential circle.

Case b): $\bar{g}=0$, i,e $Q=S^2$, a sphere.

Subcase 1: $l\geq 4$.

Let $h(c_j)=k_j\in Z_p$. $0<k_j<p$. Then $k_1+k_2\equiv 0 \bmod p$, $k_1+k_3\equiv 0 \bmod p$, k_3+k_2 $0 \bmod p$. can not all hold. Suppose $k_2+k_3\equiv 0 \bmod p$ does not hold. Set $\bar{C}=C_2\#C_3$ (See picture 2). Then $h(\bar{C})$ is a generator of Z_p, each component of $S_2-\bar{C}$ having more than one branch point. By lemma 4, $C=q^{-1}(\bar{C})$ is an invariant essential circle of f.

Subcase 2: $l\leq 3$.

Take any invariant circle C of f(if there is). $\bar{C}=q(C)$ is a circle in S^2. One component of $S^2-\bar{C}$ contains only one branch point. By lemma 4, C is not essental.

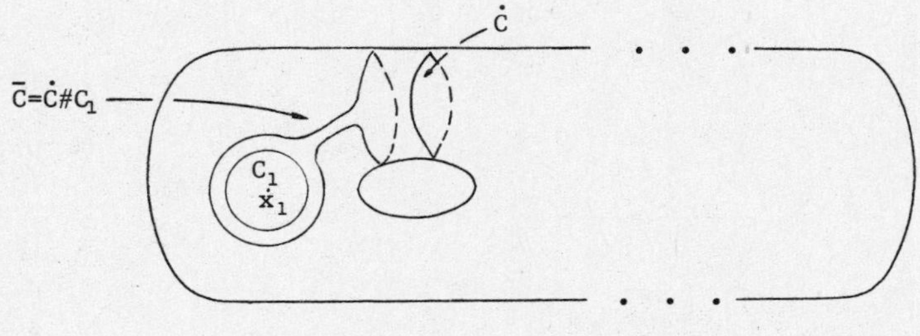

Figure 1

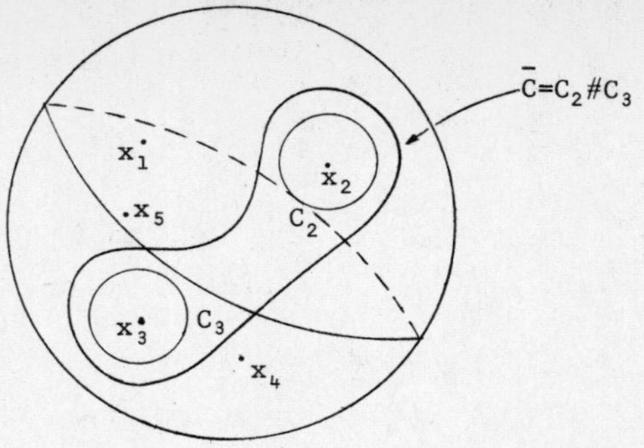

Figure 2

By discussion above, we see that f has no invariant essential circle if and only if the quotient orbifold Q is a sphere S^2 with no more then three branch points. We will prove Q is S^2 with no more then 3 branch point if and only if $p>g+1$. Hence our theorem will be proved when $p>2$.

We prove it by three steps.

Case 1): $\bar{g}\geqq 2$. We have

$$\frac{2g-2}{p} \geqq 2\times 2-2+\sum_{i=1}^{l}(1-\frac{1}{p}) \geqq 2.$$

This implies

$$p \leqq g-1.$$

Case 2): $\bar{g}=1$, i.e. $Q=F_1$. Because F_g, $g\geqq 2$ can not be a covering space of F_1, the branch set S of F_1 is not empty, i.e. $l\geqq 1$. If $l=1$, then $c^{-1}=a_1b_1a_1^{-1}b_1^{-1}$ and $h(c^{-1})=h(a_1b_1a_1^{-1}b_1^{-1})=0 \in Z_p$. This is a contradiction. Hence $l\geqq 2$. Now we have

$$\frac{2g-2}{p} = 2\times 1-2+\sum_{i=1}^{l}(1-\frac{1}{p}) = \frac{lp-l}{p}.$$

i.e.
$$2g-2 = l(p-1)$$

i.e.
$$p = \frac{2g-2}{l} + 1 \leqq g.$$

Case 3): $\bar{g}=0$. i.e. $Q=S^2$. Then

$$\frac{2g-2}{p} = -2 + \sum_{i=1}^{l}(1-\frac{1}{p}).$$

i.e.
$$2 + \frac{2g}{p-1} = l$$

and $l \geqq 4 \Leftrightarrow g \geqq p-1 \Leftrightarrow p \leqq g+1$.

We will finish the proof of theorem 1 by proving the following subtheorem.

Subthrorem: If f is of order 2, then f has an invariant essential circle.

Proof: If f is fixed point free, then $\langle f\rangle$ acts freely on F_g. $g \geqq 2$ implies $\bar{g} \geqq 2$ (Counting Eular number). Same argument as we did above shows that f has an invariant essential circle. (The purpose to set condition $p>2$ is to avoid C passing any singular point. Here fixed point free implies sigular point free.)

If f has fixed points, f has at least two fixed point y,z. (Again counting Eular number.) Choose a shortest arc A to connect them. A must be a geodisic arc. A $\cup$ f(A) is an invariant set of f. There are only finitely many intersection points between A and f(A). Let $\bar{A}$ be the subarc of A which connects y and x, x is the first intersection point between A and f(A) and after y. If $f(x) \neq x$, use the fact f fixes two ends y,z and f as an isometry preserves the length of arc, it is easy to prove there is a shorter ar c to connect y,z. It is a contridiction. So we have $f(x)=x$. Hence $C=f(\bar{A}) \cup \bar{A}$ is an invariant circle of f. If C bounds a disk, then C can be lifted to a circle in the hyperbolic plane which is formed by two geodesic arcs. It is impossible. Hence C is essential, i.e. f has an invariant essential circle-

Proof of theorem 2.

We need more lemmas.

Lemma 5: If $g>2$, there is no f of order $4g-2$ with fixed point on F_g. There is a map of order 6 with fixed point on F .

Proof: Suppose there is a f of order $4g-2$ with fixed point on F_g, $g>2$. By Riemann-Hurwitz formula, we have

$$\frac{2g-2}{4g-2} = 2\bar{g}-2+\sum_{i=1}^{l}(1-\frac{1}{\nu_i}).$$

We always assume $\nu_1 \leqq \nu_2 \leqq \cdots \leqq \nu_l$. Then the fixed point condition implies

$$\nu_l = 4g-2.$$

The D condition is $\nu_i | 4g-2$. The inequality

$$\frac{2g-2}{4g-2} < \frac{1}{2}$$

implies $\bar{g}=0$. So we have LCM condition

$$\text{l.c.m.}(\nu_1,\cdots,\check{\nu_i},\cdots\nu_l) = 4g-2.$$

and now we have

$$\frac{5g-3}{2g-1} = \sum_{i=1}^{l}(1-\frac{1}{\nu_i}).$$

Using the fact $\nu_l=4g-2$, we get

$$\frac{3}{2} = \sum_{i=1}^{l-1} (1-\frac{1}{\nu_i})$$

The possible value of l is 3 or 4. If $l=4$, then $\nu_1=\nu_2=\nu_3=2$, and the LCM condition is not satisfied. If $l=3$, then $\nu_1=3$, $\nu_2=6$ or $\nu_1=\nu_2=4$. If the first case happens, the LCM condition implies $g=2$. The second case is ruled out by the D condition.

Now we show that there is a map of order 6 with fixed point on F_2. We define the Hurwitz homomorphism

$$h\colon \pi_1(S^2-3\bar{D}) = (c_1,c_2,c_3 \mid c_1c_2c_3=1) \to Z_6$$

by $h(c_1)=h(c_2)=1$, $h(c_3)=-2$. Here $\bar{D}$ is the subdisk of S^2. Then h determines a $M-4D$ which is a convering of $S^2-3\bar{D}$ of degree 6. Here D is the subdisk of some surface M. The gerenator of deck transformation group can be extended to a map of order 6 on M with two fixed points. The Eular number of M is $(-1)\times 6+4=-2$. Hence M is F_2.

Lemma 6: There is no f of order $4g+1$ with fixed point on F_g, $g\geqq 2$.

Proof: Suppose there is a f of order $4g+1$ with fixed point on F_g, $g\geqq 2$. By Riemann-Hurwitz formula, we have

$$\frac{2g-2}{4g+1} = 2\bar{g}-2+\sum_{i=1}^{l} (1-\frac{1}{\nu_i}).$$

We always assume $\nu_1\leqq\nu_2\leqq\cdots\leqq\nu_l$. Then the fixed point condition implies

$$\nu_l = 4g+1.$$

The D condition is $\nu_i \mid 4g+1$. the inequality

$$\frac{2g-2}{4g+1} \leqq \frac{1}{2}$$

implies $\bar{g}=0$. So we have LCM condition

$$\text{l.c.m.}(\nu_1,\cdots,\check{\nu_i},\cdots\nu_l) = 4g+1.$$

and we get

$$\frac{10g}{4g+1} = \sum_{i=1}^{l} (1-\frac{1}{\nu_i}).$$

Using the fact $\nu_l=4g+1$ we get

$$\frac{3}{2} - \frac{3}{8g+2} = \sum_{i=1}^{l-1} (1-\frac{1}{\nu_i}).$$

The left side is between 1 and $\frac{3}{2}$, the right side is between $\frac{l-1}{2}$ and $l-1$. Hence $l=3$. Then we have

$$\frac{3}{8g+2} = \frac{1}{\nu_1} + \frac{1}{\nu_2} - \frac{1}{2}$$

The D condition implies that ν_i is odd. Now we have $\nu_1=3$ and $\nu_2<7$. (otherwise the right side is nagetive.) If $\nu_1=\nu_2=3$, then LCM condition implies that $4g+1=3$. If $\nu_1=3$ and $\nu_2=5$, then LCM condition implies that $4g+1=15$. Neither of them has integer solution.

Lemma 7: Any f on F_g of order p, $p>2g-2$, has no invariant nonseparating circle. There is a map on F_g of order $2g-2$ which has an invariant nonseparating circle.

Proof: We may assume $p>2$, otherwise $p\leqq 2g-2$. Suppose some f of order p has an invariant nonseperating circle. Let us wrute down Riemann-Hurwitz formula

$$J = \frac{2g-2}{p} = 2\bar{g}-2+\sum_{i=1}^{l}(1-\frac{1}{\nu_i}).$$

By lemma 3, we know $\bar{g}\geqq 1$. We first calculate J and then determine p.

Case 1) $\bar{g}>1$, then $J\geqq 2$.

Case 2) $\bar{g}=1$, then $l\geqq 2$ by the proof of case 2 of theorem 1. Now we have

$$J = \frac{2g-2}{p} = \sum_{i=1}^{l}(1-\frac{1}{\nu_i}) \geqq (1-\frac{1}{\nu_1})+(1-\frac{1}{\nu_2}) \geqq \frac{1}{2}+\frac{1}{2} = 1$$

That is $\frac{2g-2}{p} = J \geqq 1$. Hence $p\leqq 2g-2$.

Now we will construct an example f of order $2g-2$ which has an invariant nonsaperating circle on F_g for each $g\geqq 2$. For concreteness, we set $g=4$. The way to generalize to arbitrary $g\geqq 2$ is obvious.

We thought F_4 as the boundary of a wheel with three disjoint symmetry axes shown in picture 3:

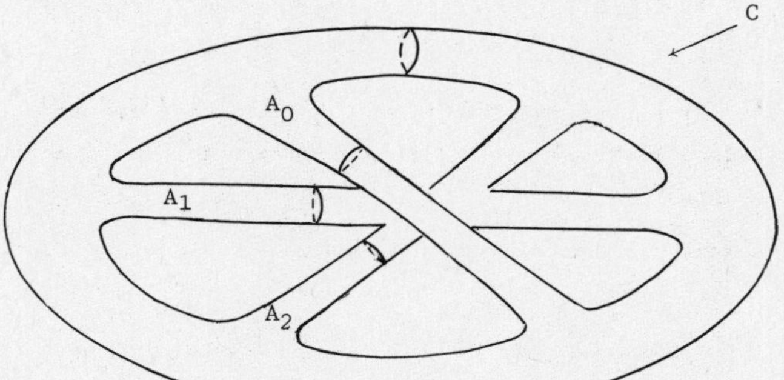

Figure 3

The map f of order $2g-2$ is formed by first rotating the wheel with angle

$\frac{2\pi}{2g-2} = \frac{\pi}{3}$ then sending A_0 to A_1, A_1 to A_2 and A_2 to A_0. It is easy to see the circle C on the wheel is nonseparating and invariant under f.

Lemma 8: For every $g>0$, there is a map of order $4g+2$ with fixed point.

Proof: It is derived easily from the work of W.J.Harvey. However we would like to give an elementary and constructive proof. For concreteness, we set $g=2$. The way to generalize it to arbitrary $g>0$ is obvious.

We thought the fundamental octagon of F_2 as the union of two regular pentagons which lies in the plane shown as the following picture 4. The oriented edges presented by same letter are identified. All corner points are identified to a point.

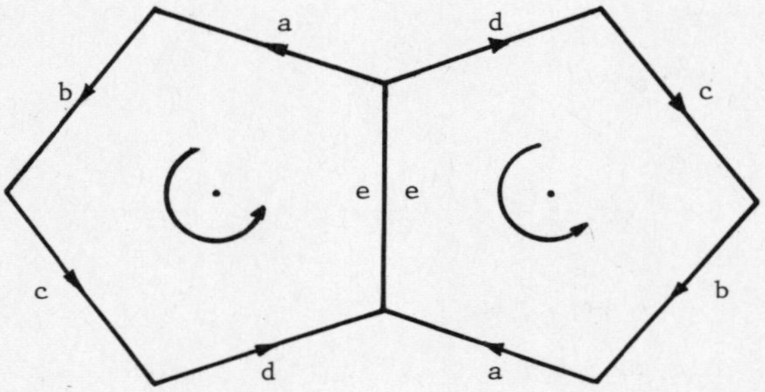

Figure 4

Let η be the map of order 2×2+1 whose restriction on each regular pentagon is a self rotation of angle $\frac{2\pi}{5}$. Let τ be a rotation of the plane with angle π which switch the two regular pentagons. The composition of these two maps is a map of order $4g+2=10$. It induces a map on F_2 with the same order and one fixed point (The image of corner points).

Proof of theorem 2: First Suppose f of order p has an invariant essential circle C. We prove that $p \leqq 0(g)$.

If C is nonseperating, by lemma 7, $p \leqq 2g-2 \leqq 0(g)$ for any $g \geqq 2$. So we may suppose C is separating. Let $F_g - C = F^\circ_{g_1} \cup F^\circ_{g_2}$. Here $F^\circ_{g_i} = F_{g_i} - D_i$, D_i is a disk. i=1,2. $g=g_1+g_2$. Each component is invariant under f. Let f°_i be the restriction of f on $F^\circ_{g_i}$. F°_i can be extended to F_{g_i}, denoted by f_i, in the standard way. f_i is again an orientation preserving periodic map of order p and f_i has a fixed point (the center of disk D_i).

We assume $g_1 \leqq g_2$.

Case 1) $g=2k$, then $g_1 \leq k$. So $p \leq 4g_1+2 \leq 4k+2=2g+2=0(g)$.

Case 2) $g=2k+1$. $g>3$. then $g_1 \leqq k$.

Subcase 1) $g_1 \leqq k-1$, then $p \leq 4g_1+2 \leq 4k-4+2=4k-2<4k=2g-2$.

Subcase 2) $g_1=k>1$, then $g_2=k+1>2$. First we have $p \leq 4g_1+2=4k+2\ 2g$. So the possible values of p, $p>2g-2$, are $2g$ and $2g-1$. If $p=2g$, then f_2 is a map of order $p=2g=4k+2=4g_2-2$ on F_{g_2} with fixed points. This possibility is ruled out by lemma 5. If $p=2g-1$, then f_1 is a map of order $p=2g-1=4g_1+1$ on F_{g_1} with fixed points. This possibility is ruled out by lemma 6. Therefore $p \leqq 2g-2=0(g)$.

Case 3) $g=3$. then $g_1=1$ and f_1 is a map of order p on F_1 which is not isotopic to identity. So $p \leqq 6$.

Now we construct a map of order $0(g)$ which has an invariant essential circle.

If $g=2k$, let f be the map of order $4k+2=2g+2$ on F_k with fixed point provided by lemma 8. f has a fixed point implies that f has an invariant disk D. Take the double of F_k-D, we get $F_{2k}=F_g$. The action of $\langle f \rangle$ on each copy of F_k-D can be matched up to give a f' on F_g of the same order $2g+2$. $C=\partial D$ is an invariant essential circle.

If $g=$ odd, $g>3$, then lemma 7 provides the required map.

If $g=3$, take the map of order 6 with fixed point on F_1 provided by lemma 8 and the map of order 6 with fixed point on F_2 provided by lemma 5, then do the same surgery as in the case $g=$ even.

We have finished the proof of theorem 2.

Reference:

[FLP] A.Fathi F.Laudenbach V.Poenaru : Travaux de Thurston sur les surface. *Aserisque* 66-67, 1979.

[H] W.J.Harvey : Cyclic group of automorphisms of a compact Riemann surface. *Quart. J. Math. Oxford*, Ser.(2) 17(1966), 86-97.

[He] J.Hempel : *3-Manifold*, Princeton University Press 1976.

[M] W.H.Meeks : Circle invariant under diffeomorphisms of finite order. *J. Diff. Geom.* 14(1979), 377-383.

[ZVG] H.Zieschang E.Vogt H.D.Goldewey : *Surfaces and Planar Discontinuous Groups*. Lecture Notes in Math. 835, Springer-verlag 1980.

JONES POLYNOMIAL AND THE CROSSING NUMBER OF LINKS

Wu Ying-qing
Peking University

This paper is based on Kauffman's paper [1], to which the reader is referred for definitions of state, universe, bracket polynomial, simple projection, and other concepts and notations which are not defined in this paper.

Let L be an oriented link and K a projection diagram of L. Denote by $\langle K\rangle$ the bracket polynomial of K. The bracket polynomial is related to the Jones polynomial $V_L(t)$ of L in the form

$$V_L(t^4) = (-t)^{-3w(K)}\langle K\rangle(t^{-1}),$$

where $w(K)$ is the twist number of K defined in [1]. For a Laurent polynomial $f(x)$, define Span $f = \max\deg f(x) - \min\deg f(x)$. Thus Span $\langle K\rangle = 4$ Span V_L. Theorem 2.10 of [1] can be rewritten as

Theorem 1 (Kauffman). If K is a simple alternating diagram of L, then Span $V_L = v(K)$, where $v(K)$ is the crossing number of the diagram K.

Also contained in [1] is a proof of the following

Theorem 2 (Murasugi [2] and Thistlethwaite [3]). If K is an arbitrary diagram of a link L, then Span $V_L \leq v(K)$.

A diagram is called simple if it is reduced, connected and not a diagram sum. The following theorem is also due to Murasugi and Thistlethwaite:

Theorem 3 ([2], [3]). If K is a simple non-alternating diagram of a link L, then Span $V_L < v(K)$.

These theorems have many important corollaries. They give solutions to some outstanding classical conjectures in knot theory, and also a partial solution to the additivity conjecture for crossing number of knots. One is referred to [2] or [3] for details.

In this paper we will continue Kauffman's work to give a simple proof of Theorem 3. First we need some definitions.

Let S be a state of a universe U. An edge E of S is called a special edge, if the two vertices $\{P_1, P_2\} = \partial E$ are not alternating, i.e. they have splitting markers as shown in Fig. 1. A region A of S is called a special region if it contains a special edge and there are not two edges of A on the boundary of a region other than A. The state S is called a special state if it is connected (i.e. the underlying universe U is connected) and contains a special region.

Proof of Theorem 3.

Let S be the state obtained by splitting every crossing in the simple non-alternating diagram K along the A-direction. Being simple, K is not a diagrammatic connected sum (i.e. no two regions of S can have two edges in common). So by definition, S is a special state. In his proof of Theorem 2, Kauffman showed that

Figure 1

$$\max \deg\langle K\rangle \le v(S) + 2(|S| - 1),$$
$$\min \deg\langle K\rangle \ge -v(S) - 2(|S^\wedge| - 1).$$

Therefore,

$$4\text{Span}\, V_L = \text{Span}\langle K\rangle \le 2v(S) + 2(|S| + |S^\wedge|) - 4,$$

where $v(S)$, the number of vertices in the underlying universe U of S, is equal to $v(K)$, the crossing number of K. The theorem now follows from the following Lemma.

Lemma. If S is a special state, then

$$|S| + |S^\wedge| \le v(S).$$

Proof. It is easily seen that the lemma is true for $v(S) = 2$. (The only possible S with $v(S) = 2$ is shown in Fig. 2). Thus we assume $v(S) > 2$ and suppose that the lemma is true for every special state S_1 with $v(S_1) < v(S)$. It is implied in the proof of Lemma 2.11 in [1] that if S' is a connected state resulting from splitting S at a vertex P (the splitting is not necessarily according to the state marker), then $|S| + |S^\wedge| \le |S'| + |S'^\wedge| + 1$. So by induction we need only to show that we can split S at a suitable vertex in such a way that the new state S' is still a special state.

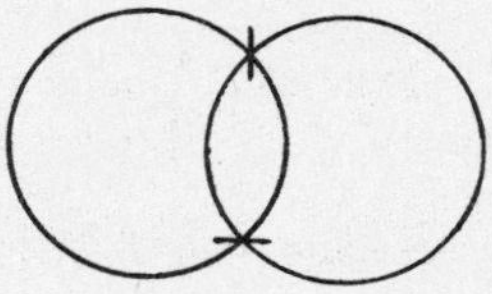

Figure 2

Let A be a special region of S and E a special edge on A with $\partial E = \{P_1, P_2\}$. If A has more than two vertices, we choose $P \ne P_1$, P_2 on A and split S at P in such a way that the region A remains "unchanged" (cf. Fig. 3). The new state S' is connected (otherwise A will not be a special region) and the region A' in S' is a special region because the regions neighbouring to the edges of A' are still different. Hence S' is a special state.

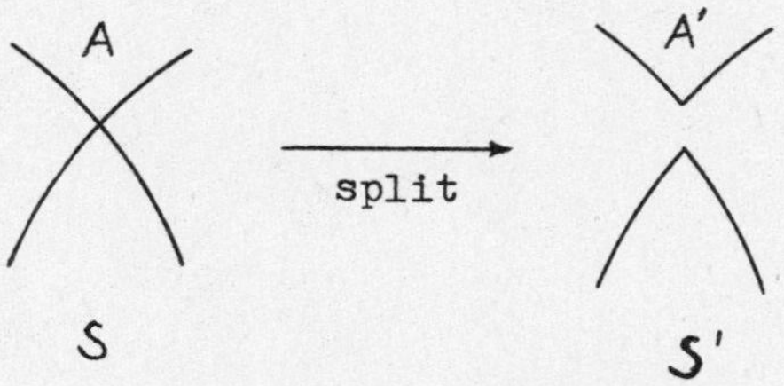

Figure 3

Now suppose A has only two vertices P_1, P_2. There are two regions B_1 and B_2 neighbouring to A, one of which, say B_1, must have more than two vertices because S is connected and $v(S) > 2$. Choose $P \ne P_1, P_2$ on B_1. Splitting at P, we obtain two possible states S_1 and S_2 (cf. Fig. 4).

If S_1 is connected, then S_1 is a special state since the two regions neighbouring to A' are still different. If S_1 is not connected, then S_2 is connected and the region C in the Figure is

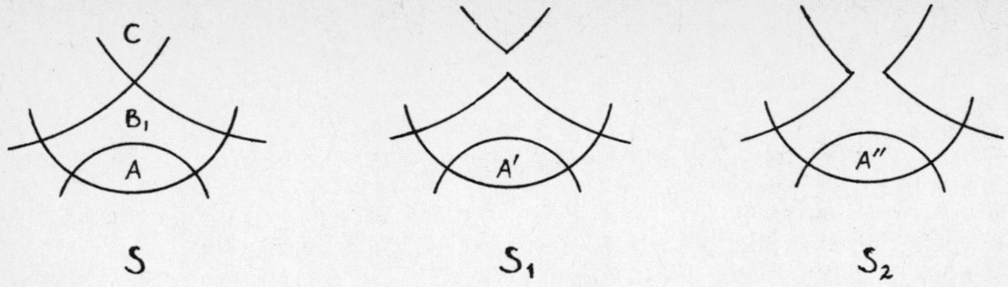

Figure 4

different from the region B_2. Thus the two regions neighbouring to A'' are different and hence S_2 is a special state. This completes the proof of the lemma, as well as that of Theorem 3.

I should like to thank Professor Boju Jiang for his help and guidance in my study period, and thank Professor Louis H. Kauffman for sending me his works, without which this paper would never have been written.

References

1. L.H.Kauffman, State models and the Jones polynomial, Topology 26(1987), 395–407.

2. K.Murasugi, Jones polynomials and classical conjectures in knot theory, I and II, Topology 26(1987), 187–194.

3. M.Thistlethwaite, A spanning tree expansion of the Jones polynomial, Topology 26(1987), 297–309.

On Complete Minimal Surfaces With Parallel and Flat Ends

Xiao Liang
Nankai Institute of Mathematics and Graduate School of
University of Science and Technology of China

Abstract

In this paper we have proved the non-existence of minimal surfaces with 5 flat ends and presented some new examples of minimal surfaces with parallel and flat ends.

§1. Introduction

A complete minimal surface M in R^3 with finite total curvature is conformally equivalent to a compact Riemann surface $\bar{M}$ from which r points, $p_1, \cdots, p_r$ have removed, $r \geq 1$. Moreover the Gauss mapping, defined on M as a meromorphic function, extends to $\bar{M}$. This is a theorem of Osserman [3]. In this paper we are only concerned with this kind of surfaces with genus 0.

In paper [2], Jorge ang Meeks proved the following theorem.

Theorem 1 [2] Let M be an embedded complete minimal surface with finite total curvature in R^3. Then the ends of M satisfy the following:

(1) All of the ends are parallel;

(2) Let K_+ and K_- be respectively the numbers of ends that normal vectors have same directions, then

$$K_+ = K_-, \quad \text{if} \quad K_+ + K_- = \text{even};$$

$$|K_+ - K_-| = 1, \quad \text{if} \quad K_+ + K_- = \text{odd}.$$

We called this kind of surfaces pseudo-embedded minimal surfaces, which are well studied in papers [1], [2] and [4]. Jorge and Meeks proved in paper [2] that they can have one or two ends and non-existence of pseudo-embedded minimal surfaces having 3, 4 and 5 ends. In paper [1], Peng gave one example with 6 ends. Xiao mentioned in paper [4] that a similar situation holds for $K > 6$, but he presented no explicit examples. Now we can provide expricit examples having $4m + 2$ ends.

Let M be the complete minimal surface with finite total curvature and embedded ends. After a suitable rotation of the coordinates, each end of M can be written as

$$X_3 = a \log(X_1^2 + X_2^2) + b + \frac{cX_1 + dX_2}{X_1^2 + X_2^2} + 0(|X|^{-2})$$

for suitable constans a, b, c and d [5].

Definition 1 [1]. Let M be complete minimal surfaces with finite total curvature and embedded ends. The end E_i is called flat if $a = 0$ at E_i and M flat-type if all the ends of M are flat.

In paper [1], Peng proved that there exist flat-type minimal surfaces in R^3 with K ends where K is any integer other than 2,3,5 and 7. He also proved that $K = 2, 3$ cannot occur. We are now to prove that $K = 5$ cannot occur either.

Here I wish to express my gratitude to Professor Peng C.K. for his encouragement.

§2. Some examples of complete minimal surfaces with parallel ends

The principal method for constructing complete minimal surfaces is the formula of Enneper-Weierstrass:

Let D be a Riemann surface, f an analytic function on D, and g a meromorphic function on D. Suppose further that g has a pole of order m at $z \in D$ iff f has zero of order $2m$ at z. From the C^3-valued function

$$\Phi(z) = (f/2)(1-g^2, i(1+g^2), 2g) = (\Phi_1, \Phi_2, \Phi_3)$$

Then

$$X(z) = Re \int_{z_0}^{z} \Phi(z)dz \tag{1}$$

is a regular conformal minimal immersion, well-defined on the universal covering space of D. (If the components of Φdz have no real periods, then X is well defined on D.) Furthermore, g is the stereographic projection of the Gauss normal mapping of X.

It can be derived from (1) that the metric on D is given by $\lambda^2 = |f^2|(1+|g^2|)^2/2$. To show that a minimal surface is a complete one, it is sufficient to establish that for every divergent curve γ on D

$$\int_\gamma |f|(1+|g|^2)dt = \infty.$$

Thus we can translate the geometrical properties of minimal surfaces in R^3 into analytical data. Now we will list some without proof.

(1) M is immersion iff $\{\Phi_j\}$ have no common zero points.

(2) M has finite curvature iff $\{\Phi_j\}$ are the mereomorphic functions on S^2.

(3) The ends of M are embedded iff $\{\Phi_j\}$ have poles whose orders are at most 2.

(4) M is regular (i .e ., single valued) iff $\{\Phi_j\}$ have no real periods for all $\Gamma \in H_1(M, Z)$.

From these relations it is not difficult to see that if M is a complete minimal surface with finite curvature and conformal to $S^2 - \{P_1, P_2, \cdots, P_k\}$, then the ends of M are embedded if and only if total curvature $C(M) = -4\pi(k-1)$.

Furthermore, if M is pseudo-embedded, then after a suitable rotation of the coordinates, we can assume

$$\begin{aligned} g &= \frac{Q}{P}, \\ f &= \frac{P^2}{\prod_{j=1}^{K_+}(z-a_j)^2 \prod_{i=1}^{K_-}(z-b_i)^2}, \\ K_+ &+ K_- = K, \end{aligned} \tag{2}$$

where P and Q are polynomials satisfying

$$\max(\deg P, \deg Q) = K - 1, \tag{3}$$

$$\begin{aligned} g(a_j) &= \frac{Q(a_j)}{P(a_j)} = 1 \quad j = 1, \cdots, K_+, \\ g(b_i) &= \frac{Q(b_i)}{P(b_i)} = -1 \quad i = 1, \cdots, K_-. \end{aligned} \tag{4}$$

Substituting it into Weierstrass representation (1), we have

$$\Phi_1 = \frac{1}{2} \frac{P^2 - Q^2}{\prod(z-a_j)^2 \prod(z-b_i)^2}, \tag{5}$$

$$\Phi_2 = \frac{1}{2} \frac{P^2 + Q^2}{\prod(z-a_j)^2 \prod(z-b_i)^2}, \tag{6}$$

$$\Phi_3 = \frac{PQ}{\prod(z-a_j)^2\prod(z-b_i)^2}. \tag{7}$$

It follows from (4) that $P(a_j) = Q(a_j)$ and $P(b_i = -Q(b_i)$. Hence

$$\begin{aligned} P - Q &= G\prod(z-a_j), \\ P + Q &= H\prod(z-b_i), \end{aligned} \tag{8}$$

where G and H are polynomials. Substituting (8) into the expression of Φ_j we have

$$\Phi_1 = \frac{1}{2}\frac{GH}{\prod(z-a_j)\prod(z-b_i)}, \tag{9}$$

$$\Phi_2 = \frac{1}{4}\left[\frac{G^2}{\prod(z-b_i)^2} + \frac{H^2}{\prod(z-a_j)^2}\right], \tag{10}$$

$$\Phi_3 = \frac{1}{4}\left[\frac{G^2}{\prod(z-b_i)^2} - \frac{H^2}{\prod(z-a_j)^2}\right]. \tag{11}$$

Since Φ_1, Φ_2, Φ_3 do not have real periods, it follows that

$$\frac{G^2}{\prod(z-b_i)^2}dz \tag{12}$$

and

$$\frac{H^2}{\prod(z-a_j)^2}dz \tag{13}$$

are exact, i.e., have no periods.

From (3) and (4) it is easily seen that

$$\deg G \leq K_- - 1 \text{ and } \deg H \leq K_+ - 1 \tag{14}$$

(at least one of these inequalities is equality),

$$\begin{aligned} G(b_i) &\neq 0 \quad i = 1, \cdots, K_-, \\ H(a_j) &\neq 0 \quad j = 1, \cdots, K_+. \end{aligned} \tag{15}$$

It follows that

$$\frac{G}{\prod(z-b_i)} = \sum_{i=1}^{K_-}\frac{d_i}{z-b_i}, \tag{16}$$

$$\frac{H}{\prod(z-a_j)} = \sum_{j=1}^{K_+}\frac{c_j}{z-a_j}, \tag{17}$$

and

$$\sum_{i=1}^{K_-} d_i \neq 0 \quad \text{or} \quad \sum_{j=1}^{K_+} c_j \neq 0. \tag{18}$$

Lemma 1 [1] If

$$\left[\sum_{i=1}^{m}\frac{A_i}{z-z_i}\right]^2 dz$$

has no periods, then

$$\sum_{\substack{j=1 \\ j\neq i}} \frac{A_j}{z_j - z_i} = 0 \quad i = 1, \cdots, m.$$

For any numbers $z_1, \cdots, z_m \in C$, $z_i \neq z_j (i \neq j)$, we define $A(z_1, \cdots, z_m)$ as the matrix (a_{ij}) where

$$a_{ij} = \begin{cases} \frac{1}{z_i - z_j} & i \neq j, \\ 0 & i = j. \end{cases}$$

The matrix A is well studied in paper [4]. If $(x_1, \cdots, x_m)$ is a solution of A (i.e. $(x_1, \cdots, x_m)A$ 0), then after a linear fractional transformation $L = \frac{az+b}{cz+d}(cz_i \neq -d$ for any $i)$, $(\frac{x_1}{cz_1+d}, \cdots, \frac{x_m}{cz_m+d})$ is a solution of $A(Lz_1, \cdots, Lz_m)$. It is easy to see that the solution space is invariant under Euclidean motions and scale transformations of complex plane. If σ is a permutation of n elements, $\sigma(x_1, \cdots, x_m)$ is a solution of $A(\sigma(z_1, \cdots, z_m))$.

Let $c = e^{\frac{2\pi}{2m+1}\sqrt{-1}}$. It is easily seen that $(1, c^m, \cdots, c^{2m^2})$ is a solution of $A(c, c^m, \cdots, c^{2m+1})$. Thus $(1, \frac{c+1}{c^2+1}c^m, \cdots)$ is a solution of $A(i\frac{c-1}{c+1}, \cdots, 0)$. Let $a_i = \sqrt{-1}\frac{c^i-1}{c^i+1}$, $M = \max\{a_i\}$, $b_i = a_i - M - 1$, $A_i = \frac{c^{(i-1)m}(c+1)}{c^i+1}$. We can easily see that $\sum A_i \neq 0$. Next we take

$$G = (\sum \frac{A_i}{z - b_i}) \prod (z - b_i),$$

$$H = (\sum \frac{A_i}{z - a_j}) \prod (z - a_j).$$

Lemma 1 implies that $\Phi_2 dz, \Phi_3 dz$ do not have real periods. That all the a_j, b_j and A_j are real implies $\Phi_1 dz$ has not real periods. Now we have already given explicit examples of pseudoembedded minimal surfaces which have $4m \pm 2$ ends.

Remark We know that there exist pseudo-embedded minimal surfaces of genus 0 which have $4m \pm 1, 4m$ ends, but we can't given explicit examples.

§3. Non-existence of flat-typed minimal surfaces of genus 0 with 5 ends

Lemma 1 [1] Let M be conformal to $S^2 - \{P_1, \cdots, P_k\}$. Then M is a flat-typed minimal surface if and only if $\Phi_j, j = 1, 2, 3$ satifying

(1) $\{\Phi_j\}$ have the poles whose order ≤ 2,

(2) $\{\Phi_j\}$ are exact, i.e., have no periods.

Lemma 2 [1] $\Phi_j dz$ $j = 1, 2, 3$ are exact if and only if $f dz, fg dz, fg^2 dz$ are exact.

Since M has finite total curvature and embedded ends, we can assume

$$g = \frac{Q}{P}$$

and

$$f = \frac{P^2}{\prod(z - a_j)^2}$$

where P and Q are polynomials without common zero points and $\max(\deg P, \deg Q) = k - 1$.

From lemma 2 it follows that the exactness of $\Phi_j dz$ is equivalent to that of the following holomorphic differentials

$$\frac{P^2}{\prod(z - a_j)^2} dz, \quad \frac{Q^2}{\prod(z - a_j)^2} dz$$

and

$$\frac{PQ}{\prod(z - a_j)^2} dz.$$

And this is equivalent to the exactness of the differentials

$$\frac{(aP+bQ)^2}{\prod(z-a_j)^2}dz \quad \forall a, \quad b \in C.$$

Since P and Q have no common zero point, we can find two $(c_m, d_m)m = 1, 2$ such that $c_mP + d_mQ$ and $\prod(z - a_i)$ have no common zero point and $c_1d_2 - c_2d_1 \neq 0$. This implies the dimension of solution space of $A(a_1, \cdots)$ is at least 2.

For any $c_1, c_2, d_1, d_2, c_1d_2 - c_2d_1 \neq 0$, let $P_1 = c_1P + d_1Q, Q_1 = c_2P + d_2Q$

$$f_1 = \frac{P_1^2}{\prod(z-a_j)^2}dz \qquad g_1 = \frac{Q_1}{P_1}$$

$$X = (Re\frac{1}{2}\int f_1(1-g_1^2), Re\frac{1}{2}\int f_1(1+g_1^2), Re\int f_1g_1)$$

is also a flat-typed minimal surface. Using this method we can construct many new examples.

In paper [1], Peng gave many examples of flat-typed minimal surfaces.

For K =even, we assume $\prod(z - a_j) = z(z^{2n+1} - 1), P = z^{n+1}, Q = z^{2n+1} + \frac{n}{n+1}$.

For K =odd, we assume $\prod(z-a_j) = z(z^n-1)(z^n-\lambda), n \geq 4, P = z^m(z^n-c), 2 \leq m \leq n-1$, $n+1 \neq 2m$

$$Q = (z^n - a)(z^n - b)$$

$\{lambda, a, b, c\}$ satisfying

$$m-1+\frac{n}{1-c} = \frac{n-1}{2}+\frac{n}{1-\lambda},$$
$$m-1+\frac{n\lambda}{\lambda-c} = \frac{n-1}{2}+\frac{n}{\lambda-1}$$

and

$$\frac{n}{1-a}+\frac{n}{1-b} = 1+\frac{n-1}{2}+\frac{n}{1-\lambda},$$
$$\frac{n}{\lambda-a}+\frac{n}{\lambda-b} = 1+\frac{n-1}{2}+\frac{n}{\lambda-1}.$$

Now we prove the non-existence for $K = 5$.

If there exists a flat-typed minimal surface for $K = 5$, then there exist $z_1, \cdots, z_5$ *s.t.* rank $A(z_1, \cdots, z_5) \leq 3$. By anti-symmtry of A, rank $A = 2$. If (c_1, c_2, c_3) is a solution of $A(z_1, z_2, z_3)$, then

$$\frac{c_1}{z_i-a_1}+\frac{c_2}{z_i-a_2}+\frac{c_3}{z_i-a_3} = 0 \quad i = 4, 5.$$

Case 1: All z_i are concyclic.

It is easy to prove that all confactor $A_{ii} \neq 0$. Hence rank $A(z_1, \cdots, z_5) = 4$

Case 2: The other case

It is easy to prove that rank A is invariant under a linear fractional transformation. Without loss of generalities, we may assume $z_1 = -1, z_2 = 0, z_3 = 1$. Then $z_4 = \frac{1}{\sqrt{3}}, z_5 = -z_4$. But rank $A(z_1, \cdots, z_5) = 4$

This leads to a contradiction. Thus we have

<u>Theorem</u> There do not exist flat-typed minimal surfaces with 5 ends.

<u>Remark</u> Similarly we can discuss the existence of flat- typed minimal surfaces with 7 ends.

References

[1] Peng C.K. Some new examples of minimal surfaces in R^3 and its applications, MSRI 07510-85.

[2] L.P.Jorge and W.H.Meeks The topology of complete minimal surfaces of finite total Gaussian curvature. Topology, 22(1983) 203-221.

[3] R.Osserman Assurvey of minimal surfaces, Van Nostrand Reinhold, New York, (1969).

[4] Xiao L. Some Results on pseudo-embedded minimal surfaces in R^3. To appear in Acta Mathematica Sinica.

[5] R.Schoen Uniqueness, symmetry, and embeddedness of minimal surfaces, J.Differential Geometry, 18(1983), 791-809.

REGULARITY OF HARMONIC MAPS INTO CERTAIN HOMOGENEOUS SPACES

Xin Yuan-Long
Institute of Mathematics,
Fudan University, China.

1. Introduction

In [SU1] R.Schoen and K.Uhlenbeck gave us a rather general regularity theorem. Then they obtained more concrete results when the target manifold is Euclidean sphere [SU2]. The key point is to treat stablity inequality. In the case of the sphere the variation cross-sections are conformal vector fields along the image of the map. When the sphere is embedded canonically in the Euclidean space the conformal vector field can be expressed as the gradiant vector fields of the height functions. Thus, the result in [SU2] can be generalized to the submanifolds in the Euclidean space as was done in [x]. On the other hand, conformal vector fields in the sphere can also be viewed as the gradiant vector fields of the eigenfunctions for the first nonzero eigenvalue of the Laplace operator in the sphere.

In the present paper we consider the case when target manifolds are compact irreduceble homogeneous spaces. The average process in this case for second variation can also be done. Thus we obtained a regularity condition involving the scalar curvature, the first eigenvalue of the target manifold and the dimensions of the both manifolds which generalized a result in [SU1]. This is Theorem 3 of this paper.

We also consider some related nonexistence of harmonic maps from complete noncompact manifolds into compact irreduceble homogeneous spaces in Section 4 of this paper.

This work was carried out during the author visiting Nankai Institute of Mathematics. He would like to express his sincere thanks for its hospitallity.

2. Preliminaries

Let M and N be Riemannian manifolds of dimension m and n respectively, $\phi:M\to N$ a harmonic map which is a critical point of the energy functional

$$E(\phi) = \int_M e(\phi)*1 = \frac{1}{2}\int_M <\phi_* e_i, \phi_* e_i>*1 , \qquad (2.1)$$

where $\{e_i\}$ is a local orthonormal frame field in M. Here and in the sequel we use the summation convention. In this paper we agree the following range of indices:

$$1 \leqq i,j, \ldots \leqq m,$$
$$1 \leqq s,t, \ldots \leqq m-1,$$
$$1 \leqq \alpha,\beta, \ldots \leqq n.$$

A harmonic map ϕ is called stable if its index is zero corresponding to the index form [EL]

$$I(v,v) = \int_M \langle -\nabla^2 v - R^N(\phi_* e_i, v)\phi_* e_i,\ v\rangle *1 \tag{2.2}$$

for $v \in T(\phi^{-1}TN)$ with compact support, where ∇^2 stands for trace Laplace operator on vector bundle $\phi^{-1}TN$ over M.

As is known, $d\phi$ is a harmonic 1-form with values in $\phi^{-1}TN$. From the Weitzenböck formula it follows that

$$(\nabla^2 d\phi)(x) = \phi_* \mathrm{Ric}\ x - R^N(f_* e_i, f_* x) f_* e_i \tag{2.3}$$

for any $x \in TM$. By using (2.3) the following Bochner type formula can be derived

$$\Delta e(\phi) = |\nabla\phi|^2 + \langle \phi_* \mathrm{Ric}\ e_i, \phi_* e_i\rangle - \langle R^N(\phi_* e_i, \phi_* e_j)\phi_* e_i, \phi_* e_j\rangle \tag{2.4}$$

Let f be a smooth function on M. Then for a local orthonormal frame field $\{e_i\}$

$$\begin{aligned} v &= \nabla f = (\nabla_{e_i} f) e_i \\ \nabla_{e_i} v &= \mathrm{Hess}(f)(e_i, e_j) e_j \\ \nabla^2 v &= \nabla\Delta f + \mathrm{Ric}\ v \end{aligned} \tag{2.5}$$

3. A regularity result

Let f be an eigenfunction in N corresponding to the first nonzero eigenvalue λ of Laplace operator on N, namely

$$\Delta f + \lambda f = 0$$

Take cross-section in vector bundle $\phi^{-1}TN$ over M

$$v = u\nabla f,$$

where u is any function on M with compact support. we have

$$\nabla_{e_i} v = (\nabla_{e_i} u)(\nabla_{\varepsilon_\alpha} f)\varepsilon_\alpha + u\langle \phi_* e_i, \varepsilon_\beta\rangle (\nabla_{\varepsilon_\beta} \nabla_{\varepsilon_\alpha} f)\varepsilon_\alpha, \tag{3.1}$$

where $\{\varepsilon_\alpha\}$ is a local orthonormal frame field with $\nabla_{\varepsilon_\alpha} \varepsilon_\beta = 0$ at a given point $y_\circ \in N$. Thus

$$\begin{aligned} |\nabla_{e_i} v|^2 = {} & |\nabla u|^2 (\nabla_{\varepsilon_\alpha} f)(\nabla_{\varepsilon_\alpha} f) + u^2 \langle \phi_* e_i, \varepsilon_\beta\rangle\langle \phi_* e_i, \varepsilon_\gamma\rangle (\nabla_{\varepsilon_\beta}\nabla_{\varepsilon_\alpha} f)(\nabla_{\varepsilon_\gamma}\nabla_{\varepsilon_\alpha} f) \\ & + 2u(\nabla_{e_i} u)\langle \phi_* e_i, \varepsilon_\beta\rangle (\nabla_{\varepsilon_\alpha} f)(\nabla_{\varepsilon_\beta}\nabla_{\varepsilon_\alpha} f) \end{aligned} \tag{3.2}$$

Substituting (3.2) into (2.2) yields

$$I(u\nabla f,u\nabla f) = \int_M\{|\nabla u|^2(\nabla_{\varepsilon_\alpha} f)(\nabla_{\varepsilon_\alpha} f)+u^2(\langle f_* e_i,\varepsilon_\beta\rangle\langle f_* e_i,\varepsilon_\gamma\rangle(\nabla_{\varepsilon_\beta}\nabla_{\varepsilon_\alpha} f)(\nabla_{\varepsilon_\gamma}\nabla_{\varepsilon_\alpha} f)$$

$$- \langle R^N(\phi_* e_i,\varepsilon_\alpha)\phi_* e_i,\varepsilon_\beta\rangle(\nabla_{\varepsilon_\alpha} f)(\nabla_{\varepsilon_\beta} f))$$

$$+ 2u(\nabla_{e_i} u)\langle\phi_* e_i,\varepsilon_\beta\rangle(\nabla_{\varepsilon_\alpha} f)(\nabla_{\varepsilon_\beta}\nabla_{\varepsilon_\alpha} f)\}*1, \qquad (3.3)$$

which is valid for any $f\varepsilon E_\lambda$, the eigenspace of dimension ℓ for the first nonzero eigenvalue λ. Let $f^1,\ldots, f^a,\ldots, f^\ell$ be an orthonormal basis in E_λ with respect to L^2 norm on E_λ. In the case of compact irreducible homogeneous space N=G/H, G has a natural action on E_λ given by $g\cdot f=f\cdot g^{-1}$ for $f\varepsilon E_\lambda$, $g\varepsilon G$. Furthermore, since the Laplaian Δ commutes with isometry, this action leaves eigenspace E_λ invariant and preserves the L^2 norm. With respect to the orthonormal basis f^a, the action of each $g\varepsilon G$ is expressed by an orthogonal matrix $O_{ab}\varepsilon SO(\ell)$ such that

$$g\cdot f^a = O_{ab}f^b \qquad (3.4)$$

Now we consider a symmetric bilinear form $(\nabla f^a)(\nabla f^a)$ defined on $T(G/H)$ by

$$(\nabla f^a)(\nabla f^a)(x,y) = (\nabla_x f^a)(\nabla_y f^a)$$

for any $x,y\varepsilon T(G/H)$. By (3.4) this is G-invariant. It defines a endmorphism $A: T(G/H)\to T(G/H)$ as follows

$$(\nabla_x f^a)(\nabla_y f^a) = \langle AX,Y\rangle$$

It is easy to see that A is symmetric and any eigenspace of A with respect to the Riemannian metric on T(G/H) is G-invariant. From the irreducibility of H it follows that $A=C\cdot id$ for constant $C>0$. Thus, we have

$$(\nabla_x f^a)(\nabla_y f^a) = C\langle X,Y\rangle. \qquad (3.5)$$

Hence we have

$$(\nabla_{\varepsilon_\alpha} f^a)(\nabla_{\varepsilon_\beta}\nabla_{\varepsilon_\alpha} f^a) = 0 \qquad (3.6)$$

Similarlly,

$$(\nabla_x\nabla_{\varepsilon_\alpha} f)(\nabla_y\nabla_{\varepsilon_\alpha} f) = \bar{C}\langle X,Y\rangle \qquad (3.7)$$

From (3.6),(3.7) and (2.5) it follows that

$$\begin{aligned}\bar{C} &=- \frac{1}{n}(\nabla_{\varepsilon_\beta}\nabla_{\varepsilon_\beta}\nabla_{\varepsilon_\alpha} f^a)(\nabla_{\varepsilon_\alpha} f^a)\\ &=- \frac{1}{n}\langle\nabla\Delta f^a+\mathrm{Ric}\nabla f,\varepsilon_\alpha\rangle\langle\nabla_{\varepsilon_\alpha} f^a\rangle\\ &=- \frac{1}{n}(-\lambda+\frac{s}{n})(\nabla_{\varepsilon_\alpha} f^a)(\nabla_{\varepsilon_\alpha} f^a)\end{aligned}$$

$$= C(\lambda - \frac{s}{n}), \tag{3.8}$$

where s is the scalar curvature of G/H.

(3.3) defines a quadratic form on E_λ. From (3.3),(3.5),(3.6),(3.7)and (3.8) we have

$$\text{trace } I = C\int_M (n|\nabla u|^2 + u^2(\lambda - \frac{2s}{n})|d\phi|^2)*1 \tag{3.9}$$

Therefore, we have the stability inequality

$$\int_M (-u\Delta u - \frac{1}{n^2}(2s-n\lambda)u^2|d\phi|^2)*1 \geqq 0 \tag{3.10}$$

for any stable harmonic map from M into any compact irreducible homogeneous space G/H, where u is any function on M with compact support. To prove the reqularity it suffices to show any minimizing tangent map $\phi: \mathbb{R}^m \to G/H$ is constant. For any tangent map ϕ there is an associate harmonic map $f_1 : \to S^{m-1}$ G/H. (3.10) becomes in this case

$$\int_{S^{m-1} \times (0,\infty)} (-\bar{\Delta}u - \frac{1}{n^2}(2s-n\lambda)u|df_1|^2 - r\frac{2\partial^2 u}{\partial r^2} - r(m-1)\frac{\partial u}{\partial r})r^{m-3}u*1 \geqq 0, \tag{3.11}$$

where $\bar{\Delta}$ denotes Laplacian on S^{m-1}. We then consider a strongly elliptic operator on S^{m-1}

$$L_1 = \bar{\Delta} + \frac{1}{n^2}(2s-n\lambda)|df_1|^2$$

and ordinary differential operator on $(0,\infty)$

$$L_2 = r^2\frac{d^2}{dr^2} + r(m-1)\frac{d}{dr} .$$

The eigenvalues of L_1 and L_2 are

$$\mu_1 \leqq \mu_2 \leqq \ldots \leqq \mu_i \leqq \to \infty$$

and

$$\delta_1 \leqq \delta_2 \leqq \ldots \leqq \delta_i \leqq \to \infty ,$$

respectively. As the way done in [SI], the stability condition becomes

$$\mu_1 + \delta_1 \geqq 0 . \tag{3.12}$$

By a direct computation

$$\delta_1 = \frac{(m-2)^2}{4} . \tag{3.13}$$

On the other hand, by using the Bochner type formula (2.4) we can estimate μ_1 as follows. From (2.4) it follows

$$\frac{1}{2}\bar{\Delta}|df_1|^2 \geqq |\tilde{\nabla} df_1|^2 + (m-2)|df_1|^2 - \frac{k(m-2)}{m-1} |df_1|^4, \tag{3.14}$$

where k is the upper bound of the sectional curvature of G/H.

noting

$$\frac{1}{2}\bar{\Delta}|df_1|^2 = |df_1|\bar{\Delta}|df_1| + |\nabla|df_1||^2$$

$$\leq |df_1|\bar{\Delta}|df_1| + |\tilde{\nabla}df_1|^2 ,$$

(3.14) becomes

$$|df_1|\bar{\Delta}|df_1| \geq (m-2)|df_1|^2 - \frac{k(m-2)}{m-1}|df_1|^4. \tag{3.15}$$

Let

$$\phi_\varepsilon = (|df_1|^2 + \varepsilon)^{\frac{1}{2}}.$$

We have then

$$\phi_\varepsilon\bar{\Delta}\phi_\varepsilon \geq (m-2)|df_1|^2 - \frac{k(m-2)}{m-1}|df_1|^4$$

and

$$-\phi_\varepsilon\bar{\Delta}\phi_\varepsilon - \frac{k(m-2)}{m-1}\phi_\varepsilon^2|df_1|^2 \leq -(m-2)|df_1|^2 \tag{3.16}$$

If

$$k\,\frac{m-2}{m-1} < \frac{2s-n\lambda}{n^2}$$

and f_1 is not a constant map, then from (3.16) we have

$$\mu_1 \leq \inf \frac{\int_{S^{m-1}}(-\bar{\Delta}\phi_\varepsilon - \frac{1}{n^2}(2s-n\lambda)|df_1|^2\phi_\varepsilon)\phi_\varepsilon *1}{\int_{S^{m-1}}\phi_\varepsilon^2 *1}$$

$$< \inf \frac{\int_{S^{m-1}}(-\bar{\Delta}\phi_\varepsilon - k\frac{m-1}{m-2}|df_1|^2\phi_\varepsilon)\phi_\varepsilon *1}{\int_{S^{m-1}}\phi_\varepsilon^2 *1}$$

$$\leq 2-m . \tag{3.17}$$

(3.12),(3.13) and (3.17) imply m>6. Set

$$d(n) = \begin{cases} 2 & \text{when } \dfrac{kn^2}{kn^2-(2s-n\lambda)} \leq 1 , \\ 5 & \text{when } \dfrac{kn^2}{kn^2-(2s-n\lambda)} = 5 , \\ [\min(1+\dfrac{kn^2}{kn^2-(2s-n\lambda)}, 6)] & \text{in other cases,} \end{cases}$$

where [•] denotes the greatest integer in a number.

The above discusion show that $m \leq d(n)$ forces f_1 to be constant. By the regularity theorem of Schoen-Uhlenbeck [SU1], we have the following result.

Theorem 3 Let M be a compact Riemannian manifold of dimension m, G/H a compact irreducible homogeneous space of dimension n. If $m \leq d(n)$, then every minimizing map from M into G/H is smooth in the interior of M. If m=d(n)+1, such a map has at most isolated singularities and in general the singular set is a closed set of Hausdorff dimension at most m-d(n)-1.

4. Some nonexistence results

Let M be a compact Riemannian manifold, f a smooth function on M, N any Riemennian manifold. let ϕ: M→N be a harmonic map. Take a cross-section $\phi_*\nabla f$ in the induced vector bundle $f^{-1}TN$ over M. From (2.5) it follows that

$$\nabla_{e_i}(\phi_*\nabla_{e_i}\nabla f) = (\tilde{\nabla}_{e_i}d\phi)(\nabla_{e_i}\nabla f) + \phi_*\nabla^2\nabla f$$

$$= \mathrm{Hess}(f)(e_i,e_j)(\nabla_{e_i}d\phi)e_j + \phi_*(\nabla\Delta f) + \phi_*\mathrm{Ric}\nabla f$$

$$= \mathrm{Hess}(f)(e_i,e_j)B_{e_ie_j}(\phi) + \phi_*(\nabla\Delta f) + \phi_*\mathrm{Ric}\nabla f$$

Thus, by using the Weitzenböck formula (2.3)

$$-\nabla^2\phi_*\nabla f = -(\nabla^2 d\phi)\nabla f - 2\nabla_{e_i}(\phi_*\nabla_{e_i}\nabla f) + \phi_*(\nabla^2\nabla f)$$

$$= R^N(f_*e_i,\phi_*\nabla f)f_*e_i - 2\phi_*\mathrm{Ric}\nabla f - \phi_*(\nabla\Delta f) - 2\mathrm{Hess}(f)(e_ie_j)B_{e_ie_j}(\phi) \tag{4.1}$$

Substituting (4.1) into (2.2) gives

$$I(\phi_*\nabla f,\phi_*\nabla f) = \int_M\{-2\langle\phi_*\mathrm{Ric}\nabla f,\phi_*\nabla f\rangle - 2\mathrm{Hess}(f)(e_ie_j)(\nabla_{e_k}f)\langle B_{e_ie_j}(\phi),\phi_*e_k\rangle$$

$$-\langle\phi_*\nabla\Delta f,\phi_*\nabla f\rangle\}*1 \tag{4.2}$$

If f is an eigenfunction of Laplace operator on M with respect to the nonzero first eigenvalue λ, then (4.2) becomes

$$I(\phi_*\nabla f,\phi_*\nabla f) = \int_M\{\lambda\langle\phi_*\nabla f,\phi_*\nabla f\rangle - 2\mathrm{Hess}(f)(e_ie_j)(\nabla_{e_k}f)\langle B_{e_ie_j}(\phi),\phi_*e_k\rangle$$

$$-2\langle\phi_*\mathrm{Ric}\nabla f,\phi_*\nabla f\rangle\}*1 \tag{4.3}$$

In addition, ϕ is relative affine immersion and $\mathrm{Ric}>\frac{\lambda}{2}$, then we have

$$I(\phi_*\nabla f,\phi_*\nabla f) = \int_M\{\lambda\langle\phi_*\nabla f,\phi_*\nabla f\rangle - 2\langle\phi_*\mathrm{Ric}\nabla f,\phi_*\nabla f\rangle\}*1<0$$

Thus, we obtain

Theorem 4.1 Let M be a compact Riemannian manifold with $\mathrm{Ric}>\frac{\lambda}{2}$, where λ is the first nonzero eigenvale of Laplace operator on M, N any Riemannian manifold, f: M→N a relative affine harmonic immersion. Then f can not be stable.

Remark By Lichnerowicz' theorem $\mathrm{Ric}>\frac{\lambda}{2}$ means m=dim M>2.

If M=G/H is compact irreducible homogeneous space we can treat (4.3) as the same as before and obtain

$$\mathrm{trace}\ I = C\int_M(\lambda-\frac{2s}{n})|d\phi|^2*1 \tag{4.4}$$

Thus, the following results in [HW] and [O] can be obtained immediately:

Theorem 4.2 Let G/H be a compact irreducible n-dimensional homogeneous space with the nonzero first eigenvalue $\lambda<\frac{2s}{n}$, where s is the scalar curvature of G/H. Then any stable harmonic map f from G/H into any Riemannian manifold has to be constant. Furthermore, if G/H is symmetric space it can be one of the following spaces:

i) Simply connected compact Lie group of type $A_n (n\geq 2)$, B_2 and $C_n (n\geq 3)$;

ii) $SU(2n)/Sp(n)$ $(n\geq 3)$;

iii) S^n $(n\geq 3)$;

iv) $Sp(p+q)/Sp(p)\times Sp(q)$ $(1\leq p\leq q,\ p+q\geq 3)$;

v) E_6/F_4;

vi) $P^2(cay)=F_4/Spin(9)$

A Riemannian manifold is called strongly parabolic if it adimits no nonconstant positive superharmonic function. As is known, $\mathbb{R}^2$ is strongly parabolic as well as a minimal surface with finite total curvature in Euclidean space and a minimal surface in Euclidean space which is an entire graph, while $\mathbb{R}^n (n\geq 3)$ is not. In fact, dimension plays no role here, and the decisive criterion is the rate of volume growth. L.Karp [k] introduced that a complete noncompact Riemannian manifold has moderate volume growth if there is $F\in\mathcal{F}$ such that $\limsup_{r\to\infty} \frac{1}{r^2F(r)}\operatorname{vol} B_r(x_0)<\infty$ for some $x_0\in M$, here $\mathcal{F}=\{F: (0,\infty)\to(0,\infty); F$ is increasing on $(0,\infty)$ and $\int_1^\infty \frac{dr}{rF(r)}=+\infty\}$ and $B_r(x_0)$ is a geodesic ball of radius r and centered at x_0 in M. He also showed that if M has moderate volume growth then it is strongly parabolic.

Thus we have the following result.

Theorem 4.3 If M is a compact or complete noncompact Riemannian manifold with moderate volume growth and G/H is n-dimensional compact irreducible homogeneous space with $\lambda<\frac{2s}{n}$, where λ is the first eigenvalue of Laplace operator of G/H and s is its scalar curvature. Then any stable harmonic map ϕ from M into G/H has to be constant.

Proof If M is compact without boundary by choosing $u\equiv 1$ in (3.10) ϕ has to be constant which is the result of [HW] and [O].

If M is complete and noncompact, we consider a strongly elliptic operator

$$L = \Delta + \frac{1}{n^2}(2s-n\lambda)|d\phi|^2$$

on any domain $D\subset M$ with $\bar{D}$ compact. Let μ be the first eigenvalue of L in D with Dirichlet boundary condtion. From (3.10) it follows that

$$\mu = \inf \frac{\int_D -uLu *1}{\int_D u^2 *1} = \inf \frac{\int_D(-u\Delta u - \frac{1}{n^2}(2s-n\lambda)u^2|d\phi|^2)*1}{\int_D u^2 *1}$$

$$\geqq 0$$

By using a theorem in [FCS] there is a positive solution $u>0$ to $Lu=\Delta u+\frac{1}{n^2}(2s-n\lambda)\cdot$ $\cdot|d\phi|^2u=0$ in M, namely

$$\Delta u = -\frac{1}{n^2}(2s-n\lambda)|d\phi|^2 u \leqq 0, \tag{4.5}$$

which means that u is a positive superharmonic function on M. Since M has moderate volume growth which is strongly parabolic, u has to be constant. Thus (4.5) implies $|d\phi|\equiv 0$ Q.E.D.

For compact irreducible symmetric space, the first eigenvalue λ of Laplace operator has been computed [N]. Thus, we have

Theorem 4.4 Let M be a manifold as that in Theorem 4.3 and N be one of the following spaces

i) Simply connected compact Lie group of type An ($n\geqq 2$), B_2 and C_n ($n\geqq 3$);
ii) $SU(2n)/Sp(n)$ ($n\geqq 3$);
iii) S^n ($n\geqq 3$);
iv) $Sp(p+q)/Sp(p)\times Sp(q)$ ($1\leqq p\leqq q$, $p+q\geqq 3$);
v) E_6/F_4 ;
vi) $p^2(cay) = F_4/Spin(9)$.

Then any stable haumonic map ϕ: $M\to N$ has to be constant.

We also can do L^P estimate for the energy density of harmonic maps into compact irreducible homogeneous space as was done in [SSY][SU2].

Theorem 4.5 Let M be a complete Riemannian manifold with Ricci curvature bounded below by a nonpositive constant $-A(A\geqq 0)$ and G/H a n-dimensional compact irreduciple homogeneous space with sectional curvature bounded above by a positive constant k. Let ϕ: $M\to G/H$ be a stable harmonic map with rank $f\leqq r$. If $r\leqq\frac{n^2k}{n^2k-(2s-n\lambda)}$ then for any nonnegative function u with compact support in M the following inequality is valid

$$\int_M |d\phi|^4 u^4 *1 \leqq C\int_M (A^2u^4 + |\nabla u|^4) *1, \tag{4.6}$$

where C is a constant depending on n, m, s and λ.

Proof Replacing u by $|d\phi|u$ in (3.10) gives

$$\frac{1}{n^2}(2s-n\lambda)\int_M |d\phi|^4u^2 *1 \leqq \int_M [u^2|\nabla|d\phi||^2 + |d\phi|^2|\nabla u|^2 + 2u|d\phi|\nabla u\cdot\nabla|d\phi|] *1 \tag{4.7}$$

From (2.4) it follows that

$$\tfrac{1}{2}\Delta|d\phi|^2 \geqq |\nabla d\phi|^2 - A|d\phi|^2 - \tfrac{r-1}{r}k|d\phi|^4$$

and

$$|d\phi|\Delta|d\phi| = |\nabla d\phi|^2 - |\nabla|d\phi||^2 - A|d\phi|^2 - \tfrac{r-1}{r}k|d\phi|^4 \tag{4.8}$$

By a computation in [SU2]

$$|\nabla d\phi| - |\nabla|d\phi||^2 \geqq \frac{1}{2mn}|\nabla|d\phi||^2 \tag{4.9}$$

Substituting (4.9) into (4.8), we have

$$|d\phi|\Delta|d\phi| \geqq \frac{1}{2mn}|\nabla|d\phi||^2 - A|d\phi|^2 - \tfrac{r-1}{r}k|d\phi|^4 \tag{4.10}$$

Multiplying (4.10) by u^2 and then integrating it over M we obtain

$$\begin{aligned}\frac{1}{2mn}\int_M u^2|\nabla|d\phi||^2{*1} &\leqq -\int_M u^2|\nabla|d\phi||^2{*1} + \int_M A|d\phi|^2u^2{*1}\\ &\quad + \int_M \tfrac{r-1}{r}k|d\phi|^4u^2{*1} - 2\int_M u|d\phi|\nabla|d\phi|\cdot\nabla u{*1}\end{aligned} \tag{4.11}$$

By adding (4.7) and (4.11)

$$\begin{aligned}\frac{1}{2mn}\int_M u^2|\nabla|d\phi||^2{*1} &\leqq \int_M (|d\phi|^2|\nabla u|^2 + A|d\phi|^2u^2){*1}\\ &\quad + \left(\tfrac{r-1}{r}k - \tfrac{1}{n^2}(2s-n\lambda)\right)\int_M|d\phi|^4u^2{*1}\\ &\leqq \int_M(|d\phi|^2|\nabla u|^2 + A|d\phi|^2u^2){*1}\end{aligned} \tag{4.12}$$

By using the Cauchy inequality for any $\varepsilon>0$

$$2u|d\phi|\nabla u\cdot\nabla|d\phi| \leqq \varepsilon u^2|\nabla|d\phi||^2 + \varepsilon^{-1}|d\phi|^2|\nabla u|^2.$$

Thus (4.7) becomes

$$\tfrac{1}{n^2}(2s-n\lambda)\int_M|d\phi|^4u^2{*1} \leqq (1+\varepsilon)\int_M u^2|\nabla|d\phi||^2{*1} + (1+\varepsilon^{-1})\int_M|d\phi|^2|\nabla u|^2{*1} \tag{4.13}$$

Substituting (4.12) into (4.13) and then replacing u by u^2 we have

$$(2s-n\lambda)\int_M|d\phi|^4u^4{*1} \leqq C_1\int_M(4|d\phi|^2u^2|\nabla u|^2 + A|d\phi|^2u^4){*1}, \tag{4.14}$$

where C_1 is dependent on m and n. By using the Cauchy inequality again we have then for any $\varepsilon>0$

$$\begin{aligned}|d\phi|^2u^2|\nabla u|^2 &\leqq \tfrac{\varepsilon}{2}u^4|d\phi|^4 + \tfrac{\varepsilon^{-1}}{2}|\nabla u|^4\\ A|d\phi|^2 &\leqq \tfrac{\varepsilon}{2}|d\phi|^4 + \tfrac{1}{2}\varepsilon^{-1}A^2\end{aligned} \tag{4.15}$$

Thus (4.14) becomes (4.6).

Q.E.D.

Corollary 4.6 Let M be a complete Riemannian manifold with nonnegative Ricci curvature and volume of geodesic ball $B_R(x_\circ)$ in M satisfying

$$\frac{\text{vol } B_R(x_\circ)}{R^4} \to 0$$

as R going to infinity. Let G/H and ϕ: $M \to G/H$ be as the same as the above theorem. Then f has to be constant. So does f in the case when M is Euclidean space of dimension 4.

Proof From (4.6) we now have

$$\int_M |d\phi|^4 u^4 *1 \leq C\int_M |\nabla u|^4 *1 \tag{4.16}$$

Choose a cut off function

$$u = \begin{cases} 1 & \text{in } B_{R/2}(x_\circ) \\ 0 & \text{out of } B_R(x_\circ) \end{cases}$$

with $|\nabla u| \leq \frac{C'}{R}$. Then from (4.16) it follows that

$$\int_{B_{R/2}(x_\circ)} |d\phi|^4 *1 \leq \int_M |d\phi|^4 u^4 *1 \leq \frac{C''}{R^4} \text{ vol } B_R(x_\circ) \to 0$$

as R goes to infinity. This mean that ϕ is constant.

When $M = \mathbb{R}^4$ by choosing.

$$u(x) = \begin{cases} 1 & \text{in } B_R(0) \\ \dfrac{\log\left(\frac{R^2}{|x|}\right)}{\log R} & \text{for } R \leq |x| \leq R^2 \\ 0 & \text{out of } B_{R^2}(0) \end{cases}$$

in (4.16) we have

$$\int_{B_R(0)} |d\phi|^4 *1 \leq C(\log R)^{-3} .$$

Letting R go to infinity we complete the proof of the corollary.

REFERENCES

[EL] J. Eells and L. Lemaire : "Selected topics in harmonic maps" CBMS n.50, 1983.

[FCS] D. Fischer-Colbrie and R. Schoen : "The structure of complete stable minimal surfaces in 3-manifolds of non-negative scalar curvature" Comm. Pure Appl. Math. 33(1980), 199-211.

[HW] R. Howard and S.W. Wei : "Nonexistence of stable harmonic maps to and from certain homogeneous spaces and submanifolds of Euclidean space" Trans. A.M.S. 294(1) (1986), 319-331.

[K] L. Karp : "Subharmonic function, harmonic mappings and isometric immersions" Seminar on differential geometry, Ann. Math. Studies 102, Princeton Univ. Press.

[N] T. Nagano : "Stability of harmonic maps between symmetric spaces" Lecture Notes Math. 949,Springer-Verlag. (1982), 130-137.

[O] Y. Ohnita : "Stability of harmonic maps and standard minimal immersions" Tôhoku Math. J. 38(1986), 259-267.

[SI] J. Simons : "Minimal varietes in Riemannian manifolds" Ann. Math. 88(1968), 62-105.

[SSY] R. Schoen, L. Simon and S.T. Yau : "Curvature estimates for minimal hypersurfaces" Acta Math. 134(1975),275-288.

[SU1] R. Schoen and K. Uhlenbeck : "A regularity theory for harmonic maps" J. Diff. Geom. 17(1982), 307-335-

[SU2] R. Schoen and K. Uhlenbeck : "Regularity of minimizing harmonic maps into the sphere" Invent. Math. 78(1984), 89-100.

[X] Y.L. Xin : "Liouville type theorems and regularity of harmonic maps" Proc. DD6 Symp. Lecture Notes Math. 1255. Springer-Verlag (1987).

On Infinitesimal Deformations of Surfaces in E^3

Yang Wenmao†

§0. Introduction

In 1876, *O.Bonnet*[1] studied the isometric deformations of surfaces in E^3, which preserve mean curvature H, and showed that the surfaces with H =constant admit such deformations. This study was continued by *W.C.Graustein*[2] (1924), *E.Cartan*[3] (1942) and others. Later, in 1985, *S.S.Chern*[4] studied this problem. Chern's formulation was taken up by *I.M.Roussos*[5] (1986) to get further results. Afterwards, the present author generalized the above concept and defined BII-isometry (resp. BIII-isometry), which preserves the second fundamental form (resp. the third fundamental form) and two principal curvatures $\kappa_i, i = 1,2$ (for simplicity, we call them O. Bonnet deformations in short), and got some results[6],[7]. Up on this very base, in the present paper, he continues his study of every kind of infinitesimal BI, BII, BIII-isometry, and gets and necessary and sufficient conditions for surfaces to admit non- trivial infinitesimal BI, BII, BIII-isometry (see Theorems 1-4).

The author would like to thank Professor S.S.Chern for his constant help and Nankai Institute of Mathematics for their hospitality during my stay there.

§1. Infinitesimal deformation of surface

Set a surface M in E^3 into a one-parametric family of surfaces $\{M_t\}(|t| \le \delta, \delta > 0)$, $M_0 = M, x(t)e_1(t)e_2(t)e_3(t)$ is the orthonormal frame field of $M_t, x(0) = x, e_i(0) = e_i, xe_1e_2e_3$ is the orthonormal frame field of M, and in E^3 into a one-parametric family of surfaces $\{M_t\}(|t| \le \delta, \delta > 0)$, $M_0 = M, x(t)e_1(t)e_2(t)e_3(t)$ is the orthonormal frame field of M_t, $x(0) = x, e_i(0) = e_i, xe_1e_2e_3$ is the orthonormal frame field of M, and

$$dx(t) = \omega_i(t)e_i(t), \quad i = 1,2 \tag{1}$$

$$de_i(t) = \omega_{ij}(t)e_j(t), \quad i,j = 1,2,3 \tag{2}$$

where $\omega_i(0) = \omega_i$ and $\omega_{ij}(0) = \omega_{ij}$ are 1-forms of M. Set

$$\omega_i(t) = \omega_i + t\varphi_i + \cdots, \quad i = 1,2 \tag{3}$$

$$\omega_{ij}(t) = \omega_{ij} + t\varphi_{ij} + \cdots, \quad i,j = 1,2,3 \tag{4}$$

We call $\phi = \{\varphi_1, \varphi_2, \varphi_{12}, \varphi_{13}, \varphi_{23}\}$ an infinitesimal deformation of M, satisfying the following integlable conditions introduced from the structure equations of M_t:

$$\omega_1 \wedge \varphi_{13} + \varphi_1 \wedge \omega_{13} + \omega_2 \wedge \varphi_{23} + \varphi_2 \wedge \omega_{23} = 0; \tag{5}$$

$$d\varphi_1 = -\omega_2 \wedge \varphi_{12} - \varphi_2 \wedge \omega_{12}, \quad d\varphi_2 = \omega_1 \wedge \varphi_{12} + \varphi_1 \wedge \omega_{12}; \tag{6}$$

$$d\varphi_{12} = -\omega_{13} \wedge \varphi_{23} - \varphi_{13} \wedge \omega_{23}; \tag{7}$$

$$d\varphi_{13} = -\varphi_{23} \wedge \omega_{12} - \omega_{23} \wedge \varphi_{12}, \quad d\varphi_{23} = \varphi_{13} \wedge \omega_{12} + \omega_{13} \wedge \varphi_{12}. \tag{8}$$

Under the deformation, the invariations of three fundamental forms and mean curvature and Gauss curvature of the surface are respectively given by

$$\frac{1}{2}\delta I = \omega_1\varphi_1 + \omega_2\varphi_2, \tag{9}$$

† This work is supported by the National Foundations of Science.

$$\delta II = \omega_1\varphi_{13} + \varphi_1\omega_{13} + \omega_2\varphi_{23} + \varphi_2\omega_{23}, \tag{10}$$

$$\frac{1}{2}\delta III = \omega_{13}\varphi_{13} + \omega_{23}\varphi_{23}, \tag{11}$$

$$\begin{aligned} 2\delta H * 1 =& \omega_{13} \wedge \varphi_2 + \varphi_{13} \wedge \omega_2 - \omega_{23} \wedge \varphi_1 - \varphi_{23} \wedge \omega_1 \\ & - 2H(\omega_1 \wedge \varphi_2 + \varphi_1 \wedge \omega_2), \end{aligned} \tag{12}$$

$$\delta K * 1 = \omega_{13} \wedge \varphi_{23} + \varphi_{13} \wedge \omega_{23} - K(\omega_1 \wedge \varphi_2 + \varphi_1 \wedge \omega_2) \tag{13}$$

$$*1 = \omega_1 \wedge \omega_2.$$

If $\delta I = 0$ (resp. $\delta II = 0, \delta III = 0$), we call ϕ I (resp. II, III) -isometry. If again $\delta\kappa_i = 0$ (or $\delta H = 0, \delta K = 0$), we call ϕ O.Bonnet I (resp. II, III,) -isometry. We shall denote the O.Bonnet isometry by $BI, BII, BIII$-isometry. If infinitesimal deformation ϕ is both I- and II-isometry, we call ϕ a trivial or a rigid deformation of surface M. If a surface M admits a non-trivial infinitesimal BI-isometric (resp. BII, BIII-isometric) ϕ, we call M is a Bonnet I-surface or simply BI-surface (resp. BII, BIII-surface).

Finally, let us consider the possible changes of the frames. Now set

$$\begin{aligned} e_1^* =& e_1(t)\cos(\rho + t\sigma + \cdots) + e_2(t)\sin(\rho + t\sigma + \cdots) \\ e_2^*(t) =& - e_1(t)\sin(\rho + t\sigma + \cdots) + e_2(t)\cos(\rho + t\sigma + \cdots) \\ e_3^*(t) =& e_3(t) \end{aligned} \tag{14}$$

where

$$\begin{aligned} \cos(\rho + t\sigma + \cdots) &= \cos\rho - t\sin\sigma + \cdots \\ \sin(\rho + t\sigma + \cdots) &= \sin\rho + t\cos\sigma + \cdots \end{aligned} \tag{15}$$

Substituting (14) and (15) into (1), (2), and analogous equations $(1)^*, (2)^*$, we have

$$\begin{gathered} \omega_1^* = \omega_1\cos\rho + \omega_2\sin\rho, \ \omega_2^* = -\omega_1\sin\rho + \omega_2\cos\rho \\ \omega_{12}^* = \omega_{12} + d\rho \\ \omega_{13}^* = \omega_{13}\cos\rho + \omega_{23}\sin\rho, \ \omega_{13}^*\sin\rho + \omega_{23}\cos\rho \end{gathered} \tag{16}$$

and

$$\begin{aligned} \varphi_1^* =& \sigma(-\omega_1\sin\rho + \omega_2\cos\rho) + \varphi_1\cos\rho + \varphi_2\sin\rho \\ \varphi_2^* =& - \sigma(\omega_1\cos\rho + \omega_2\sin\rho) - \varphi_1\sin\rho + \varphi_2\cos\rho \\ \varphi_{12}^* =& \varphi_{12} + d\sigma \\ \varphi_{13}^* =& \sigma(-\omega_{13}\sin\rho + \omega_{23}\cos\rho) + \varphi_{13}\cos\rho + \varphi_{23}\sin\rho \\ \varphi_{23}^* =& - \sigma(\omega_{13}\cos\rho + \omega_{23}\sin\rho) - \varphi_{13}\sin\rho + \varphi_{23}\cos\rho. \end{aligned} \tag{17}$$

We call ω_1, ω_2 the coframe of the metric

$$I = (\omega_1)^2 + (\omega_2)^2 \tag{18}$$

and ω_{12} the connection form associated to I, which is determined by the structure equations

$$d\omega_1 = -\omega_2 \wedge \omega_{12}, \quad d\omega_2 = \omega_1 \wedge \omega_{12}. \tag{19}$$

We shall denote the complex structure of (18) by

$$\omega = \omega_1 + i\omega_2, \quad i^2 = -1.$$

Thus (19) becomes

$$d\omega = i\omega \wedge \omega_{12}. \tag{20}$$

We need the following lemma about changes of the coframe.

Lemma 1 If the coframe undergoes the transformations (i), (ii) and (iii), then the associated connection forms are respectively given as follows:

(i) if $\omega^* = \bar{\omega}$, then $\omega_{12}^* = -\omega_{12}$;

(ii) if $\omega^* = e^{i\tau}\omega$, then $\omega_{12}^* = \omega_{12} - d\tau$;

(iii) if $\omega^* = A\omega$, then $\omega_{12}^* = \omega_{12} + *d\log A$.

Where τ, A are functions, and "$*$" the Hodge $*$- operator, such that $*\omega_1 = \omega_2, *\omega_2 = -\omega_1$.

§2. Infinitesimal BI-isometry

Lemma 2 Let ϕ be an infinitesimal I-isometry of a surface M in Euclidean space E^3. We may properly choose a coframe, such that

$$\varphi_1 = \varphi_2 = \varphi_{12} = 0. \tag{1}$$

Proof For the deformation ϕ, we write

$$\varphi_1 = a_1\omega_1 + a_2\omega_2, \quad \varphi_2 = a_3\omega_1 + a_4\omega_2. \tag{2}$$

Using (1.9) and (2), we have

$$\frac{1}{2}\delta I = a_1(\omega_1)^2 + (a_2 + a_3)\omega_1\omega_2 + a_4(\omega_2)^2.$$

$\delta I = 0$ implies that $a_1 = a_2 + a_2 = a_4 = 0$, hence

$$\varphi_1 = a_2\omega_2, \quad \varphi_2 = -a_2\omega_1 \tag{3}$$

Inserting (3) into $(1.16)_{1,2}$, we get

$$\begin{aligned}\varphi_1^* &= (\sigma + a_2)(-\omega_1 \sin\rho + \omega_2 \cos\rho),\\ \varphi_2^* &= -(\sigma + a_2)(\omega_1 \cos\rho + \omega_2 \sin\rho).\end{aligned}$$

We may choose a new frame such that $\sigma = -a_2$. Using the equations above, we have $\varphi_1^* = \varphi_2^* = 0$. It follows from (1.16) that $\varphi_{12} = 0$. Q.E.D.

For the surface M, we have

$$dx = \omega_1 e_1 + \omega_2 e_2; \tag{4}$$

$$de_i = \omega_{ij} e_j, \quad i, j = 1, 2, 3; \tag{5}$$

$$\omega_{ij} + \omega_{ji} = 0. \tag{6}$$

Write

$$\omega_{12} = h\omega_1 + k\omega_2, \tag{7}$$

$$\omega_{13} = a\omega_1 + b\omega_2, \quad \omega_{23} = b\omega_1 + c\omega_2. \tag{8}$$

The mean curvature H and Gaussian curvature K are respectively

$$2H = a + c, \qquad K = ac - b^2 \tag{9}$$

and the Gaussian equation,

$$d\omega_{12} = -K * 1 \tag{10}$$

Taking exterior differentiation of (8), we get the existence of functions $\alpha, \beta, \gamma, \delta$ such that

$$\begin{aligned} da - ab\omega_{12} &= \alpha\omega_1 + \beta\omega_2, \\ db + (a-c)\omega_{12} &= \beta\omega_1 + \gamma\omega_2, \\ dc + 2b\omega_{12} &= \gamma\omega_1 + \delta\omega_2. \end{aligned} \tag{11}$$

Furthemore, taking exterior differentiation of (11), we have the existence of function $A, \cdots, E$ such that

$$\begin{aligned} d\alpha - 3\beta\omega_{12} &= A\omega_1 + (B - bK)\omega_2, \\ d\beta + (\alpha - 2\gamma)\omega_{12} &= (B + bK)\omega_1 + (C + aK)\omega_2, \\ d\gamma + (2\beta - \delta)\omega_{12} &= (C + cK)\omega_1 + (D + bK)\omega_2, \\ d\delta + 3\gamma\omega_{12} &= (D - bK)\omega_1 + E\omega_2. \end{aligned} \tag{12}$$

Now consider a piece of oriented surface M of sufficient smoothness and containing no umbilics, and write

$$f = a - c > 0, \qquad g = 2b, \tag{13}$$

$$F = \sqrt{f^2 + g^2} = 2\sqrt{H^2 - K} > 0. \tag{14}$$

We determine the first and second covariant derivatives of f and g as follows:

$$df - 2g\omega_{12} = f_i\omega_i, \quad dg + 2f\omega_{12} = g_i\omega_i; \tag{15}$$

$$\begin{aligned} df_1 - (2g_1 + f_2)\omega_{12} &= f_{1i}\omega_i \\ df_2 - (2g_2 - f_1)\omega_{12} &= f_{2i}\omega_i \\ dg_1 + (2f_1 - g_2)\omega_{12} &= g_{1i}\omega_i \qquad i = 1, 2. \\ dg_2 + (2f_2 + g_1)\omega_{12} &= g_{2i}\omega_i \end{aligned} \tag{16}$$

Using (11)-(13), we get

$$\begin{aligned} f_1 = \alpha - \gamma, \quad & f_2 = \beta - \delta, \\ g_1 = 2\beta, \quad & g_2 = 2\gamma; \end{aligned} \tag{17}$$

$$\begin{aligned} f_{11} = A - C - cK, \quad & f_{12} = B - D - 2bK; \\ f_{21} = B - D + 2bK, \quad & f_{22} = C - E + aK; \\ g_{11} = 2(B + bK), \quad & g_{12} = 2(C + aK); \\ g_{21} = 2(C + cK), \quad & g_{22} = 2(D + bK). \end{aligned} \tag{18}$$

Let ϕ be the infinitesimal deformation, satisfying the euqation (1), and write

$$\varphi_{13} = b_1\omega_1 + b_2\omega_2, \quad \varphi_{23} = b_3\omega_1 + b_4\omega_2. \tag{19}$$

From (1.5) and using (1) and (19), we have

$$b_2 = b_3 = S. \tag{20}$$

From $\delta H = \delta K = 0$, using (1), (13), (19), (20), (1.12) and (1.13), we have

$$b_1 = -b_4 = R, \tag{21}$$

$$fR + gS = 0. \tag{22}$$

From (1.3) and (14), we introduce the angle ψ by writing

$$\cos\psi = gF^{-1}, \quad \sin\psi = -fF^{-1}. \tag{23}$$

The equation (22) can be rewritten as

$$R \sin \psi - S \cos \psi = 0.$$

We introduce a function L,

$$R = L \cos \psi, \qquad S = L \sin \psi. \tag{24}$$

By (20), (21) and (24), (19) becomes

$$\begin{aligned} \varphi_{13} &= L(\omega_1 \cos \psi + \omega_2 \sin \psi), \\ \varphi_{23} &= L(\omega_1 \sin \psi - \omega_2 \cos \psi). \end{aligned} \tag{25}$$

Now introduce the complex structure

$$\varphi = \varphi_{13} + i\varphi_{23}. \tag{26}$$

By (1.17), (26) we rewrite (25) as

$$\varphi = L e^{i\psi} \bar{\omega}. \tag{27}$$

From (1.8) it follows that

$$d\varphi = i\varphi \wedge \omega_{12}. \tag{28}$$

Because of (27), we obtain from Lemma 2 the connection form of φ:

$$-\omega_{12} - d\psi - *d \log L$$

Using (28), we know that the connection form of the complex structure φ is ω_{12}, thus, we get

$$d\varphi + 2\omega_{12} + *d \log L = 0. \tag{29}$$

It follows from (23) that

$$f \cos \psi + g \sin \psi = 0 \tag{30}$$

Taking differentiation of (30) and using (15), we get

$$d\psi + 2\omega_{12} - \theta = 0 \tag{31}$$

where the 1-form θ is

$$F^2 \theta = (fg_1 - gf_1)\omega_1 + (fg_2 - gf_2)\omega_2 \tag{32}$$

From (29) and (31) it follows that

$$\theta = - * d \log L \tag{33}$$

or

$$d \log L = *\theta. \tag{33'}$$

This is a total differential equation for determining of the function L, whose integrable condition is

$$d * \theta = 0. \tag{34}$$

Using (32), (15) and (16), we have

$$F^2 d * \theta = \{f\Delta g - g\Delta f \tag{35}$$

$$-2F^{-2}[(f^2 - g^2)\langle \nabla f, \nabla g \rangle - fg(|\nabla f|^2 - |\nabla g|^2)]\} * 1$$

where

$$\Delta f = \sum f_{ii}, \ \Delta g = \sum g_{ii},$$
$$|\nabla f|^2 = \sum (f_i)^2, \ |\nabla g|^2 = \sum (g_i)^2, \tag{36}$$
$$\langle \nabla f, \nabla g \rangle = \sum f_i g_i, \ i = 1,2.$$

Taking exterior differentiation of (31), we have

$$d\theta = -2K * 1 \tag{37}$$

Taking exterior differentiation of (33) and using (37) yields

$$\Delta \log L = 2K \tag{38}$$

where $\Delta \log L = \sum (\log L)_{ii}, d * d \log L = \Delta \log L * 1$, $d * d$ is the Laplace-Beltrami operator on M, and

$$d \log L = \sum \omega_i (\log L)_i,$$
$$d(\log L)_1 - (\log L)_2 \omega_{12} = \sum \omega_i (\log L)_{1i}, \tag{39}$$
$$d(\log L)_2 + (\log L)_1 \omega_{12} = \sum \omega_i (\log L)_{2i}.$$

Taking the $*$-operator and exterior differentiation of (31) and using (34) gives us

$$\Delta \psi = -2(h_1 + k_2) \tag{40}$$

where

$$\begin{aligned} dh - k\omega_{12} &= h_1 \omega_1 + (l + \frac{1}{2}K)\omega_2, \\ dk + h\omega_{12} &= (l - \frac{1}{2}K)\omega_1 + k_2 \omega_2, \end{aligned} \tag{41}$$

and l is the function.

From the discussion above it can be seen that, to detirmine an infinitesimal BI-isometry of M, first we get ψ by (23), then we obtain a solution L of (33), which is solvable when the integrable condition (34) is satisfied, finally we attain the deformation ϕ determined by (1) and (2).

Theorem 1 The surface M is an infinitesimal BI- surface if and only if (34) is satisfied, i .e .,

$$d * \theta = 0 \tag{34}$$

where θ is an 1-form defined by (32).

Corollary 1 Condition (34) may be rewritten as one of the following:

(i)

$$\begin{vmatrix} f & f \\ \Delta f & \Delta g \end{vmatrix} = 2 \begin{vmatrix} f^2 - g^2 & fg \\ |\nabla|^2 - |\nabla g|^2 & \langle \nabla f, \nabla g \rangle \end{vmatrix}; \tag{42}$$

(ii)

$$\begin{aligned} &F^2[2(B+D)f - (A-E)g + fgK] \\ =&2\{2(\alpha\beta - \nu\delta)(f^2 - g^2) - [\alpha^2 + \delta^2 - 3(\beta^2 + \nu^2) - 2(\alpha\nu + \beta\delta)]fg\}; \end{aligned} \tag{43}$$

(iii)

$$\text{If } \ b = 0 \ \ (\text{in } (8)), \ \ f \neq 0, g = 0,$$
$$B + D = 2(h\alpha - k\delta). \tag{44}$$

Corollary 2 Let M be a surface with constant mean curvature (H=const.), then M is an infinitesimal BI- surface. This is similar to the classical case about finite Bonnet I-isometry.

Corollary 3 Let M be an infinitesimal BI-surface with non-constant mean curvature ($H \neq$cons and a metric conformal with the metric I

$$\hat{I} = (\varphi_{13})^2 + (\varphi_{23})^2 = L^2 I. \tag{45}$$

Then the Gaussian curvature of the metric $\hat{I}$ is

$$\hat{K} = -KL^{-2}. \tag{46}$$

Using (25), we get $\varphi_{13} \wedge \varphi_{23} = -L^2\omega_1 \wedge \omega_2$. Taking exterior differentiation of ω_{12}, we get (46) immediately.

Corollary 4 Let M be a flat surface with zero Gaussian curvature ($K = 0$). Then M is an infinitesimal BI- surface.

Corollary 5 All surfaces of revolution are infinitesimal H-surfaces.

§3. Infinitesimal BIII-isometry

First of all, we will prove a lemma about III-isometry, which is similar to Lemma 2.

Lemma 3 Let ϕ be an infinitesimal III-isometry of a surface M with non-zero Gaussian curvature ($K \neq 0$). We may properly choose a coframe, such that

$$\varphi_{13} = \varphi_{23} = \varphi_{12} = 0. \tag{1}$$

Proof The proof os this lemma is similar to that of Lemma 2.

Since $K \neq 0$ and the 1-forms ω_{13} and ω_{23} are linearly independent, we may write

$$\omega_{12} = h\omega_1 + k\omega_2, \tag{2}$$

$$\omega_1 = a\omega_{13} + b\omega_{23}, \quad \omega_2 = b\omega_{13} + c\omega_{23}. \tag{3}$$

The mean curvature and Gaussian curvature are

$$2H = \frac{a+c}{ac-b^2}, \quad K = \frac{1}{ac-b^2}. \tag{4}$$

The Gaussian equation is

$$d\omega_{12} = -K\omega_1 \wedge \omega_2 = -\omega_{13} \wedge \omega_{23}. \tag{5}$$

Taking exterior differentiation of (3), we get the existence of functions $\alpha, \beta, \nu, \delta$ such that

$$\begin{aligned} da - 2b\omega_{12} &= \alpha\omega_{13} + \beta\omega_{23}, \\ db + (a-c)\omega_{12} &= \beta\omega_{23} + \nu\omega_{23}, \\ dc + 2b\omega_{12} &= \nu\omega_{13} + \delta\omega_{23}. \end{aligned} \tag{6}$$

Taking exterior differentiation of (6) gives the existence of functions $A, \cdots, E$ such that

$$\begin{aligned} d\alpha - 3\beta\omega_{12} &= \omega_{13} + (B-b)\omega_{23}, \\ d\beta + (\alpha - 2\nu)\omega_{12} &= (B+b)\omega_{13} + (C+a)\omega_{23}, \\ d\nu + (2\beta - \delta)\omega_{12} &= (C+c)\omega_{13} + (D+b)\omega_{23}, \\ d\delta + 3\nu\omega_{12} &= (D+b)\omega_{13} + E\omega_{23}. \end{aligned} \tag{7}$$

Now consider a piece of oriented surface M of sufficiant sommthness and containing no umbilit points, and we write

$$f = a - c > 0, \quad g = 2b, \tag{8}$$

$$F = (f^2 + g^2)^{1/2} = 2(H^2K^2 - K^{-1})^{1/2} > 0. \tag{9}$$

We ditermine the first and second covariant derivatives of f and g as follows:

$$\begin{aligned} df - 2g\omega_{12} &= f_i\omega_{i3} \\ dg + 2f\omega_{12} &= g_i\omega_{i3} \qquad i = 1, 2; \end{aligned} \tag{10}$$

$$\begin{aligned} df_1 - (2g_1 + f_2)\omega_{12} &= f_{1i}\omega_{i3} \\ df_2 - (2g_2 - f_1)\omega_{12} &= f_{2i}\omega_{i3} \qquad i = 1, 2. \\ dg_1 + (2f_1 - g_2)\omega_{12} &= g_{1i}\omega_{i3} \\ dg_2 + (2f_2 + g_1)\omega_{12} &= g_{2i}\omega_{i3} \end{aligned} \tag{11}$$

Using (6) and (7), we get

$$\begin{aligned} f_1 = \alpha - \gamma, \quad & f_2 = \beta - \delta \\ g_1 = 2\beta, \quad & g_2 = 2\gamma, \end{aligned} \tag{12}$$

$$\begin{aligned} f_{11} = A - C - c, \quad & f_{12} = B - D - 2b \\ f_{21} = B - D + 2b, \quad & f_{22} = C - E + a \\ f_{11} = 2(B + b), \quad & g_{12} = 2(C + a) \\ g_{21} = 2(C + c), \quad & g_{22} = 2(D + b). \end{aligned} \tag{13}$$

Let ϕ be an infinitesimal deformation satisflying the equation (1), and write

$$\varphi_1 = a_1\omega_{13} + a_2\omega_{23}, \quad \varphi_2 = a_3\omega_{13} + a_3\omega_{23}. \tag{14}$$

From (1.5) and using (1) and (14), we have

$$a_2 = a_3 = S. \tag{15}$$

From $\delta H = \delta K = 0$, using (1), (8), (14), (15) and (1.12), (1.13), we have

$$a_1 = -a_1 = R, \tag{16}$$

$$fR + gS = 0. \tag{17}$$

Using (8) and (9), we introduce the angle ψ by writing

$$\cos\psi = gF^{-1}, \quad \sin\psi = -fF^{-1}. \tag{18}$$

From (17), we get

$$R = L\cos\psi, \quad S = L\sin\psi \tag{19}$$

We note that two complex structures

$$\omega = \omega_{13} + i\omega_{23}, \quad \varphi = \varphi_1 + i\varphi_2 \tag{20}$$

Using (15), (16), (19) and (20), we write (14) as

$$\varphi = Le^{i\psi}\bar{\omega}. \tag{21}$$

We get from (1.6)

$$d\varphi = i\varphi \wedge \omega_{12}. \tag{22}$$

By (21) and from Lemma 2, it follows that the connection form of φ is

$$-\omega_{12} - d\psi - *d\log L.$$

From (22), we know that the connection form of φ is ω_{12}, thus we get

$$d\psi + 2\omega_{12} + *d\log L = 0. \tag{23}$$

We have similar results to Section 2.

Theorem 2 The surface M with non-zero Gaussian curuature is an infinitesimal BIII-surface if and only if it satisfies

$$d * \theta = 0 \tag{24}$$

where θ is an 1-form defined by

$$F^2\theta = (fg_1 - gf_1)\omega_{13} + (fg_2 - gf_2)\omega_{23}. \tag{25}$$

Corollary 1 Condition (24) may be rewritten as one of the following:

(i)

$$F^2 \begin{vmatrix} f & g \\ \Delta f & \Delta g \end{vmatrix} = 2 \begin{vmatrix} f^2 - g^2 & fg \\ |\nabla f|^2 - |\nabla g|^2 & \langle \nabla f, \nabla g \rangle \end{vmatrix}; \tag{26}$$

(ii)

$$\begin{aligned} &F^2[2(B+D)f - (A-E)g + fg] \\ =&2\{2(\alpha\beta - \nu\delta)(f^2 - g^2) - [\alpha^2 + \delta^2 - 3(\beta^2 + \nu^2) - 2(\alpha\nu + \beta\delta)]fg\}; \end{aligned} \tag{27}$$

(iii) If $b = 0$ (in (3)), $f \neq 0, g = 0$, then

$$B + D = 2(h\alpha - k\delta). \tag{28}$$

Corollary 2 Let M be a surface with H/K=constant. Then M is an infinitesimal BIII-surface.

Corollary 3 Let M be an infinitesimal BIII-surface with $H/K \neq$const., and the metric of the Gaussian image of M be

$$I_g = (\omega_{13})^2 + (\omega_{23})^2. \tag{29}$$

Then the Gaussian curvature of the metric

$$\hat{I} = (\varphi_1)^2 + (\varphi_2)^2 = L^2 I_g, \tag{30}$$

which is conformal to I_g, is

$$\hat{K} = -KL^{-2}. \tag{31}$$

Corollary 4 Let M be an infinitesimal BIII-surface. Then the following equations hold:

$$\theta = - * d\log L, \tag{32}$$

$$d\theta = -2K * 1, \tag{33}$$

$$\Delta \log L = 2. \tag{34}$$

Note: The left side of (34) is $\sum(\log L)_{ii}$, where the subscript "i" denotes the covariant derivative with respect to the fundamental forms ω_{13}, ω_{23}.

§4. Infinitesimal BII-siometry

Let ϕ be an infinitesimal BII-isometry of a surface M, and

$$\varphi_1 = a_1\omega_1 + a_2\omega_2, \quad \varphi_2 = a_3\omega_1 + a_4\omega_2; \tag{1}$$

$$\varphi_{13} = b_1\omega_1 + b_2\omega_2, \quad \varphi_{23} = b_3\omega_1 + b_4\omega_2. \tag{2}$$

Inserting (1), (2) into (1.10), we get

$$\begin{aligned}\delta II =&(b_1 + aa_1 + ba_3)(\omega_1)^2 + (b_4 + ba_2 + ca_4)(\omega_2)^2 \\ &+ (b_2 + b_3 + ba_1 + aa_2 + ca_3 + ba_4)\omega_1\omega_2\end{aligned}$$

Where a, b and c are defined by (2.8). From $\delta II = 0$ and above equation it follows that

$$\begin{aligned} b_1 + aa_1 + ba_3 = 0, \\ b_2 + b_3 + ba_1 + aa_2 + ca_3 + ba_4 = 0, \\ b_4 + ba_2 + ca_4 = 0. \end{aligned} \tag{3}$$

Inserting (1), (2) into (1.5), we get

$$b_2 - b_3 + ba_1 - aa_2 + ca_3 - ba_4 = 0. \tag{4}$$

Using (3) and (4), we have

$$\begin{bmatrix} b_1 & b_2 \\ b_3 & b_4 \end{bmatrix} = - \begin{bmatrix} a_1 & a_3 \\ a_2 & a_4 \end{bmatrix} \begin{bmatrix} a & b \\ b & c \end{bmatrix}. \tag{5}$$

Substituting (5) into (2), we have by (2.8)

$$\varphi_{13} = -a\omega_{13} - a_3\omega_{23}, \quad \varphi_{23} = -a\omega_{13} - a_4\omega_{23}. \tag{6}$$

By (1) and (6), from (1.12) and (1.13) it follows that

$$\begin{aligned} \delta H &= - aa_1 - b(a_2 + a_3) - ca_4, \\ \delta K &= - 2K(a_1 + a_4). \end{aligned}$$

Since ϕ is a BII-isometry, $\delta H = \delta K = 0$, we get

$$aa_1 + b(a_2 + a_3) + ca_4 = 0, \tag{7}$$

$$K(a_1 + a_4) = 0. \tag{8}$$

We need a lemma of II-isometry, similar to Lemmas 1 and 2.

Lemma 4 Let ϕ be an infinitesimal II-isometry of a surface M, we may properly choose a coframe such that

$$\varphi_{12} = 0. \tag{9}$$

Proof It is easy to see that (see Section 1), under the change (1.14) of coframe, we have $\varphi^*_{12} = \varphi_{12} + d\sigma$. To show that $\varphi^*_{12} = 0$, it is sufficient to prove that $\varphi_{12} = d\lambda$ is locally an exact 1-form, that is, φ_{12} is closed:

$$d\varphi_{12} = 0. \tag{10}$$

In fact, inserting (1) and (6) into (1.7), we have

$$\begin{aligned} d\varphi_{12} &= (a_1 + a_4)\omega_{13} \wedge \omega_{23} \\ &= K(a_1 + a_4) * 1 = 0 \end{aligned}$$

because of (8). Thus (10) is established, hence there is function λ such that $\varphi_{12} = d\lambda, \varphi_{12}^* = d(\lambda + \sigma)$. So long as we choose $\sigma = -\lambda$, then $\varphi_{12}^* = 0$. *Q.E.D.*

From (9), the integrable conditions (1.6) and (1.8) of ϕ become

$$d\varphi_1 = -\varphi_2 \wedge \omega_{12}, \quad d\varphi_2 = \varphi_1 \wedge \omega_{12}; \tag{11}$$

$$d\varphi_{13} = -\varphi_{23} \wedge \omega_{12}, \quad d\varphi_{23} = \varphi_{13} \wedge \omega_{12}. \tag{12}$$

(I) In case of $K \neq 0$. We choose coframe such that $b = 0$, thus in (2.13),

$$f = a - c \neq 0, \quad g = 0, \quad K = ac \neq 0 \tag{13}$$

From $K \neq 0$, using (8), (7), we get

$$a_1 = -a_4 = R, \qquad fR = 0$$

Since $f \neq 0$, $a_1 = a_4 = R = 0$ by the above equations, and we write $a_2 = \lambda, a_3 = \mu$.

From (1) and (6) it follows that

$$\varphi_1 = \lambda\omega_2, \quad \varphi_2 = \mu\omega_1, \tag{14}$$

$$\varphi_{13} = -\mu\omega_{23}, \quad \varphi_{23} = -\lambda\omega_{13}. \tag{15}$$

Taking exterior differentiation of (14) and using (11), we get

$$\begin{aligned} d\lambda + (\lambda + \mu)k\omega_1 &= \rho\omega_2 \\ d\mu - (\lambda + \mu)h\omega_2 &= \sigma\omega_1, \end{aligned} \tag{16}$$

$$d(\lambda + \mu) = -(\lambda + \mu)(k\omega_1 - h\omega_2) + \sigma\omega_1 + \rho\omega_2. \tag{17}$$

Taking exterior differentiation of (15) and using (12) and (16) yields

$$\begin{aligned} a\rho - ch(\lambda + \mu) &= 0, \\ c\sigma + ak(\lambda + \mu) &= 0. \end{aligned} \tag{18}$$

We find ρ, σ by (18) and substitute them into (16), then we get

$$\begin{aligned} d\lambda &= (\lambda + \mu)pa^{-1}\omega_2, \\ d\mu &= (\lambda + \mu)pc^{-1}\omega_1, \end{aligned} \tag{19}$$

$$p = ch - ak. \tag{20}$$

It follows from (19) that

$$d\log(\lambda + \mu) = p(c^{-1}\omega_1 + a^{-1}\omega_2). \tag{21}$$

Taking exterior differentiation of (19) and using (21), we have

$$\begin{aligned} (pa^{-1})_1 + p^2K^{-1} + pa^{-1}k &= 0, \\ (pc^{-1})_2 - p^2K^{-1} - pc^{-1}h &= 0 \end{aligned} \tag{22}$$

where

$$d(pa^{-1}) = (pa^{-1})_i\omega_i$$
$$d(pc^{-1}) = (pc^{-1})_i\omega_i \qquad \subset = 1,2 \tag{23}$$

From (1.11), (20) and

$$dh = h_i\omega_i, \quad dk = k_i\omega_i, \quad i = 1,2$$

we have

$$\begin{aligned}(pa^{-1})_1 &= a^{-1}(ch_1 - ak_1) + a^{-2}(a\gamma - c\alpha)h\\ (pa^{-1})_2 &= a^{-1}(ch_2 - ak_2) + a^{-2}(a\delta - c\beta)h\\ (pc^{-1})_1 &= c^{-1}(ch_1 - ak_1) + c^{-2}(a\gamma - c\alpha)k\\ (pc^{-1})_2 &= c^{-1}(ch_2 - ak_2) + c^{-2}(a\delta - c\beta)k.\end{aligned} \tag{24}$$

Inserting (20) and (24) into (22), we get

$$\begin{aligned}ch_1 - ak_1 - ca^{-1}\alpha h + ch^2 - ahk - a^2(a^{-1} - c^{-1})k^2 &= 0\\ ch_2 - ak_2 + ac^{-1}\delta k - c^2(a^{-1} + c^{-1})h^2 + chk - ak^2 &= 0.\end{aligned} \tag{25}$$

This is the integrable condition of (16), i.e., the necessary and sufficient condition for a surface M to be an infinitesimal BII-surface.

From above discussion we obtain the following:

Theorem 3 Let M be a surface with a non-zero Gaussian curvature, the necessary and sufficient condition for the surface to be an infinitesimal BII-surface, is that (25) are satisfied.

If (25) holds, then the equation system (19) of λ, μ is integrable, hence the deformation of M is given by (14) and (15).

Taking exterior differentiation of (21), we get:

Corollary Let M be an infinitesimal BII-surface with a non-zero Gaussian curvature. Then the right-hand side of (21)

$$\theta = (ch - ak)(c^{-1}\omega_1 + a^{-1}\omega_2) \tag{26}$$

is a closed 1-form, i.e., $d\theta = 0$.

(II) In case of $K = 0$. Let M be a non-plane surface, choose the frame $e_1e_2e_3$ such that one of the principal curvatures $a \neq 0$, and the other $c = 0$, thus

$$a \neq 0, \qquad b = c = 0; \tag{27}$$

$$\omega_{13} = a\omega_1, \qquad \omega_{23} = 0. \tag{28}$$

From the structure equations and (28), we have

$$d\omega_{13} = 0, \qquad \omega_{13} \wedge \omega_{12} = 0$$

which implies

$$\omega_{13} = a\omega_1 = du, \qquad \omega_{12} = h\omega_1 \tag{29}$$

where u is a function. It is easy to see that all the curvature curves except u-curves ($\omega_2 = 0$) in M are lines. Choose arclength of these lines as parametre v, thus

$$\omega_1 = a^{-1}du, \qquad \omega_2 = dv. \tag{30}$$

Taking exterior differentiation of (30), we get

$$h = (\log a)'_v. \tag{31}$$

According to Lemma 4, (9), (11) and (12) hold, too. Using (27) and from (7) we get

$$a_1 = 0 \tag{32}$$

From (28) and (32) it follows that

$$\varphi_{13} = 0, \qquad \varphi_{23} = -a_2 du. \tag{33}$$

Taking exterior differentiation of (1) and (33), we have

$$\begin{gathered}(da_2 - ha_4\omega_1) \wedge \omega_2 = 0 \\ (da_3 - ha_2\omega_2) \wedge \omega_1 + (da_4 + ha_3\omega_1) \wedge \omega_2 = 0. \\ da_2 \wedge du = 0\end{gathered} \tag{34}$$

It follows from (34) that

$$a_2 = \rho(u). \tag{35}$$

From $(34)_{1,2}$ we introduce the functions A_1, A_2, A_3:

$$\begin{aligned} da_2 &= ha_4\omega_1 \\ da_3 &= A_1\omega_1 + (A_2 + ha_2)\omega_2 \\ da_4 &= (A_2 - ha_3)\omega_1 + A_3\omega_2 \end{aligned} \tag{36}$$

Using (35) and (36) gives

$$ha_4 = a\rho'_u \tag{37}$$

Denote

$$\begin{aligned} da &= a_{,i}\omega_i \\ da_{,i} &= a_{,ij}\omega_j \\ dh &= h_i\omega_i \qquad\qquad i,j = 1,2 \\ dh_i &= h_{ij}\omega_j \end{aligned} \tag{38}$$

where

$$\begin{aligned} a_{,1} &= a^2(\log a)'_u, \quad a_{12} = ah = a(\log a)'_v \\ h_1 &= (a^{-1}(\log a)''_{uu}, \quad h_2 = (\log a)''_{vv}. \end{aligned} \tag{39}$$

Taking differentiation of (37) and using (36), we have

$$\begin{aligned} h_1a_4 + h(A_2 - ha_3) &= a_{,1}\rho'_u + a^2\rho''_{uu} \\ h_2a_4 + hA_3 &= a_{,2}\rho'_u. \end{aligned} \tag{40}$$

Taking exterior differentiation of $(36)_{2,3}$ gives us

$$dA_1 \wedge \omega_1 + dA_2 \wedge \omega_2 + (hA_1 + h_1a_2 + h^2a_4)\omega_1 \wedge \omega_2 = 0,$$

$$dA_2 \wedge \omega_1 + dA_3 \wedge \omega_2 + [2hA_1 + h_2a_3 + h^2(a_2 - a_3)]\omega_1 \wedge \omega_2 = 0.$$

Using E.Cartan's Lemma, we know that there are functions $B_1, \cdots, B_4$ such that

$$\begin{aligned} dA_1 &= B_1\omega_1 + (B_2 + hA_1 + h_1a_2 + h^2a_4)\omega_2 \\ dA_2 &= B_2\omega_1 + B_3\omega_2 \\ dA_3 &= [B_3 + 2hA_2 + (h_2 - h^2)a_3 + h^2a_2]\omega_1 + B_4\omega_2. \end{aligned} \tag{41}$$

Differentiate (40) and use (36) and (41), then we have

$$\begin{aligned} hB_2 - h^2A_1 + 2h_1A_2 - 3hh_1a_3 + h_{11}a_4 &= a_{,11}\rho'_u + 3aa_{,1}\rho''_{uu} + a^3\rho'''_{uuu} \\ hB_3 + (h_2 - h^2)A_2 + h_1A_3 - 2hh_2a_2 + h_{12}a_4 &= h^3\rho + a_{,12}\rho'_u + 2aa_{12}\rho''_{uu} \\ hB_3 + (h_2 + 2h^2)A_2 + h_1A_3 - h^3a_3 + h_{21}a_4 &= -h^3\rho + a_{,21}\rho'_u + aa_{,2}\rho''_{uu} \\ hB_4 + 2h_2A_3 + h_{22}a_4 &= a_{,22}\rho'_u \end{aligned} \tag{42}$$

It follows from $(42)_{2,3}$ that

$$3h^2A_2 + h(2h_2 - h^2)a_3 + 2h^3\rho + aa_{,2}\rho''_{uu} = 0 \tag{43}$$

From (40) and (43), we have

$$2h(h_2 - h^2)a_3 = 4ha^2\rho''_{uu} + 3(ha_{,1} - h_1a)\rho'_u + 2h^3\rho, \tag{44}$$

$$\begin{aligned} 2h^2A_2 + 3ha^2\rho''_{uu} + \frac{3}{2}(ha_{,1} - h_1a)\rho'_u + 4h^3\rho &= 0 \\ h^2A_3 + a(h_2 - ah)\rho'_u &= 0. \end{aligned} \tag{45}$$

From the above discussion it can be seen that, to determine the infinitesimal BII-isometry ϕ, we find a_1 (by (32)), a_2(by (35)), a_3(by (37)), a_4(by (44)), where $a_{,i}, a_{,ij}, h_i, h_{ij}$ are defined on surface M. There exist three cases as follows:

(i) Cylinder M: $m(s,z) = m(s) + kz$

$$m(s) = x(s)i + y(s)j,$$

where $0ijk$ is a frame in E^3, $m(s)$ a plane curve parametrized by its arclength s, and

$$m' = \alpha, \qquad \alpha' = k\beta,$$

Choose the frame of M, by

$$e_1 = \alpha, \qquad e_2 = k, \qquad e_3 = -\beta$$

we have

$$dm = \omega_1e_1 + \omega_2e_2$$

$$\begin{aligned} \omega_1 &= ds, & \omega_2 &= dz \\ \omega_{12} &= 0, & h &= 0 \\ \omega_{13} &= a\omega_1, & a &= -k \neq 0 \\ \omega_{23} &= 0, & b &= c = 0. \end{aligned}$$

From (32), (35) and (37), we conclude that $a_1 = 0, a_2 = \rho(s), \rho'_s = 0, \rho =$const, (37) and (44) identically hold, and a_3 and a_4 are two arbitrary functions of s. We can see that the deformation ϕ of the cylinder depends on two drbitrary functions a_3 and a_4 and one arbitrary constant ρ.

(ii) Cone M: $m(s,v) = vm(s), m^2(s) = 1, v > 0$
where $m(s)$ is a curve on the unit sphere centered at the origin and parametrized by its arclength s. We have

$$m' = \alpha, \qquad \alpha' = \kappa\beta.$$

Choose the frame of M by

$$e_1 = \alpha, \quad e_2 = m, \quad e_3 = \alpha \times m$$

and we have

$$dm = \omega_1 e_1 + \omega_2 e_2$$

$$\begin{aligned} \omega_1 &= \nu ds, & \omega_2 &= d\nu \\ \omega_{12} &= h\omega_1, & h &= \nu^{-1} \\ \omega_{13} &= a\omega_1, & a &= \kappa h \\ \omega_{23} &= 0, & b &= c = 0 \end{aligned}$$

$$dh = h_1\omega_1 + h_2\omega_2, \quad h_1 = 0, \quad h_2 = -\nu^{-2}$$

$$da = a_{,1}\omega_1 + a_{,2}\omega_2, \quad a_{,1} = \kappa'_s \nu^{-2}.$$

From (32), (35), (37) and (44) it follows that

$$a_1 = 0, \quad a_2 = \rho(s), \quad a_4 = \kappa\rho'_s$$

$$a_3 = \kappa^2 \rho''_{ss} + \frac{3}{4}(\kappa'_s \rho'_s + \rho).$$

We can see that the deformation ϕ of the cone depends on one arbitrary function ρ.

(iii) Tangential developable surface of non-plane curve

$$M: \ m(s, \bar{v}) = m(s) + \bar{v}\alpha(s), \qquad \bar{v} > 0$$

where $m(s)$ is an non-plane curve with parametre by its arclength s, we hve

$$m' = \alpha, \quad \alpha' = \kappa\beta, \quad \kappa \neq 0$$

$$\beta' = -\kappa\alpha + \tau\nu, \quad \nu' = -\tau\beta, \quad \tau \neq 0$$

$$dm = (\kappa\bar{v}ds)\beta + (ds + d\bar{v})\alpha.$$

Let $v = \bar{v} + s, v - s > 0$, then

$$e_1 = \beta, \quad e_2 = \alpha, \quad e_3 = -\nu$$

$$dm = \omega_1 e_1 + \omega_2 e_2$$

$$\begin{aligned} \omega_1 &= \kappa(v - s)ds, & \omega_2 &= dv \\ \omega_{12} &= h\omega_1, & h &= -(v - s)^{-1} \\ \omega_{13} &= a\omega_1, & a &= -\tau\kappa^{-1}(v - s)^{-1} \\ \omega_{23} &= 0, & b &= c = 0. \end{aligned}$$

By (32), (35), (37) and (44),

$$\begin{aligned} a_1 &= 0, & a_2 &= \rho(s) \\ a_3 &= \sigma(s), & a_4 &= \tau\kappa^{-1}\rho' \end{aligned}$$

$$4(\tau\kappa^{-1})^2\rho'' - 3\kappa^{-1}(\tau\kappa^{-1})'\rho' + 2\rho = 0 \tag{*}$$

We can see that the deformation of the tangential developable surface depends on one arbitrary function σ, where the function ρ is determined by the second order ordinary differential equation (*).

Now, we are in the position to get the following theorem.

Theorem 4 Let M be a developable surface ($K = 0$). Then it is an infinitesimal BII-surface. Such deformation ϕ is determined by

$$\varphi_1 = a_2\omega_2, \qquad \varphi_2 = a_3\omega_1 + a_4\omega_2$$

$$\varphi_{12} = 0, \quad \varphi_{13} = 0, \quad \varphi_{23} = -aa_2\omega_1$$

where the functions a_2, a_3, a_4 are given by (35), (37), (44), resp., In detail, for the three kinds of developable surface, we have

(i) if M is a cylinder, then ϕ depends on two arbitrary functions and one arbitrary constant;

(ii) if M a cone, then ϕ depends on one arbitrary function;

(iii) if M is a tangential developable of curve, then ϕ depends on one arbitrary function.

Example For a cylinder

$$M: \qquad m(s,z) = m(s) + zk$$

we define a family of the cylinders

$$M_t: \qquad m_t(s,z) = m(s) + tzk$$

where t is the family of parameter.

Let a map $F: M \to M_t, m(s,z) \mapsto m_t(s,z)$. Geometrically it is easy to see that this map F is a BII-isometry from M into M_t.

References

1. O.Bonnet: Memoire sur la theorie des surfaces applicables, Journal Ecole Polytechnique, 42(1867), 72-92.

2. W.C.Graustein: Applicability with preservation of both curvatures, Bull. Amer. Math. Soc., 30(1924), 19-27.

3. E.Cartan: Couples des surfaces applicables avec conservation des courbures principales, Bull. Sciences Math., 66(1942), 55-85.

4. S.S.Chern: Deformation of surfaces preserving principal curvatures, Differential geometry and complex analysis, 1985, 155-163.

5. I.M.Roussos: Mean-curvature-preserving isometries of surfaces in ordinary space, 1986, to appear.

6. Yang Wenmao: On III-isometric deformations of surfaces preserving the principal curvatures, 1986, to appear.

7. Yang Wenmao: Deformations of Codazzi tensors on Riemannian surface, 1986, to appear.

8. A.Svec: Global differential geometry of surfaces, VEB Deutscher Verlag der Wissenschaften, Berlin, 1981.

Yang Wenmao
Department of Mathematics
Wuhan University
Wuhan, China

Local Expressions of Classical Geometric Elliptic Operators

Yu Yanlin

In [1] Atiyah, Bott and Patodi pointed out that all the classical elliptic operators are, locally, twisted Dirac operators. This is a very beautiful observation, which has already become almost a common opinion. But until now neither of its natural statement nor the proof is available. This paper will give a full description of this observation.

§1. Operator $d+\delta$

Let M be an oriented Riemannian manifold of dim n. As far as the local problem is concerned in this paper, we always assume that M is diffeomorphic to an open contractible set of $\mathbf{R}^n$. As usual we have a operator $d+\delta : \wedge^*(M) \to \wedge^*(M)$, where $\wedge^*(M)$ is the set of differential forms, and $\delta = d^*$. Let us choose an oriented orthonormal basis $\{E_1, \cdots, E_n\}$, whose dual basis is denoted by $\{\omega_1, \cdots, \omega_n\}$. In [4] the following operators were introduced.

$$e_i^+ = \omega_i + i(E_i) : \ \wedge^*(M) \to \wedge^*(M),$$

$$e_i^- = \omega_i - i(E_i) : \ \wedge^*(M) \to \wedge^*(M).$$

They are subject only to the relations

$$\begin{cases} e_i^+ e_j^+ + e_j^+ e_i^+ = 2\delta_{ij}, \\ e_i^- e_j^- + e_j^- e_i^- = -2\delta_{ij}, \\ e_i^+ e_j^- + e_j^- e_i^+ = 0. \end{cases}$$

Moreover, we have

$$\begin{aligned} d+\delta &= \sum_i \omega_i \nabla_{E_i} - \sum_i i(E_i) \nabla_{E_i} \\ &= \sum_i e_i^- [E_i + \frac{1}{4} \sum_{j,k} \Gamma_{ij}^k (e_j^- e_k^- - e_j^+ e_k^+)], \end{aligned}$$

Where ∇ is the Levi-Civita connection and Γ_{ij}^k defined by

$$\nabla_{E_i} E_j = \sum_k \Gamma_{ij}^k E_k$$

Now we introduce some new operators as follows.

$$\begin{aligned} \gamma_0 =& (\sqrt{-1})^{\frac{n(n-1)}{2}} e_1^+ \cdots e_n^+, \\ L_i =& e_i^- \gamma_0, \\ R_i =& e_i^+ \gamma_0. \end{aligned}$$

Lemma 1. The following hold true:

(i) $\gamma_0^2 = 1$;

(ii) $\gamma_0 = (\sqrt{-1})^{\frac{n(n-1)}{2} + p(2n-p-1)} * : \wedge^p(M) \to \wedge^{n-p}(M)$,

where $* : \wedge^p(M) \to \wedge^{n-p}(M)$ is the Hodge star homomorphism; (As a corollary, when n is even we have

$$\gamma_0 = (-1)^p \alpha : \ \wedge^p(M) \to \wedge^{n-p}(M),$$

where α is a well-known homomorphism, used to give the splitting

$$\wedge^*(M) = \wedge_+(M) \oplus \wedge_-(M).)$$

(iii) $L_i\gamma_0 = \gamma_0 L_i$, $R_i\gamma_0 = -\gamma_0 R_i$.

Proof. (i) and (iii) are trivial due to the relations satisfied by e_i^+, e_i^-. Now we are going to prove (ii). Suppose $\omega_{i_1} \wedge \cdots \wedge \omega_{i_p} \in \wedge^*(M)$, without loss of generality we assume $i_1 < \cdots < i_p$, then

$$\begin{aligned} e_1^+ \cdots e_n^+(\omega_{i_1} \wedge \cdots \wedge \omega_{i_p}) = & \sum_{j_1 < \cdots < j_{n-p}} \in (j_1, \cdots, j_{n-p}, i_1, \cdots, i_p) e_{j_1}^+ \cdots e_{j_{n-p}}^+ e_{i_1}^+ \cdots e_{i_p}^+ \\ & \cdot (\omega_{i_1} \wedge \cdots \wedge \omega_{i_p}) = (-1)^{\frac{p(p-1)}{2}} \in (j_1, \cdots, j_{n-p}, i_1, \cdots, i_p) e_{j_1}^+ \cdots e_{j_{n-p}}^+ \cdot 1 \\ = & (-1)^{\frac{p(p-1)}{2} + p(n-p)} \sum_{j_1 < \cdots < j_{n-p}} \in (i_1, \cdots, i_p, j_1, \cdots, j_{n-p}) \omega_{j_1} \wedge \cdots \wedge \omega_{j_{n-p}} \\ = & (-1)^{\frac{1}{2}p(2n-p-1)} * (\omega_{i_1} \wedge \cdots \wedge \omega_{i_p}). \end{aligned}$$

completing the proof.

Lemma 2. The algebra generated by $\{L_i, R_i | i = 1, \cdots, n\}$ is subject only to the relations

$$\begin{cases} L_i L_j + L_j L_i = -2\delta_{ij}, \\ R_i R_j + R_j R_i = -2\delta_{ij}, \\ L_i R_j = R_j L_i. \end{cases}$$

Proof. Since the relations satisfied by $\{e_i^+, e_i^-\}$ are equivalent to those above, hence the lemma is true.

Remark. Lemma 2 tells us that the algebra generated by $\{L_i, R_i | i = 1, \cdots, n\}$ is $C_n \oplus C_n$, where C_n is the Clifford algebra. Thus the action of L_i, R_i on $\wedge^*(M)$ makes $\wedge^*(M)$ a $C_n \oplus C_n$ module.

Theorem A. Fix an oriented othonormal basis $\{E_1, \cdots, E_n\}$ and let L_i, R_i be defined as above, then we have

$$d + \delta = \sum_i L_i \{\check{E}_i + \frac{1}{4} \sum_{j,k} \Gamma_{ij}^k (L_j L_k + R_j R_k)\} \gamma_0,$$

Where $\check{E}_i$ is defined, as in [4], by

$$\check{E}_i(\sum f \omega_{i_1} \wedge \cdots \wedge \omega_{i_p}) = \sum (E_i f) \omega_{i_1} \wedge \cdots \wedge \omega_{i_p}, \ \forall \text{ function } f.$$

Proof. In [4], we have

$$d + \delta = \sum_i e_i^- \{\check{E}_i + \frac{1}{4} \sum_{j,k} \Gamma_{ij}^k (e_j^- e_k^- - e_j^+ e_k^+)\},$$

hence the theorem is true by Lemma 1.

§2. Operator $\bar{\partial} + \bar{\partial}^{\#}$

Given an almost complex manifold M of real dim $2n$, which is diffeomorphic to an open set of $\mathbf{R}^{2n}$, and let J be the almost complex structure, i.e., $J^2 = -1$. Then we have

$$TM \oplus \mathbb{C} = T \oplus \bar{T},$$

Where

$$T = \{X \in TM \oplus \mathbb{C} | JX = \sqrt{-1}X\},$$

$$\bar{T} = \{X \in TM \oplus \mathbb{C} | JX = -\sqrt{-1}X\}.$$

So we have

$$\wedge^*(M) \equiv \wedge^*(TM \oplus \mathbb{C}) = \wedge^*(T \oplus \bar{T}) = \wedge^*(T) \oplus \wedge^*(\bar{T}).$$

Define $\wedge^{(p,q)}(M)$ by

$$\wedge^{(p,q)}(M) = \wedge^p(T) \oplus \wedge^q(\bar{T}).$$

The notion of a connection D on M is defined as follows. For the complex vector fields X and Y, $D_X Y$ is a complex vector field such that the following conditions are satisfied

$$D_{fX_1+X_2}Y = fD_{X_1}Y + D_{X_2}Y,$$

$$D_X(fY_1 + Y_2) = fD_X Y_1 + D_X Y_2 + (Xf)Y_1$$

Where X_i and Y_i are complex vector fields and f, any complex valued function on M.

Definition. A connection is called real, if it satisfies

$$\overline{D_X Y} = D_{\bar{X}}\bar{Y} \quad \forall X, Y,$$

Where $\bar{X}$ stands for the complex conjugate of X.

Definition. A connection D is said to be almost complex, if

$$D_X(JY) = JD_X Y \quad \forall X, Y.$$

It is easy to get the following lemma.

Lemma 3. For a connection D on an almost complex manifold M, the conditions below are equivalent.

(a) D is almost complex;

(b) $(D_X J) = 0 \ \forall X$;

(c) $D_X \wedge^{(p,q)}(M) \subset \wedge^{(p,q)}(M) \quad \forall X, p, q$.

A Hermitian metric on an almost complex manifold M^{2n} is a Riemannian metric $\langle , \rangle$ invariant by the almost complex structure J, i.e.,

$$\langle JX, JY \rangle = \langle X, Y \rangle \quad \forall X, Y.$$

An almost complex manifold with a fixed Hermitian metric is known as a Hermitian manifold.

Let M^{2n} be a Hermitian manifold, and ∇ the Levi- Civita connection. It is easy to see that

(i) ∇ is real;

(ii) $\langle \nabla_X Y, Z \rangle + \langle Y, \nabla_X Z \rangle = X\langle , Z \rangle$;

(iii) $\nabla_X Y - \nabla_Y X = [X, Y]$.

Unfortunately, ∇ is not almost complex in general. A classical result says: ∇ is almost complex if and only if M is a Kahler manifold.

Now we will make a slight change of ∇ so as to get an almost complex connection. Define a connection $\tilde{\nabla}$ by

$$\tilde{\nabla}_X Y = \frac{1}{2}(\nabla_X Y - J\nabla_X(JY)) \quad \forall X, Y,$$

and we have

(i) $\tilde{\nabla}$ is indeed a connection,

(ii) $\tilde{\nabla}$ is real,

(iii) $\tilde\nabla$ preserves the metric, i.e.

$$\langle \tilde\nabla_X, Y, Z\rangle + \langle Y, \tilde\nabla_X Z\rangle = X\langle Y, Z\rangle,$$

(iv) $\tilde\nabla$ is almost complex,

(v) $\tilde\nabla = \nabla$ iff M is a Kahler manifold.

(Note: (i)-(iv) are easy to check.)

For a Hermitian manifold M^{2n}, there always exists locally a basis $\{E_1, \cdots, E_n, E_{\bar 1}, \cdots, E_{\bar n}\}$ satisfying the following conditions

(i) $\overline{E_a} = E_{\bar a}$,

(ii) $JE_i = E_{\bar i}$, $JE_{\bar i} = -E_i$,

(iii) $\langle E_a, E_b\rangle = \delta_{ab}$.

Where the indices $a, b, \cdots$ belong to the set $\{1, \cdots, n, \bar 1, \cdots, \bar n\}$, and the notation $\bar a$ means that

$$\bar a = \begin{cases} \bar i & \text{if } a = i, \\ i & \text{if } a = \bar i. \end{cases}$$

A basis satisfying the above conditions (i)-(iii) is called an S- basis.

Let $Z_i = \frac{1}{\sqrt 2}(E_i - \sqrt{-1}E_{\bar i})$, $Z_{\bar i} = \frac{1}{\sqrt 2}(E_i + \sqrt{-1}E_{\bar i})$, then $\{Z_1, \cdots, Z_n, Z_{\bar 1}, \cdots, Z_{\bar n}\}$ is a basis also, which satisfies the following conditions

(iv) $\overline{Z_a} = Z_{\bar a}$,

(v) $JZ_i = \sqrt{-1}Z_i$, $JZ_{\bar i} = -\sqrt{-1}Z_{\bar i}$,

(iv) $\langle Z_a, Z_b\rangle = \delta_{a\bar b}$.

A basis satisfying the above condition (iv)-(vi) is called a U-basis. We say that the S-basis $\{E_a\}$ and the U-basis $\{Z_a\}$, defined as above, are related to each other. It is easy to see that each one of them determines the other.

Let $\{\omega_1, \cdots, \omega_n, \omega_{\bar 1}, \cdots, \omega_{\bar n}\}$ and $\{\Omega_1, \cdots, \Omega_n, \Omega_{\bar 1}, \cdots, \Omega_{\bar n}\}$ be the dual bases of $\{E_a\}$ and $\{Z_a\}$ respectively. Then we have

$$\begin{cases} \overline{\omega_a} = \omega_a \\ J\omega_i = -\omega_{\bar i}, \quad J\omega_{\bar i} = \omega_i \end{cases}$$

$$\begin{cases} \overline{\Omega_a} = \Omega_{\bar a}, \\ J\Omega_i = \sqrt{-1}\Omega_i, \quad J\Omega_{\bar i} = -\sqrt{-1}\Omega_{\bar i}, \end{cases}$$

$$\begin{cases} \omega_i = \frac{1}{\sqrt 2}(\Omega_i + \Omega_{\bar i}) \\ \omega_{\bar i} = \frac{1}{\sqrt{-2}}(\Omega_i - \Omega_{\bar i}) \end{cases}$$

$$\begin{cases} \Omega_i = \frac{1}{\sqrt 2}(\omega_i + \sqrt{-1}\omega_{\bar i}) \\ \Omega_{\bar i} = \frac{1}{\sqrt 2}(\omega_i - \sqrt{-1}\omega_{\bar i}). \end{cases}$$

In this case the Kahler form is

$$\Phi = -2\sqrt{-1}\sum_i \Omega_i \wedge \Omega_{\bar i}.$$

Let Γ^c_{ab} be the components of the Levi-Civita connection ∇ with respect to the U-basis $\{Z_a\}$, i.e.,

$$\nabla_{Z_a} Z_b = \sum_c \Gamma^c_{ab} E_c.$$

Similarly $\tilde\Gamma^c_{ab}$ is defined by

$$\tilde\nabla_{Z_a} Z_b = \sum_c \tilde\Gamma^c_{ab} Z_c.$$

The fact that ∇ and $\tilde{\nabla}$ are real means that

$$\overline{\Gamma^c_{ab}} = \Gamma^{\bar{c}}_{\bar{a}\bar{b}}, \quad \overline{\tilde{\Gamma}^c_{ab}} = \tilde{\Gamma}^{\bar{c}}_{\bar{a}\bar{b}}.$$

The fact that ∇ and $\tilde{\nabla}$ preserve the metric means

$$\Gamma^c_{ab} + \Gamma^{\bar{b}}_{a\bar{c}} = 0,$$

$$\tilde{\Gamma}^c_{ab} + \tilde{\Gamma}^{\bar{b}}_{a\bar{c}} = 0.$$

And that $\tilde{\nabla}$ is almost complex refers to

$$\tilde{\Gamma}^{\bar{j}}_{ai} = \tilde{\Gamma}^{j}_{a\bar{i}} = 0.$$

Lemma 4. There hold

$$\Gamma^j_{ai} = \tilde{\Gamma}^j_{ai}, \quad \Gamma^{\bar{j}}_{a\bar{i}} = \tilde{\Gamma}^{\bar{j}}_{a\bar{i}},$$
$$\Gamma^k_{i\bar{j}} = \frac{-\sqrt{-1}}{4} d\Phi(Z_i, Z_{\bar{j}}, Z_{\bar{k}}),$$
$$\Gamma^{\bar{k}}_{\bar{i}j} = \frac{\sqrt{-1}}{4} d\Phi(Z_{\bar{i}}, Z_j, Z_k).$$

Proof. It is well known that $d = \sum_a \omega_a \nabla_{E_a}$, so $d = \sum_a \Omega_a \nabla_{Z_a}$, hence

$$d\Omega_c = \sum_a \Omega_a \nabla_{Z_a} \Omega_c = -\sum_{a,b} \Gamma^c_{ab} \Omega_a \Omega_b,$$
$$d\Phi = -2\sqrt{-1} \sum_{i,a,b} \{-\Gamma^i_{ab} \Omega_a \Omega_b \Omega_{\bar{i}} + \Gamma^{\bar{i}}_{ab} \Omega_i \Omega_a \Omega_b\}.$$

By the property $\Gamma^c_{ab} + \Gamma^{\bar{b}}_{a\bar{c}} = 0$, a little complicated computation shows

$$\begin{aligned} d\Phi = 2\sqrt{-1}\{ & -\sum_{j<k} \sum_i 2\Gamma^k_{\bar{i}j} \Omega_{\bar{i}} \Omega_j \Omega_k \\ & + \sum_{j<k} \sum_i 2\Gamma^k_{i\bar{j}} \Omega_i \Omega_{\bar{j}} \Omega_{\bar{k}} \\ & + \sum_{i,j,k} \Gamma^i_{\bar{j}\bar{k}} \Omega_{\bar{i}} \Omega_{\bar{j}} \Omega_{\bar{k}} \\ & - \sum_{i,j,k} \Gamma^i_{jk} \Omega_i \Omega_j \Omega_k \}, \end{aligned}$$

so

$$\frac{\sqrt{-1}}{4} d\Phi(Z_{\bar{i}}, Z_j, Z_k) = \frac{\sqrt{-1}}{4} \cdot (2\sqrt{-1}) \cdot (-2)\Gamma^{\bar{k}}_{\bar{i}j} = \Gamma^{\bar{k}}_{\bar{i}j},$$
$$-\frac{\sqrt{-1}}{2} d\Phi(Z_i, Z_{\bar{j}}, Z_{\bar{k}}) = (-\frac{\sqrt{2}}{4}) \cdot (2\sqrt{-1}) \cdot 2\Gamma^k_{i\bar{j}} = \Gamma^k_{i\bar{j}},$$

the proof is complete.

We introduce some linear operators as follows

$$\begin{aligned}
B_i^+ &= \Omega_i + i(Z_i), \quad B_i^- = \Omega_i - i(Z_i),\\
B_{\bar i}^+ &= \Omega_{\bar i} + i(Z_{\bar i}), \quad B_{\bar i}^- = \Omega_{\bar i} - i(Z_{\bar i}),\\
\gamma &= B_{\bar 1}^+ \cdots B_{\bar n}^+ B_{\bar 1}^- \cdots B_{\bar n}^-,\\
R_i &= B_i^+ \gamma, \quad R_{\bar i} = -\sqrt{-1} B_i^- \gamma,\\
L_i &= B_{\bar i}^+ \gamma, \quad L_{\bar i} = \sqrt{-1} B_{\bar i}^- \gamma.
\end{aligned}$$

It follows that

$$\begin{aligned}
\Omega_i &= \frac{1}{2}(R_i + \sqrt{-1}R_{\bar i})\gamma, \ i(Z_i) = \frac{1}{2}(R_i - \sqrt{-1}R_{\bar i})\gamma,\\
\Omega_{\bar i} &= \frac{1}{2}(L_i - \sqrt{-1}L_{\bar i})\gamma, \ i(Z_{\bar i}) = -\frac{1}{2}(L_i + \sqrt{-1}L_{\bar i})\gamma.
\end{aligned}$$

We present the following lemmas without proofs, for these proofs are similar to the corresponding ones in [4] or in §1 of this paper.

Lemma 5. There hold

$$\begin{aligned}
B_a^+ B_b^+ + B_b^+ B_a^+ &= 2\delta_{ab},\\
B_a^- B_b^- + B_b^- B_a^- &= -2\delta_{ab},\\
B_a^+ B_b^- + B_b^- B_a^+ &= 0.
\end{aligned}$$

Lemma 6.
(i) $\gamma^2 = 1$,
(ii) $\gamma = (-1)^{\frac{n(n-1)}{2}+p} : \wedge^{(p,q)}(M) \to \wedge^{(p,q)}(M)$,
(iii) $L_a \gamma = \gamma L_a$, $R_a \gamma = -\gamma R_a$.
Lemma 7. The algebra generated by $\{L_a, R_a\}$ is subject only to the relations

$$\begin{aligned}
L_a L_b + L_b L_a &= -2\delta_{ab},\\
R_a R_b + R_b R_a &= -2\delta_{ab},\\
L_a R_b &= R_b L_a.
\end{aligned}$$

Proposition. As operators acting on $\wedge^*(M)$, there holds

$$\tilde\nabla_{E_a} = \check E_a + \frac{1}{4}\sum_{b,c}\langle \tilde\nabla_{E_a} E_b, E_c\rangle (L_b L_c + R_b R_c),$$

where $\{E_a\}$ is an S-basis for the Hermitian manifold M^{2n}.
Proof. Since $\tilde\nabla$ is almost complex, we have

$$\begin{aligned}
\tilde\nabla_{Z_a} &= \check Z_a - \sum_{b,c} \tilde\Gamma_{ab}^c \Omega_b i(Z_c)\\
&= \check Z_a - \sum_{i,j} \tilde\Gamma_{ai}^j \Omega_i i(Z_j) - \sum_{i,j} \tilde\Gamma_{a\bar i}^{\bar j} \Omega_{\bar i} i(Z_{\bar j})\\
&= \check Z_a - \sum_{i,j} \langle \tilde\nabla_{Z_a} Z_i, Z_{\bar j}\rangle \Omega_i i(Z_j)\\
&\quad - \sum_{i,j} \langle \tilde\nabla_{Z_a} Z_{\bar i}, Z_j\rangle \Omega_{\bar i} i(Z_{\bar j}).
\end{aligned}$$

Hence

$$\tilde{\nabla}_{E_a} = \check{E}_a - \sum_{i,j}\{\langle\tilde{\nabla}_{E_a}Z_i, Z_{\bar{j}}\rangle\Omega_i i(Z_j) + \langle\tilde{\nabla}_{E_a}Z_{\bar{i}}, Z_j\rangle\Omega_{\bar{i}} i(Z_{\bar{j}})\}.$$

Let us deal with the term $\sum_{i,j}\langle\tilde{\nabla}_{E_a}Z_i, Z_{\bar{j}}\rangle\Omega_i i(Z_j)$ first.

$$\begin{aligned}\langle\tilde{\nabla}_{E_a}Z_i, Z_{\bar{j}}\rangle =&\frac{1}{2}\langle\tilde{\nabla}_{E_a}(E_i - \sqrt{-1}E_{\bar{i}}), (E_j + \sqrt{-1}E_{\bar{j}})\rangle\\ =&a_{ij} + \sqrt{-1}b_{ij},\end{aligned}$$

where

$$\begin{aligned}a_{ij} =&\frac{1}{2}\{\langle\tilde{\nabla}_{E_a}E_i, E_j\rangle + \langle\tilde{\nabla}_{E_a}E_{\bar{i}}, E_{\bar{j}}\rangle\},\\ b_{ij} =&\frac{1}{2}\{\langle\tilde{\nabla}_{E_a}E_i, E_{\bar{j}}\rangle - \langle\tilde{\nabla}_{E_a}E_{\bar{i}}, E_j\rangle\}.\end{aligned}$$

Since $\tilde{\nabla}$ is real and preserves the metric, a_{ij} and b_{ij} are real numbers and satisfy

$$a_{ij} = -a_{ji}, \qquad b_{ij} = b_{ji}.$$

Further more, since $\tilde{\nabla}$ is almost complex, we have

$$\begin{aligned}\langle\tilde{\nabla}_{E_a}E_{\bar{i}}, E_{\bar{j}}\rangle =&\langle\tilde{\nabla}_{E_a}JE_i, JE_j\rangle = \langle J\tilde{\nabla}_{E_a}E_i, JE_j\rangle = \langle\tilde{\nabla}_{E_a}E_i, E_j\rangle,\\ \langle\tilde{\nabla}_{E_a}E_i, E_{\bar{j}}\rangle =&\langle\tilde{\nabla}_{E_a}(-JE_{\bar{i}}), JE_j\rangle = -\langle\tilde{\nabla}_{E_a}E_{\bar{i}}, E_j\rangle.\end{aligned}$$

Hence

$$\begin{aligned}a_{ij} =&\langle\tilde{\nabla}_{E_a}E_i, E_j\rangle = \langle\tilde{\nabla}_{E_a}E_{\bar{i}}, E_{\bar{j}},\rangle\\ b_{ij} =&\langle\tilde{\nabla}_{E_a}E_i, E_{\bar{j}}\rangle = -\langle\tilde{\nabla}_{E_a}E_{\bar{i}}, E_j\rangle.\end{aligned}$$

On the other hand,

$$\begin{aligned}\Omega_i i(Z_j) =&\frac{1}{4}(R_i + \sqrt{-1}R_{\bar{i}})\gamma(R_j - \sqrt{-1}R_{\bar{j}})\gamma\\ =&-\frac{1}{4}(R_i + \sqrt{-1}R_{\bar{i}})(R_j - \sqrt{-1}R_{\bar{j}})\\ =&-\frac{1}{4}\{(R_iR_j + R_iR_j + 2\delta_{ij}) - [2\delta_{ij} + \sqrt{-1}(R_iR_j - R_iR_j)]\}.\end{aligned}$$

Note: $R_iR_j + R_{\bar{i}}R_{\bar{j}} + 2\delta_{ij}$ and a_{ij} are antisymmetric on the indices i, j while $2\delta_{ij} + \sqrt{-1}(R_iR_{\bar{j}} - R_{\bar{i}}R_j)$ and b_{ij} are symmetric, so

$$\begin{aligned}&\sum_{i,j}\langle\tilde{\nabla}_{E_a}Z_i, Z_{\bar{j}}\rangle\Omega_i i(Z_j)\\ =&(-\frac{1}{4})\sum_{i,j}\{a_{ij}(R_iR_j + R_{\bar{i}}R_{\bar{j}} + 2\delta_{ij}) - \sqrt{-1}b_{ij}[2\delta_{ij} + \sqrt{-1}(R_iR_{\bar{j}} - R_{\bar{i}}R_j)]\}\\ =&-\frac{1}{4}\sum_{i,j}\{a_{ij}R_iR_j + a_{ij}R_{\bar{i}}R_{\bar{j}} + b_{ij}R_iR_{\bar{j}} - b_{ij}R_{\bar{i}}R_j\} + \frac{\sqrt{-1}}{2}\sum_i b_{ii}\\ =&-\frac{1}{4}\sum_{b,c}\langle\tilde{\nabla}_{E_a}E_b, E_c\rangle R_bR_c + \frac{\sqrt{-1}}{2}\sum_i b_{ii}.\end{aligned}$$

Similarly,

$$\sum_{i,j}\langle\tilde{\nabla}_{E_a}Z_{\bar{i}}, Z_j\rangle\Omega_{\bar{i}}i(Z_{\bar{j}})$$
$$= -\frac{1}{4}\sum_{b,c}\langle\tilde{\nabla}_{E_a}E_b, E_c\rangle L_b L_c - \frac{\sqrt{-1}}{2}\sum_i b_{ii},$$

thus proving the proposition.

Suppose $\mathbf{L} : \wedge^*(M) \to \wedge^*(M)$ is a linear operator, we define three linear operators $\bar{\mathbf{L}}, \mathbf{L}^*, \mathbf{L}^\#$ by

$$\begin{aligned} \bar{\mathbf{L}}(\omega) &= \overline{\mathbf{L}(\bar{\omega})} && \forall\omega \in \wedge^*(M), \\ \langle \mathbf{L}^*\omega, \Omega\rangle_0 &= \langle\omega, \mathbf{L}\Omega\rangle_0 && \forall\omega, \Omega \in \wedge^*(M), \\ \ll \mathbf{L}^\#\omega, \Omega \gg_0 &= \ll \omega, \mathbf{L}^\#\Omega \gg_0 && \forall\omega, \Omega \in \wedge^*(M) \end{aligned}$$

where $\langle,\rangle_0$ and $\ll,\gg_0$ are defined by

$$\langle\omega, \Omega\rangle_0 = \int_M \langle\omega, \Omega\rangle d\ vol(M),$$
$$\ll \omega, \Omega \gg_0 \int_M \langle\omega, \bar{\Omega}\rangle d\ vol(M),$$

in which $\langle\omega, \Omega\rangle$ is induced by the Riemannian metric in a standard way. It is easy to see that

$$\mathbf{L}^\# = \overline{\mathbf{L}^*} = (\bar{\mathbf{L}})^*, \qquad d = \bar{d},$$

thus

$$d^\# = d^*.$$

Given an 1-form ω, we define a vector field ω^* by

$$\omega(Y) = \langle\omega^*, Y\rangle \qquad \forall \text{ vector field } Y.$$

Then it is easy to check that $\omega_a^* = E_a$ and $\Omega_a^* = Z_{\bar{a}}$. It is well known that

$$d = \sum_a \omega_a \nabla_{E_a}$$

and

$$d^* = -\sum_a i(E_a)\nabla_{E_a},$$

so by the Lemmas above we have

$$\begin{aligned} d &= \sum_a \omega_a \nabla_{E_a} = \sum_a \Omega_a \nabla_{Z_a} \\ &= \sum_a \Omega_a\{\tilde{\nabla}_{Z_a} - \sum_{i,j}\Gamma_{ai}^{\bar{j}}\Omega_i i(Z_{\bar{j}}) \\ &\quad - \sum_{i,j}\Gamma_{a\bar{i}}^{j}\Omega_{\bar{i}} i(Z_j)\}. \end{aligned}$$

Thus

$$\bar{\partial} = \sum_i \Omega_{\bar{i}} \tilde{\nabla}_{Z_{\bar{i}}} - \sum_{i,j,k} \Gamma^j_{k\bar{i}} \Omega_k \Omega_{\bar{i}} i(Z_j)$$
$$= \sum_i \Omega_{\bar{i}} \tilde{\nabla}_{Z_{\bar{i}}} + \frac{\sqrt{-1}}{4} \sum_{i,j,k} d\Phi(Z_k, Z_{\bar{i}}, Z_{\bar{j}}) \Omega_k \Omega_{\bar{i}} i(Z_j).$$

Similarly,

$$d^{\#} = d^* = -\sum_a i(E_a) \nabla_{E_a} = -\sum_a i(\omega_a^*) \nabla_{E_a}$$
$$= -\sum_a i(\Omega_a^*) \nabla_{Z_a} = -\sum_a i(Z_{\bar{a}}) \nabla_{Z_a}$$
$$= -\sum_a i(Z_{\bar{a}}) \{ \tilde{\nabla}_{Z_a} - \sum_{i,j} \Gamma^{\bar{j}}_{ai} \Omega_i i(Z_{\bar{j}})$$
$$- \sum_{i,j} \Gamma^j_{a\bar{i}} \Omega_{\bar{i}} i(Z_j) \},$$

thus

$$\bar{\partial}^{\#} = -\sum_i i(Z_{\bar{i}}) \tilde{\nabla}_{Z_i} + \sum_{i,j,k} \Gamma^j_{\bar{k}i} i(Z_k) \Omega_i i(Z_{\bar{j}})$$
$$= -\sum_i i(Z_{\bar{i}}) \tilde{\nabla}_{Z_i} + \frac{\sqrt{-1}}{4} \sum_{i,j,k} d\Phi(Z_{\bar{k}}, Z_i, Z_j) i(Z_k) \Omega_i i(Z_{\bar{j}}).$$

<u>Lemma 8</u>. There holds

$$\sum_i \Omega_{\bar{i}} \tilde{\nabla}_{Z_{\bar{i}}} - \sum_i i(Z_{\bar{i}}) \tilde{\nabla}_{Z_i}$$
$$= \frac{1}{\sqrt{2}} \sum_a L_a \{ \check{E}_a + \frac{1}{4} \sum_{b,c} \langle \tilde{\nabla}_{E_a} E_b, E_c \rangle (L_b L_c + R_b R_c) \} \gamma.$$

Proof. Since

$$\sum_i \Omega_{\bar{i}} \tilde{\nabla}_{Z_{\bar{i}}} - \sum_i i(Z_{\bar{i}}) \tilde{\nabla}_{Z_i} = \sum_i \{ \frac{1}{2\sqrt{2}} (L_i - \sqrt{-1} L_{\bar{i}}) \gamma \tilde{\nabla}_{E_i + \sqrt{-1} E_{\bar{i}}}$$
$$- (-\frac{1}{2\sqrt{2}}) (L_i + \sqrt{-1} L_{\bar{i}}) \gamma \tilde{\nabla}_{E_i - \sqrt{-1} E_{\bar{i}}} \}$$
$$= \frac{1}{\sqrt{2}} \sum_a L_a \gamma \tilde{\nabla}_{E_a}$$
$$= \frac{1}{\sqrt{2}} \sum_a L_a \{ \check{E}_a + \frac{1}{4} \sum_{b,c} \langle \tilde{\nabla}_{E_a} E_b, E_c \rangle (L_b L_c + R_b R_c) \} \gamma,$$

the lemma holds true.

<u>Theorem B</u>. Given a Hermitian almost complex manifold M^{2n} and choose an S-basis $\{E_a\}$, then

$$\bar{\partial} + \bar{\partial}^{\#} = \frac{1}{\sqrt{2}} \sum_a L_a \{ \check{E}_a + \frac{1}{4} \langle \tilde{\nabla}_{E_a} E_b, E_c \rangle (L_b L_c + R_b R_c) \} \gamma$$
$$+ \sum_a L_a \{ \frac{1}{32} \sum_{b,c} [d\Phi(JE_a, E_b, E_c) + d\Phi(JE_a, JE_b, JE_c)] R_b R_c$$
$$+ \frac{1}{16} \sum_b d\Phi(E_a, JE_b, E_b) \} \gamma.$$

In particular, if M^{2n} is a Kahler manifold, then

$$\bar{\partial}+\bar{\partial}^{\#}=\frac{1}{\sqrt{2}}\sum_a L_a\{\check{E}_a+\frac{1}{4}\langle\nabla_{E_a}E_b,E_c\rangle(L_bL_c+R_bR_c)\}\gamma,$$

where ∇ is the Levi-Civita connection.

Proof. What we need to prove is the equality

$$\frac{\sqrt{-1}}{4}d\Phi(Z_k,Z_{\bar{i}},Z_{\bar{j}})\Omega_k\Omega_{\bar{i}}i(Z_j)+\frac{\sqrt{-1}}{4}d\Phi(Z_{\bar{k}},Z_i,Z_j)i(Z_k)\Omega_i i(Z_{\bar{j}})$$
$$=\sum_a L_a\{\frac{1}{32}\sum_{b,c}[d\Phi(JE_a,E_b,E_c)+d\Phi(JE_a,JE_b,JE_c)]R_bR_c+\frac{1}{16}\sum_b d\Phi(E_a,JE_b,E_b)\}\gamma.$$

It is sufficient to check by a straightforward computation, hence we omit the details.

§3. $C_{2n}\oplus C_{2n}$ modules.

Now we are going to discuss the $C_{2n}\oplus C_{2n}$ modules $\wedge^*(M^{2n})$'s appearing in §1 and §2. We note that, $\wedge^*(M^{2n})$ appearing in §1 and §2 are the same linear space, however, as $C_{2n}\oplus C_{2n}$ modules they are different. Hence we denote $\wedge^*(M)$ by $\wedge_1^*(M)$ or $\wedge_2^*(M)$ according as $\wedge^*(M)$ was defined in §1 or §2.

Chosen a proper basis $\{E_a\},\wedge^*(M^{2n})$ may be regarded as $C^\infty(M)\otimes\wedge^*(2n)$, where $C^\infty(M)$ is the set of C^∞ functions on M, and

$$\wedge^*(2n)=\{\sum_{i_1<\cdots<i_p} c_{i_1}\cdots{}_{i_p}\,\omega_{i_1}\wedge\cdots\wedge\omega_{i_p}|c_{i_1}\cdots{}_{i_p}\in\mathbf{C}\},$$

For any $\mathbf{L}\in C_{2n}\oplus C_{2n}, f\in C^\infty(M),\omega\in\wedge^*(2n)$, there holds

$$\mathbf{L}(f\omega)=f\mathbf{L}(\omega),$$

hence we can discuss the $C_{2n}\oplus C_{2n}$ modules $\wedge_1^*(2n)$ and $\wedge_2^*(2n)$ in instead of the $C_{2n}\oplus C_{2n}$ modules $\wedge_1^*(M^{2n}),\wedge_2^*(M^{2n})$.

We introduce another $C_{2n}\oplus C_{2n}$ module $\wedge_0^*(2n)$ as follows: Let Ψ be the irreducible C_{2n} module of dim 2^n, we make $\Psi\otimes\Psi$ into a $C_{2n}\oplus C_{2n}$ module by setting

$$(C_{2n}\oplus C_{2n})\times(\Psi\otimes\Psi)\to(\Psi\otimes\Psi):$$
$$(\mathbf{L}_\alpha\oplus\mathbf{L}_\beta,f_1\otimes f_2)\mapsto(\mathbf{L}_\alpha f_1)\otimes(\mathbf{L}_\beta f_2).$$

According to [2], $\Psi\otimes\Psi$ may be expressed in the following way. Let

$$\mathbf{L}_i=L_i-\sqrt{-1}L_{\bar{i}},\quad \hat{\mathbf{L}}_i=L_i+\sqrt{-1}L_{\bar{i}},$$
$$\mathbf{R}_i=R_i-\sqrt{-1}R_{\bar{i}},\quad \hat{\mathbf{R}}_i=R_i+\sqrt{-1}R_{\bar{i}},$$

then $\Psi\otimes\Psi$ may be treated as the linear space spanned by the set $\{\mathbf{L}_{i_1}\cdots\mathbf{L}_{i_p}\hat{\mathbf{L}}\otimes\mathbf{R}_{j_1}\cdots\mathbf{R}_{j_q}\hat{\mathbf{R}}|\forall p,q,$ and $i_1<\cdots<i_p,j_1<\cdots<j_q\}$, where $\hat{\mathbf{L}}=\hat{\mathbf{L}}_1\cdots\hat{\mathbf{L}}_n,\hat{\mathbf{R}}=\hat{\mathbf{R}}_1\cdots\hat{\mathbf{R}}_n$. We denote the $C_{2n}\oplus C_{2n}$ module $\Psi\otimes\Psi$ by $\wedge_0^*(2n)$.

Lemma 9. For $\alpha=0,1,2,\wedge_\alpha^*(2n)$ has the following properties:

(i) dim $\wedge_\alpha^*(2n)=2^{2n}$;

(ii) $\wedge_\alpha^*(2n)$ is a irreducible $C_{2n}\oplus C_{2n}$ module;

(iii) there exists an element $\theta_\alpha\in\wedge_\alpha^*(2n)$ such that

$$\hat{\mathbf{L}}_i\theta_\alpha=0,\quad \hat{\mathbf{R}}\theta_\alpha=0.$$

Proof. (i) and (ii) are trivial. We choose $\theta_0 = \hat{\mathbf{L}} \otimes \hat{\mathbf{R}}$, $\theta_1 = (\frac{\omega_1 + \sqrt{-1}\omega_1}{2}) \wedge \cdots \wedge (\frac{\omega_n + \sqrt{-1}\omega_n}{2})$, $\theta_2 = \Omega_1 \wedge \cdots \wedge \Omega_n$, then it is easy to check (iii).

Lemma 10. There exists a unique $C_{2n} \oplus C_{2n}$ module isomorphism $\Phi_\alpha : \wedge_0^*(2n) \to \wedge_\alpha^*(2n)$ such that $\Phi_\alpha(\theta_0) = \theta_\alpha$.

Proof. Define Φ_α by

$$\begin{aligned}\Phi_\alpha =&(\mathbf{L}_{i_1} \cdots \mathbf{L}_{i_p} \hat{\mathbf{L}} \otimes \mathbf{R}_{j_1} \cdots \mathbf{R}_{j_q} \hat{\mathbf{R}}) \\ =&\mathbf{L}_{i_1} \cdots \mathbf{L}_{i_p} \mathbf{R}_{j_1} \cdots \mathbf{R}_{j_q} \theta_\alpha .\end{aligned}$$

It can be proved by Lemma 9 that Φ_α is a $C_{2n} \oplus C_{2n}$ module isomorphism. The uniqueness is trivial, hence the lemma holds true.

§4. A remark

Given a Riemannian manifold M^n, a C_n module Δ, and a complex vector space V, we have two vector bundles $\pi : M \times \Delta \times V \to M$ and $\pi_0 : M \times V \to M$. Let D be a connection on π_0, then by [3] the twisted Dirac operator can be expressed as

$$\sum_i L_i[(\check{E}_i + \frac{1}{4}\sum_{j,k} \Gamma_{ij}^k L_j L_k) \otimes 1 + 1 \otimes D_{E_i}]$$

where $\{E_i\}$ is an orthonormal basis and Γ_{ij}^k's the components of the Levi-Civita connection. Hence, by Theorems A and B the operators $d + \delta, \bar{\partial} + \bar{\partial}^\#$ may be regarded as twisted Dirac operators with a slight modification caused by γ_0 and γ.

References

[1] Atiyah, M.Bott, R. and Patodi, V.K., On the Heat equation and Index Theorem, Invent. Math., 19(1973), 279-330.

[2] Wu, H., The Bochner technique, Proceedings of 1980 Beijing Symposium on Differential Geometry and Differential Equations (S.S.Chern and Wu Wen-tsun eds), Science Press and Gordon and Breach, New York, 2(1982), 929-1072.

[3] Yu, Y.L., Local index theorem for Dirac operator, Acta Math. Sinica, New Series 3(1987), 153-169.

[4] Yu, Y.L., Local index theorem for Signature operators, Acta Math. Sinica, New Series 3(1987), 363-372.

Yu Yanlin
Institute of Mathematics, Academia Sinica
and
Nankai Institute of Mathematics

Volume of Geodesic Balls

Yu Yanlin
Institute of Mathematics, Academia Sinica
and Nankai Institute of Mathematics

Given a Riemannian manifold M of dim n, and $p \in M$; let $B_p(r)$ be a geodesic ball with p and r as its center point and radius respectively, and denote its volume by $V_p(r)$. In this paper we will examine the Taylor's expansion of $V_p(r)$ at p with respect to the variable r, so that the coefficients in the expansion can be expressed in terms of the components of the covariant derivatives of the Riemannian curvature at p. So far as we know, there are only a few first terms in the expansion that have been examined this way (see [1]).

§1 Expansions relative to normal coordinates

Fix an orthonormal frame $\{E_1(p), \cdots, E_n(p)\}$ at p, and for each point $q \in M$ around p define its normal coordinates $(y_1, \cdots, y_n)$ by

$$q = Exp_p(y_1 E_1(p) + \cdots + y_n E_n(p)),$$

where $Exp_p : T_pM \to M$ is the exponential map. Let $\{E_1, \cdots, E_n\}$ be the orthonormal frame field around p, which is parallel along geodesics passing through p, such that $\{E_1, \cdots, E_n\}_p = \{E_1(p), \cdots, E_n(p)\}$. Here is a well known lemma below.

<u>Lemma 1.</u>

$$\sum_s y_s \frac{\partial}{\partial y_s} = \sum_s y_s E_s,$$
$$\sum_s y_s \Gamma^j_{si} = 0 \qquad \forall i, j,$$

where $\{\Gamma^k_{ij}\}$ are the Christoffel coefficients of the Levi-Civita connection ∇, defined by

$$\nabla_{E_i} E_j = \sum_k \Gamma^k_{ij} E_k.$$

For a function f, define

$$\hat{f}(m) = \frac{1}{m!} \sum_{i_1, \cdots, i_m} \frac{\partial^m f}{\partial y_{i_1} \cdots \partial y_{i_m}}(p) \cdot y_{i_1} \cdots y_{i_m}.$$

Sometimes we denote $\hat{f}(m)$ by $f(\hat{m})$. As usual $\sum_{m=0}^{\infty} \hat{f}(m)$ is called the Taylor's expansion of f at p with respect to the variables $y_1, \cdots, y_n$.

Let us introduce an example to describe the task of this section. Suppose that a vector field X is given, we now define functions f_i, and $\tilde{f}_i$ by

$$X = \sum f_i E_i,$$
$$X = \sum \tilde{f}_i \frac{\partial}{\partial y_i}.$$

Our task is to solve the following problem.

Problem. How to express $\hat{f}_i(m), \hat{\tilde{f}}_i(m)$ in terms of the covariant derivatives of X and the Riemannian curvature?

For a general tensor T, we may pose the same problem also.

Given a tensor T of type (r, s), and let $\{T^{i_1\cdots i_r}_{j_1\cdots j_s}\}$ be its components with respect to $\{E_1, \cdots, E_n\}$. The covariant differential ∇T of T is a tensor of type $(r, s+1)$, whose components relative to $\{E_1, \cdots, E_n\}$ are $\{(\nabla T)^{i_1\cdots i_r}_{j_1\cdots j_s i}\}$. It is well known that

$$\begin{aligned}&(\nabla T)^{i_1\cdots i_r}_{j_1\cdots j_s i}\\ =&E_i T^{i_1\cdots i_r}_{j_1\cdots j_s} + \Gamma^{i_1}_{il} T^{l i_2\cdots i_r}_{j_1\cdots j_s} + \cdots + \Gamma^{i_r}_{il} T^{i_1\cdots i_{r-1} l}_{j_1\cdots j_s}\\ &- \Gamma^{l}_{ij_1} T^{i_1\cdots i_r}_{l j_2\cdots j_s} - \cdots - \Gamma^{l}_{ij_s} T^{i_1\cdots i_r}_{j_1\cdots j_{s-1} l}.\end{aligned}$$

Here we employ the summation convention where use has been made of that repeated indices implies addition. Here after we will follow this notation. As usual $(\nabla T)^{i_1\cdots i_r}_{j_1\cdots j_s i}$ is still denoted by $T^{i_1\cdots i_r}_{j_1\cdots j_s, i}$.

Proposition 2. Given a tensor T of type (r, s), and let $\{T^{i_1\cdots i_r}_{j_1\cdots j_s}\}$ be its components with respect to $\{E_1, \cdots, E_n\}$, then

$$T^{i_1\cdots i_r}_{j_1\cdots j_s}(\hat{m}) = \frac{1}{m!} \sum_{\alpha_1, \cdots, \alpha_m} T^{i_1\cdots i_r}_{j_1\cdots j_s, \alpha_1\cdots\alpha_m}(p) \cdot y_{\alpha_1} \cdots y_{\alpha_m}.$$

Proof. We are going to prove the following stronger equalities

$$\begin{aligned}&\sum_{\alpha_1, \cdots, \alpha_m} y_{\alpha_1} \cdots y_{\alpha_m} \frac{\partial^m}{\partial y_{\alpha_1} \cdots \partial y_{\alpha_m}} T^{i_1\cdots i_r}_{j_1\cdots j_s} =\\ &= \sum_{\alpha_1\cdots\alpha_m} y_{\alpha_1} \cdots y_{\alpha_m} T^{i_1\cdots i_r}_{j_1\cdots j_s, \alpha_1\cdots\alpha_m} \qquad (F_m),\end{aligned}$$

which implies the proposition immediately. If $m = 1$, using Lemma 1 we have

$$\begin{aligned}&\sum_k y_k \frac{\partial}{\partial y_k} T^{i_1\cdots i_r}_{j_1\cdots j_s} = \sum_k y_k E_k T^{i_1\cdots i_r}_{j_1\cdots j_r}\\ =&y_k E_k T^{i_1\cdots i_r}_{j_1\cdots j_s} + y_k \Gamma^{i_1}_{kl} T^{l i_2\cdots i_r}_{j_1\cdots j_s} + \cdots\\ &- y_k \Gamma^{l}_{kj_1} T^{i_1\cdots i_r}_{l j_1\cdots j_s} - \cdots\\ =&\sum_k y_k T^{i_1\cdots i_r}_{j_1\cdots j_s, k}.\end{aligned}$$

Hence (F_1) is true. If the equality (F_m) is true, then

$$\begin{aligned}&\sum_{\alpha_1\cdots\alpha_{m+1}} y_{\alpha_1} \cdots y_{\alpha_{m+1}} \frac{\partial^{m+1}}{\partial y_{\alpha_1} \cdots \partial y_{\alpha_{m+1}}} T^{i_1\cdots i_r}_{j_1\cdots j_s} =\\ =&\sum_k y_k \frac{\partial}{\partial y_k} \Big(\sum_{\alpha_1\cdots\alpha_m} y_{\alpha_1} \cdots y_{\alpha_m} \frac{\partial^m}{\partial y_{\alpha_1} \cdots \partial y_{\alpha_m}} T^{i_1\cdots i_r}_{j_1\cdots j_s} \Big)\\ &- \sum_{\alpha_1\cdots\alpha_m} \Big(\sum_k y_k \frac{\partial}{\partial y_k} (y_{\alpha_1} \cdots y_{\alpha_m}) \Big) \frac{\partial^m}{\partial y_{\alpha_1} \cdots \partial y_{\alpha_m}} \cdot T^{i_1\cdots i_r}_{j_1\cdots j_s}\\ =&\sum_k y_k \frac{\partial}{\partial y_k} \Big(\sum_{\alpha_1\cdots\alpha_m} y_{\alpha_1} \cdots y_{\alpha_m} T^{i_1\cdots i_r}_{j_1\cdots j_s, \alpha_1\cdots\alpha_m} \Big)\\ &- m \sum_{\alpha\cdots\alpha_m} y_{\alpha_1} \cdots y_{\alpha_m} T^{i_1\cdots i_r}_{j_1\cdots j_s, \alpha_1\cdots\alpha_m}\\ =&\sum_{\alpha_1\cdots\alpha_m} y_{\alpha_1} \cdots y_{\alpha_m} \Big(\sum_k y_k \frac{\partial}{\partial y_k} T^{i_1\cdots i_r}_{j_1\cdots j_s, \alpha_1\cdots\alpha_m} \Big).\end{aligned}$$

By the equality (F_1) we have

$$\sum_k y_k \frac{\partial}{\partial y_k} T^{i_1\cdots i_r}_{j_1\cdots j_s,\alpha_1\cdots\alpha_m} = \sum_k y_k T^{i_1\cdots i_r}_{j_1\cdots j_s,\alpha_1\cdots\alpha_m k}.$$

Consequently

$$\begin{aligned}&\sum y_{\alpha_1}\cdots y_{\alpha_{m+1}} \frac{\partial^{m+1}}{\partial y_{\alpha_1}\cdots \partial y_{\alpha_{m+1}}} T^{i_1\cdots i_r}_{j_1\cdots j_s}\\ &= \sum y_{\alpha_1}\cdots y_{\alpha_m} y_k T^{i_1\cdots i_r}_{j_1\cdots j_s,\alpha_1\cdots\alpha_m k},\end{aligned}$$

thus proving the proposition.

In [3] we defined H_{ij}, H_{ijk} by

$$\begin{aligned}H_{ij} &= \langle \frac{\partial}{\partial y_i}, E_j\rangle,\\ H_{ijk} &= \sum_l H_{il}\Gamma^k_{lj},\end{aligned}$$

and had the following lemma.

Lemma 3. There holds

$$\begin{cases} \hat{d}H_{ij} = \delta_{ij} - H_{ij} + y_\alpha H_{i\alpha j} \\ \hat{d}H_{ijk} = -H_{ijk} + y_\beta H_{i\alpha} R_{\alpha\beta jk}, \end{cases}$$

where $\hat{d} = \sum y_s \frac{\partial}{\partial y_s}$, $\{R_{ijkl}\}$ are the components of the Riemannian curvature, defined by

$$R_{ijkl} = -\langle (\nabla_{E_i}\nabla_{E_j} - \nabla_{E_j}\nabla_{E_i} - \nabla_{[E_i,E_j]})E_k, E_l\rangle.$$

Let $R_{ij} = \sum_{\alpha,\beta} y_\alpha y_\beta R_{i\alpha\beta j}$, $R_{ijk} = \sum_\alpha y_\alpha R_{i\alpha jk}$, then by Proposition 2 we have

$$\widehat{R_{ij}}(m) = \begin{cases} 0 & \text{if } m = 0, 1, \\ \frac{1}{(m-2)!}\sum R_{i\alpha_1\alpha_2 j,\alpha_3\cdots\alpha_m}(p) y_{\alpha_1}\cdots y_{\alpha_m} & \text{if } m \geq 2; \end{cases}$$

$$\widehat{R_{ijk}}(m) = \begin{cases} 0 & \text{if } m = 0, \\ \frac{1}{(m-1)!}\sum R_{i\alpha_1 jk,\alpha_2\cdots\alpha_m}(p) y_{\alpha_1}\cdots y_{\alpha_m} & \text{if } m \geq 1. \end{cases}$$

Proposition 4. There hold

$$\widehat{H_{ij}}(m) = \widehat{\delta_{ij}}(m) + \sum_{s\geq 1}\ \sum_{\substack{k_1,\cdots,k_s\geq 2\\ k_1+\cdots+k_s=m}}\ \sum_{j_1,\cdots,j_{s-1}} \frac{\widehat{R_{ij_1}}(k_1)\widehat{R_{j_1j_2}}(k_2)\cdots \widehat{R_{j_{s-1}j}}(k_s)}{[k_1,\cdots,k_s]},$$

and

$$\widehat{H_{ijk}}(m) = \frac{1}{m+1}\{\widehat{R_{ijk}}(m) + \sum_{s\geq 1}\ \sum_{\substack{k_1,\cdots,k_s\geq 2\\ m_0+k_1+\cdots+k_s=m}}\ \sum_{j_1,\cdots,j_s} \frac{\widehat{R_{ij_1}}(k_1)\cdots \widehat{R_{j_{s-1}j_s}}(k_s)\widehat{R_{j_sjk}}(m_0)}{[k_1,\cdots,k_s]}\}$$

where $[k_1,\cdots,k_s] = k_1(k_1+1)(k_1+k_2)(k_1+k_2+1)\cdots(k_1+\cdots+k_s)(k_1+\cdots+k_s+1)$.

Proof. For any function f, $(\hat{d}f)(\hat{m}) = mf(\hat{m})$. Hence by Lemma 3 we have

$$(m+1)\widehat{H_{ij}}(m) = \widehat{\delta_{ij}}(m) + y_\alpha \widehat{H_{i\alpha j}}(m-1),$$
$$(m+1)\widehat{H_{ijk}}(m) = \sum_{m_1+m_2=m} \widehat{H_{i\alpha}}(m_1)\widehat{R_{\alpha jk}}(m_2).$$

Then the proposition can be easily proved by induction on m.

Let the matrix (H^{ij}) be the inverse of (H_{ij}), it is easy to see that $\widehat{H^{ij}}(m) = \delta_{ij}$ if $m = 0$, and if $m \geq 1$, we have

$$\widehat{H^{ij}}(m) = \sum_{s\geq 1} \sum_{\substack{l_1,\cdots,l_s\geq 1 \\ l_1+\cdots+l_s=m}} \sum_{\alpha_1,\cdots,\alpha_{s-1}} (-1)^s \widehat{H_{i\alpha_1}}(l_1)\widehat{H_{\alpha_1\alpha_2}}(l_2)\cdots \widehat{H_{\alpha_{s-1}j}}(l_s).$$

Now let us solve the problem of this section. By proposition 2, we can get $\widehat{f_i}(m)$. Since $\tilde{f}_i = \sum H^{ij} f_j$, $\tilde{f}_i(m) = \sum_{m_1+m_2=m} H^{ij}(\widehat{m_1}) f_j(\widehat{m_2})$, hence we can get $\tilde{f}_i(\hat{m})$ also.

§2 Volume of a geodesic ball

Let us give the following definitions first.

Definition 1. Let T and S be tensors of the types $(1, 1+r)$ and $(1, 1+s)$ respectively, where $r, s \geq 0$, and we define a tensor $T\hat{\otimes}S$ of the type $(1, 1+r+s)$ by

$$(T\hat{\otimes}S)^i_{jj_1\cdots j_r k_1\cdots k_s} = \sum_l T^i_{kj_1\cdots j_r} S^k_{jk_1\cdots k_s}.$$

Definition 2. Let $T(1), \cdots, T(n)$ be tensors of the types $(1, 1+r_1), \cdots, (1, 1+r_n)$ respectively, where $r_i \geq 0$, and we define a tensor $\det(T(1), \cdots, T(n))$ of the type $(0, r_1 + \cdots + r_n)$ by

$$(det(T(1), \cdots, T(n)))_{\cdots} = \sum_{\substack{i_1,\cdots,i_n \\ j_1,\cdots,j_n}} \delta^{i_1\cdots i_n}_{j_1\cdots j_n} (T(1))^{i_1}_{j_1\cdots} \cdots (T(n))^{i_n}_{j_n\cdots}$$

Definition 3. Let T be a tensor of the type (o, r), and we define a tensor $AS(T)$ of the type $(0, 0)$ as follows:

$$AS(T) = \begin{cases} \sum_\pi T_{i_1,\cdots i_r} g^{i_{\pi(1)}i_{\pi(2)}} \cdots g^{i_{\pi(r-1)}i_{\pi(r)}}, \text{if } r = \text{even}, \\ 0 \quad \text{if } r = \text{odd} \end{cases}$$

where $\sum_\pi$ means the summation running over the pemutations of r elements.

Definition 4. For $m \geq 2$, let $K(m)$ be a tensor of the type $(1, 1+m)$, whose components relative to $\{E_1, \cdots, E_n\}$ are

$$(K(m))^i_{jj_1\cdots j_m} = R_{ij_1j_2j,j_3\cdots j_m}.$$

And for $s \geq 0$, and $m_1, \cdots, m_s \geq 2$ we define

$$K(m_1, \cdots, m_s) = \begin{cases} K(m_1)\hat{\otimes}\cdots\hat{\otimes}K(m_s) & \text{if } s \geq 1, \\ \{\delta^i_j\} & \text{if } s = 0 \end{cases}$$

which is of the type $(1, 1 + m_1 + \cdots + m_s)$.

Let

$$M = \begin{pmatrix} m^1_1 & \cdots & m^1_{s_1} \\ \cdots & \cdots & \cdots \\ m^n_1 & \cdots & m^n_{s_n} \end{pmatrix},$$

where $s_1, \cdots, s_n \geq 0$, $m^i_j \geq 2$. We define

$$\|M\| = (\prod_{i=1}^{n}[m^i_1, \cdots, m^i_{s_i}])(\prod_{i=1}^{n}\prod_{j=1}^{s_i}(m^i_j - 2)!),$$

$$|M| = \sum_{i=1}^{n}\sum_{j=1}^{s_i} m^i_j,$$

$$\Gamma_m = \{M \mid |M| = m\}.$$

Let $D^n(r)$ and $S^{n-1}(r)$ be respectively the ball and the sphere in the Euclidean space of dim n, which have the same radius r. We denote their volumes respectively by $\mathrm{Vol}(D^n(r))$ and $\mathrm{Vol}(S^{n-1}(r))$. It is well known that

$$\mathrm{Vol}(D^n(r)) = \frac{r^n}{n}\mathrm{Vol}(S^{n-1}(1)),$$
$$\mathrm{Vol}(S^{n-1}(r)) = r^{n-1}\mathrm{Vol}(S^{n-1}(1)).$$

Theorem. Let $V_p(r)$ be the volume of the geodesic ball with p and r as its center point and radius respectively, then we have

$$\frac{V_p(r)}{\mathrm{Vol}(D^n(r))} = \sum_{m=\text{even}} \frac{n}{m+n} \cdot \frac{(n-2)!!}{(n-2+m)!!m!!} \cdot \left(\sum_{M\in\Gamma_m} \frac{AS\det(K(M))}{\|M\|}(p) \right) r^m$$

where

$$K(M) = (K(m^1_1, \cdots, m^1_{s_1}), \cdots, K(m^n_1, \cdots, m^n_{s_n}))$$

Proof. Define $x_1, \cdots, x_n$ and ρ by

$$\begin{cases} \rho = |y| = \sqrt{y_1^2 + \cdots + y_n^2}, \\ y_i = \rho x_i, i = 1, \cdots, n \end{cases}$$

then $x_1^2 + \cdots + x_n^2 = 1$. Let dx denote the volume element of $S^{n-1}(1)$. Thus we have

$$\begin{aligned} V_p(r) &= \int \cdots \int_{y_1^2+\cdots+y_n^2\leq r^2} \sqrt{det((\frac{\partial}{\partial y_i}, \frac{\partial}{\partial y_j}))} dy_1 \cdots dy_n \\ &= \int_{o\leq\rho\leq r}\int_{|x|=1} det(H_{ij}) \cdot \rho^{n-1} dx \cdot d\rho \\ &= \sum_{m=0}^{\infty}\left(\int_{|x|=1}\frac{det\widehat{(H_{ij})}(m)}{\rho^m}\right)\int_0^r \rho^{m+n-1}d\rho \\ &= \sum_{m=0}^{\infty}\frac{1}{n+m}\left(\int_{S^{n-1}(1)}\frac{det(H_{ij})(\hat{m})}{\rho^m}\right) r^{m+n} \end{aligned}$$

It is known that

$$det(H_{ij})(\hat{m}) = \left(\sum_{\substack{i_1 - i_n \\ j_1\cdots j_n}} \frac{1}{n!} \in (i_1, \cdots, i_n) \in (j_1 \cdots j_n) H_{i_1 j_1} \cdots H_{i_n j_n} \right)(\hat{m}).$$

By Proposition 4, we have

$$\widehat{H_{ij}}(m)=\sum_{s\geq 0}\sum_{\substack{m_1,\cdots,m_s\geq 2\\ m_1+\cdots+m_s=m}}\sum_{y_{\alpha_1}\cdots\alpha_m}\frac{(K(m_1,\cdots,m_s)(p))^i_{j\alpha_1\cdots\alpha_m}y_{\alpha_1}\cdots y_{\alpha_m}}{[m_1,\cdots,m_s]\cdot(m_1-2)!\cdots(m_s-2)!}$$

Hence

$$\int_{S^{n-1}(1)}\frac{det(H_{ij})(\hat{m})}{\rho^m}dx$$
$$=\sum_{M\in\Gamma_m}\frac{det(K(m_1^1,\cdots,m_{s_1}^1),\cdots,K(m_1^n,\cdots,m_{s_n}^n))_{\alpha_1\cdots\alpha_m}}{\|M\|}\cdot$$
$$\int_{S^{n-1}(1)}x_{\alpha_1}\cdots x_{\alpha_m}dx.$$

Thus the rest of the proof can be reduced to that of the following proposition.

Proposition 5. There holds

$$\int_{S^{n-1}(1)}x_{\alpha_1}\cdots x_{\alpha_m}dx$$
$$=\begin{cases}0 & \text{if } m=\text{odd.}\\ \frac{(n-2)!!}{(n-2+m)!!m!!}vol(S^{n-1}(1))\sum_\pi\delta_{\alpha_{\pi(1)}\alpha_{\pi(2)}}\cdots\delta_{\alpha_{\pi(m-1)}\alpha_{\pi(m)}} & \text{if } m=\text{even}\end{cases}$$

where π runs over permutations of the set $\{1,\cdots,m\}$.

Proof. if $\{\alpha_1,\cdots,\alpha_m\}=\{\underbrace{1,\cdots,1}_{e_1},\underbrace{2,\cdots,2}_{e_2},\cdots,\underbrace{n,\cdots,n}_{e_n}\}$ then $e_1+\cdots+e_n=m$. By Weyl's result [2] we have

$$\int_{S^{n-1}(1)}x_{\alpha_1}\cdots x_{\alpha_m}dx=\int_{S^{n-1}(1)}x_1^{e_1}x_2^{e_2}\cdots x_n^{e_n}dx$$
$$=\begin{cases}\frac{(n-2)!!}{(n-2-m)!!}[(e_1-1)!!]\cdots[(e_n-1)!!]vol(S^{n-1}(1)), & \text{if all } e_1,\cdots,e_n \text{ are even}\\ 0, & \text{otherwise.}\end{cases}$$

When m is even and one of $e_1,\cdots,e_n$ is odd, it is trivial $\sum_\pi\delta_{i_{\pi(1)}i_{\pi(2)}}\cdots\delta_{i_{\pi(m-1)}i_{\pi(m)}}=0$. When all $e_1,\cdots,e_n$ are even, by a combinatorial consideration we have

$$\sum_\pi\delta_{i_{\pi(1)}i_{\pi(2)}}\cdots\delta_{i_{\pi(m-1)}i_{\pi(m)}}=\frac{(\frac{e_1+\cdots+e_n}{2})!}{(\frac{e_1}{2}!)\cdots(\frac{e_n}{2}!)}(e_1!)\cdots(e_n!)$$
$$=\frac{m!!}{(e_1!!)\cdots(e_n!!)}(e_1!)\cdots(e_n!)=(m!!)[(e_1-1)!!]\cdots[(e_n-1)!!]$$

thus proving 5.

References

[1] Gray A., Vanhecke L., Riemannian geometry as determined by the volumes of small geodesic ball, Acta Math., 142 (1979) 157-198.

[2] Weyl H., On the volume of tubes, Amer. J. Math., 61(1939), 461-472.

[3] Yu.Yanlin, Local index theorem for Dirac operator, Acta Mathematica Sinica, New Series, 3(1987), 152-169.

On Complete Minimal Immersion $\chi : RP^2 - \{a, b\} \to R^3$ With Total Curvature -10π

Shaoping Zhang
Nankai Institute of Mathematics

Introduction

The existence of complete immersed minimal surfaces in R^3 is one of the fundamental problems in the classical theory of minimal surfaces. W.H.Meeks, III [3] has shown that all complete minimal surfaces immersed in R^3 with total curvature K greater than -8π are plane, cateniod, Enneper's surfaces and Meeks minimal Mobius strip. For $K \leq -8\pi$, the classification seems to be more complicated. We restrict ourselves to the case of non-orientable immersions $\chi : M \to R^3$, in particular, with $K = -10\pi$. Then the Osserman's theorem [6] implies that there exists a compact surface $\hat{M}$, a finite number of points $p_1, \cdots, p_n$ of $\hat{M}$ and a conformal mapping from M to $\hat{M} - \{p_1, \cdots, p_n\}$. Moreover, by Chern-Osserman's inequality [2] we have $2n - \chi(\hat{M}) \leq 5$, where $\chi(\hat{M})$ represents the Euler characteristic of $\hat{M}$. Meeks' theorem [3] says $\chi(\hat{M})$ needs to be odd. Therefore $\chi(\hat{M})$ must be 1, -1 or -3. But one can exclude the last alternative by virtue of Rado's theorem [7]. In 1984, Maria Elisa G.G. Oliveira [4] established an example of non-orientable complete minimal surface with $K = -10\pi$, ends 2 and genus 1. In the present paper we try to determine all immersions in this special case, i.e., when $\chi(\hat{M}) = 1$ and $n = 2$.

In the first section we recall some useful preliminaries. A uniqueness theorem, one of our main results, is given in section 2. Some new examples are established in section 3. In the last section we give some remarks towards the classification.

I am very grateful to Professor C.K. Peng for raising this problem to me and teaching me a great deal of the background material, and to Professor S.S. Chern for his continued help, advice and stimulus.

§1. Preliminaries And Notations

The Weierstrass representation plays an important role in studying minimally immersed orientable surface $\chi : M \to R^3$. It can be stated as following

Theorem 1 *If $\phi : M \to R^3$ is a minimally immersed orientable surface, then there exists a meromorphic function $g : M \to R^3$ which is the Gauss map for the surface, and a meromorphic 1-form ω which can be expressed as $\omega = f(z)dz$ in local coordinates, satisfying*

i) ω has a zero of order exactly $2m$ just at a point where g has a pole of order m;

ii) ω_j has no real periods, i.e., the real part of the integral of ω_j along any closed path is zero, where $j = 1, 2, 3$ and $\omega_1 = \frac{1}{2}(1 - g^2)\omega$, $\omega_2 = \frac{i}{2}(1 + g^2)\omega$ and $\omega_3 = g\omega$, such that

$$(*) \qquad \phi(p) = \mathrm{Re}\left(\int_{p_0}^{p} \omega_1, \int_{p_0}^{p} \omega_2, \int_{p_0}^{p} \omega_3\right) + \phi(p_0).$$

Conversely $\phi : M \to R^3$ defined by (*) gives a minimally immersed surface.

Remark: We call f and g Weierstrass functions.

We are concerned with a complete minimal immersion (CMI) $\chi : RP^2 - \{a, b\} \to R^3$. For the completeness we have the following

Lemma 2 *Let $\phi : C \cup \{\infty\} - \{x_1, \cdots, x_s\} \to R^3$ be a minimal immersion with associated g and ω as above. Then ϕ is complete if and only if max $\{m_{1j}, m_{2j}, m_{3j}\} \geq 2$ at each x_j, where m_{jk} = the pole order of ω_j at x_k.*

For nonorientable χ a natural method is to consider its 2-sheet orientable covering (that is why we mentioned $C \cup \{\infty\}$ in Lemma 2), and then study the two associated Weierstrass functions. Meeks [3] gave the following necessary and sufficient condition to check whether ϕ is such a lifting or not.

Lemma 3 *Let $\phi : U \subseteq C \cup \{\infty\} \to R^3$ be a minimal immersion and f, g be the Weierstrass functions associated to ϕ. Then $\phi(I(z)) = \phi(z)$ for all $z \in U$ if and only if the following occurs*

i) $g(I(z)) = I(g(z))$

ii) $f(I(z)) = -\overline{f(z)}\,\overline{(zg(z))}^2$

where $I : C \cup \{\infty\} \to C \cup \{\infty\}$ is the transformation given by $z \to -\frac{1}{\bar{z}}$.

We omit the proofs of both lemmas (cf.[1]) here.

Remark: In particular, after changing coordinates in $C \cup \{\infty\}$ if necessary, we may denote by $\phi : C - \{0, a, -\frac{1}{a}\} \to R^3$ the lifting of our immersion $\chi : RP^2 - \{a, b\} \to R^3$, where a is a positive real number.

Immediately we have the following

Corollary 4 *Let ϕ be the lifting mapping of a CMI χ as above. Then we may assume*

(a)
$$g = cz^k \prod_{j=1}^{r} \frac{\bar{a}_j z + 1}{z - a_j}$$

where k, r are non-negative integers and a_j's, c are complex numbers such that $a_j \bar{a}_l \neq -1$ for any j, l;

(b)
$$f = i \frac{\prod_{j=1}^{r} (z - a_j)^2}{z^{n_1} (z - a)^{n_2} (az + 1)^{n_2}}$$

with $k + r + 1 = n_1 + n_2$ and $n_1, n_2 \geq 2$.

In particular, when $k + r = 5$, then $(n_1, n_2) = (2, 4), (3, 3)$ or $(4, 2)$.

For our purpose the following lemma is important. It is slightly different from what was stated in [1] and will be used repeatedly.

Lemma 5 *Suppose that $\phi : C - \{0, a, -\frac{1}{a}\} \to R^3$ is a minimal immersion with the extended Weierstrass functions f and g. Assume $\phi(z) = \phi(I(z))$ for all z in the domain. Then for all z we have*

$$\alpha(z) + \overline{\beta(I(z))} = 0, \quad \gamma(z) = \overline{\gamma(I(z))}.$$

Moreover, denote $Res_{w=z} f(w), Res_{w=z} fg^2(w), Res_{w=z} fg(w)$ by $\alpha(z), \beta(z)$ and $\gamma(z)$ respectively. Then the condition ii) in Theorem 1 is satisfied if and only if

$$\alpha(0) = \alpha(a) = 0, \quad \alpha(0) + \alpha(-\frac{1}{a}) = 0;$$

$$\beta(0)+\beta(a)=0,\quad \beta(0)+\beta(-\frac{1}{a})=0;$$

$$\gamma(0)+\gamma(a)=0,\quad \gamma(0)+\gamma(-\frac{1}{a})=0.$$

Proof: Let Γ be a circle around z of sufficiently small radius. A direct computation shows that $\omega_j=\overline{I^*\omega_j}$. Then $\int_\Gamma \omega_j = -\overline{\int_{I(\Gamma)}\omega_j}$, $j=1,2,3$. It follows that $\alpha(z)-\beta(z)=\overline{\alpha(I(z))-\beta(I(z))}$, $\alpha(z)+\beta(z)=-\overline{\alpha(I(z))+\beta(I(z))}$ and $\gamma(z)=\overline{\gamma(I(z))}$ by virtue of the Residue Theorem. The first assertion then follows.

Similarly, let a' be a or $-\frac{1}{a}$, then $\int_\Gamma\omega_j=\int_{I(\Gamma)}\omega_j=\int_\Gamma I^*\omega_j=\overline{\int_\Gamma\omega_j}$, $j=1,2,3$, where Γ is an I-invariant closed curve which is the boundary of a domain D containing 0 and a', Γ is positively oriented as the boundary of S^2-D.

We then obtain that the conditions $\int_\Gamma\omega_j=0$, $j=1,2,3$, are equivalent to the conditions ω_j, $j=1,2,3$ have no real periods. It implies the second conclusion.

Corollary 6 *If one of $\alpha(0),\beta(0),\alpha(a'),\beta(a')$ is zero, then both α and β are exact. Similarly, if either $\gamma(0)$ or $\gamma(a')$ is zero then γ is exact.*

Notations: $k(z)=$ the branch number of g at z. Denote $a\bar{a}, b\bar{b},\cdots$, by $A,B,\cdots$, respectively.

§2. A Uniqueness Theorem

In this section we want to show the following

Theorem A *There exists no other CMI $\chi: RP^2-\{a,b\}\to R^3$ with $K=-10\pi$ and $\max[k(a),k(b)]\geq 3$ besides the example established by Oliveira.*

Proof: W.L.O.G. we assume $k(a)\leq k(b)$. Lift χ to $\phi: C-\{0,a,-\frac{1}{a}\}\to R^3$, where a is a real positive number. Also write $g: C\cup\{\infty\}\to C\cup\{\infty\}$ for the extension of Gauss map of ϕ. Then g is a 5-sheet covering map since $K=-10\pi$. It follows that $k(0)\leq 5$. By Corollary 4,

$$g=z^k\prod_{j=1}^{5-k}\frac{\bar{a}_jz+1}{z-a_j}\qquad \text{with } 3\leq k\leq 5$$

and

$$f=i\frac{\prod_{j=1}^{5-k}(z-a_j)^2}{z^{n_1}(z-a)^{n_2}(az+1)^{n_2}}\qquad \text{with } n_1+n_2=6,\ n_1,n_2\geq 2$$

In any case we find $\beta(0)=0$. It follows $\alpha(*)\equiv\beta(*)\equiv 0$ by Corollary 6. We are going to consider the various alternatives for the functions f and g.

Case 1 $k(0)=5$

Then we would have

$$g=z^5$$

and

$$f=i\frac{1}{z^{n_1}(z-a)^{n_2}(az+1)^{n_2}}$$

with $(n_1,n_2)=(2,4),(3,3)$ or $(4,2)$ possibly. It is easy to derive a contradiction in this case.

Case 2 $k(0)=4$

First, we would have

$$g=\frac{z^4(\overline{p}z+1)}{z-p} \text{ for some } p\neq 0,\infty.$$

Moreover, by a series of direct computation we would find $p\neq a,-\frac{1}{a}$. It would follow

$$f=i\frac{(z-p)^2}{z^{n_1}(z-a)^{n_2}(az+1)^{n_2}}$$

with $(n_1,n_2)=(2,4),(3,3)$ or $(4,2)$ possibly.

It is easy to exclude the alternative $(n_1,n_2)=(4,2)$. Indeed, $\alpha(0)=0$ and $\alpha(-\frac{1}{a})=0$ would imply $\frac{a}{\overline{p}}>0$ and $\frac{a}{\overline{p}}<0$, respectively.

Moreover, if $(n_1,n_2)=(2,4)$, by $\alpha(0)=0$, we would find $\frac{a}{\overline{p}}=2-2A$. It would follow $\beta(a)=i\frac{20\overline{p}^2(4A^2-A+4)}{a^5(A+1)^7}$, which contradicts $\beta(a)=0$.

Similarly, it is impossible that $(n_1,n_2)=(3,3)$. Otherwise, by $\beta(0)=0$ we would have

$$(A^2+5A+10)x^2+6A(A^2+4A+5)x+3A^2(2A^2+7A+7)=0 \tag{1}$$

where $x=\frac{a}{\overline{p}}$. On the other hand, $\beta(-\frac{1}{a})=0$ would imply

$$(10A^2+5A+1)x^2-6(5A^2+4A+1)x+3(7A^2+7A+2)=0. \tag{2}$$

From equations (1) and (2) we can derive a contradiction $A^2+A+1=0$.

Case 3 $k(0)=3$

Now the Gauss map can be expressed as

$$g=\frac{z^3(\overline{p}z+1)(\overline{q}z+1)}{(z-p)(z-q)} \qquad \text{where } p,q\neq 0,\infty \text{ and } p\overline{q}\neq -1.$$

We can easily find that neither p nor q are in $\{a,-\frac{1}{a}\}$ and then get

$$f=\frac{i(z-p)^2(z-q)^2}{z^{n_1}(z-a)^{n_2}(az+1)^{n_2}} \text{ with } (n_1,n_2)=(2,4),(3,3) \text{ or } (4,2) \text{ possibly.}$$

We will complete our proof by showing that there exists a unique expected immersion in this case. The calculation is rather lengthy and complicated.

Let $p_0=\frac{a}{p}, q_0=\frac{a}{q}, x=p_0+q_0, y=p_0q_0$.

Step 1 If $(n_1,n_2)=(4,2)$,

by $\alpha(a) = 0$, and $\alpha(-\frac{1}{a}) = 0$, we would obtain $\frac{1}{(p_0-1)} + \frac{1}{(q_0-1)} - \frac{A}{(A+1)} = 0$ and $\frac{1}{(p_0+A)} + \frac{1}{(q_0+A)} + \frac{1}{A(A+1)} = 0$, which are equivalent to $(1+2A)p_0 - Aq_0 = 3A+2$ and $(A^2+2A)p_0 + q_0 = -2A^3+3A^2$ respectively, and then q_0 is a non-zero real number. However, a direct computation shows that $\gamma(0) = i\frac{q_0}{a}$. It contradicts $\mathrm{Im}\gamma(0) = 0$.

Step 2 If $(n_1, n_2) = (3,3)$,
then we would have

$$f = \frac{i(z-p)^2(z-q)^2}{z^3(z-a)^3(az+1)^3}$$

and

$$g = \frac{z^3(\overline{p}z+1)(\overline{q}z+1)}{(z-p)(z-q)}$$

and then

$$fg^2 = \frac{z^3(\overline{p}z+1)^2(\overline{q}z+1)^2}{(z-a)^3(az+1)^3}, \quad fg = i\frac{(z-p)(z-q)(\overline{p}z+1)(\overline{q}z+1)}{(z-a)^3(az+1)^3}.$$

Observation: We would have $\alpha(*), \beta(*)$ and $\gamma(*) = 0$ by virtue of Corollary 6 since $\beta(0) = \gamma(0) = 0$ now.

By definition,

$$\beta(a) = Res_{z=a} fg^2 = \frac{1}{2}F''(a), \text{ where } F = \frac{z^3(\overline{p}z+1)^2(\overline{q}z+1)^2}{(az+1)^3}$$

After a staightforward calculation, we would get

$$\begin{aligned} &A^2(A^2+5A+10)x^2 + 12Axy - 3(A-1)y^2 \\ &+ 6A^3(A^2+4A+1)x + 2A^2(A^2+5A+10)y + 3A^4(A^2+7A+7) = 0 \end{aligned} \tag{3}$$

Similarly, by $\beta(-\frac{1}{a}) = 0$, we would find

$$\begin{aligned} &(10A^2+5A+1)x^2 - 12A^2xy + 3A(A-1)y^2 \\ &- 6(5A^2+4A+1)x + 2(10A^2+5A+1)y + 3(7A^2+7A+2) = 0 \end{aligned} \tag{4}$$

Furthermore, by $\gamma(a) = 0$, we would get

$$\begin{aligned} &(A^2-4A+1)Y - 3A(A-1)(x\overline{y}+y\overline{x})A(A^2-4A+1)X \\ &+ 3A^2(A-1)(x+\overline{x}) + 6A^2(y+\overline{y}) + A^2(A^2-4A+1) = 0 \end{aligned} \tag{5}$$

We want to derive a contradiction from the system of (3), (4) and (5), which can be simplified as following

$$m^2 + 2n - 3(A^2+A+1) = 0 \tag{6}$$

$$4Amn - (A-1)n^2 - 4A^2m - 4A^2(A-1) = 0 \tag{7}$$

$$\begin{aligned}&(A^2-4A+1)(N-AM) - 3A(A-1)(m\bar{n}+n\bar{m}) + 3A(A^2+4A+1)(n+\bar{n})\\ &+3A(A-1)(A^2+3A+1)(m+\bar{m}) - A(9A^4+35A^3+58A^2+35A+9) = 0\end{aligned} \tag{8}$$

where $m = x + 3(A-1)$ and $n = y + 6A$.

Write $m = re^{i\theta}$, where $r \in R$ and $-\frac{1}{2}\pi \le \theta < \frac{1}{2}\pi$. Applying (7) to (8) we would get

$$\begin{aligned}&(A-1)r^4\cos 4\theta + 8Ar^3\cos 3\theta - 6(A^3-1)r^2\cos 2\theta\\ &- A(3A^2+A+3)r\cos\theta + (A-1)(9A^6+1A^5+43A^2+18A+9) = 0\end{aligned} \tag{9}$$

and

$$(A-1)r^4\sin 4\theta + 8Ar^3\sin 3\theta - 6(A^3-1)r^2\sin 2\theta - 8A(3A^2+A+3)r\sin\theta = 0. \tag{10}$$

We also apply (7) to (9) to get

$$\begin{aligned}&(A^2-4A+1)r^4 + 12A(A-1)r^3\cos\theta - [6(A^4-A^3+6A^2-A+1)\cos 2\theta\\ &+4A(A^2-4A+1)]r^2 - 12A(A-1)(A^2-3A+1)r\cos\theta\\ &+(9A^6-18A^5+4A^4-88A^3+4A^2-1A+9) = 0\end{aligned} \tag{11}$$

By (10) $\times \cos 2\theta$ $-$ (9) $\times \sin 2\theta$, we would get

$$\begin{aligned}&(A-1)r^4\sin 2\theta - (A-1)(9A^4+18A^3+43A^2+18A+9)\sin 2\theta\\ &+8Ar^3\sin\theta + 8A(3A^2+A+3)r\sin\theta = 0.\end{aligned} \tag{12}$$

By (10) $\times \sin 2\theta$ + (9) $\times \cos 2\theta$, we would get

$$\begin{aligned}&(A-1)r^4\cos 2\theta + 8Ar^3\cos\theta - 6(A^3-1)r^2 - 8A(3A^2+A+3)r\cos\theta\\ &+(A-1)(9A^4+18A^3+43A^2+18A+9)\cos 2\theta = 0\end{aligned} \tag{13}$$

Now, there are two possibilities, $\theta = 0$ and $\theta \ne 0$. If $\theta = 0$, then, by using (11) and (13), one would easily derive

$$\begin{aligned}&7(A^2+A+1)r^2 - 2(A-1)(6A^2+A+6)r\\ &+(9A^4-6A^3+43A^2-6A+9) = 0.\end{aligned}$$

This is impossible since the discriminant is equal to

$$-(27A^6+81A^5+237A^4+339A^3+237A^2+81A+27) < 0.$$

It is most difficult to derive a contradiction in the case $\theta \ne 0$. We do it as follows.

First, from (12), we would solve out

$$\cos\theta = -\frac{4Ar(r^2+v)}{(A-1)(r^4-u)} \tag{14}$$

where $u = 9A^4 + 18A^3 + 43A^2 + 18A + 9$ and $v = 3A^2 + A + 3$.

Then, substituting (14) to (13) we would get

$$64A^2R^2(R+v)(vR+u) = (A-1)^2(R^2-u)^2[R^2 + 6(A^2+A+1)R + u]. \tag{15}$$

Let $\lambda = A + \dfrac{1}{A} \geq 2$ and $\sigma = \dfrac{R}{A} \geq 0$. By (14) and $\cos^2\theta \leq 1$, one would get

$$16A^2R(R+v)^2 \leq (A-1)^2(R^2-u)^2$$

and then, noticing (15), one would have

$$(R+v)(R^2 + 6(A^2+A+1)R + u) \leq 4R(vR+u)$$

or equivalently

$$\psi(h) \equiv h^3 + 3(2\lambda+1)h^2 - 3(4\lambda+23)h - 114\lambda + 25 \leq 0$$

where $h = \sigma - 3\lambda \geq -3\lambda$.

By studying the function ψ on $[-3\lambda, \infty)$, one would conclude $h \geq -4$. On the other hand, carrying (14) to (11), one would have

$$\begin{aligned}
&(A^2 - 4A + 1)R^3(vR+u) \\
&- 48A^2[R - (A^2 - 3A + 1)](R^2-u)(A-1)^2\frac{R^2 + 6(A^2+A+1)R + u}{64A^2} \\
&- 3R(R+v)(R^2 + (A^2+A+1)R + u)(A^4 - A^3 + 6A^2 - A + 1) \\
&+ (6A^4 - 10A^3 + 52A^2 - 10A + 6)R^2(vR+u) \\
&+ (9A^6 - 18A^5 + 4A^4 - 88A^3 + 4A^2 - 18A + 9)R(vR+u) = 0.
\end{aligned}$$

It can be rewritten in the terms of h as follows

$$\begin{aligned}
&3(\lambda-2)h^5 + (60\lambda^2 - 61\lambda + 10)h^4 \\
&+ (432\lambda^3 + 24\lambda^2 + 198\lambda + 468)h^3 \\
&+ (1296\lambda^4 + 2016\lambda^3 + 1632\lambda^2 + 5790\lambda - 1404)h^2 \\
&+ (1296\lambda^5 + 6912\lambda^4 + 7344\lambda^3 + 31080\lambda^2 + 7703\lambda + 12850)h \\
&+ (6480\lambda^5 + 12960\lambda^4 + 50688\lambda^3 + 38052\lambda^2 + 45375\lambda + 11250) = 0.
\end{aligned}$$

However since $\lambda \geq 2$ and $h \geq -4$ the left hand side would be greater than

$$\begin{aligned}
&114[(9\lambda^4 + 14\lambda^3 + 11\lambda^2 + 40\lambda - 10)h^2 \\
&+ (48\lambda^4 + 51\lambda^3 + 216\lambda^2 + 53\lambda + 89)h \\
&(90\lambda^4 + 352\lambda^3 + 260\lambda^2 + 315\lambda + 78)]
\end{aligned}$$

which is positive since the discriminant is

$$-(936\lambda^8 + 12816\lambda^7 + 9791\lambda^6 + 28770\lambda^5 + 24433\lambda^4 + 13880\lambda^3 - 9554\lambda - 8701) < 0.$$

So we arrive at a contradiction again.

The remaining case is $(n_1, n_2) = (2, 4)$. Since the result is quite interesting we will give a somewhat detailed proof here.

Now

$$f = \frac{i(z-p)^2(z-q)^2}{z^2(z-a)^4(az+1)^4} \text{ and } g = \frac{z^3(\overline{p}z+1)(\overline{q}z+1)}{(z-p)(z-q)}$$

where $p, q \overline{\in} \{a, -\frac{1}{a}\}$ will be determined later. We use the same notations as above. First, a direct computation shows

$$\alpha(0) = \frac{2ip^2q^2}{A^2(A+1)^4}\left[-\frac{1}{p} - \frac{1}{q} + \frac{2}{a} - 2a\right].$$

It follows $x = 2 - 2A$.

Then investigate $\beta(*)$.

$$\beta(a) = \frac{i\overline{p}^2\overline{q}^2}{3!A^2}F'''(a), \text{ where } F = \frac{z^4(z+\frac{1}{\overline{p}})^2(z+\frac{1}{\overline{q}})^2}{(z+\frac{1}{a})^4}.$$

Set $w = z + \frac{1}{a}$ and express $\overline{F}(w) = F(w - \frac{1}{a})$ in terms of w. Then we find

$$\begin{aligned}\overline{F}'''(w) =& 4!w - 2 \times 3!\frac{4-x}{a} \\ &+ 2 \times \frac{3!}{w^4}\frac{28 - 35x + 20\overline{y} + 10x^2 - 10x\overline{y} + 2\overline{y}^2}{aA^2} \\ &- \frac{4!}{w^5}\frac{28 - 42x + 30\overline{y} + 15x^2 - 20x\overline{y} + 6\overline{y}^2}{A^3} + \frac{5!}{w^6}\frac{4 - 7x + 6\overline{y} + 3x^2 - 5x\overline{y} + 2\overline{y}^2}{aA^3} \\ &- \frac{6!}{3!w^7}\frac{1 - 2x + 2\overline{y} + x^2 - 2x\overline{y} + \overline{y}^2}{A^4}.\end{aligned}$$

Therefore, by $\beta(a) = 0$ and the data of x, we get

$$(A^2 - 3A + 1)y^2 + 10A(A^2 - A + 1)y + 5A^2(4A^2 - A + 4) = 0. \tag{16}$$

It follows that y is real since the discriminant is equal to

$$20A^2(A^4 + 3A^3 + 4A^2 + 3A + 1) > 0.$$

In the same way, by $\gamma(a) = 0$, we have

$$\begin{aligned}&(A-1)[(A^2 - 8A + 1)y^2 + 4A(4A^2 - 7A + 4)y \\ &- (4A^4 - 27A^3 + 30A^2 - 27A + 4)] = 0.\end{aligned} \tag{17}$$

If $A \neq 1$ one would obtain the following immediately from (16) and (17)

$$5y^2 - 2(3A^2 - 9A + 3)y + (4A^4 - 7A^3 + 15A^2 - 7A + 4) = 0$$

whose discriminant $-(11A^4 + 19A^3 - 24A^2 + 19A + 11) < 0$. It contradicts the result that y is real.
Hence $A = 1$ and then, solving (16), we find $y = 5 \pm 2\sqrt{15}$. It follows that $p = -q = \pm\frac{1}{\sqrt{-5 \pm 2\sqrt{15}}}$, coinciding with those in the example given by Oliveira. This proves the theorem.

§3. New Examples

Next, we turn to establish new examples. The process is the same as above. By the previous theorem we know that they, if any, satisfy $k \equiv \max(k(a_1), k(a_2)) \leq 2$, where a_1 and a_2 are the ends.

W.L.O.G., we assume the associated Weierstrass functions f and g taking the form shown in the previous section and assume $k_2 \equiv k(a_2) \leq k(a_1)$. Namely, we assume $k(a_2) \leq k(a_1)$, $a_1 = 0$, $a_2 = -\frac{1}{a}$, a is a positive real number, and

$$f = i\frac{(z-a)^{2k'} \prod_{j=1}^{5-k'} (z - p_j)^2}{z^{n_1}(z-a)^{n_2}(az+1)^{n_2}}, \text{ and } g = \frac{z^k(az+1)^{k'} \prod_{j=1}^{5-k'} (\overline{p_j}z + 1)}{(z-a)^{k'} \prod_{j=1}^{5-k'} (z - p_j)}$$

here k' is a non-negative integer, $(n_1, n_2) = (2,3), (3,3)$ or $(4,2)$, and $p_j \neq 0, \infty, a, -\frac{1}{a}$.

a) all CMIs with $k = 2$ and $g(a_1) = g(a_2)$

First we claim $k' = k_2$ since $g(-\frac{1}{a})$ needs to be zero by the hypothesis, g taking the same value at the ends, and $g(0) = 0$ now. $g(-\frac{1}{a}) = 0$ also implies $k' \geq 1$. Moreover, some simple computation shows that $k' \neq 2$. Then we have $k' = 1$. Secondly, one concludes $(n_1, n_2) \neq (3,3)$. Otherwise $\beta(0) = 0$ would follow $\alpha(a) = 0$ by Corollary 6. On the other hand, a direct computation would show $\alpha(0) = \frac{i(a-p)^2(a-q)^2}{a^3(A+1)^3}$ which would never vanish.

Hence we need only to deal with the cases $(n_1, n_2) = (2,4)$ or $(4,2)$. In both cases we have $\beta(0) = 0$. Let $x = \frac{a}{p} + \frac{a}{q}$ and $y = \frac{a}{p}\frac{a}{q}$ again.

<u>Case 1</u> $(n_1, n_2) = (2,4)$

We will show that there exists a unique CMI in this case. Indeed, $\alpha(0) = 0$ and $\alpha(a) = 0$ are equivalent to $x = 1 - 2A$ and $y = \frac{4A^2 + A + 1}{1 - A}$ respectively. Moreover, by use of Corollary 6, one has $\gamma(a) = 0$ since $\gamma(0) = 0$ clearly. And then one gets, noticing the data of x and y

(18) $$4A^7 + 15A^6 + 25A^5 + 23A^4 + 10A^3 - A^2 - 3A - 1 = 0.$$

This equation has a unique positive solution, say A_0. We choose $a = \sqrt{A_0}$ and p, q such that they satisfy the following system

$$\frac{a}{p} + \frac{a}{q} = 1 - 2A_0$$
$$\frac{a}{p}\frac{a}{q} = \frac{4A_0^2 + A_0 + 1}{1 - A_0}.$$

It is also clear that $p, q \neq a, -\frac{1}{a}$. Hence we get a good immersion.

Case 2 $(n_1, n_2) = (4, 2)$

As before one needs to determine a, p, and q such that ω_j, $j = 1, 2, 3$, has no real periods. After some linear algebra, one finds $\beta(a) = 0$ is equivalent to $x = -2A$. Similarly, by $\mathrm{Im}\gamma(0) = 0$, one gets $\mathrm{Re}[(x - a\overline{p} - a\overline{q} + A - 1)pq] = 0$ and then solves out, noticing the data of x,

$$\cos\theta = \frac{2Ar}{A+1} \tag{18}$$

here one writes $pq = re^{i\theta}$.

Finally, $\mathrm{Im}\gamma(a) = 0$ gives

$$\mathrm{Re}\{[A + (2A + 1)pq](1 - Are^{-i\theta})\} = 0.$$

By substituting (18) to the above one solves out

$$R \equiv r^2 = \frac{A+1}{4A^2 - A - 1}. \tag{19}$$

Therefore one needs only to choose p and q such that

$$p + q = -2are^{i\theta} \qquad pq = re^{i\theta} \qquad \text{where } r \text{ and } \theta \text{ satisfy (18) and (19).}$$

It is easy to see that there exists one family of CMIs in this case parametrized by a. Finally one needs to determine the domain of a. The restriction on a is caused by the condition $p, q \neq a, -\frac{1}{a}$ and the fact $|\cos\theta| \leq 1$. It turns out that $a \in (1, \infty)$.

b) examples of CMI with $k = 1$ and $\gamma(a_1) = \gamma(a_2)$

Now we can write

$$g = \frac{z(az+1)(\overline{p}z+1)(\overline{q}z+1)(\overline{s}z+1)}{(z-a)(z-p)(z-q)(z-s)} \text{ where } p, q, s \neq a, -\frac{1}{a}, 0, \infty.$$

And we consider the case in which $(n_1, n_2) = (4, 2)$, i.e.

$$f = i\frac{(z-p)^2(z-q)^2(z-s)^2}{z^4(az+1)^2}.$$

Then $\beta(-\frac{1}{a}) = 0$ clearly. Let $x = ap + aq + as$, $y = apaq + aqas + asap$ and $t = apaqas$.

<u>Computation</u> of $\beta(0)$

$$\beta(0) = \frac{2i}{A}\left[\overline{p} + \overline{q} + \overline{s} - \frac{1}{-a}\right].$$

It follows

(20)
$$x = -1.$$

<u>Computation</u> of $\beta(a)$

$$\beta(a) = \frac{(2i(\overline{p}a+1)^2(\overline{q}a+1)^2(\overline{s}a+1)^2)}{A}\left[\frac{\overline{p}}{\overline{p}a+1} + \frac{\overline{q}}{\overline{q}a+1} + \frac{\overline{s}}{\overline{s}a+1} - \frac{1}{a}\right].$$

By virtue of (20), it follows

(21)
$$y = 1 - 2t.$$

<u>Computation</u> of $\gamma(a)$

$$\gamma(a) = i\frac{(t+x+y)(aA - ax + \frac{y}{a} - \frac{t}{aA})}{aA(A+1)}.$$

Since $\mathrm{Im}\gamma(a) = 0$, by substituting (20) and (21) to the above, we get

(22)
$$\mathrm{Re}(1-\overline{t})[A^3 + A^2 + A - (2A+1)t] = 0.$$

<u>Computation</u> of $\gamma(0)$

$$\gamma(0) = \frac{i}{2}F''(0) \text{ where } F = \frac{(\overline{p}z+1)(\overline{q}z+1)(\overline{s}z+1)(z-p)(z-q)(z-s)}{(z-a)(az+1)}.$$

By using (20) and (21), we can find $\mathrm{Im}\gamma(0) = 0$ implies

(23)
$$\mathrm{Re}[2T + (A^2 - 1)t] = 0.$$

Set $t = m + in$, where m and n are real. Then we solve out, from (22) and (23), that

(24)
$$m = \frac{2A^3 + 2A^2 + 2A}{4A^3 + 3A^2 + 4A + 1} \quad \text{and} \quad T = A(1-2A)\frac{A^2 + A + 1}{4A^3 + 3A^2 + 4A + 1}.$$

We need to check whether there exists any positive A such that $T \geq m^2$ or not. It turns out that if $A \in (0, A_0)$, where A_0 is the unique positive solution of $4w^3 + 3w^2 - 1 = 0$, then the answer is affirmative. The reason for $A \neq A_0$ is $p, q, s \neq a, -\frac{1}{a}$.

§4. Remarks

We have shown that there exist so rich CMIs from $RP^2 - \{a, b\}$ to R^3, for $k = 1, 2$, in the previous section. Analogously we can establish many CMIs from $RP^2 - \{p\}$ to

R^3, even when $k = 4$. However it seems too complicated to determine immersions in those cases since we need to consider all alternatives unless we can develop something to exclude some of them prior to our computation.

Furthermore, if we want to give the classification for $K = -10\pi$ we have to deal with the case in which the number of ends happens to be 3, that is, the equality in $2n - \chi(\hat{M}) \leq m$ occurs since now $m = 5, n = 3$ and $\chi(\hat{M}) = 1$.

We recall the following [3]

Theorem *There exists no CMI from $RP^2 - \{a, b\}$ to R^3 with total curvature -6π.*

Noticing that these two cases are analogous in the sense that all the ends are embeded, we have the following

Question: Does there exist any CMI from $RP^2 - \{a_1, a_2, a_3\}$ to R^3 with total curvature -10π?

The answer is "negative" for $k = 5$ and open for the remaining cases.

Recently Rob Kusner [8] established an example in this case with $k = 2$. Indeed for each odd $n \geq 3$ he gave an example with total curvature $-2(2n-1)\pi$ and ends n. I would like to thank him for informing me this result.

REFERENCE

[1] J.L.Barbosa & A.G.Colares, *Lecture Notes in Mathematics* 1195, Springer-Verlag, 1986.

[2] S.S.Chern & R.Osserman, *Complete Minimal Surfaces in Euclidean n-space*, J.d'Analyse Math., 19(1967).

[3] W.H.Meeks, *The Classification of Complete Minimal Surfaces with Total Curvature Greater Than -8π*, Duke Math.J. 48 (1981).

[4] Maria Elisa G.G.de Oliveira, *Superficies Minimas nao- orientaveis no R^n*, Ph.D Thesis IMEUSP, 1984.

[5] R.Osserman, *A Survey of Minimal Surfaces*, Van Nostrand, 1969.

[6] R.Osserman, *Global Properties of Minimal Surfaces in E^3 & E^n*, Ann.of Math. 80(1964).

[7] Rado, *On The Problem of Plateau* Springer, Berlin, 1933.

[8] Rob Kusner, *Conformal Geometry And Complete Minimal Surfaces*, Bulletin (New Series) of the AMS Vol.17 (1987).

LOCAL ATIYAH-SINGER INDEX THEOREM FOR FAMILIES OF DIRAC OPERATORS

Zhang Wei-ping

§0. Introduction

The first success of proving the Atiyah-Singer index theorem directly by heat kernel method was achieved by Patodi [10], who carried out the "fantastic cancellation" (cf. [9]) for the Laplace operators and for the first time proved a local version of the Gauss-Bonnet-Chern theorem. In recent years, several different direct heat kernel proofs of the Atiyah-Singer index theorem for Dirac operators have appeared independently: Bismut [3], Getzler [6], [7] and Yu [13] or [14], see also Berline- Vergne [2]. All the proofs have their own advantages.

Motivated by the problem of generalizing the heat kernel proofs of the index theorem to prove a local index theorem for families of elliptic operators, Quillen [12] introduced the concept of superconnections, and this was developed by Bismut to give a heat kernel representation for the Chern character of families of first order elliptic operators. Then using his probabilitistic method, Bismut [4] obtained a proof of the local index theorem for families of Dirac operators.

In this paper, we will use the method of Yu [14] to give another proof of the local index theorem for families of Dirac operators. The key point is Yu's idea of comparing the corresponding terms in the Taylor expansion series of functions, thus avoiding probability and some complicated estimates.

It seems that the method in [7] can be also generalized to give a proof of the local index theorem for families of Dirac operators.

Note also that Yu [15] has presented a direct proof of the local index theorem for signature operators explicitly in the same spirit as [13]. We found that our proof can be modified immediatly to give a proof of the local index theorem for families of signature operators.

For simplicity, we write out our proof only for classical Dirac operators, but our proof works also for twisted Dirac operators, as pointed out in [13]. For a brief account cf. the Appendix.

We take Yu [13] and Chap. I-III of Bismut [4] as our basic references.

We are deeply grateful to Professor Yu Yanlin who introduced this subject to us and kindly explained to us the key points of his work [13], without his encouragement, the present paper could never have been finished. We would also like to thank Nankai Institute of Mathematics for hospitality and some other services.

§1. Clifford module and supertrace

Let $V = V^0 \oplus V^1$ be a super (or Z_2-graded) complex vector space. As in [12], we use ε to denote the involution giving the grading: $\varepsilon(v) = (-1)^{\deg(v)}v$. Then the space End (V) of endomorphisms of V is a super algebra. The even (resp.odd) elements of End (V) commute (resp. anticommute) with ε.

Definition 1.1 The supertrace tr_s of $k \in \mathrm{End}(V)$ is defined by

$$\mathrm{tr}_s k = \mathrm{tr}\varepsilon k.$$

It is easy to verify that

$$\begin{aligned} &\mathrm{tr}_s k = \mathrm{tr}(k|_{V^0}) - \mathrm{tr}(k|_{V^1}), \ k \text{ even} \\ &\mathrm{tr}_s k = 0, \ k \text{ odd} \end{aligned} \tag{1.2}$$

Now let H be a Grassmann algebra, then H is naturally Z_2- graded. Let K be the graded tensor product

$$K = \mathrm{End}(V)\hat{\otimes} H$$

then K possesses a nature Z_2-grading. The supertrace concept can be naturally extended to K and takes its value in H (cf. [8], [4]).

Definition 1.3 If $h \in H$, $k \in K$, define

$$\mathrm{tr}_s(hk) = h(\mathrm{tr}_s k).$$

Now let $N = 2n$ be an even integer. $c(R^{2n})$ denotes the Clifford algebra of R^{2n} generated by $1, e_1, \cdots, e_{2n}$ with the following commutative relations:

$$e_i^2 = -1, \quad e_i e_j + e_j e_i = 0, \quad i \neq j.$$

As well known, Spin $(2n)$ acts unitarily on the 2^n-dimensional complex vector space of spinors S and that $c(R^{2n}) \otimes_R C$ can be identified with End(S), so that Spin $(2n) \subset \mathrm{End}(S)$. Set

$$\tau = (\sqrt{-1})^n e_1 \cdots e_{2n} \tag{1.4}$$

Then τ acts unitarily on S and moreover

$$\tau^2 = 1 \tag{1.5}$$

Set

$$S_+ = \{s \in S : \tau s = s\}, S_- = \{s \in S : \tau s = -s\}. \tag{1.6}$$

Then S_+, S_- are 2^{n-1}-dimensional vector subspaces of S such that

$$S = S_+ \oplus S_- \tag{1.7}$$

Spin(2n) acts unitarily and irreducibly on S_+ and S_-.

S can be naturally regarded as a super vector space with obvious grading and involution τ. Let W be a usual vector space, then $S \otimes W$ also carries a naural grading.

Lemma 1.8

$$\mathrm{tr}_s e_{i_1} \cdots e_{i_p} = 0, \quad p < 2n$$
$$\mathrm{tr}_s e_1 \cdots e_{2n} = (\frac{2}{\sqrt{-1}})^n.$$

Let m be another integer, $f_1, \cdots, f_\alpha, \cdots, f_m$ denote the canonical oriented Euclidean basis, and $dy^1, \cdots, dy^\alpha, \cdots, dy^m$ the basis of the corresponding dual space.

Definition 1.9 ϵ denote the graded tensor product of the Z_2-graded algbra $c(R^{2n})$ and $\Lambda(R^m)$, i.e.

$$\epsilon = c(R^{2n}) \hat{\otimes} \Lambda(R^m).$$

For simplicity, we will use no "$\hat{\otimes}$" sign to indicate the product in ϵ. e.g. we have

$$e_i dy^\alpha + dy^\alpha e_i = 0. \tag{1.10}$$

§2. Fibration of manifolds

A large portion of this section, which is adapted from [4], is included here only for completeness.

Let B be an m-dimensional connected compact Riemannian manifold. Denote its Riemannian metric by g_B.

X is a connected compact orientable manifold of even dimension 2n. We assume that X is a spin manifold, so that $w_2(X) = 0$.

M is a $2n+m$ dimensional compact connected manifold. We assume that $\pi : M \to B$ is a submersion of M onto B, which defines a fibration of M with fiber X. Namely, we assume that there is an open covering _U of B such that for every U∈_U, there exists a C^∞-diffeomorphism $\varphi_U : \pi^{-1}(U) \to U \times X$, and if $U \bigcap V \neq \phi$, $\varphi_U \circ \varphi_V^{-1} : (U \bigcap V) \times X \to (U \bigcap V) \times X$ is given by $(y,x) \mapsto (y, f_{U,V}(y,x))$ where $f_{U,V}(y,\cdot)$ is a C^∞-diffeomorphism of X which is C^∞ in both variables y and x.

For $y \in B$, $\pi^{-1}(y)$ is then a submanifold G_y of M, and π defines a fibration G of M. TG denote the 2n- dimensional vector bundle on M whose fiber T_xG is the tangent space at x to the fiber $G_{\pi(x)}$. And we assume that M is oriented.

By using any Riemannian structure on M, we can obtain an m-dimensional smooth sub-vector bundle T^HM of TM such that

$$TM = T^HM \oplus TG \tag{2.1}$$

In particular, $\forall x \in M$, T^H_xM and $T_{\pi(x)}B$ are isomorphic under π.

Recall that B is Riemannian, so we can lift the Euclidean scalar product g_B of TB to T^HM. And we assume that TG is endowed with a scalar product g_G. Thus we can introduce in TM a new scalar product $g_B \oplus g_G$, and denote by ∇^L the Levi-Civita connection on TM with respect to this metric.

Let O be the SO(2n) bundle of oriented orthonormal frames in TG.

We now do the assumption that the bundle TG is spin. Namely we assume that the $SO(2n)$ bundle $O \to M$ can be lifted to a Spin(2n) bundle O' so that the projection $O' \to O$ induces the covering projection Spin(2n)→SO(2n) on each fiber.

Let F, F_+, F_- be the bundle of spinors:

$$\begin{aligned} F &= O' \times_{\mathrm{Spin}(2n)} S \\ F_\pm &= O' \times_{\mathrm{Spin}(2n)} S_\pm \end{aligned} \tag{2.2}$$

Recall that if $e \in TG$, $\|e\| = 1$, e acts unitarily by Clifford multiplication on F, and interchanges F_+ and F_-.

Definition 2.3 ϵ is the bundle defined by

$$\epsilon_x = c(T_xG) \hat{\otimes} \Lambda_{\pi(x)}(B).$$

The supertrace construction of §1 can be extended obviously to give a supertrace on the superalgebra bundle ϵ.

Next we construct a connection on TG.

Definition 2.4 ∇ denotes the connection on TG defined by the following relation:

$$\nabla_Y Z = P_G(\nabla^L_Y Z), \quad Y \in TM, \quad Z \in TG$$

where P_G denote the orthogonal projection of TM on TG. ∇ obviously preserves the scalar product in TG.

∇ can be lifted to give a connection on F, F_+ and F_- respectively. Let $H^\infty, H^\infty_+, H^\infty_-$ denote the set of C^∞ sections of F, F_+, F_- over M.

Clearly, we may regard $H^\infty, H^\infty_\pm$, as the set of C^∞ sections over B of infinite dimensional vector bundles, which we still denote by $H^\infty, H^\infty_\pm$. The fibers $H^\infty_y, H^\infty_{\pm,y}$ are the sets of C^∞ sections over G_y of $F, F_\pm$.

Also note that since F is an Hermitian bundle, and since the fibers G_y carry a natural volume element dx, if $h, h' \in H_y$, we can define the scalar product

$$<h,h'>=\int_{G_y} <h(x),h'(x)> dx. \tag{2.5}$$

We now define a connection on $H_\pm$.

Definition 2.6 $\tilde{\nabla}$ denotes the connection on $H_\pm$ such that if $Y \in TB$, $h \in H_\pm$, then

$$\tilde{\nabla}_Y h = \nabla_{Y^H} h$$

where Y^H is the (unique) lifting of $Y \in TB$ in $T^H M$.

§3. Dirac operators and a heat kernel formula for the index

Definition 3.1 D denotes the operator acting on H

$$D = \sum_1^{2n} e_i \nabla e_i$$

Of course the operator D acts fiberwise in the fibers G_y.

For $y \in B$, D_y denotes the restriction of D to the fiber H_y^∞. D interchanges H_+^∞ and H_-^∞. D_+, D_- are the restrictions of D to H_+^∞, H_-^∞.

By Atiyah and Singer [1], the difference bundle over B

$$\ker D_{+,y} - \ker D_{-,y} \tag{3.2}$$

is well defined in the sense of K-theory.

The aim of this paper is to present a calculation of the Chern character of (3.2) as a differential form over B explicitly.

First, we will give a brief review of the heat kernel representation of this Chern character given by Bismut [4].

In what follows, to simplify the notations, we will use the following conventions:

(1) All the summation signs will be omitted;

(2) The subscripts α, β will be used for horizontal variables and the subscripts i, j for vertical ones (i.e. the variables in TG);

(3) We omit H in $(f_\alpha)^H$, i.e., we identify the orthonormal basis $f_1, \cdots, f_m$ of $T_y B$ with their lift in $T_x^H M$ (for $x \in G_y$). Also, $dy^1, \cdots, dy^m$ are now considered as differential forms on M;

(4) We omit the exterior product sign $\wedge$;

(5) We omit the "$\sim$" in $\tilde{\nabla}$.

We extend $e_1, \cdots, e_{2n}; f_1, \cdots, f_m$ to give an orthonormal frame $E_1, \cdots, E_{2n}; F_1, \cdots, F_m$ in the way of [13] and pick a fixed spin frame as in [13].

Denote

$$\Gamma_{IJ}^K = <\nabla_{Z_I}^L Z_J, Z_K>$$

where Z_I is the total notation for E_i, F_α, etc.

Definition 3.3 Define

$$\begin{aligned} H =& e_i\left(E_i + \frac{1}{4}\Gamma_{ij}^k e_j e_k + \frac{1}{2}\Gamma_{ij}^\alpha e_j dy^\alpha + \frac{1}{4}\Gamma_{i\alpha}^\beta dy^\alpha dy^\beta\right) + \\ & dy^\alpha\left(F_\alpha + \frac{1}{4}\Gamma_{\alpha i}^j e_i e_j + \frac{1}{2}\Gamma_{\alpha i}^\beta e_i dy^\beta + \frac{1}{4}\Gamma_{\alpha\beta}^\gamma dy^\beta dy^\gamma\right), \\ I =& H^2. \end{aligned}$$

Let I^y be the restriction of I to G_y. For a given $y \in B$, the operator I^y acts on $H_y\hat{\otimes}\Lambda_y(B)$ in the following sense: if $h \in \mathrm{End}(F), \eta, \eta' \in \Lambda(B)$, $e \in F$, then the action of $h\eta \in \mathrm{End}(F)\hat{\otimes}\Lambda(B)$ on $e\eta' \in F\hat{\otimes}\Lambda(B)$ is given by

$$(h\eta)(e\eta') = (-1)^{\deg\eta\cdot\deg e} h(e)\cdot\eta\eta' \tag{3.4}$$

further more, if $h \in H_y^\infty$, $\eta \in \Lambda_y(B)$,

$$I^y(h\eta) = (I^y h)\eta. \tag{3.5}$$

As indicated in [4], using standard results on elliptic equations, we can construct the "heat kernel" semi-group e^{-tI^y} which also acts in the fiber. For any $t > 0, e^{-tI^y}$ is given by a kernel $P_t^y(x,x')$ (for $x, x' \in G_y$) which is C^∞ in $(t,x,x') \in (0,+\infty)\times G_y \times G_y$.

Since the fibration $M \to B$ is locally trivial, there is an open neighborhood of y in B such that $\pi^{-1}(U)$ is diffeomorphic to $U \times X$. In what follows, we will not distinguish $\pi^{-1}(U)$ and $U \times X$.

In particular, since I^y is a smooth family of second order elliptic differential operators, it is not difficult to prove that $P_t^y(x,x')$ is C^∞ in $(t,x,x',y) \in (0,\infty)\times X\times X\times U$, cf. [4], Proposition 2.8.

For $x, x' \in G_y$, $P_t^y(x,x')$ is a linear mapping from F_x, into $F_x\hat{\otimes}\Lambda_y(B)$.

Let τ_x be the involution defining the grading in $F_{x'}$ then $\mathrm{Hom}(F_{x'},F_x)$ has a nature grading. The even (resp.odd) elements commute (resp. anticommute) with τ_x. Thus, $P_t^y(x,x')$ is an even element of the graded tensor product $\mathrm{Hom}(F_{x'},F_x)\hat{\otimes}\Lambda_y(B)$. In particular, $P_t^y(x,x)$ is an even element in the graded algebra $\mathrm{End}(F_x)\hat{\otimes}\Lambda_y(B)$, and $\mathrm{tr}_s P_t^y(x,x)$ is an even element in $\Lambda_y(B)$.

As in [12], we change the normalization constant in the definition of the Chern character. Namely, for a vector bundle V with connection form μ and curvature C, we set

$$\mathrm{Ch}(V) = \mathrm{Tr}exp(-C). \tag{3.6}$$

In [4], Bismut proved the following fundamental result:

<u>Theorem 3.7</u> (Bismut [4]). Let H^t be given by

$$\begin{aligned} H^t =& e_i(E_i + \frac{1}{4}\Gamma_{ij}^k e_i e_j + \frac{1}{2}\Gamma_{ij}^\alpha e_j \frac{dy^\alpha}{\sqrt{t}} + \frac{1}{4t}\Gamma_{i\alpha}^\beta dy^\alpha dy^\beta) \\ &+ \frac{dy^\alpha}{\sqrt{t}}(F_\alpha + \frac{1}{4}\Gamma_{\alpha i}^j e_i e_j + \frac{1}{2}\Gamma_{\alpha i}^\beta e_i \frac{dy^\beta}{\sqrt{t}} + \frac{1}{4}\Gamma_{\alpha\beta}^\gamma \frac{dy^\beta dy^\gamma}{t}) \\ I^t =& t(H^t)^2 \end{aligned} \tag{3.8}$$

Then

$$\int_{G_y} \mathrm{tr}_s(P_1^{L,t,y}(x,x))dx \tag{3.9}$$

is a C^∞ form over B which is a representative of $\mathrm{Ch}(\ker D_{+,y} - \ker D_{-,y})$, where $P_1^{L,t,y}(x,x')$ is the C^∞ kernel over G_y of $e^{-I_y^t}$.

The goal of this paper is to calculate out

$$\mathrm{tr}_s(P_1^{L,t,y}(x,x))dx \tag{3.10}$$

§4. A local parametrix and Minakshsundaram-Pleijel equations

In this section, we shall deduce (3.10) to a calculable form, and in the next section we will carry out the explicit calculation.

In all what follows, we may keep in mind that we are fixing a typical G_y, so that the subscript y will be omitted unless necessary.

First, as in [5], $\forall t > 0$, let φ_t be the homomorphism

$$(4.0)\qquad \begin{aligned} &\varphi_t : \mathrm{Hom}(F_{x'}, F_x)\hat{\otimes}\Lambda_y(B) \to \mathrm{Hom}(F_{x'}, F_x)\hat{\otimes}\Lambda_y(B) \\ &\varphi_t : h dy^\alpha \mapsto \frac{1}{\sqrt{t}} h dy^\alpha, h \in \mathrm{Hom}(F_{x'}, F_x) \end{aligned}$$

then clearly,

$$H^t = \varphi_t(H).$$

Hence

$$\begin{aligned} \mathrm{tr}_s e^{-I^t} &= \int_{G_y} \mathrm{tr}_s P_1^{L,t,y}(x,x)dx \\ &= \mathrm{tr}_s e^{-t(\varphi_t(H))^2} = \mathrm{tr}_s e^{-t\varphi_t H^2} \\ &= \mathrm{tr}_s \varphi_t e^{-tH^2} \\ &= \int_{G_y} \varphi_t \mathrm{tr}_s P_t^y(x,x)dx \end{aligned}$$

Thus we get a corollary of Theorem 3.7:

Proposition 4.1 $\forall t > 0$,

$$\varphi_t \int_{G_y} \mathrm{tr}_s P_t^y(x,x)dx$$

is a representative of $\mathrm{Ch}(\ker D_{+,y} - \ker D_{-,y})$.

We wish to calculate out

$$(4.2)\qquad \lim_{t\to 0} \int_{G_y} \mathrm{tr}_s(\varphi_t P_t^y(x,x))dx$$

Now we note that $P_t(x,x')$ is uniquely characterized by the following properties:

$$(4.3)\qquad \lim_{t\to 0}\left(\frac{\partial}{\partial t} + I_{x'}\right)P_t(x,x') = 0$$

and $\forall V(x') \in H_y$,

$$(4.4)\qquad \lim_{t\to 0} \int_{G_y} P_t(x,x')V(x')dx' = V(x)$$

Proof: Recalling that

$$e^{-tI}V(x) = \int_{G_y} P_t(x,x')V(x')dx'$$

so (4.3) and (4.4) are clearly hold. Now let $G_t(x,x')$ be another C^∞ function satisfying (4.3) and (4.4), we have

$$(4.5a)\qquad \left(\frac{\partial}{\partial t} + I_{x'}\right)(P_t(x,x') - G_t(x,x')) = 0$$

and $\forall V \in H_y$,

(4.5b)
$$\lim_{t\to 0}\int_{G_y}(P_t(x,x')-G_t(x,x'))V(x')dx'=0$$

From (4.5a,b) it is obvious that we should have

$$P_t(x,x')=G_t(x,x').$$

Definition 4.6 $\forall t>0$, set

(4.7)
$$H_N(x,x';t)=\frac{e^{-\rho^2/4t}}{(4\pi t)^n}\sum_{i=1}^{N}t^iU^i(x,x'),$$

where $N\geq n+[\frac{1}{2}m]$ and $\rho=d(x,x')$, $(x,x')\in\Delta(\varepsilon)=\{(x,x')\in M\times M|d(x,x')<\varepsilon\}$ for some sufficiently small $\varepsilon>0$, and each $U^{(i)}$ is a .

(4.8)
$$U^{(i)}(x,x'):F_{x'}\to F_x\hat{\otimes}\Lambda_y(B)$$

If $H_N(x,x';t)$ satisfies the following two conditions:
(1)

(4.9)
$$(\frac{\partial}{\partial t}+I_{x'})H_N(x,x';t)=\frac{e^{-\rho^2/4t}}{(4\pi t)^n}t^N h(x,x';t)$$

(2)
$$U^{(0)}(x,x)=Id:F_x\to F_x$$

where h is a continuous function, then we call H_N a local parametrix for I.

We now show that this H_N does exist.

Lemma 4.10 (Compare Yu [13])
If $\phi\in C^\infty(M)$, $S\in H^\infty$, then

$$\begin{aligned}I(\phi S)=&-\phi_{ii}S-2\phi_i\nabla_{E_i}S+\phi I(S)\\&+(a_i\phi_i\Gamma^k_{ij}e_je_k+b_i\phi_i\Gamma^\alpha_{ij}e_jdy^\alpha+c_i\phi_i\Gamma^\beta_{i\alpha}dy^\alpha dy^\beta)\end{aligned}$$

for some constants a_i,b_i,c_i.

Proof. First, as in [13], we easily deduced that

(4.11)
$$H(\phi S)=e_i\phi_iS+dy^\alpha\phi_\alpha S+\phi HS$$

(4.12)
$$\begin{aligned}H(e_i\phi_iS)=&e_i(e_i\phi_l)e_lS+e_ie_l\phi_l(E_iS)+\frac{1}{4}\Gamma^k_{ij}e_ie_je_k\phi_le_lS\\&+\frac{1}{2}\Gamma^\alpha_{ij}e_ie_jdy^\alpha\phi_le_lS+\frac{1}{4}\Gamma^\beta_{i\alpha}e_idy^\alpha dy^\beta e_l\phi_lS\\&+dy^\alpha(F_\alpha\phi_l)e_lS+dy^\alpha\phi_le_l(F_\alpha S)+\frac{1}{4}\Gamma^j_{\alpha i}dy^\alpha e_ie_j\phi_lS\\&+\frac{1}{2}\Gamma^\beta_{\alpha i}dy^\alpha e_idy^\beta e_l\phi_lS+\frac{1}{4}\Gamma^\gamma_{\alpha\beta}dy^\alpha dy^\beta dy^\gamma e_l\phi_lS\\=&-\phi_{ii}-\phi_ie_iHS+\phi_{\alpha i}dy^\alpha e_iS-2\phi_i\nabla_{E_i}S\\&+a_i\phi_i\Gamma^k_{ij}e_je_k+b_i\phi_i\Gamma^\alpha_{ij}e_jdy^\alpha+c_i\phi_i\Gamma^\beta_{i\alpha}dy^\alpha dy^\beta\end{aligned}$$

$$H(\phi_\alpha dy^\alpha S) = \phi_{i\alpha} e_i dy^\alpha S - \phi_\alpha dy^\alpha (HS) \tag{4.13}$$

and from (4.11) it follows that

$$H(\phi HS) = e_i\phi_i(HS) + dy^\alpha \phi_\alpha (HS) + \phi(H^2 S) \tag{4.14}$$

Now from (4.11) to (4.14), we obtain the Lemma by summation.

Set $\phi = \frac{e^{-\rho^2/4t}}{(4\pi t)^n}$ as in (4.9). Recall from [13] that in the local normal coordinate system, if $\rho^2(x,x') = x_1^2 + \cdots + x_{2n}^2$, we have

$$\phi_i = -\phi \frac{x_i}{2t} \tag{4.15}$$

$$\phi_{ii} = \phi\left(\frac{\rho^2}{4t} - \frac{n}{t} - \frac{1}{t}\sum_i x_i B_i\right) \tag{4.16}$$

for some functions B_i.

Set $S = \sum_{i=0}^N t^i U^{(i)}(x,x')$, then from (4.15), (4.16) and (4.10) we have

$$\frac{\partial}{\partial t} H_N(x';t) = \phi \sum_{i=0}^N \left(\frac{\rho^2}{4t^2} - \frac{n}{t} + \frac{i}{t}\right) t^i U^{(i)}(x')$$

$$\begin{aligned} I(H_N(x';t)) = &-\phi\left(\frac{\rho^2}{4t^2} - \frac{n}{t} - \frac{x_i B_i}{t}\right)S - 2\phi\left(\frac{-x_i}{2t}\right)\nabla_{E_i} S + \phi I S \\ &+ (a_i\phi_i \Gamma_{ij}^k e_j e_k + b_i \phi_i \Gamma_{ij}^\alpha e_j dy^\alpha + c_i \phi_i \Gamma_{i\alpha}^\beta dy^\alpha dy^\beta) S \\ = &-\sum_{i=0}^N \left(\frac{\rho^2}{4t^2} - \frac{n}{t} - \frac{x_i B_i}{t}\right) t^i U^{(i)} + \sum_{i=0}^n t^{i-1} \hat{d} U^{(i)} \\ &+ \sum_{i=0}^N I(U^{(i)}) t^i - \frac{1}{2}(a_i x_i \Gamma_{ij}^k e_j e_k + b_i x_i \Gamma_{ij}^\alpha e_j dy^\alpha \\ &+ c_i x_i \Gamma_{i\alpha}^\beta dy^\alpha dy^\beta) t^{i-1} U^{(i)}. \end{aligned}$$

Hence we obtain from (4.9) the following analogue of the Minakshsundaram-Pleijel equations given by Yu [13]:

$$\begin{aligned} &(\hat{d} + x_i(a_i\Gamma_{ij}^k e_j e_k + b_i \Gamma_{ij}^\alpha e_j dy^\alpha + c_i \Gamma_{i\alpha}^\beta dy^\alpha dy^\beta))U^{(i)} + (x_i B_i + i)U^{(i)} \\ &= -IU^{(i-1)}, \qquad i \le N \\ &U^{(0)}(x,x) = Id : F_x \to F_x. \end{aligned} \tag{4.17}$$

<u>Proposition 4.18</u> The local parametrix H_N exists iff $\forall i \le N$, $U^{(i)}$ satisfies the equations in (4.17).

In the next section, we will calculate out the local index throughout these equations.

§5. The local index theorem

First we make it explicit what to be calculate. Recall from [14] that if H_N is the local parametrix, we have

$$P_t(x,x') - H_N(x,x';t) = O(t^{1+N-n}) \tag{5.1}$$

while from (4.0) it is clear that

$$\lim_{t\to 0} t^{[\frac{1}{2}m]+\frac{2}{3}}\varphi_t = 0 \tag{5.2}$$

so when $N \geq [\frac{1}{2}m] + n$, we have

$$\lim_{t\to 0} \varphi_t(O(t^{1+N-n})) = 0 \tag{5.3}$$

From (5.1) and (5.3), it follow that

$$\lim_{t\to 0} \varphi_t P_t(x,x) = \lim_{t\to 0} \varphi_t H_N(x,x;t) \tag{5.4}$$

The supertrace of the right hand of (5.4) is precisely what we procceed to calculate out.

Now we take a convension similar to that in [13]:

Let φ be a C^∞ function defined locally in a neighborhood of x, denote the degree of zero of φ at x by $\nu(\varphi)$, to every

$$\begin{aligned}\alpha(x') = &\varphi_{l_1}(x')\frac{\partial}{\partial x_{i_1}}\varphi_{l_2}(x')\cdots\frac{\partial}{\partial x_{i_m}}\varphi_{l_{m+1}}(x')dy^{\alpha_1}\cdots dy^{\alpha_p}\cdot\\ &\cdot e_{j_1}\cdots e_{j_s} : F'_x \to F_x; \alpha_i \neq \alpha_j\ (i\neq j); j_a \neq j_t\ (a\neq t)\end{aligned} \tag{5.5a}$$

We define

$$\chi(\alpha) = m + p + s - \nu(\varphi_1\cdots\varphi_{m+1}) \tag{5.5b}$$

and we denote $\{\chi < m\}$ the linear space generated by all the elements α for which $\chi(\alpha) < m$, etc. and denote $(\chi < m)$ an element of $\{\chi < m\}$, e.g. $\omega = \eta + (\chi < m)$ means that there exists a $\beta \in \{\alpha < m\}$ such that $\omega = \eta + \beta$, we can also write it as

$$\omega \equiv \eta \mod\{\chi < m\} \tag{5.6}$$

<u>Lemma 5.7</u>

$$\begin{aligned}\Gamma^k_{ij} &= \frac{1}{2}\sum_l R_{iljk}x_l + (\chi < -1),\\ \Gamma^\alpha_{ij} &= \frac{1}{2}\sum_l R_{ilj\alpha}x_l + (\chi < -1),\\ \Gamma^\beta_{i\alpha} &= \frac{1}{2}\sum_l R_{il\alpha\beta}x_l + (\chi < -1).\end{aligned}$$

<u>Proof.</u> Comparing [13], we only need to note that we are working on a fixed G_y.

<u>Proposition 5.8</u>

$$\begin{aligned}I = &-\frac{\partial^2}{\partial x_i^2} + \frac{1}{4}R_{ijst}x_i\frac{\partial}{\partial x_j}e_s e_t + \frac{1}{2}R_{ijs\alpha}x_i\frac{\partial}{\partial x_j}e_s dy^\alpha + \frac{1}{4}R_{ij\alpha\beta}x_i\frac{\partial}{\partial x_j}dy^\alpha dy^\beta\\ &+\frac{1}{64}x_i x_j R_{irlk}R_{rjst}e_l e_k e_s e_t + \frac{1}{16}x_i x_j R_{irk\alpha}R_{rjst}e_k e_s e_t dy^\alpha +\\ &+\frac{1}{32}x_i x_j R_{irlk}R_{rj\alpha\beta}e_l e_k dy^\alpha dy^\beta + \frac{1}{16}x_i x_j R_{irk\alpha}R_{rjl\beta}e_k dy^\alpha e_l dy^\beta +\\ &+\frac{1}{16}x_i x_j R_{irl\alpha}R_{rj\lambda\mu}e_l dy^\alpha dy^\lambda dy^\mu +\\ &+\frac{1}{64}x_i x_j R_{ir\alpha\beta}R_{rj\lambda\mu}dy^\alpha dy^\beta dy^\lambda dy^\mu + (\chi < 2).\end{aligned}$$

Proof. It follows directly from Lemma 5.7 and the generalized Lichnerowicz formula given by Bismut [4], Theorem 3.5.

As in [13], we denote

$$a_0 = -\frac{1}{4}R_{ijst}x_i\frac{\partial}{\partial x_j}e_se_t - \frac{1}{2}R_{ijs\alpha}x_i\frac{\partial}{\partial x_j}e_s dy^\alpha - \frac{1}{4}R_{ij\alpha\beta}x_i\frac{\partial}{\partial x_j}dy^\alpha dy^\beta;$$

$$a_2 = \sum_i \frac{\partial^2}{\partial x_i^2};$$

$$\begin{aligned} a_{-2} = -\Big(&\frac{1}{64}x_ix_jR_{irkl}R_{rjst}e_le_ke_se_t + \frac{1}{16}x_ix_jR_{irk\alpha}R_{rjst}e_ke_se_tdy^\alpha + \\ &+\frac{1}{32}x_ix_jR_{irlk}R_{rj\alpha\beta}e_le_kdy^\alpha dy^\beta + \frac{1}{16}x_ix_jR_{irk\alpha}R_{rjl\beta}e_kdy^\alpha e_ldy^\beta + \\ &\frac{1}{16}x_ix_jR_{irl\alpha}R_{rj\lambda\mu}e_ldy^\alpha dy^\lambda dy^\mu + \\ &+\frac{1}{64}x_ix_jR_{ir\alpha\beta}R_{rj\lambda\mu}dy^\alpha dy^\beta dy^\lambda dy^\mu\Big). \end{aligned}$$

And we set

$$A_i = a_ix_lR_{iljk}e_je_k + b_ix_lR_{ilj\alpha}e_jdy^\alpha + c_ix_lR_{il\alpha\beta}dy^\alpha dy^\beta.$$

Obviously,

$$x_i(a_i\Gamma^k_{ij}e_je_k + b_i\Gamma^\alpha_{ij}e_jdy^\alpha + c_i\Gamma^\beta_{i\alpha}dy^\alpha dy^\beta) = x_iA_i + (\chi < 0). \tag{5.9}$$

Lemma 5.10 $\forall 1 \le j \le 2n$,

$$\frac{\partial}{\partial x_j}(x_iA_i) = 0.$$

Proof.

$$\frac{\partial}{\partial x_j}(x_ix_lR_{ilst}e_se_t) = x_i(R_{jist} + R_{ijst})e_se_t = 0,$$

the other two can be proved in the same way.

Corollary 5.11 For $S \in (Hom(F,F)\hat{\otimes} \wedge_y (B))$,

$$\begin{aligned} a_0((\sum_i x_iA_i)S) =&(\sum_i x_iA_i)a_0S + (\chi < \chi(a_0S)),\\ a_2((\sum x_iA_i)S) =&(\sum x_iA_i)a_2S. \end{aligned}$$

Lemma 5.12

$$a_{-2}A_i \equiv A_ia_{-2}, \quad \mathrm{mod}\{\chi < 3\}.$$

Proof. Direct calculations.

Now we recall the basic idea of Yu[13] of comparing the corresponding terms of the Taylor expansion series: let f be a function in a neighborhood of x, $f : U \to R$ or F. We expand f by its Taylor series:

$$f = \sum_{m=0}^{\infty} \hat{f}(m) \tag{5.13}$$

where $\hat{f}(m)$ is the m-th degree homogeneous polynomial in $x_1, \cdots, x_{2n}$. We know that, for $V \in F_{x'}, U^{(i)}(x,x') : F_{x'} \to F_x, U^{(i)}V$ can be viewed as a spinor field, which under the fixed spin frame, can be viewed as a function with values in F which we still denote by $U^{(i)}$.

Notice that $\hat{d}\hat{U}(m) = m\hat{U}(m)$, and denote $\sum_i x_i B_i = h$ for some h, comparing the corresponding terms of Taylor series in (4.17), we get

$$\begin{aligned}&(m+i)\hat{U}^{(i)}(m) + x_i A_i \hat{U}^{(i)}(m-2)\\&+ \sum_{\substack{m_1+m_2=m\\ m_1>0}} \hat{h}(m_1)\hat{U}^{(i)}(m_2)\\=&a_2\hat{U}^{(i-1)}(m+2) + a_0\hat{U}^{(i-1)}(m) + a_{-2}\hat{U}^{(i-1)}(m-2)\\&+\sum f_j \hat{U}^{(i-1)}(m_j)\end{aligned}$$

where $\chi(f_j) < 2$. Rewrite it as

$$\begin{aligned}\hat{U}^{(i)}(m) =&\frac{1}{(m+i)}\sum_{\alpha}\hat{U}^{(i-1)}(m+\alpha) + \sum_j g_j\hat{U}^{(i-1)}(m_j)+\\&+(\sum_i x_i A_i)\sum_j s_{m_j}\hat{U}^{(i-1)}(m_j)\end{aligned} \tag{5.14}$$

with $\chi(g_j) < 2$, $\chi(S_{m_j}) \le 2$. From (5.14), it can be easily deduced that

$$\begin{aligned}\hat{U}^{(i)}(m) = &\sum_{\alpha_1,\cdots,\alpha_i} \frac{a_{\alpha_1}\cdots a_{\alpha_i}}{\Gamma(\alpha_1,\cdots,\alpha_i;m)}(\hat{U}^{(0)}(m+\alpha_1+\cdots+\alpha_i))+\\&\sum_j f_j\hat{U}^{(0)}(m_j) + (\sum_i x_i A_i)\sum_j \tilde{S}_j\hat{U}^{(0)}(m_j)\end{aligned} \tag{5.15}$$

with $\chi(f_j) < 2i$, $\chi(\tilde{S}_j) < 2i$. Note that in the deducing, Lemma 5.12 and Corollary 5.11 are freely used.

Proposition 5.16 For $i < n$,

$$\mathrm{tr}_s U^{(i)}(x) = 0.$$

Proof. c.f. [13] or compare with the following proof of Lemma 5.20.

Corollary 5.17

$$\lim_{t\to 0}\varphi_t(\sum_{i=0}^{n-1}\mathrm{tr}_s U^{(i)}\frac{t^i}{(4\pi t)^n}) = 0$$

So what we really ought to calculate out is

$$\begin{aligned}&\lim_{t\to 0}\varphi_t(\frac{1}{(4\pi t)^n}\sum_{k=0}^{[\frac{1}{2}m]}\mathrm{tr}_s U^{(n+k)}(x,x)t^{n+k})\\=&\lim_{t\to 0}\frac{1}{(4\pi)^n}\sum_{k=0}^{[\frac{1}{2}m]}t^k\varphi_t(\mathrm{tr}_s U^{(n+k)}(x,x)).\end{aligned} \tag{5.18}$$

Let us take a look at the one

$$\lim_{t\to 0}t^k\varphi_t(\mathrm{tr}_s U^{(n+k)}(x,x)),\quad 0\le k\le [\frac{1}{2}m] \tag{5.19}$$

Lemma 5.20 If $\chi(\alpha) < 2n+2k$, then

$$\lim_{t\to 0}t^k\varphi_t(\mathrm{tr}_s\alpha) = 0,\quad 0\le k\le [\frac{1}{2}m].$$

Proof. We can assume that α can be written as

$$\alpha = \varphi(x')dy^{\alpha_1}\cdots dy^{\alpha_p}e_{j_1}\cdots e_{j_s}$$

If $\chi(\alpha) < 2n+2k$, then either $\varphi(x) = 0$ or $p+s < 2n+2k$. In the former case, $\mathrm{tr}_s\alpha = 0$ is trivial and in the latter case, if $s < 2n$, then $\mathrm{tr}_s(\varphi(x)dy^{\alpha_1}\cdots dy^{\alpha_p}e_{j_1}\cdots e_{j_s}) = \varphi(x)dy^{\alpha_1}\cdots dy^{\alpha_p}\cdot \mathrm{tr}_s(e_{j_1}\cdots e_{j_s}) = 0$ and if $s \geq 2n$, then $p < 2k$ so

$$\begin{aligned}\lim_{t\to 0} t^k\varphi_t(\mathrm{tr}_s\alpha) &= \lim_{t\to 0}\varphi(x)\varphi_t(dy^{\alpha_1}\cdots dy^{\alpha_p})t^k\mathrm{tr}_s(e_{i_1}\cdots e_{i_s})\\ &= \lim_{t\to 0}\varphi(x)\mathrm{tr}_s(e_{i_1}\cdots e_{i_s})dy^{\alpha_1}\cdots dy^{\alpha_p}t^{k-\frac{1}{2}p} = 0\end{aligned}$$

Lemma 5.21 If $\varphi(x,x') \in \mathrm{Hom}(F_{x'},F_x)\hat{\otimes}\wedge_y(B)$, and some $\alpha_k = 0$, then

$$(a_{\alpha_1}\cdots a_{\alpha_l}\varphi) < 2l + \chi(\varphi)$$

Proof. cf. [13].

Now we can easily see from (4.17), (5.15), (5.19), (5.20) and the above Lemma 5.21 that $\sum_i x_iA_i$, a_0, $h = \sum_i x_iB_i$ and the term $(\chi < 2)$ in Propotion 5.18 are really irrelavent for the calculation of the supertrace (5.19). We write this result as follows:

Proposition 5.22 If $V^{(i)}(x,x') \in \mathrm{Hom}(F_{x'},F_x)\hat{\otimes}\wedge_y(B)$ satisfies the following equations:

$$x_i\frac{\partial}{\partial x_i}V^{(i)} + iV^{(i)} = (\sum_i\frac{\partial^2}{\partial x_i^2} + a_{-2})V^{(i-1)} \tag{5.23}$$

$$V^{(0)}(x,x) = Id : F_x \to F_x$$

then

$$\lim_{t\to 0}\varphi_t(\mathrm{tr}_sV^{(n+k)}(x,x))t^k = \lim_{t\to 0}\varphi_t(\mathrm{tr}_sU^{(n+k)}(x,x))t^k \tag{5.24}$$

where $0 \leq k \leq [\frac{1}{2}m]$.

Now the analogue of (5.15) is

$$V^{(i)}(m) = \sum_{\alpha_1,\cdots,\alpha_i}\frac{a_{\alpha_1}\cdots a_{\alpha_i}}{\Gamma(\alpha_1,\cdots,\alpha_i;0)}(V^{(0)}(m+\alpha_1+\cdots+\alpha_i)) \tag{5.25}$$

with each $\alpha_i \neq 0$. So

$$\chi(V^{(n+k)}) \leq 2n+2k. \tag{5.26}$$

$V^{(n+k)}$ can be expressed as a sum of the following terms:

$$\alpha = \varphi(x')dy^{\alpha_1}\cdots dy^{\alpha_p}e_{i_1}\cdots e_{i_s} \tag{5.27}$$

where in the process, we only take the interchanging between $dy^{\alpha}{}'s$ and $e_i's$, e.g., $e_idy^{\alpha} = -dy^{\alpha}e_i$, and has not interchanged the order of $e_i's$. It may be happened to α the following cases:

(1) $\varphi(x) = 0$, then $\alpha(x) = 0$;

(2) $\varphi(x) \neq 0$, but $\exists r \neq q \ni \alpha_r = \alpha_q$, then $\alpha(x) = 0$;

(3) $\varphi(x) \neq 0$, and $\forall r \neq q$, $\alpha_r \neq \alpha_q$, then $p+s = 2k+2n$, if $p < 2k$, then $\lim_{t\to 0}\varphi_t(\varphi(x)dy^{\alpha_1}\cdots$ 0, and if $p > 2k$, then $s < 2n, \mathrm{tr}_s(e_{i_1}\cdots e_{i_s}) = 0$. So in this case we must have $p = 2k$ and $s = 2n$ for a possible non-zero contribution to the supertrace, but if we have some $i_r = i_q$, $r \neq q$, then $\chi(e_{i_1}\cdots e_{i_{2n}}) < 2n$ which implies that $\mathrm{tr}_s(e_{i_1}\cdots e_{i_s}) = 0$.

Summarizing these, we get

Proposition 5.28 The only terms having nontrivial contributions to (5.24) are thouse of α's such that

$$\alpha = \varphi(x')dy^{\alpha_1}\cdots dy^{\alpha_p}e_{i_1}\cdots e_{i_s}$$

where $\varphi(x) \neq 0$, $\alpha_r \neq \alpha_q$ $(r \neq q)$, $e_{i_t} \neq e_{i_l}$ $(t \neq l)$ and $p + s = 2n + 2k$. In particular, if in the original expression of α, there have some e_i's resppeared, then $\lim_{t\to 0} t^k \varphi_t(\mathrm{tr}_s \alpha) = 0$.

Proof. All that we need to notice is the following:

$$e_i e_j \equiv -e_j e_i \qquad (\mathrm{mod}\{\chi < 2\}).$$

From this proposition and Lemma 1.8, we immediately have:

Proposition 5.29 If $W^{(i)}(x,x') \in \wedge_{x'}(TG)\hat{\otimes} \wedge_y (B)$ satisfies the equations

$$\begin{aligned}&x_i\frac{\partial}{\partial x_i}W^{(i)} + iW^{(i)}\\ &=(\sum_i \frac{\partial^2}{\partial x_i^2} - \frac{1}{64}x_i x_j R_{irIJ}R_{rjST}dz^I dz^J dz^S dz^T)W^{(i-1)},\\ &W^{(0)}(x,x) = 1\end{aligned} \tag{5.30}$$

where we denote $z = (x, y)$ and use I, J etc., to denote the total subscripts i, j, α, etc. Then

$$\lim_{t\to 0}\varphi_t(\mathrm{tr}_s(P_t(x,x)))dx = (\frac{1}{2\pi\sqrt{-1}})^n \sum_{k=0}^{[\frac{1}{2}m]}(W^{(n+k)}(x))_{2n} \tag{5.31}$$

where $(\cdot)_{2n}$ stands for the term which is a multiple of $dx_1\cdots dx_{2n}$.

Proof. This follows directly from (5.4), (5.18), (5.19), (5.22), (5.28) and (1.8). Notice that $dx \cdot dx = 0$.

Denote by $\Omega = -\frac{1}{2}R_{ijIJ}dz^I dz^J$ the matrix of two forms over M. Clearly, Ω is the curvature matrix for the connection ∇ in (2.4) of the vector bundle TG over M. As in the usual computations of characteristic classes, we take the identification of Ω to its Chern root matrix:

$$\begin{pmatrix} 0 & u_1 & & & \\ -u_1 & 0 & & & \\ & & \ddots & & \\ & & & 0 & u_n \\ & & & -u_n & 0 \end{pmatrix} \tag{5.32}$$

then

$$x_i x_j \Omega_{ir}\Omega_{rj} = -\sum_{l=1}^{n}(x_{2l-1}^2 + x_{2l}^2)u_l^2 \tag{5.33}$$

and (5.30) becomes

$$x_i\frac{\partial}{\partial x_i}W^{(i)} + iW^{(i)} = (\sum_{i=0}^{n}\frac{\partial^2}{\partial x_i^2} + \frac{1}{16}\sum_{l=1}^{n}(x_{2l-1}^2 + x_{2l}^2)u_l^2)W^{(i-1)} \tag{5.34}$$

$$W^{(0)}(x,x) = 1$$

As in [13], set

$$H(x_1,\cdots,x_{2n};t)=\frac{e^{-\rho^2/4t}}{(4\pi t)^n}\sum_{i=0}^{\infty}W^{(i)}(x_1,\cdots,x_{2n})t^i \tag{5.35}$$

By (5.34) and (5.35), we have:

$$\begin{gathered}\frac{\partial H}{\partial t}=\sum_i\frac{\partial^2}{\partial x_i^2}H+\frac{1}{16}\sum_{l=1}^{n}(x_{2l-1}^2+x_{2l}^2)u_l^2H\\ \lim_{t\to 0}(4\pi t)^nH(0,\cdots,0;t)=1\end{gathered} \tag{5.36}$$

Solving this equation as in [13], we find

$$H(x_1,\cdots,x_{2n};t)=(\frac{1}{4\pi})^n\prod_{l=1}^{n}(\frac{\sqrt{-1}u_l/2}{\sinh\sqrt{-1}u_lt/2}e^{(x_{2l-1}^2+x_{2l}^2)\frac{\sqrt{-1}u_l}{8}\coth\frac{\sqrt{-1}u_lt}{2}}) \tag{5.37}$$

So

$$H(0,\cdots,0;t)=\frac{1}{(4\pi)^n}\prod_{l=1}^{n}\frac{\sqrt{-1}u_l/2}{\sinh\sqrt{-1}u_lt/2} \tag{5.38}$$

Combining with (5.35), we get

$$\sum_{i=0}^{\infty}W^{(i)}(0)t^i=\prod_{l=1}^{n}\frac{\sqrt{-1}u_lt/2}{\sinh\sqrt{-1}u_lt/2} \tag{5.39}$$

Thus

$$\sum_{i=0}^{\infty}W^{(i)}(0)=\prod_{l=1}^{n}\frac{\sqrt{-1}u_l/2}{\sinh\sqrt{-1}u_l/2}=\hat{A}(\sqrt{-1}\Omega) \tag{5.40}$$

Notice that when $N>n+[\frac{1}{2}m]$, $W^{(N)}(0)=0$ and when $N<n$, $W^{(N)}(0)$ is not a multipul of $dx^1\cdots dx^{2n}$. Hence from (5.31) and (5.40) we finally get

$$\lim_{t\to 0}\varphi_t(\mathrm{tr}_sP_t(x,x))dx=(\frac{1}{2\pi\sqrt{-1}})^n(\hat{A}(\sqrt{-1}\Omega))_{2n} \tag{5.41}$$

This is what we call the local Atiyah-Singer index theorem for families of Dirac operators. As a direct corollary, we get

Theorem (Atiyah-Singer [1]):

$$(\frac{1}{2\pi\sqrt{-1}})^n\int_{G_y}\hat{A}(\sqrt{-1}\Omega) \tag{5.42}$$

is a representative of $\mathrm{Ch}(\ker D_{+,y}-\ker D_{-,y})$.

APPENDIX

In this appendix, we will briefly outline how our method works for twisted Dirac operators. For simplicity (and without loss of generality), we only carry out the single operator case. Now, by Lichnerowicz formula, we can deduce that (Compare [13])

$$\begin{aligned}D^2=&-\frac{\partial^2}{\partial x_i^2}+\frac{1}{4}R_{ijst}x_i\frac{\partial}{\partial x_j}e_se_t+\frac{1}{64}x_ix_jR_{irst}R_{rjpq}e_se_te_pe_q\\&+\frac{1}{2}e_ie_j\otimes F(e_i,e_j)+(\chi<2)\end{aligned} \tag{a.1}$$

where F is the curvature matrix of the connection of the vector bundle ξ over G. Now doing similarly as in §5, we see that if $V^{(i)}$ is the solution of the following equations

(a.2)
$$x_i \frac{\partial}{\partial x_i} V^{(i)} + iV^{(i)} = [\sum_i \frac{\partial^2}{\partial x_i^2} - \frac{1}{16} x_i x_j \Omega_{ir} \Omega_{rj} - \frac{1}{2} dx^s dx^t \otimes F(e_s, e_t)] V^{(i-1)}$$
$$V^{(-1)} \equiv 0, \quad V^{(0)}(0) = Id : (F \otimes \xi)_x \to (F \otimes \xi)_x$$

then the local index is

(a.3)
$$\tau(D) = \frac{1}{2\pi(\sqrt{-1})^n} \mathrm{tr}_s V^{(n)}(0)$$

Aside (5.32), we pick another identification

(a.4)
$$-\frac{1}{2} dx^i dx^j \otimes F(e_i, e_j) \leftrightarrow 1 \otimes \begin{pmatrix} v_1 & & \\ & \ddots & \\ & & v_N \end{pmatrix}$$

where $N = \dim \xi$. Set $H = \frac{e^{-\rho^2/4t}}{(4\pi t)^n} \sum_i t^i V^{(i)}$ then by solving an equation similar to (5.36), we find

$$H(x_1, \cdots, x_{2n}; t) = \prod_{l=1}^{n} \left(\frac{\sqrt{-1} u_l}{8\pi \sinh(\sqrt{-1} u_l t/2)} e^{(x_{2l-1}^2 + x_{2l}^2) \frac{\sqrt{-1} u_l}{8} \coth \frac{\sqrt{-1} u_l t}{2}} \right) \cdot \begin{pmatrix} e^{v_1 t} & & \\ & \ddots & \\ & & e^{v_N t} \end{pmatrix}$$

So

$$\tau(D)$$
$$= \lim_{t \to 0} \left(\frac{1}{2\pi\sqrt{-1}}\right)^n \frac{1}{n!} \cdot \frac{\partial}{\partial t^n} \left((4\pi t)^n \left(\prod_1^n \frac{\sqrt{-1} u_l}{8\pi \sinh(\sqrt{-1} u_l t/2)}\right) (e^{v_1 t} + \cdots + e^{v_N t})\right)$$
$$= \left(\frac{1}{2\pi}\right)^n \left(\hat{A}(\Omega) \mathrm{ch}\left(\frac{F}{\sqrt{-1}}\right)\right)_{2n}$$

this is the local Atiyah-Singer index theorem for twisted Dirac operators.

REFERENCES

[1] M.F.Atiyah & I.M.Singer: The index of elliptic operators IV. Annals of Math. 93 119-138 (1971).

[2] N.Berline & M.Vergne: A computation of the equivariant index of the Dirac operators. Bull. Soc. Math. France 113 305-345 (1985).

[3] J.-M.Bismut: The Atiyah-Singer theorems: A probabilistic approach. J.Functional Analysis 57 56-99 (1984).

[4] J.-M.Bismut: The Atiyah-Singer index theorem for families of Dirac operators: Two heat equation proofs. Invent. Math. 83 91-151 (1986).

[5] J.-M.Bismut & D.Freed: The analysis of elliptic families I. Commun. Math. Phys. 106 159-176 (1986).

[6] E.Getzler: Pseudodifferential operators on supermanifolds and the Atiyah-Singer index theorem. Commun. Math. Phys. 92 163-178 (1983).

[7] E.Getzler: A short proof of the local Atiyah-Singer index theorem. Topology 25 111-117 (1986).

[8] V.Mathai & D.Quillen: Super connections, Thom classes and equivariant differential forms. Topology 25 85-110 (1986).

[9] H.P.Mckean, Jr. & I.M.Singer: Curvature and the eigenvalue of the Laplacian J.Differentia Geometry 1 43-69 (1967).

[10] V.K.Patodi: Curvatures and the eigenforms of the Laplace operator. J.Differential Geometry 5 233-249 (1971).

[11] V.K.Patodi: An analytic proof of Riemann-Roch-Hirzebruch theorem for Kaehler manifolds. J.Diff. Geom. 5 251-283 (1971).

[12] D.Quillen: Superconnections and the Chern character. Topology 24 89-95 (1985).

[13] Yu Yan-lin: Local index theorem for Dirac operators. Acta. Math. Sinica. (New Sieries) vol.3, No.2, pp.152-169 (1987).

[14] Yu Yan-lin: Lectures on Atiyah-Singer index theorem (in Chinese) Nankai Institute of Mathematics, meomographic notes (1986).

[15] Yu Yan-lin: Local index theorem for Signature operators. Acta Math. Sinica (New Series) vol.3, No.4, pp.363-372.

Institute of Mathematics
Academia Sinica
Beijing, China
April, 1987

Note added in proof: We learned after completing this work that Berline-Vergne, in a preprint dating Sept. 1986, had also given a differential geometric proof of this Bismut local index theorem. (Topology 26 No.4 (1987))

LECTURE NOTES IN MATHEMATICS
Edited by A. Dold and B. Eckmann

Some general remarks on the publication of proceedings of congresses and symposia

Lecture Notes aim to report new developments - quickly, informally and at a high level. The following describes criteria and procedures which apply to proceedings volumes. The editors of a volume are strongly advised to inform contributors about these points at an early stage.

§1. One (or more) expert participant(s) of the meeting should act as the responsible editor(s) of the proceedings. They select the papers which are suitable (cf. §§ 2, 3) for inclusion in the proceedings, and have them individually refereed (as for a journal). It should not be assumed that the published proceedings must reflect conference events faithfully and in their entirety. Contributions to the meeting which are not included in the proceedings can be listed by title. The series editors will normally not interfere with the editing of a particular proceedings volume - except in fairly obvious cases, or on technical matters, such as described in §§ 2, 3. The names of the responsible editors appear on the title page of the volume.

§2. The proceedings should be reasonably homogeneous (concerned with a limited area). For instance, the proceedings of a congress on "Analysis" or "Mathematics in Wonderland" would normally not be sufficiently homogeneous.

One or two longer survey articles on recent developments in the field are often very useful additions to such proceedings - even if they do not correspond to actual lectures at the congress. An extensive introduction on the subject of the congress would be desirable.

§3. The contributions should be of a high mathematical standard and of current interest. Research articles should present new material and not duplicate other papers already published or due to be published. They should contain sufficient information and motivation and they should present proofs, or at least outlines of such, in sufficient detail to enable an expert to complete them. Thus resumes and mere announcements of papers appearing elsewhere cannot be included, although more detailed versions of a contribution may well be published in other places later.

Surveys, if included, should cover a sufficiently broad topic, and should in general not simply review the author's own recent research. In the case of surveys, exceptionally, proofs of results may not be necessary.

"Mathematical Reviews" and "Zentralblatt für Mathematik" require that papers in proceedings volumes carry an explicit statement that they are in final form and that no similar paper has been or is being submitted elsewhere, if these papers are to be considered for a review. Normally, papers that satisfy the criteria of the Lecture Notes in Mathematics series also satisfy this

.../...

requirement, but we would strongly recommend that the contributing authors be asked to give this guarantee explicitly at the beginning or end of their paper. There will occasionally be cases where this does not apply but where, for special reasons, the paper is still acceptable for LNM.

§4. Proceedings should appear soon after the meeeting. The publisher should, therefore, receive the complete manuscript within nine months of the date of the meeting at the latest.

§5. Plans or proposals for proceedings volumes should be sent to one of the editors of the series or to Springer-Verlag Heidelberg. They should give sufficient information on the conference or symposium, and on the proposed proceedings. In particular, they should contain a list of the expected contributions with their prospective length. Abstracts or early versions (drafts) of some of the contributions are very helpful.

§6. Lecture Notes are printed by photo-offset from camera-ready typed copy provided by the editors. For this purpose Springer-Verlag provides editors with technical instructions for the preparation of manuscripts and these should be distributed to all contributing authors. Springer-Verlag can also, on request, supply stationery on which the prescribed typing area is outlined. Some homogeneity in the presentation of the contributions is desirable.

Careful preparation of manuscripts will help keep production time short and ensure a satisfactory appearance of the finished book. The actual production of a Lecture Notes volume normally takes 6 -8 weeks.

Manuscripts should be at least 100 pages long. The final version should include a table of contents and as far as applicable a subject index.

§7. Editors receive a total of 50 free copies of their volume for distribution to the contributing authors, but no royalties. (Unfortunately, no reprints of individual contributions can be supplied.) They are entitled to purchase further copies of their book for their personal use at a discount of 33.3 %, other Springer mathematics books at a discount of 20 % directly from Springer-Verlag. Contributing authors may purchase the volume in which their article appears at a discount of 33.3 %.

Commitment to publish is made by letter of intent rather than by signing a formal contract. Springer-Verlag secures the copyright for each volume.

Vol. 1201: Curvature and Topology of Riemannian Manifolds. Proceedings, 1985. Edited by K. Shiohama, T. Sakai and T. Sunada. VII, 336 pages. 1986.

Vol. 1202: A. Dür, Möbius Functions, Incidence Algebras and Power Series Representations. XI, 134 pages. 1986.

Vol. 1203: Stochastic Processes and Their Applications. Proceedings, 1985. Edited by K. Itô and T. Hida. VI, 222 pages. 1986.

Vol. 1204: Séminaire de Probabilités XX, 1984/85. Proceedings. Edité par J. Azéma et M. Yor. V, 639 pages. 1986.

Vol. 1205: B.Z. Moroz, Analytic Arithmetic in Algebraic Number Fields. VII, 177 pages. 1986.

Vol. 1206: Probability and Analysis, Varenna (Como) 1985. Seminar. Edited by G. Letta and M. Pratelli. VIII, 280 pages. 1986.

Vol. 1207: P.H. Bérard, Spectral Geometry: Direct and Inverse Problems. With an Appendix by G. Besson. XIII, 272 pages. 1986.

Vol. 1208: S. Kaijser, J.W. Pelletier, Interpolation Functors and Duality. IV, 167 pages. 1986.

Vol. 1209: Differential Geometry, Peñíscola 1985. Proceedings. Edited by A.M. Naveira, A. Ferrández and F. Mascaró. VIII, 306 pages. 1986.

Vol. 1210: Probability Measures on Groups VIII. Proceedings, 1985. Edited by H. Heyer. X, 386 pages. 1986.

Vol. 1211: M.B. Sevryuk, Reversible Systems. V, 319 pages. 1986.

Vol. 1212: Stochastic Spatial Processes. Proceedings, 1984. Edited by P. Tautu. VIII, 311 pages. 1986.

Vol. 1213: L.G. Lewis, Jr., J.P. May, M. Steinberger, Equivariant Stable Homotopy Theory. IX, 538 pages. 1986.

Vol. 1214: Global Analysis – Studies and Applications II. Edited by Yu.G. Borisovich and Yu.E. Gliklikh. V, 275 pages. 1986.

Vol. 1215: Lectures in Probability and Statistics. Edited by G. del Pino and R. Rebolledo. V, 491 pages. 1986.

Vol. 1216: J. Kogan, Bifurcation of Extremals in Optimal Control. VIII, 106 pages. 1986.

Vol. 1217: Transformation Groups. Proceedings, 1985. Edited by S. Jackowski and K. Pawalowski. X, 396 pages. 1986.

Vol. 1218: Schrödinger Operators, Aarhus 1985. Seminar. Edited by E. Balslev. V, 222 pages. 1986.

Vol. 1219: R. Weissauer, Stabile Modulformen und Eisensteinreihen. III, 147 Seiten. 1986.

Vol. 1220: Séminaire d'Algèbre Paul Dubreil et Marie-Paule Malliavin. Proceedings, 1985. Edité par M.-P. Malliavin. IV, 200 pages. 1986.

Vol. 1221: Probability and Banach Spaces. Proceedings, 1985. Edited by J. Bastero and M. San Miguel. XI, 222 pages. 1986.

Vol. 1222: A. Katok, J.-M. Strelcyn, with the collaboration of F. Ledrappier and F. Przytycki, Invariant Manifolds, Entropy and Billiards; Smooth Maps with Singularities. VIII, 283 pages. 1986.

Vol. 1223: Differential Equations in Banach Spaces. Proceedings, 1985. Edited by A. Favini and E. Obrecht. VIII, 299 pages. 1986.

Vol. 1224: Nonlinear Diffusion Problems, Montecatini Terme 1985. Seminar. Edited by A. Fasano and M. Primicerio. VIII, 188 pages. 1986.

Vol. 1225: Inverse Problems, Montecatini Terme 1986. Seminar. Edited by G. Talenti. VIII, 204 pages. 1986.

Vol. 1226: A. Buium, Differential Function Fields and Moduli of Algebraic Varieties. IX, 146 pages. 1986.

Vol. 1227: H. Helson, The Spectral Theorem. VI, 104 pages. 1986.

Vol. 1228: Multigrid Methods II. Proceedings, 1985. Edited by W. Hackbusch and U. Trottenberg. VI, 336 pages. 1986.

Vol. 1229: O. Bratteli, Derivations, Dissipations and Group Actions on C*-algebras. IV, 277 pages. 1986.

Vol. 1230: Numerical Analysis. Proceedings, 1984. Edited by J.-P. Hennart. X, 234 pages. 1986.

Vol. 1231: E.-U. Gekeler, Drinfeld Modular Curves. XIV, 107 pages. 1986.

Vol. 1232: P.C. Schuur, Asymptotic Analysis of Soliton Problems. VIII, 180 pages. 1986.

Vol. 1233: Stability Problems for Stochastic Models. Proceedings, 1985. Edited by V.V. Kalashnikov, B. Penkov and V.M. Zolotarev. VI, 223 pages. 1986.

Vol. 1234: Combinatoire énumérative. Proceedings, 1985. Edité par G. Labelle et P. Leroux. XIV, 387 pages. 1986.

Vol. 1235: Séminaire de Théorie du Potentiel, Paris, No. 8. Directeurs: M. Brelot, G. Choquet et J. Deny. Rédacteurs: F. Hirsch et G. Mokobodzki. III, 209 pages. 1987.

Vol. 1236: Stochastic Partial Differential Equations and Applications. Proceedings, 1985. Edited by G. Da Prato and L. Tubaro. V, 257 pages. 1987.

Vol. 1237: Rational Approximation and its Applications in Mathematics and Physics. Proceedings, 1985. Edited by J. Gilewicz, M. Pindor and W. Siemaszko. XII, 350 pages. 1987.

Vol. 1238: M. Holz, K.-P. Podewski and K. Steffens, Injective Choice Functions. VI, 183 pages. 1987.

Vol. 1239: P. Vojta, Diophantine Approximations and Value Distribution Theory. X, 132 pages. 1987.

Vol. 1240: Number Theory, New York 1984–85. Seminar. Edited by D.V. Chudnovsky, G.V. Chudnovsky, H. Cohn and M.B. Nathanson. V, 324 pages. 1987.

Vol. 1241: L. Gårding, Singularities in Linear Wave Propagation. III, 125 pages. 1987.

Vol. 1242: Functional Analysis II, with Contributions by J. Hoffmann-Jørgensen et al. Edited by S. Kurepa, H. Kraljević and D. Butković. VII, 432 pages. 1987.

Vol. 1243: Non Commutative Harmonic Analysis and Lie Groups. Proceedings, 1985. Edited by J. Carmona, P. Delorme and M. Vergne. V, 309 pages. 1987.

Vol. 1244: W. Müller, Manifolds with Cusps of Rank One. XI, 158 pages. 1987.

Vol. 1245: S. Rallis, L-Functions and the Oscillator Representation. XVI, 239 pages. 1987.

Vol. 1246: Hodge Theory. Proceedings, 1985. Edited by E. Cattani, F. Guillén, A. Kaplan and F. Puerta. VII, 175 pages. 1987.

Vol. 1247: Séminaire de Probabilités XXI. Proceedings. Edité par J. Azéma, P.A. Meyer et M. Yor. IV, 579 pages. 1987.

Vol. 1248: Nonlinear Semigroups, Partial Differential Equations and Attractors. Proceedings, 1985. Edited by T.L. Gill and W.W. Zachary. IX, 185 pages. 1987.

Vol. 1249: I. van den Berg, Nonstandard Asymptotic Analysis. IX, 187 pages. 1987.

Vol. 1250: Stochastic Processes – Mathematics and Physics II. Proceedings 1985. Edited by S. Albeverio, Ph. Blanchard and L. Streit. VI, 359 pages. 1987.

Vol. 1251: Differential Geometric Methods in Mathematical Physics. Proceedings, 1985. Edited by P.L. García and A. Pérez-Rendón. VII, 300 pages. 1987.

Vol. 1252: T. Kaise, Représentations de Weil et GL_2 Algèbres de division et GL_n. VII, 203 pages. 1987.

Vol. 1253: J. Fischer, An Approach to the Selberg Trace Formula via the Selberg Zeta-Function. III, 184 pages. 1987.

Vol. 1254: S. Gelbart, I. Piatetski-Shapiro, S. Rallis. Explicit Constructions of Automorphic L-Functions. VI, 152 pages. 1987.

Vol. 1255: Differential Geometry and Differential Equations. Proceedings, 1985. Edited by C. Gu, M. Berger and R.L. Bryant. XII, 243 pages. 1987.

Vol. 1256: Pseudo-Differential Operators. Proceedings, 1986. Edited by H.O. Cordes, B. Gramsch and H. Widom. X, 479 pages. 1987.

Vol. 1257: X. Wang, On the C*-Algebras of Foliations in the Plane. V, 165 pages. 1987.

Vol. 1258: J. Weidmann, Spectral Theory of Ordinary Differential Operators. VI, 303 pages. 1987.

Vol. 1259: F. Cano Torres, Desingularization Strategies for Three-Dimensional Vector Fields. IX, 189 pages. 1987.

Vol. 1260: N.H. Pavel, Nonlinear Evolution Operators and Semigroups. VI, 285 pages. 1987.

Vol. 1261: H. Abels, Finite Presentability of S-Arithmetic Groups. Compact Presentability of Solvable Groups. VI, 178 pages. 1987.

Vol. 1262: E. Hlawka (Hrsg.), Zahlentheoretische Analysis II. Seminar, 1984–86. V, 158 Seiten. 1987.

Vol. 1263: V.L. Hansen (Ed.), Differential Geometry. Proceedings, 1985. XI, 288 pages. 1987.

Vol. 1264: Wu Wen-tsün, Rational Homotopy Type. VIII, 219 pages. 1987.

Vol. 1265: W. Van Assche, Asymptotics for Orthogonal Polynomials. VI, 201 pages. 1987.

Vol. 1266: F. Ghione, C. Peskine, E. Sernesi (Eds.), Space Curves. Proceedings, 1985. VI, 272 pages. 1987.

Vol. 1267: J. Lindenstrauss, V.D. Milman (Eds.), Geometrical Aspects of Functional Analysis. Seminar. VII, 212 pages. 1987.

Vol. 1268: S.G. Krantz (Ed.), Complex Analysis. Seminar, 1986. VII, 195 pages. 1987.

Vol. 1269: M. Shiota, Nash Manifolds. VI, 223 pages. 1987.

Vol. 1270: C. Carasso, P.-A. Raviart, D. Serre (Eds.), Nonlinear Hyperbolic Problems. Proceedings, 1986. XV, 341 pages. 1987.

Vol. 1271: A.M. Cohen, W.H. Hesselink, W.L.J. van der Kallen, J.R. Strooker (Eds.), Algebraic Groups Utrecht 1986. Proceedings. XII, 284 pages. 1987.

Vol. 1272: M.S. Livšic, L.L. Waksman, Commuting Nonselfadjoint Operators in Hilbert Space. III, 115 pages. 1987.

Vol. 1273: G.-M. Greuel, G. Trautmann (Eds.), Singularities, Representation of Algebras, and Vector Bundles. Proceedings, 1985. XIV, 383 pages. 1987.

Vol. 1274: N. C. Phillips, Equivariant K-Theory and Freeness of Group Actions on C*-Algebras. VIII, 371 pages. 1987.

Vol. 1275: C.A. Berenstein (Ed.), Complex Analysis I. Proceedings, 1985–86. XV, 331 pages. 1987.

Vol. 1276: C.A. Berenstein (Ed.), Complex Analysis II. Proceedings, 1985–86. IX, 320 pages. 1987.

Vol. 1277: C.A. Berenstein (Ed.), Complex Analysis III. Proceedings, 1985–86. X, 350 pages. 1987.

Vol. 1278: S.S. Koh (Ed.), Invariant Theory. Proceedings, 1985. V, 102 pages. 1987.

Vol. 1279: D. Ieşan, Saint-Venant's Problem. VIII, 162 Seiten. 1987.

Vol. 1280: E. Neher, Jordan Triple Systems by the Grid Approach. XII, 193 pages. 1987.

Vol. 1281: O.H. Kegel, F. Menegazzo, G. Zacher (Eds.), Group Theory. Proceedings, 1986. VII, 179 pages. 1987.

Vol. 1282: D.E. Handelman, Positive Polynomials, Convex Integral Polytopes, and a Random Walk Problem. XI, 136 pages. 1987.

Vol. 1283: S. Mardešić, J. Segal (Eds.), Geometric Topology and Shape Theory. Proceedings, 1986. V, 261 pages. 1987.

Vol. 1284: B.H. Matzat, Konstruktive Galoistheorie. X, 286 pages. 1987.

Vol. 1285: I.W. Knowles, Y. Saitō (Eds.), Differential Equations and Mathematical Physics. Proceedings, 1986. XVI, 499 pages. 1987.

Vol. 1286: H.R. Miller, D.C. Ravenel (Eds.), Algebraic Topology. Proceedings, 1986. VII, 341 pages. 1987.

Vol. 1287: E.B. Saff (Ed.), Approximation Theory, Tampa. Proceedings, 1985–1986. V, 228 pages. 1987.

Vol. 1288: Yu. L. Rodin, Generalized Analytic Functions on Riemann Surfaces. V, 128 pages, 1987.

Vol. 1289: Yu. I. Manin (Ed.), K-Theory, Arithmetic and Geometry. Seminar, 1984–1986. V, 399 pages. 1987.

Vol. 1290: G. Wüstholz (Ed.), Diophantine Approximation and Transcendence Theory. Seminar, 1985. V, 243 pages. 1987.

Vol. 1291: C. Mœglin, M.-F. Vignéras, J.-L. Waldspurger, Correspondances de Howe sur un Corps p-adique. VII, 163 pages. 1987

Vol. 1292: J.T. Baldwin (Ed.), Classification Theory. Proceedings, 1985. VI, 500 pages. 1987.

Vol. 1293: W. Ebeling, The Monodromy Groups of Isolated Singularities of Complete Intersections. XIV, 153 pages. 1987.

Vol. 1294: M. Queffélec, Substitution Dynamical Systems – Spectral Analysis. XIII, 240 pages. 1987.

Vol. 1295: P. Lelong, P. Dolbeault, H. Skoda (Réd.), Séminaire d'Analyse P. Lelong – P. Dolbeault – H. Skoda. Seminar, 1985/1986. VII, 283 pages. 1987.

Vol. 1296: M.-P. Malliavin (Ed.), Séminaire d'Algèbre Paul Dubreil et Marie-Paule Malliavin. Proceedings, 1986. IV, 324 pages. 1987.

Vol. 1297: Zhu Y.-l., Guo B.-y. (Eds.), Numerical Methods for Partial Differential Equations. Proceedings. XI, 244 pages. 1987.

Vol. 1298: J. Aguadé, R. Kane (Eds.), Algebraic Topology, Barcelona 1986. Proceedings. X, 255 pages. 1987.

Vol. 1299: S. Watanabe, Yu.V. Prokhorov (Eds.), Probability Theory and Mathematical Statistics. Proceedings, 1986. VIII, 589 pages. 1988.

Vol. 1300: G.B. Seligman, Constructions of Lie Algebras and their Modules. VI, 190 pages. 1988.

Vol. 1301: N. Schappacher, Periods of Hecke Characters. XV, 160 pages. 1988.

Vol. 1302: M. Cwikel, J. Peetre, Y. Sagher, H. Wallin (Eds.), Function Spaces and Applications. Proceedings, 1986. VI, 445 pages. 1988.

Vol. 1303: L. Accardi, W. von Waldenfels (Eds.), Quantum Probability and Applications III. Proceedings, 1987. VI, 373 pages. 1988.

Vol. 1304: F.Q. Gouvêa, Arithmetic of p-adic Modular Forms. VIII, 121 pages. 1988.

Vol. 1305: D.S. Lubinsky, E.B. Saff, Strong Asymptotics for Extremal Polynomials Associated with Weights on $\mathbb{R}$. VII, 153 pages. 1988.

Vol. 1306: S.S. Chern (Ed.), Partial Differential Equations. Proceedings, 1986. VI, 294 pages. 1988.

Vol. 1307: T. Murai, A Real Variable Method for the Cauchy Transform, and Analytic Capacity. VIII, 133 pages. 1988.

Vol. 1308: P. Imkeller, Two-Parameter Martingales and Their Quadratic Variation. IV, 177 pages. 1988.

Vol. 1309: B. Fiedler, Global Bifurcation of Periodic Solutions with Symmetry. VIII, 144 pages. 1988.

Vol. 1310: O.A. Laudal, G. Pfister, Local Moduli and Singularities. V, 117 pages. 1988.

Vol. 1311: A. Holme, R. Speiser (Eds.), Algebraic Geometry, Sundance 1986. Proceedings. VI, 320 pages. 1988.

Vol. 1312: N.A. Shirokov, Analytic Functions Smooth up to the Boundary. III, 213 pages. 1988.

Vol. 1313: F. Colonius, Optimal Periodic Control. VI, 177 pages. 1988.

Vol. 1314: A. Futaki, Kähler-Einstein Metrics and Integral Invariants. IV, 140 pages. 1988.

Vol. 1315: R.A. McCoy, I. Ntantu, Topological Properties of Spaces of Continuous Functions. IV, 124 pages. 1988.

Vol. 1316: H. Korezlioglu, A.S. Ustunel (Eds.), Stochastic Analysis and Related Topics. Proceedings, 1986. V, 371 pages. 1988.

Vol. 1317: J. Lindenstrauss, V.D. Milman (Eds.), Geometric Aspects of Functional Analysis. Seminar, 1986–87. VII, 289 pages. 1988.

Vol. 1318: Y. Felix (Ed.), Algebraic Topology – Rational Homotopy. Proceedings, 1986. VIII, 245 pages. 1988

Vol. 1319: M. Vuorinen, Conformal Geometry and Quasiregular Mappings. XIX, 209 pages. 1988.